INTERTIDAL DEPOSITS

DEPOSITS
River Mouths, Tidal Flats, and Coastal Lagoons

Marine Science Series

The CRC Marine Science Series is dedicated to providing state-of-the-art coverage of important topics in marine biology, marine chemistry, marine geology, and physical oceanography. The Series includes volumes that focus on the synthesis of recent advances in marine science.

CRC MARINE SCIENCE SERIES

SERIES EDITORS

Michael J. Kennish, Ph.D.
Peter L. Lutz, Ph.D.

PUBLISHED TITLES

The Biology of Sea Turtles, Peter L. Lutz and John A. Musick
Chemical Oceanography, Second Edition, Frank J. Millero
Coastal Ecosystem Processes, Daniel M. Alongi
Ecology of Estuaries: Anthropogenic Effects, Michael J. Kennish
Ecology of Marine Bivalves: An Ecosystem Approach, Richard F. Dame
Ecology of Marine Invertebrate Larvae, Larry McEdward
Intertidal Deposits: River Mouths, Tidal Flats, and Coastal Lagoons,
    Doeke Eisma
Morphodynamics of Inner Continental Shelves, L. Donelson Wright
Ocean Pollution: Effects on Living Resources and Humans, Carl J. Sindermann
Physical Oceanographic Processes of the Great Barrier Reef, Eric Wolanski
The Physiology of Fishes, David H. Evans
Pollution Impacts on Marine Biotic Communities, Michael J. Kennish
Practical Handbook of Estuarine and Marine Pollution, Michael J. Kennish
Practical Handbook of Marine Science, Second Edition, Michael J. Kennish

FORTHCOMING TITLES

Benthic Microbial Ecology, Paul F. Kemp
Chemosynthetic Communities, James M. Brooks and Charles R. Fisher
Environmental Oceanography, Second Edition, Tom Beer
Handbook of Marine Mineral Deposits, D.S. Cronan
Seabed Instability, M. Shamin Rahman

INTERTIDAL DEPOSITS

River Mouths, Tidal Flats, and Coastal Lagoons

Doeke Eisma

Netherlands Institute for Sea Research (NIOZ)
Texel, The Netherlands and
Institute of Earth Sciences
Utrecht University

with
P. L. de Boer
G. C. Cadée
K. Dijkema
H. Ridderinkhof
and
C. Philippart

CRC Press
Boca Raton   Boston   London   New York   Washington, D.C.

Library of Congress Cataloging-in-Publication Data

Eisma, D., Dr.
    Intertidal deposits: river mouths, tidal flats, and coastal
lagoons / by Doeke Eisma, with P.L. de Boer ... [et al.].
      p. cm. -- (Marine science series)
    Includes bibliographical references and index.
    ISBN 0-8493-8049-9 (alk. paper)
    1. Estuarine sediments. I. Title. II. Series.
GC97.7.E38   1997
551.46'09--dc21
                                            97-28683
                                                CIP

No claim to original U.S. Government works
International Standard Book Number 0-8493-8049-9
Library of Congress Card Number 97-28683
Printed in the United States of America 1 2 3 4 5 6 7 8 9 0
Printed on acid-free paper

Author

Doeke Eisma, born in Haarlem, The Netherlands, in 1932, majored in physical geography, soil science, and sedimentology at Utrecht University. In 1961 he joined the Netherlands Institute for Sea Research, Texel, The Netherlands, where he became head of the Department of Marine Geology and Geochemistry in 1971. He retired from the Institute in 1997.

In 1968 he obtained his Ph.D. at Groningen University with a thesis on the composition and distribution of Dutch coastal sands. He has studied coastal and shelf deposits, mollusk shells, and the composition, transport, deposition, and flocculation of suspended matter in different coastal and shelf areas around the world.

In 1970 he became a lecturer in oceanography at Leiden University and in 1980 at Utrecht University, where he became full professor in marine sedimentology in 1990. He has published over 100 research papers, several review papers and chapters, and recently a book entitled *Suspended Matter in the Aquatic Environment* (1993). He edited a book on *Climate Change: Impact on Coastal Habitation* (1995) and also published three popular books (on the Wadden Sea, the North Sea, and the Dutch-Belgian coast), as well as numerous articles in newspapers and periodicals on planning, energy, housing, and environmental problems.

Table of Contents

Preface

This book is the outcome of experience, first with the Wadden Sea and later with other intertidal areas in the world visited by the author and co-authors, partly to do research, partly because of more personal interest and fascination. Intertidal areas can be important for many reasons besides scientific or economic ones, as the many activities testify that aim to keep intertidal areas in a (more or less) natural state. This book is concerned primarily with intertidal deposits and the conditions of their formation, which include a strong influence of water movement (the tides and, less obviously, waves), climate, sediment particle characteristics, vegetation, fauna, and man. This influence of various factors becomes increasingly clear when visiting intertidal areas and when reading the voluminous literature. The shear volume of the literature, even when it is restricted to what is relevant for intertidal deposits, as well as the widespread occurrence of intertidal deposits along the world's coasts in very different conditions, make it difficult to reach an overview. Our main purpose in writing this book therefore was to put together relevant knowledge on intertidal deposits worldwide, based to a large extent on, besides personal experience, papers and books that are generally accessible and only to a limited extent on less accessible reports and other grey literature.

While writing this book we had much assistance from the libraries at the Netherlands Institute of Sea Research (NIOZ) on Texel, at the Institute of Earth Sciences of Utrecht University, and the Utrecht University Library Centrum Uithof, who were very helpful in tracing and providing relevant publications. We thank Dr. P. Castaing (Bordeaux), Prof. T. T. Healy (Waikato), and Dr. Z. Shi (Shanghai) for additional information and sending copies. To Bob Verschuur, Arco de Graaf, and Bert Aggenbach at NIOZ and the Audiovisuele Dienst at the Institute of Earth Sciences, Utrecht University, we are much indebted for making the many copyproofs and drawings. To J. van de Kam, J. Abrahamse, M. Hemminga, Th. Piersma, S. Tougaard, KLM Aerocarto, and NIOZ we are much indebted for the use of photographs of intertidal areas, and to Henk Hobbelink of NIOZ for digitizing most of the photographs. We thank several reviewers for reading (parts of) the text. To Joke Hart, Anneke Bol, Hans Malschaert, and Carmen Blaauboer of NIOZ, Marnella van der Tol of the Institute of Earth Sciences (Utrecht), Machteld Vervloet (Abcoude), and my daughter Roos, we are very grateful for all their typing and computer work. This book could not have been made without their help.

D. Eisma
Texel/Utrecht

P. L. de Boer
Institute of Earth Sciences
Utrecht University

H. Ridderinkhof
G. C. Cadée
C. Philippart
Netherlands Institute for Sea Research
Texel

K. Dijkema
Institute for Forestry and
Nature Research
Texel

Introduction

Intertidal sediments occur along all tidal coasts with loose sediment and range from bare flats to sediments covered with salt marsh or mangrove. They are formed along tidal channels, in bays and lagoons, in estuaries, and along coastal seas and inlets. No intertidal deposits occur on rocky coasts except in rias. They are also absent at high latitudes above approximately 70° to 73°, where the ice-free period is too short for intertidal deposits to be formed or where tidal sediments and vegetation are frequently disturbed by moving ice. Although the tidal range influences the kind of sediment that is deposited, intertidal sediments occur at all tidal ranges and even where the tidal range is so small that the coast is regarded as tideless. In such areas, storm surges can result in variations in water level and sediment deposits that can be very similar to intertidal sediments.

There is a large variation in intertidal deposits, ranging from extensive tidal flats and giant ripples to small local deposits. This book is concerned with all deposits that are flooded at high tide and fall dry at the lowest tides, as well as those that may be called "supratidal" but nevertheless are occasionally flooded by the highest tides and may be strongly influenced by storm winds (storm surges). Subdivisions of tidal areas are usually made on the basis of the tidal range as defined by Davies (1964), Bird (1970), and Hayes (1975). This also implies a morphological classification: river dominated deltas and wave-built features are mainly present along microtidal coasts; barrier islands, tidal flats, and salt marshes mainly on mesotidal coasts; and funnel-shaped estuaries, extensive tidal flats, and salt marshes on macrotidal coasts. Based on geological criteria, Dalrymple et al. (1992) and Boyd et al. (1992) distinguished wave-dominated and tide-dominated coasts, with tidal flats in sheltered areas where the fluvial input is small, and flats along open coasts where the coast is prograding. Other subdivisions have been made for estuaries and tidal areas separately, with the main distinction between them being that estuaries have a strong river (freshwater) influence, and that in some estuaries the tides are small or almost absent. Fairbridge (1980) made a subdivision (of estuaries) based on the dynamic factors that are at the basis of regional variations: river characteristics, wave energy, tidal range, biological sedimentary factors, sediment characteristics, and tectonics, while McCann (1980) and Dionne (1988) suggested a classification (of tidal areas) based on sediment composition, position (intertidal or subtidal), tidal range, and location (in an estuary or delta, on an exposed coast, etc.). Perillo (1995) pointed out that the existing classifications (of estuaries) may place different types within the same category and proposed a more complex classification based on genetic and morphological considerations. The subdivisions used in this book are largely descriptive and cover tide-dominated intertidal areas with some additional adjacent phenomena such as supratidal flats and wind flats. Barrier islands, tidal beaches, subtidal deposits, and nontidal deposits in deltas will not be discussed, or only in a subordinate way, as they are dominated by waves, longshore currents, and river flow, or are tidal without becoming intertidal. Most intertidal environments are well described by the terminology that is commonly used, but some misunderstanding may come from the use of the term "lagoon." A lagoon is "an expanse of water separating an offshore bar from the shore," but also "an enclosed or nearly enclosed body of water separated from the sea by a low, narrow, elongated strip of land" (Visser ed. 1980). This leaves a wide range of application so that both the Wadden Sea in northwest Europe and the Laguna Madre in Mexico/Texas can be called lagoons. Here the use of the term "lagoon" is limited to "an enclosed or nearly enclosed body of water where salt and freshwater interact, that is separated from the coastal sea by a low barrier and connected with it by one or a few restricted inlets," which comes near to the definition of a lagoon given by Kjerve and Magill (1989) and the characteristics given by Isla (1995).

Intertidal deposits are not isolated units: they are part of larger systems that include subtidal areas (and the processes operating there) and, probably to a lesser extent, supratidal areas. In tidal inlets and deltas the intertidal deposits often have a very limited extension compared to subtidal and supratidal units. Also along beaches the intertidal area usually is only a limited fringe. Because of the interrelation between intertidal areas and adjacent subtidal and supratidal areas, the intertidal deposits will be discussed within a wider context.

The aim of this book is to bring together both descriptive knowledge on the intertidal deposits around the world and insight into the processes that are instrumental in forming them: sediment transport and deposition under the influence of tides and waves, bottom fauna, and vegetation. Emphasis will be on the more important intertidal areas, which are distinguished on the basis of their size (area) and morphology, the tidal range, the degree and type of vegetational cover, the amount and type of benthic fauna, the human use of the area, and the degree to which they have been studied. The latter varies considerably: although the coast is more accessible than the open sea or the ocean, and less cost is involved in studying it, the best-known intertidal areas are those near large centers of study, and those where human use of the coast is most intensive.

The first part of this book (Chapters 1 through 5) is descriptive, mainly covering the intertidal areas that are relatively well known, but also including short sections on tidal areas that have been studied less. The following three chapters deal with intertidal morphology (Chapter 6), sediment composition and structure (Chapter 7), and sediment transport and deposition (Chapter 8), which are focused on intertidal processes. The fauna and flora in relation to intertidal sedimentation is divided into five sections in Chapter 9: benthic fauna and microflora (diatoms), salt marshes which occur mainly, but not exclusively, in temperate and cold climates, mangrove (which occur along tropical and subtropical coasts), algal mats (which develop on salt flats in dry hot climates), and seagrasses. The Holocene history of intertidal areas is discussed in Chapter 10, followed by Chapter 11 on the large influence of man on present intertidal areas.

1 Worldwide Distribution of Intertidal Areas

On the basis of the data given in *The World's Coastline* (Bird and Schwartz 1985) and the literature referred to elsewhere in this book, the most important intertidal areas can be listed as follows:

Europe: The Wadden Sea, the Wash, the Bay du Mont-St. Michel, and the Bassin d'Arcachon. Along the remaining coasts of Britain, France, and Portugal small areas with intertidal deposits occur in estuaries, rias, and bays. In the White Sea the tidal flats are probably extensive but not much studied. In the fjords of Norway and the Faroer, tidal deposits are virtually absent.

Africa: The lagoons from Guinea-Bissau to Benin (with the Volta river delta), the Niger delta, the lagoons and bays in Cameroon and Equatorial Guinea, the Zaire river mouth, the lagoons and deltas in Natal and Mozambique, and the open-coast mangrove flats in Tanzania. From Morocco to Senegal along the Sahara coast, intertidal deposits are rare and small with larger deposits at the Banc d'Arguin, but become more numerous and extensive in lagoons and river estuaries from the Senegal river mouth to Sierra Leone. Also between Equatorial Guinea and the Zaire river mouth, intertidal deposits are present in bays, lagoons, and behind spits. South of the Zaire river mouth only small intertidal areas are present in Langebaan lagoon and near Port Elisabeth. In Tanzania and Kenya and along eastern Madagascar, rather large intertidal areas are present in bays, lagoons, estuaries, and deltas. Along Kenya and Somalia, they decrease in size (and number).

South Asia: The sebkhas along the southern Persian Gulf, the Indus delta with the Rann of Kutch and the Bay of Khambhat, the deltas along the Indian east coast, the Ganges delta area, the Irrawaddy delta, the Chao Phraya river delta, the Mahakam delta in Indonesia, and the open coast flats along the Mekong river mouth and the Red river delta. Smaller areas with intertidal deposits are present along the Persian Gulf and the northern shore of the Indian Ocean up to the Indus river delta. From Burma to the east the intertidal deposits are relatively large because of the large sediment supply from even small rivers. In Indonesia relatively large intertidal areas covered with mangrove are present along eastern Sumatra, south and east Kalimantan, south Irian Jaya, and the adjacent coasts of Papua, New Guinea, but their sediments have been little studied.

North and East Asia: South and West Korea, Bohai Bay (Yellow river delta), the China mainland coast between Shandong and Fujian (Jiangsu coast, Chang Jiang estuary, Hangzhou Bay, Wenzhou Bay), West Taiwan, and the Zhujiang river mouth. Small intertidal areas are present along the Shandong peninsula and along the Chinese coast from Wenzhou Bay to west of the Zhujiang river mouth. Large intertidal areas are present along the northern Gulf of Anadyr, West Kamchatka, the northern Sea of Okhotsk with the Bay of Penzhinskaya, and along Sakhalin and smaller areas along northern Siberia, but little has been published about them. No intertidal deposits are present north of 73°N.

Australia/New Zealand/Large Pacific Islands: The west and northwest coast of Australia, (Shark Bay to King Sound), Cambridge Gulf–Van Diemen Gulf–Arnhem Land, river mouths in the Gulf of Carpentaria, and Broad Sound. Numerous smaller intertidal areas occur along the other parts of Australia, New Zealand, and the larger Pacific islands as part of river deltas and in bays.

America West Coast: The Mackenzie River delta, the Yukon and Kuskokwim deltas, Bristol Bay, Cook Inlet, the Queen Charlotte Islands, San Francisco Bay, the Colorado river delta, the Colombia coast, and the Gulf of Guayaquil. There are no intertidal deposits north of 70°N. Numerous small deposits are present in bays, river mouths, and lagoons from Alaska to Ecuador. Hardly any tidal deposits are present from northern Peru to central Chile. Small intertidal deposits occur in the fjords of southern Chile.

America East Coast: San Antonio Bay, the Amazon–Guyana–Orinoco delta coast, the Mexico–Texas lagoons, the Louisiana–Mississippi delta coast, the SW coast of Florida, the *Sabellaria* reefs in western Florida, the bays and marshes in Georgia and North and South Carolina, the Bay of Fundy, and the St. Lawrence estuary–Labrador–Hudson Bay flats. Smaller intertidal flats are present along most of the American west coast from Tierra del Fuego to Greenland and, in particular, between Boston and Nova Scotia. As along the American west coast, no intertidal deposits are present north of 70°N.

The tidal range, tidal type, and wave height characteristics for these areas are given in Table 1.1, based on data given by Davies (1972). The wave characteristics are given for the open coast that is exposed to waves coming from the coastal sea or ocean. Where waves are relatively high (Bassin d'Arcachon, Bay of Fundy), it should be realized that the intertidal deposits here have been formed in well-protected bays.

Many river mouths on tidal coasts contain small areas of intertidal deposits along estuaries and deltas. They are mentioned where they have been the object of sediment studies. For a number of large coastal areas sedimentological information on intertidal deposits is limited and only a general description can be given. This concerns northern Russia from the White Sea to the Gulf of Anadyr and the Sea of Okhotsk, Alaska, Iceland, the Philippines, north Australia, Tanzania and Kenya, and parts of the South American east coast. For the sake of completeness, these are included in Chapters 2, 3, and 4. Statistics on the occurrence and extent of tidal flats, salt marshes, and mangrove are available for western Europe, the east coast of the U.S., and Australia. They are given after Table 1.2 in Tables 1.3 through 1.6.

TABLE 1.1
Characteristics of the Most Important Intertidal Areas Tidal Range

	Tidal range			
	Micro	Meso	Macro	Tidal regime
EUROPE				
Iceland	x	x	x	S
White Sea		x	x	S
Wadden Sea**	x	x		S
Ooster, Wester Schelde		x		S
The Wash*			x	S
Dyfi estuary	x			S
Baie du Mont-St. Michel			x	S
Bassin d'Arcachon			x	S
Tejo river mouth, Sado estuary		x		S
Rias northern Spain		x		S
Thames estuary (Essex coast)		x	x	S
Norfolk flats			x	S
Dee estuary			x	S
Bristol Channel/ Severn estuary			x	S
Solway Firth			x	S
France, Atlantic coast			x	S
Baltic (wind flats)	x			M
Mediterranean, Caspian Sea (wind flats)	x			M,S
AFRICA				
Northwest Africa	x			S
Senegal, Gambia	x			S
Guinea-Bissau-Guinea		x		S
Sierra Leone-Benin	x	x		S
Niger delta	x	x		S
Cameroon-Equatorial Guinea		x		S
Zaire river estuary	x			S
Langebaan lagoon	x			S
Kwazulu-Natal-Mozambique**	x	x	x	S
Tanzania-Kenya		x		S
Madagascar west coast			x	S
SOUTH ASIA				
Persian Gulf	x	x		M
Indus delta-Rann of Kutch		x		M
Gulf of Khambhat (Cambay), Gulf of Kachchh (Kutch)			x	M
(Mud banks Kerala)		x		M
Deltas east coast India	x			S
Sri Lanka	x			S
Ganges-Brahmaputra river mouth*			x	S
Irrawaddy river delta			x	S
Gulf of Thailand	x	x		D
West coast Burma, Thailand, Malaysia		x		S

TABLE 1.1 *(continued)*
Characteristics of the Most Important Intertidal Areas Tidal Range

	Tidal range			Tidal regime
	Micro	Meso	Macro	
Mekong river mouth*		x		M
Red river mouth*			x	D
The Philippines	x	x		M
Indonesia**	x	x		M,D,S
Gulf of Papua		x		M
NORTHEAST ASIA				
North Siberia	x			S(M)
East Siberia (wind flats)	x			S
Gulf of Anadyr	x			S
Sea of Okhotsk			x	M
West Korea*			x	S
Bohai Bay/Huang He delta	x	x		S
China coast**			x	S
West Taiwan*			x	S
AUSTRALIA				
South and east Australia	x			M,S
West Australia	x			M
Northwest/North Australia**			x	S
North Australia		x	x	S,M
Gulf of Carpentaria	x	x		M
Broad Sound			x	M
New Zealand	x	x		S
Pacific Islands	x			S
NORTH AMERICA				
Mackenzie river delta	x			S
Yukon river delta	x			M
Kuskokwim river delta	x			M
Bristol Bay		x		M
Cook Inlet			x	M
Queen Charlotte Islands			x	M,S
Fraser river delta		x		M
Yaquina Bay, San Francisco Bay		x		M
Colorado river delta			x	M
Lagoons Mexico/Texas (wind flats)	x			D,M
Louisiana/Mississippi delta**	x			D,M
Southwest Florida–Belize	x			M
Northwest Florida	x			S
Andros, Caicos, Grand Cayman	x			S
Georgia-Maine		x		S
Bay of Fundy			x	S
St. Lawrence river estuary			x	M

TABLE 1.1 *(continued)*
Characteristics of the Most Important Intertidal Areas Tidal Range

	Tidal range			Tidal regime
	Micro	Meso	Macro	
Hudson Bay-Labrador-Baffin island	x	x	x	S
CENTRAL/SOUTH AMERICA				
Colombia coast*			x	M
Gulf of Guayaquil			x	M
San Sebastian Bay			x	S
San Antonio Bay			x	S
Bahia Blanca		x		S
Mar Chiquita	x			S
South American east coast	x	x	x	S,M
Amazon-Orinoco delta*		x	(x)	S

* Along an open coast.

** Partially along an open coast.

Micro = microtidal; meso = mesotidal; macro = macrotidal; tidal regime (Davies 1964): S = semidiurnal; M = mixed; D = diurnal.

TABLE 1.2
Frequency of Waves >2.4 High and Largest Waves at the Most Important Intertidal Areas

	Wave frequency (%)				Largest wave (m)			
	<10	10–20	20–30	>30	<2.4	2.4–3.7	3.7–4.9	>4.9
EUROPE								
Iceland			x	x				x
White Sea	x				?	?	?	?
Wadden Sea**		x					x	
Ooster, Wester Schelde		x					x	
The Wash*		x					x	
Dyfi estuary		x					x	
Baie du Mont-St. Michel		x					x	
Bassin d'Arcachon				x				x
Tejo river mouth, Sado estuary			x				x	
Rias northern Spain				x			x	
Thames estuary (Essex coast)		x					x	
Norfolk flats		x					x	
Dee estuary		x					x	
Bristol Channel/Severn estuary		x						x
Solway Firth		x					x	
France, Atlantic coast		x		x			x	x
Baltic (wind flats)		x				x		

TABLE 1.2 *(continued)*
Frequency of Waves >2.4 High and Largest Waves at the Most Important Intertidal Areas

	Wave frequency (%)				Largest wave (m)			
	<10	10–20	20–30	>30	<2.4	2.4–3.7	3.7–4.9	>4.9
Mediterranean, Caspian Sea (wind flats)	x					x	x	
AFRICA								
Northwest Africa		x			x	x		
Senegal and Gambia	x				x			
Guinea-Bissau-Guinea	x				x			
Sierra Leone-Benin	x	x				x	x	
Niger delta	x					x	x	
Cameroon-Equatorial Guinea	x					x		
Zaire river mouth	x					x		
Langebaan lagoon			x				x	
Kwazulu-Natal-Mozambique	x	x			x	x	x	
Tanzania-Kenya	x				x	x		
Madagascar west coast	x					x		
SOUTH ASIA								
Persian Gulf	x				x			
Indus-delta-Rann of Kutch		x			x	x	x	
Gulf of Khambhat (Cambay), Gulf of Kachchh (Kutch)		x			x			
(Mud banks Kerala)		x				x		
Deltas east coast India	x				x			
Sri Lanka	x	x				x		
Ganges-Brahmaputra river mouth		x					x	
Irrawaddy river delta	x					x		
Gulf of Thailand	x					x		
West coast Burma, Thailand, Malaysia	x				x	x		
Mekong river mouth		x				x		
The Philippines	x					x	x	
Red river mouth	x					x		
Indonesia	x				x			
Gulf of Papua	x					x		
NORTHEAST ASIA								
Siberia	x				?	?	?	?
East Siberia (wind flats)	?	?	?	?	?	?	?	?
Gulf of Anadyr	x				x			
Sea of Okhotsk	x					x	x	
West Korea	x					x		
Bohai Bay/Huang He delta	x					x		
China coast		x					x	
West Taiwan		x					x	

TABLE 1.2 *(continued)*
Frequency of Waves >2.4 High and Largest Waves at the Most Important Intertidal Areas

	Wave frequency (%)				Largest wave (m)			
	<10	10–20	20–30	>30	<2.4	2.4–3.7	3.7–4.9	>4.9
AUSTRALIA								
South and east Australia		x	x				x	
West Australia			x				x	
Northwest/North Australia	x				x			
North Australia	x				x			
Gulf of Carpentaria	x					x		
Broad Sound	x					x		
New Zealand	x	x	x	x	x	x	x	x
Pacific Islands	?	?	?	?	?	?	?	?
NORTH AMERICA								
Mackenzie river delta	?	?	?	?	?	?	?	?
Yukon river delta	x				x			
Kuskokwim river delta	x				x			
Bristol Bay	x				x			
Cook Inlet		x					x	
Queen Charlotte Islands			x		x			
Fraser river delta				x			x	
Yaquina Bay, San Francisco Bay			x			x		
Colorado river delta		x			x			
Lagoons Mexico/Texas	x	x				x	x	
Louisiana/Mississippi delta	x					x		
Southwest Florida - Belize	x					x		
NW. Florida		x				x		
Andros, Caicos, Grand Cayman		x				x		
Georgia-North/South Carolina			x				x	
Bay of Fundy			x					x
St. Lawrence river	x						x	
Hudson bay-Labrador-Baffin Island	x						x	
CENTRAL/SOUTH AMERICA								
Colombia coast	x					x		
Gulf of Guayaquil	x					x		
San Sebastian Bay			x					x
San Antonio Bay		x					x	
Bahia Blanca		x				x		
South America east coast		x			x	x	x	
Amazon-Orinoco		x	x			x	x	

Note: Wave frequency (%) = Percentage frequency of occurrence in at least two quarters of the year of waves >2.4 m high. Largest wave (m) = greatest height (in m) reached by waves occurring with a frequency of 3% or more in at least two quarters of the year.

Data from Davies 1972.

TABLE 1.3
Area of Tidal Flats in Europe

	km²	%
Iceland	?	?
Murman coast, White Sea	?	?
Norway	(very small)	
Sweden	(very small)	
Wadden Sea	4800	52
(Denmark–West Germany–The Netherlands)		
The Netherlands (Eastern and Western Scheldt)	249	3
Belgium	<5	
England (The Wash: 740 km²)	1672	17
Scotland	474	5
Wales	262	3
Iceland	800	9
France (Baie du Mont St. Michel: 250 km²	700	8
Baie d'Arcachon 155 km²)		
Spain	100?	1?
Portugal (Tejo: 120 km²)	200?	2?
(Italy)	–25?	
Total area ~	9300	100

Data from W. J. Wolff, ed., 1988.

TABLE 1.4
Area of Salt Marsh on the Atlantic
(Tidal Coasts of Europe)

	km²	Number of areas > 5 km²
Svalbard	?	?
Russia	?	?
Iceland	?	?
Norway	45	1
Denmark	81	7
West Germany	188	14
The Netherlands	129	9
Belgium	4	0
United Kingdom	371	17
Iceland	180	6
France	148	6
Spain	400?	1
Portugal	92	5
Total area ~	1640	66

Data from W. J. Wolff, ed., 1988.

TABLE 1.5
Distribution of Salt Marshes and Mangrove
(Florida) Along the U.S. East Coast
(Gulf of Mexico Coasts Not Included)

State	Area (ha)	% of total	
Maine	588	0.10	
New Hampshire	151	0.03	
Massachusetts	3213	0.55	
Rhode Island	216	0.04	3.08
Connecticut	840	0.14	
New York	4666	0.79	
New Jersey	8445	1.43	
Delaware	17707	3.00	
Maryland	27899	4.73	15.34
Virginia	44839	7.61	
North Carolina	64284	10.91	
South Carolina	176442	29.93	73.50
Georgia	192508	32.66	
Florida	47631	8.08	
Total	**589,429**	**100.00**	

Adapted from Reimold, 1977.

TABLE 1.6
Areas of Bare Mud Flats and Mangrove Mud Flats along the Australian Coast

	Area nr.	Total area (km²)	Bare mud flats (km²)	(%)	Mangrove mud flats (km²)	(%)
Queensland (680 km)	(1)	10.536	3.894	(40.0)	1.206	(11.5)
	(2)	17.109	1.231	(7.2)	3.198	(18.7)
New South Wales (520 km)	(3)	7.503	27	(0.4)	87	(1.2)
Victoria (540 km)	(4)	6.210	177	(2.9)	15	(0.2)
South Australia (270 km)	(5)	12.342	486	(3.9)	204	(1.7)
West Australia (1180 km)	(6)	13.269	333	(2.5)	0	
	(7)	4.002	2.217	(55.4)	15	(0.01)
	(8)	16.704	3.537	(21.2)	1.812	(10.9)
	(9)	2.475	48	(1.9)	159	(6.4)
	(10)	3.160	1.215	(38.5)	393	(12.4)
North Territories (230 km)	(11)	23.463	4.608	(19.6)	2.929	(12.5)
Tasmania (430 km)	(12)	7.119	45	(0.6)	0	
Total		**123.892**	**17.818**	**(14.4)**	**10.018**	**(8.1)**

Adapted from Thom 1984, based on the data of Galloway et al. 1980.

1.1. TIDES

The tidal range of coastal waters ranges from less than 5 cm along the eastern part of north Siberia and in the so-called tideless enclosed seas (Mediterranean, Black Sea, Caspian Sea, the Baltic, and the Red Sea), to more than 15 meters in the Bay of Fundy in eastern Canada. Davies (1964, 1972) and Bird (1970), followed by Hayes (1975) and many others, have distinguished a microtidal regime with a tidal range between 0 and 2 meters, a mesotidal regime with ranges between 2 and 4 meters, and a macrotidal regime with tidal ranges larger than 4 meters. Barrier islands and river deltas develop particularly in microtidal areas, inlets, and tidal deltas at tidal ranges of up to about 3 meters, and tidal flats and salt marshes predominantly in meso- and macrotidal areas. Ehlers (1988) proposed a modified classification based on observations made along the North Sea: a microtidal regime at a tidal range less than 1.4 meters, and a mesotidal regime between 1.4 and 2.9 meters, but this still leaves a large tidal inlet with a large ebb tidal delta (the Texel inlet in the western Wadden Sea with a tidal range of 1.2 meters) in a microtidal area. Sha (1989) therefore pointed out that for tidal inlets the tidal prism is a better measure. The distribution of tidal range (in meters at spring tide) along the world's coasts is given in Figure 1.1 (from Davies 1972). The actual distribution of tidal range may vary more at short distances than is indicated because of regional and local changes in topography and water depth which may induce a larger or smaller tidal range than can be indicated on the scale of Figure 1.1. Figure 1.1 and Table 1.1

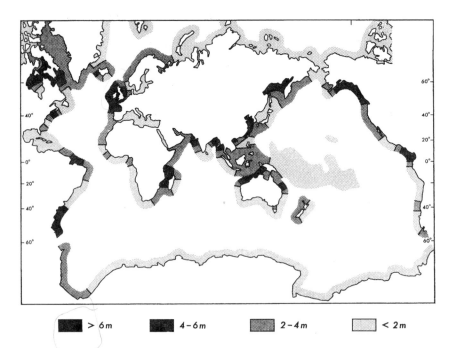

> 6 m   4 – 6 m   2 – 4 m   < 2 m

FIGURE 1.1 Distribution of tidal range along the world's coasts, in meters at spring tide. (From Davies 1972. With permission.)

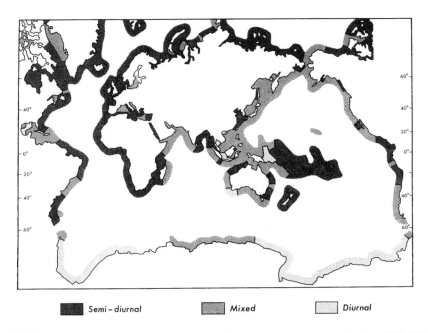

| Semi-diurnal | Mixed | Diurnal |

FIGURE 1.2 Distribution of tidal types along the world's coasts. (From Davies 1972. With permission.)

indicate that there is no relationship between the presence of intertidal deposits and the tidal range: 25 of the intertidal areas listed in Table 1.1 are along a microtidal coast, 15 along a mesotidal coast, and 28 along a macrotidal coast. Ten areas are along coasts with both micro- and mesotidal ranges, three along coasts with meso- to macrotidal ranges, and five along coasts where the tidal range varies from micro- to meso- and to macrotidal. Also the distribution of all intertidal deposits, as indicated in Chapters 2, 3, and 4, does not seem to indicate that the presence of intertidal deposits is related to the tidal range.

Nor does the tidal regime seem to be an important factor for the presence or absence of intertidal deposits: 52 of the areas listed in Tables 1.1 and 1.2 occur along coasts with a semidiurnal tide, 24 along coasts with a mixed semidiurnal/diurnal tide, and two along coasts with a diurnal tide. When the areas are added that have one tidal regime in one part and another regime in another part, the numbers become 58, 32, and 4, respectively. This distribution reflects the worldwide distribution of tidal regimes (Figure 1.2). The semidiurnal tide has two high tides and two low tides within approximately 24 hours, and there can be a relatively small daily and fort-nightly inequality in tidal range. Diurnal tides have only one high and one low tide within about 24 hours, while mixed tides show a sequence of a low low tide, a low high tide, a high low tide, and a high high tide. As pointed out by Davies (1972), the tidal regime determines the duration of drying out between tides. Differences in the length of this period may affect the degree of sediment consolidation during the dry period and, in warm climates, the salt concentration in the sediments. The tidal regime also affects the distribution of organisms in the intertidal zone and the intensity of tidal currents, which is related to the time that is available to move water masses. Ginsburg et al. (1977) introduced an "exposure index" as a distinctive parameter, which is related to the surface elevation above an arbitrary level (mean sea level or low tide level) and the tidal variations in water level. This was applied to a limited area on Andros Island in the Bahamas, but is only useful in areas where a rather large number of tidal measurements have been made that relate water level to time.

A relationship with waves seems more plausible. Relatively few intertidal deposits are found along coasts that are exposed to waves from the open sea (or ocean): of the areas listed in Tables 1.1 and 1.2, only 15 of 55 are along the open coast. The other intertidal deposits occur in rather sheltered areas: bays, estuaries, river mouths, lagoons, and behind spits or barriers. Where they occur along an open coast, they mostly occur where waves are usually low. Where intertidal deposits occur along an open coast with rather large waves (with 3% of the largest waves between 3.7 to 4.9 meters) the tide is macrotidal or nearly so, which indicates a relatively large tidal influence on sediment transport and deposition (the German Bight, the Wash, the Ganges river mouth, and the Chinese coasts). Along the Guyana coast between the Amazon river mouth and the Orinoco delta, the mud concentrations in the mud banks are very high, so that waves are strongly reduced (NEDECO 1968, Wells and Kemp 1986). Exceptions are the mud banks at the Mekong river mouth and along the western Taiwan coast, but their presence along an open coast is probably related to the direction of the larger waves, which is approximately parallel to the coast.

1.2. ZONATION

Intertidal deposits usually show a zonation related to the duration of submergence — from subtidal below (mean) low spring tide level to supratidal above (mean) high spring tide level. In-between lower flats below (mean) low neap tide level, and higher flats above (mean) high neap level have been distinguished, with middle flats around mean sea level, but in some areas only lower flats and higher flats are distinguished, approximately below and above mean sea level. These distinctions and subdivisions are not made in all intertidal areas and not in the same way. Also it should be realized that the water level can become lower than mean low spring tide level, and higher than mean high spring tide level, so that flats called supratidal or extratidal may be flooded regularly. The zonation is often based on grain size and/or vegetation or morphology as well as on tidal level, but, as shown by Amos (1995), there is a wide variation in the relation between grain size or vegetation and the tidal zone (Figure 1.3). Therefore a worldwide zonation can only be distinguished in a rather general way, roughly indicating the level in relation to the tides, and only in a more consequent and specific way for each separate intertidal area.

1.3. VEGETATION

Most intertidal deposits are partly or (almost) entirely covered with vegetation, which usually consists of salt marsh or mangrove. Salt marshes occur predominantly in

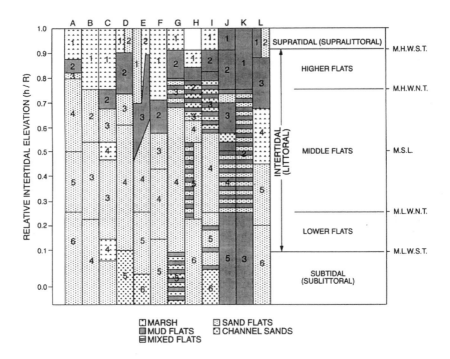

FIGURE 1.3 Summary of classifications of tidal exposure and associated intertidal flat zonations. A = The Wash, England; B = Burry Inlet, S. Wales; C = S.E. Australia; D = Cobequid Bay, Canada; E = Minas Basin, Canada; F = Hudson Bay, Canada; G = German Bight, Germany; H = The Wash, England; I = Bay du Mont St. Michel, France; J = Gulf of California, Mexico; K = Wenzhou Bay, China; L = South Australia. A and H are from different authors. (From Amos 1995. With permission.)

cold or temperate climates, although some tropical salt marshes do occur. The cold/temperate salt marshes have their southern limit at about 30°N in West Africa (Morocco), 20°N in East Africa (Sudan), 30°N in the Persian Gulf (Shatt el Arab), 26°N along the mainland of East Asia (near Fuzhou, China), 34°N in Japan, 30°N along the American west coast (northern Mexico), and 24°N along the American east coast (also in northern Mexico). In the southern hemisphere they have their northern limit in South America at about 32°S along the west coast (in central Chile) and at about 29°S along the east coast (in southern Brazil). In southern Africa their northern limit is at about 33°S, in New Zealand at about 38°S, and in southern Australia at about 30°S. Along the African west and east coasts, in the Red Sea, in the Persian Gulf, in western South America, and in western Australia, the salt marshes grade landward and along the shore into saltpans or dry desert with little or no vegetation. Along the Mexican and China coasts, in Japan and New Zealand, and along eastern Australia, there is no dry zone along the shore between the salt marshes and the tropical mangrove vegetation: salt marshes are at short distances replaced by mangrove or they intermingle. Only along the Australian east coast the change in coastal vegetation is gradual: the most northern salt marsh is at about 14°S. North of this, the saline coastal vegetation has little in common with the salt marshes farther south, while to the south the number of typical salt marsh species gradually increases.

Mangrove generally flourishes in tropical areas where the air temperature in the coldest month does not fall below 20°C and the temperature range in a year is about 10°C at most. In Florida, along the Red Sea, and in Brazil and New Zealand, some mangrove species tolerate minimum temperatures between 10° and 15°C. Mangrove occurs between 32°N (the southern tip of Japan) and 38° S (the northern tip of New Zealand. Where mangrove can grow, salt marshes tend to disappear. Salt marshes usually cover the upper tidal zone above mid-tide level and usually around high tide level or the occasionally flooded zone just above high tide level (supratidal flats); mangrove covers most of the intertidal area to below mean tide level. In Louisiana, salt marshes are fringed by mangrove; the mangrove occasionally disappears during a cold winter, but expands over the salt marshes when the winter is warm. Also in south Australia, between Spencer Gulf and Sydney, the higher flats are covered with salt marsh and the lower parts with mangrove.

Along tropical and subtropical coasts salt marshes are relatively rare. They occur around high tide level or just above high tide on supratidal flats, in East Africa (Egypt and Sudan), at the Shatt el Arab in the Persian Gulf, along Princess Charlotte Bay in northeast Australia, in the Gulf of Guyaquil and other parts of the coast of Ecuador, and in Belize. Salt marshes occur in these areas at the inner margin of mangrove forests, except in the Shatt el Arab where they border bare tidal flats because the climate is too cold there (at 30°N) for mangrove. In Cayenne, salt marshes occur as a pioneer vegetation on freshly deposited muds. In northeast Australia north of Princess Charlotte Bay, saline grass fields occur around high tide level. In the tropics and subtropics where high temperatures result in a high salinity of the water and the sediment, mangrove cannot grow (the limit is at about 90‰S), and the sediment remains without vegetation (salt flats) or becomes covered with algal mats. These consist of algal filaments and threads that form a dense layered network up to several centimeters thick that may develop into stromatoliths. They may be interlayered with sediment because of seasonal or episodic flooding. They are best developed on flats in the Bahamas and adjacent islands and along the southern fringes of Shark Bay in western Australia, and are not to be confused with the "mats" of diatoms and bacteria at the sediment surface or just below, which are weakly held together by mucus and are of millimeter thickness.

In many coastal areas intertidal deposits are not covered by vegetation. This is primarily the case in the cold and temperate regions where salt marshes usually cover only the deposits around high tide level, but this is not everywhere the case: mud flats near Port Elisabeth in South Africa are covered with a salt marsh type of vegetation down to low tide level, whereas in many temperate regions *Spartina* and *Salicornia* may grow below high tide level. In the tropics and subtropics, where mangrove is common, the tidal deposits below approximately mid-tide level are bare. Where conditions are unfavorable for mangrove, the intertidal zone may remain entirely without vegetation. This occurs where the sediment is regularly reworked by currents and waves, as at the Ganges-Brahmaputra (Meghna) river mouth, along eastern Sumatra in Indonesia, along the open coast of Mozambique, at mud banks along the Guyana coast, and at the Mekong and Red river mouths. In tropical or subtropical climates with a marked dry season, intertidal sediments around high tide level dry out and become highly saline. The vegetation is reduced to a few sparsely distributed halophytes or no vegetation at all, while regular flooding increases the salinity of the sediment, because the evaporation is usually high. Under desert conditions such salt flats around high tide level can develop into saltpans and sebkhas with salt encrustations. Saline flats with no or only sparse vegetation occur along the coasts from Senegal to Guinea-Bissau where they are called tanns, at the Indus delta and the Rann of Kutch, along the coast of Sri Lanka, in northern Australia, and along the drier parts of the coast of Venezuela. Saltpans and sebkhas occur along the Persian Gulf and along the coasts of the Red Sea, western North Africa, along the north coast of Libya (the Gulf of Syrte), and in some parts of the coast along northern Mexico. The flats along the Bay d'Arguin in West Africa are partly salt flats but are much influenced by the wind, which regularly supplies sand from the Sahara desert.

On bare intertidal sediment surfaces, in particular where they are fine-grained, benthic diatoms may form a dense cover. Through the mucus that is produced by the diatoms and which fixes them to the sediment particles, the sediment particles are glued together so that they resist resuspension/erosion to a certain extent. This effect is most clearly seen below high tide level on bare tidal flats, i.e., mainly in temperate and cold climates, but it may occur in all climates. Where mangrove dominates, however, only small areas of sediment between the trees may be covered by diatoms, which, being plants, need light for photosynthesis.

1.4. BENTHIC FAUNA

The distribution of benthic fauna in intertidal sediments is much less well known worldwide than the distribution of vegetation. Generally mollusks — bivalves as well as gastropods — occur in all intertidal areas, but burrowing bivalves are much more common where vegetation is absent or sparse, so they tend to dominate at higher latitudes in temperate or cold climates. This is also the case with burrowing worms, which also occur mostly in intertidal deposits that are bare. Burrowing crabs mostly dominate in intertidal sediments in tropical and subtropical climates as well as in some areas with a dry Mediterranean climate such as the south of Portugal and the Californian coast near Los Angeles, but burrowing crabs are also common along the American/Canadian Pacific coast farther north. In addition to crabs, a burrowing fish (the mudskipper) is also common on tropical and subtropical intertidal muds The effects on the sediment of burrowing by benthic fauna vary. Benthic fauna may fix bottom sediment by building tubes and producing mucus, even may build low reefs (like *Sabellaria*), or may form banks (like oysters and mussels) that protect the sediment and induce deposition. Other benthic organisms, however, loosen the sediment through burrowing or through crawling over the surface and thus enhance resuspension or bottom transport.

1.5. SEDIMENT

Intertidal deposits may consist of mineral grains ranging from clay-size to coarse sand, of carbonate particles of the same size range, and of other organic materials such as mollusk shells and shell fragments, coral debris, and plant remains. At high latitudes, gravel or even boulders may also be present in the sediment which have not been transported by tidal currents but by (floating) ice, or by freezing and thawing, in combination with gravity. As the sediment composition indicates, the sediment that goes into intertidal deposits may come from various sources: a nearby river, longshore transport from a source farther away, local or regional erosion or reworking of older deposits, or a biogene source of locally or regionally produced carbonate.

The total amount of suspended material brought by rivers to the sea is about 20 $\times 10^{15}$ g · y^{-1} with an additional 1 to 2 $\times 10^{15}$ g · y^{-1} of bed load (sand, gravel). About 6 $\times 10^{15}$ g · y^{-1} is supplied by the 25 rivers with the largest sediment load (not necessarily the largest rivers) (Table 1.7), and about 10 $\times 10^{15}$ g · y^{-1} is supplied by small rivers with basins smaller than 10.000 km^2 (Milliman and Meade 1983; Milliman and Syvitsky 1992). Only a few large rivers supply sediment to intertidal deposits over a wide area (the Amazon, the Yellow river, the Chang Jiang, and the Mississippi). The supply from the other large rivers is mostly transported offshore (Ganges, Zaire) and is deposited on the shelf, on the continental slope, or in the deep ocean. Usually the sediment of intertidal deposits comes from a regional source at a relatively short distance. Sediment also may come from more than one source, as in the Wadden Sea where the sand is supplied from reworked Pleistocene and Holocene river sands in the southern North Sea, and the fine-grained material has been transported in suspension mostly from the Belgian coast and the Channel.

In the next chapters the formation of intertidal deposits in the intertidal areas listed in Tables 1.1 and 1.2 will be discussed. The areas will not be grouped geographically but according to the tidal range, which is the most common denominator. They can also be grouped according to other characteristics — type of sediment, degree and type of vegetational cover, climate, presence or absence of tidal channels — but these characteristics, except climate, may vary within a single intertidal area, and for the intertidal bottom sediment, climate seems to be a less determining characteristic than tidal range. It determines, however, to a large extent whether the sediments are covered with salt marsh, mangrove, or algal mats, or are bare. Therefore a subdivision is made (in Chapter 9) according to the degree and type of vegetational cover.

TABLE 1.7
Average Sediment Discharge to the Ocean of the 25 Rivers with the Largest Sediment Load

	Average sediment discharge[a] $(10^6 \; t \cdot y^{-1})$	Average water discharge $(km^3 \cdot y^{-1})$	Average concentration $(mg \cdot l^{-1})$
Amazon	1000–1300 [1000–1300]	6300	190
Yellow River (Huang He)	1100 (100) [1200]	49	22040
Ganges-Brahmaputra	900–1200	970	1720
Chang Jiang	480	900	531
Irrawaddy	260 [260]	430	619
Magdalena	220	240	928
Mississippi	210 (400) [500]	580	362
Godavari	170	92	1140
Orinoco	150 (150) [150]	1100	136
Red River (Hung Ho)	160	120	1301
Mekong	160	470	340
Purari/Fly	110	150	1040
Salween	~100	300	300
Mackenzie	199 (100) [100]	310	327
Parana/Uruguay	100	470	195
Zhu Jiang (Pearl)	80	300	228
Copper	70 (70) [70]	39	1770
Choshui	66	6	11000
Yukon	60 (60) [60]	195	308
Amur	52	325	160
Indus	50 [250]	240	208
Zaire	43	1250	34
Liao He	41	6	6833
Niger	40	190	210
Danube	40 [70]	210	190
Rivers that Formerly Discharged Large Sediment Loads			
Nile	0 [125]	0 (was 39)	
Colorado	>1 [125]	1 (was 20)	
Other Rivers that Discharge Large Volumes of Water			
Zambesi	20	220	90
Ob	16	385	42
Yenesei	13	560	23
Lena	12	510	24
Columbia	8 [15]	250	32
St. Lawrence	3	450	7

[a] ( ) Presumed natural level; [ ] year 1890.

Adapted from Milliman and Meade (1983) and Meade (personal communication).

2 Macrotidal Deposits

In this and the following two chapters a description is given of the intertidal areas around the world, subdivided into macro-, meso-, and microtidal areas. They form the basis for the more process-oriented Chapters 6 through 9. The wind flat areas, where the tidal range is small but onshore winds can raise the water level one or two meters (with extensive flooding where the coastal area is flat), are described in Chapter 5. This chapter is concerned with macrotidal areas.

Macrotidal deposits are present where the tidal range is more than 4 m and the configuration of the coast is such that intertidal deposits can be formed with a) shallow water, b) an open coast or an open connection with the coastal sea so that the tides can come in without strong reduction in range, and c) granular material available for transport and deposition. This means that along steep and rocky coasts no macrotidal deposits (or any intertidal deposits) are found except to a limited extent at the head of bays or fjords on platforms at the bases of cliffs. For the development of intertidal deposits of some importance, a rather large shallow area needs to be present so that more than just a tidal beach or a small salt marsh is formed. Macrotidal deposits of some extension are therefore present only in a number of specific areas, while other macrotidal areas, such as the Bay of Penzhinskaya at the northern end of the Sea of Okhotsk, with a tidal range of up to 10 or 12 m, do not have extensive intertidal deposits.

The distinction between macrotidal and mesotidal is not sharp. In macrotidal areas, the spring tidal range or the mean tidal range is more than 4 m. In practice it does not make much difference which limit is taken as a basis for distinction, but the neap tidal range may be well below 4 m in macrotidal areas, since in areas with semidiurnal tides or with mixed tides, the monthly inequality can be very large. Another complication can occur where the tidal range shifts within an intertidal area. Where the macrotidal part is relatively small or otherwise not very significant, the area is considered to be mesotidal and will be discussed in the next chapter (as for instance the Amazon–Guyana–Orinoco area along the north coast of South America, where only a small area near the Amazon mouth is macrotidal). Where macrotides dominate, the area is discussed in this chapter.

2.1. THE WASH (NORTH SEA)

Along the British east coast at about 53°N, the Wash forms a roughly rectangular embayment of 20 by 30 km, which is open toward the North Sea, has a spring tidal

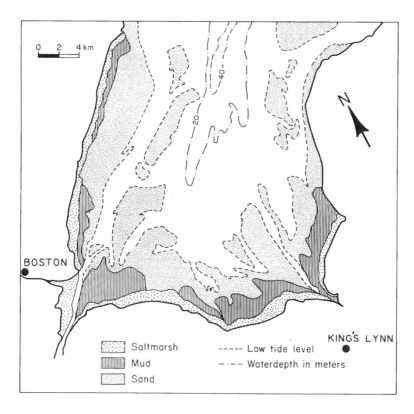

FIGURE 2.1   The Wash. (From Wingfield et al. 1978. With permission.)

range of 6.5 m and a neap tidal range of 3.5 m. The inner margin around high tide level is covered with salt marshes and about half of the remainder of the embayment is filled with bare intertidal sediments (Evans 1975, Evans and Collins 1975, McCave and Geiser 1978, Wingfield et al. 1978; Figure 2.1). The connection with the North Sea is not barred and consists of a more than 40-m-deep channel, a shallower secondary channel, and exposed intertidal flats. At the transition from the open North Sea to the inner embayment, the flats consist of well-sorted clean sands that extend outward into the subtidal zone down to about 6 m. Waves, generally 1 m high but much larger during storms from the north or northeast to east, as well as the tides, prevent fine material from being deposited here. Broad banks, up to 8 km long and 3 km wide, extend from the North Sea into the embayment with an abrupt slope toward the inshore tidal channels. The banks are flat and mostly rippled. On their outer parts below and above low tide level, regular trains of sand waves, with a wave length of 10 to 50 m and an amplitude of 0.5 to 3 m, occur with approximately parallel crests and a lateral extension of several hundred meters. They have a shoreward asymmetry, but do not seem to migrate. Mud patches are deposited in the low parts between them. Between the banks, roughly rhomboid, dominantly flood-oriented megaripples 30 to 60 cm high occur with a long axis of 5 to 15 m and a short axis of 3 to 5 m. On the flat sandy areas, small ripple marks in all directions are abundant, together with some burrowing bivalves, surface dwelling gastropods, and low mounds of *Mytilus edulis*, the edible mussel. Abundant worms that make tubes by gluing sand grains and mollusk shell fragments together (*Lanice sp.*) occur near low tide level. Landward of the low sand flats, locally a low muddy flat occurs that consists of sand with mud laminae (mud laminae <4 mm thick) and mud drapes, abundant small scour marks filled with mud balls and mollusk shell fragments, and some living mollusks. There is some bioturbation by a locally common burrowing worm (*Arenicola sp.*).

Most of the intertidal flats farther inshore consist of interlaminated sand and mud: on the seaward side, sand with mud laminae; more inward, mud with sand laminae. The sand with mud laminae consists on the average of 70% sand with some mollusk shell fragments; the mud with sand laminae, of more than 50% silty mud. Here the laminae are 2 to 7 mm thick and often occur in groups. In the more muddy sediment there is more bioturbation and where bioturbation is strong, the sediment becomes mottled. The muddy surface can be furrowed; the furrows come together and form meandering networks of shallow steep-walled creeks. The meandering creeks usually follow meander belts and do not rework wide areas of the intertidal flat. The larger creeks join the main channels, which are shallow and almost dry at low tide. In the main channels, sand and mud are eroded, and steep banks collapse in broken blocks of sediment. These are transported by the flood and transformed into mud balls and mud pebbles, which can be transported over distances up to 200 m.

On the more inward muddy flats there is abundant bioturbation by *Corophium*, a burrowing crustacean, and in the more sandy flats by *Arenicola* and some burrowing bivalves (*Scrobicularia sp., Macoma sp.*, and *Mya sp.*, more seaward *Cerastoderma sp.*). On the higher mud flats only scattered bivalves occur, and bioturbation is mainly by worms and some *Corophium sp.* Mud cracks are formed, and shallow scour features as well as large concentrations (banks) of living mussels (*Mytilus edulis*) are present and measure up to 1 km^2 in area and more than 2 m thick. *Mytilus* also forms banks on sandy sediments; they fix mud by filtering the mud that is in suspension in the water and release it in the form of pellets: fecal pellets as well as pseudofeces, which consist of the sediment particles not ingested by the animal. The pellets are dropped between the mussels on the banks and around the banks, which gives a muddy sediment. Also a dense population of benthic diatoms may live on the mud: they attach themselves with mucus to sediment particles and form a so-called algal mat. In this way they are less easily resuspended, but storm waves and strong currents can break the cohesive mat apart. Mud cracks appear higher up the intertidal flats where they are less frequently submerged. The cracks are coated with iron oxide and hydroxide.

Salt marshes occur at high tide level, often separated by a low step from the bare tidal flats farther seaward. Near the salt marsh, isolated patches of vegetation are present, which are the initial stage of a salt marsh. Inland, the salt marsh vegetation becomes more luxurious with increasing height above sea level and decreasing submergence by the tides. Steep-sided meandering creeks, 2 to 3 m deep, dissect the salt marshes. The creek bottoms consist of sand or mud and of mud breccias, which are formed by lateral erosion of the creek walls. The salt marshes almost everywhere end landward at a dike.

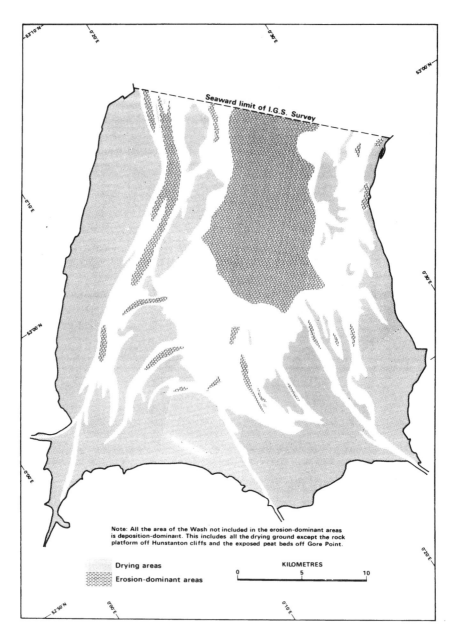

FIGURE 2.2 Areas with dominant erosion and dominant deposition in the Wash. (From Evans 1965. With permission.)

At present, erosion dominates on the outer subtidal parts of the Wash and in the larger channels (Figure 2.2). Elsewhere deposition dominates with only some local erosion in channels and creeks. Yearly 6.8×10^6 tons of suspended matter and 1.4×10^4 tons of bed load is transported into the Wash, mainly through the central channel

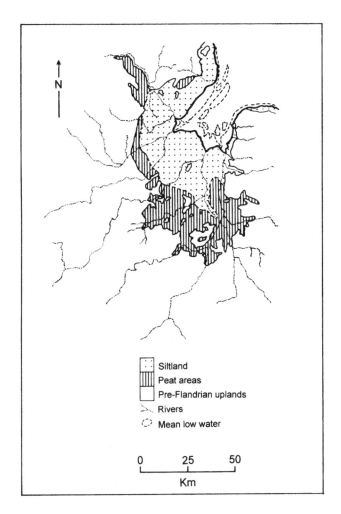

FIGURE 2.3 Land reclaimed since late medieval times around the Wash. (From Wingfield et al. 1978. With permission.)

where the residual sediment transport is landward (inward) and mostly takes place during spring tide. Along the margins, residual transport is seaward (Ke et al. 1996). The present pattern of large banks and channels has been stable over the last few hundred years, as was established from old charts, but since 1828, when more precise soundings were started, the deeper parts have become deeper and the shallower parts shallower. Sedimentation started during the (postglacial) Flandrian sea level rise, and up to 7 m of sediment has been deposited in a vertical sequence of coastal progradation. About 5000 y · BP, sea level reached its present level. The belt of intertidal flats began to form around 2000 BP, followed by the progradation that continues at present. Since the Roman period (which ended at about 450 AD), but mainly since the mid-17th century, the inner marshes and flats have been reclaimed (Figure 2.3), which led to an increase in progradation (1.5 m · y^{-1} in 1828 to 1871; 7 m · y^{-1} in 1903 to 1918; 5.7 m · y^{-1} in

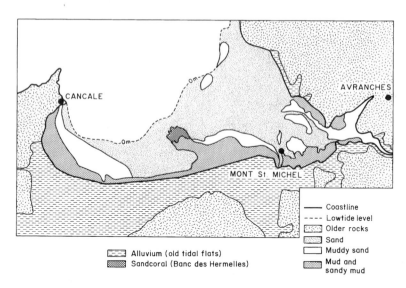

FIGURE 2.4   The Baie du Mont-St. Michel. (From Nikodic 1981. With permission.)

1917 to 1952; Steers 1967). The sediment supplied by the small rivers flowing into the Wash is not sufficient to account for the deposition that took, and takes, place; the sediment therefore is assumed to be supplied from the nearby North Sea floor and the adjacent coast (see also *The Norfolk flats* in Section 2.4).

2.2.   BAIE DU MONT-ST. MICHEL

The Baie du Mont-St. Michel is located within the Gulf of St. Malo along the northwest coast of France around 48° 40′ N on the western side of the English Channel (Larsonneur 1975, Larsonneur and co-workers 1994; Figure 2.4). The bay is open to the north and a few small rivers and streams flow into it. The southern part of the bay is filled up with intertidal deposits which have been formed at a spring tide range of 15.3 m and a neap tide range of about 5 m. The large spring tide range is probably caused by the development of a stationary wave in the bay. Water level is also influenced by the wind and by atmospheric pressure, with high levels during western storms and low air pressure. The southern part of the bay is sheltered against storm winds from the west and is mainly influenced by local winds and strong winds from the northeast. The intertidal deposits cover an area of about 200 km² and consist of bare sandy or muddy flats around high tide level on the inner side of a fringe of salt marshes. Shell banks are common and migrate landward up to spring tide level (at the west side, with a speed of up to 1 km per 10 years) and mingle with the salt marshes. They are formed and moved by the occasional swell entering the bay; the shells are mainly locally produced and reworked bivalves. In the center of the tidal flats between low spring tide level and low neap tide level a large (4 km²) bank of sand coral has been formed by colonies of *Sabellaria*, a polychaete that builds tubular structures up to one meter high with sand grains (the Banc des Hermelles).

The intertidal sediments contain a high percentage of carbonate: 50 to 90% in the coarse sands, 80 to 90% in the shell banks, and 40 to 50% in the fine muddy sands and muds. The sediments can be divided into three groups: those in the western part of the bay, those in the northeastern, more exposed part of the bay, and those in the southeastern estuarine part. Muddy sediments dominate in the western part and form a wide belt from Cancale to near the Banc des Hermelles. Mud deposition is enhanced by the presence of oyster beds in that area, which reduce the current velocity, protect the sediment against resuspension, and induce sedimentation of suspended material through the formation of fecal pellets and pseudofeces. The muddy deposits are flat, bare, and watery with small ebb-tidal creeks. The turbidity increases inward here from less than 5 mg · l^{-1} in the open bay to more than 40 mg · l^{-1} near the shore. Around spring tide level there is near Cancale a thin, narrow, sandy gravel beach (median gravel diameter around 2 mm) of local origin. Farther south and eastward a dike forms the coast with a fringe of salt marsh with shell banks in front.

Toward the east the sand flats broaden and make up the largest part of the flats. The sand is generally fine or very fine and is covered with current and oscillatory ripples. Coarse sands with *Lanice sp.* occur on the seaward side, finer sands with *Arenicola sp.* more landward, and fine sands with *Corophium sp.* on the highest tidal flats up to the salt marshes. Particularly in this eastern area, shell banks are present as well as low sand banks up to 40 cm high. Muddy patches are formed behind obstacles, at the mussel banks, and behind the Banc des Hermelles. Tidal channels cut across the flats (Figure 2.5), incised and meandering in their upper part, straight in the middle part, and in their lower part splitting up into several, frequently shifting distributary branches.

More to the east and southeast sandy levees occur along the larger channels, which are partly tidal and partly fed by river discharge. The sediment is much reworked and redistributed, which reduces the bottom fauna. Along the eastern shore north of the Bec d'Andain the coast is more exposed to storm waves and swells from the west: sandy beaches and dunes occur around spring tide level with a series of bars with low dunes more seaward. The depressions behind the bars are flooded, regularly, and mud is deposited in them which dries out and breaks up during neap tide. New bars are formed on the outside and intermingle with the high flats and the salt marshes.

More inward at the eastern side of the bay, the flats are more estuarine, with numerous shifting, meandering channels and wide salt marshes that are flooded only a few times per year. Sandy muds and muddy sands are deposited in alternating laminae only a few mm thick (locally called tangue). The sandy laminae grade upward into finer material. The fine laminae have a sharp contact with the more sandy lamina above. Besides laminated sediment, lenticular and flaser bedding occurs, as well as load structures, convolute bedding, and bioturbation. Along channels, sliding and slumping takes place. Mud cracks are common and ice cracks are formed where ice crystals occur during the winter. The sediment is not very cohesive, is thixotropic, easily resuspended and (re)deposited at a high rate of up to several cm or dm per year, with an average accretion of up to 2 cm · y^{-1}. The sediment layering reflects the tidal cycles, as well as larger tidal biweekly or annual variations. Intertidal deposition in the entire bay is estimated to be about 1.5×10^6 m^3 · y^{-1} (or about 0.7×10^{12} g · y^{-1}).

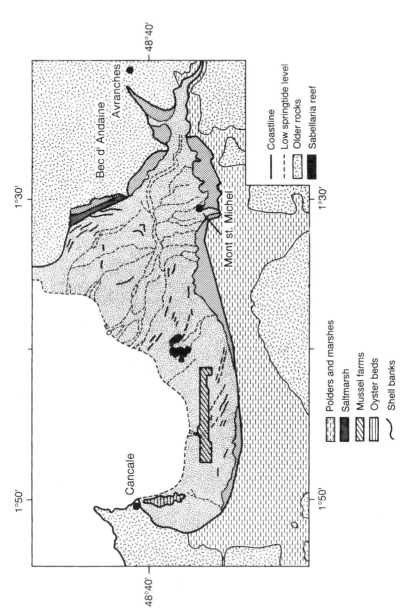

FIGURE 2.5   Channels, oyster beds, and mussel farms in the Baie du Mont-St. Michel. (Adapted from Larsonneur and co-authors 1994.)

Deposition in the bay started around 8000 BP during the Holocene transgression. Around 7500 BP the bay reached its greatest extension at a sea level 10 meters below its present level. Sedimentation has continued up to the present with approximately the same overall deposition rate as the present rate, but deposition was distributed unequally and regularly shifted from one area to another. Bare intertidal deposits were formed mainly when (relative) sea level was rising; during periods of slow sea level rise or no rise at all, salt marshes increased in area. Regularly sand and shell bars were formed around spring tide level. Around 2000 BP the salt marshes reached the level of today, completing a Holocene infill of about 15-m thickness (maximum). The Banc des Hermelles was probably formed originally in a natural depression (part of a former river outflow channel), where natural oyster banks were present (as they still were in depressions about 160 years ago). The shells probably provided a suitable substrate for the sedentary polychaetes. A similar bioherm, but much smaller, occurs at the northeastern end of the bay.

2.3. BASSIN D'ARCACHON

The Bassin d'Arcachon, with an area of 160 km², is located in southern France on the Atlantic coast at about 44° 40′ N and is largely separated from the open sea (the Bay of Biscay) by a large sand spit with dunes. An inlet about 6 km wide forms the connection between the bay and the coastal sea (Verger 1988, Carruesco 1989; Figure 2.6). The tides are semidiurnal with a mean tidal range of 3.1 meters and a maximum range of 4.9 meters. About 52% of the area of the basin is covered with intertidal deposits: bare flats, flats with seagrass (*Zostera*) in the lower tidal zone, and salt marshes around high tide level. The sediment is mostly sand with more silt and clay along the inner margins (Figure 2.7). Within the bay a series of interconnecting tidal channels has been formed that separate isolated large flat banks (Figure 2.8). Most of the channels are either flood dominated or ebb dominated. Salt marshes dominate along the inner margins of the bay, while on the central part of one of the larger banks they form an island (Ile aux Oiseaux) that usually remains dry. Oyster farms have been set up around mid-tide level and on the lower flats. Both the seagrass areas and the oyster farms have increased extensively during the past 30 to 40 years. A small river, the Leyre, flows into the basin at the southeastern end and forms a small delta there.

The sand shows a marked difference in composition between the inside and the outer margin: on the outer side it has the same composition as the sands elsewhere along the coast, while the sand along the inner margin has been supplied from other, older, inland sources. The finer sediment (clays) has a more uniform composition throughout the bay. On the flats, often dense networks of usually shallow, flat, mostly sinuous and interconnecting creeks have been formed. Such networks occur particularly more inward in the bay. The configuration of the creeks, as well as of the channels and salt marshes has been rather stable during the past centuries. On maps since 1708 and on aerial photographs from 1934 to 1954, little difference from the present situation can be seen. The changes that are observed (since the early 19th century) include, apart from the increase of seagrass and oyster farms, increased sedimentation as well as erosion, both on a local scale. Sedimentation has increased because

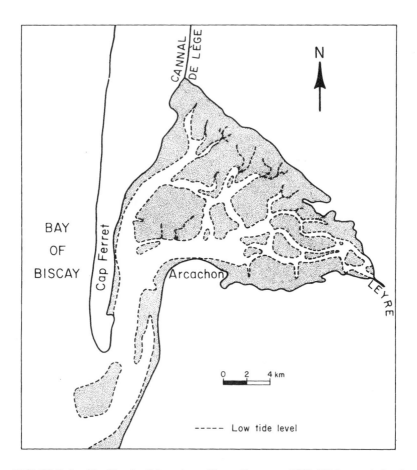

FIGURE 2.6   The Bassin d'Arcachon. (From Carruesco 1989. With permission.)

of an increased supply of sediment from streams and canals, induced by increased canalization, while erosion of salt marshes has increased as a result of the reclamation in the 18th century, which led to erosion outside the dikes and a northward shift of the channels at the Leyre river mouth (Bressolier-Bousquet 1991). Between 1964 and 1988 the salt marsh area has increased about 10% as a result of sedimentation on the delta front of the Leyre river. Since 1989 mud flats have been increasingly colonized by *Spartina* (Labourg et al. 1995).

During the Holocene the first channels connecting the estuary of the Leyre river with the open sea were formed around 5000 BP. The estuarine area was enlarged from south to north, and sands with mollusk shells were deposited on the seaward side with simultaneous deposition of fine rippled and flaser sands, which became covered with marsh deposits. A beach barrier was probably present at that time. After a rapid sea level rise between 2600 and 2200 BP, which resulted in the formation of a new embayment, this basin was subsequently filled with sandy sediment that is very similar to the present intertidal deposits. After 1500 BP the

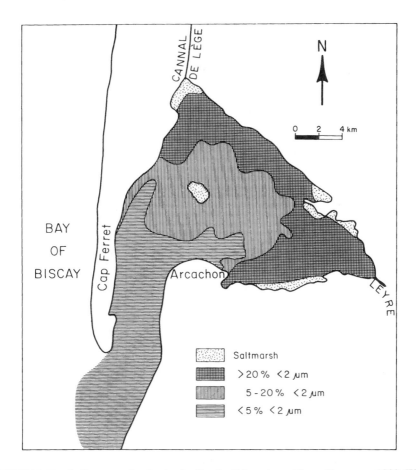

FIGURE 2.7   Sediment grain size in the Bassin d'Arcachon. (From Carruesco 1989. With permission.)

sand drifting southward along the beach pushed the inlet toward the south, while at the same time the bay was enlarged and the present situation developed. The Leyre river was able to keep the entrance open, while other, smaller rivers along the Les Landes coast during the Holocene were closed off from the sea, which resulted in the formation of a series of lakes behind the coastal barrier. Subsidence at the Bassin d'Arcachon is small — in the order of 0.5 mm · y^{-1}, or 5 cm per century.

2.4.   MACROTIDAL RIVER ESTUARIES, RIAS, AND BAYS IN NORTHWEST EUROPE

Along almost all coasts of England and Wales the mean spring tidal range is above 4 m in bays and estuaries like the Solway Firth, Morecambe Bay, Liverpool Bay, Milford Haven, the Bristol channel (Severn estuary), those along the English Channel coast, the Thames estuary, the north coast of Norfolk, the Wash, the Humber estuary,

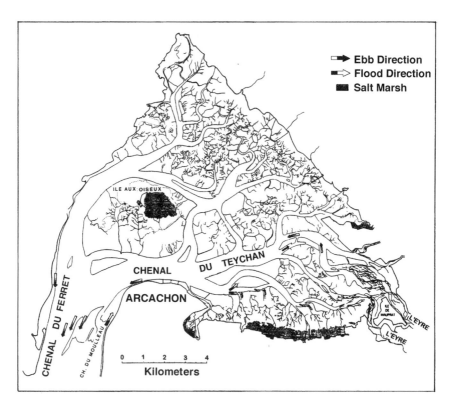

FIGURE 2.8   Channels, creeks, and salt marshes in the Bassin d'Arcachon.

and the smaller river estuaries farther north up to the Scottish border. In some bays the tidal range may reach 6 m or more, up to more than 8 m in Morecambe Bay and more than 12 m in the Severn estuary. In Scotland the tidal range reaches more than 5 m in the Firth of Forth and the Tay estuary. High tidal ranges — above 4 m — also occur along the French Channel coast (Seine estuary 8.0 m, Bay du Mont-St. Michel 15.3 m) as well as along the coast of Bretagne and at the estuaries and bays farther south as far as the Bassin d'Arcachon (4.9 m). Along the Spanish ria coast the tidal range rarely reaches above 4 m except the coast around and east of Santander (see Chapter 3, Section 3.2).

The macrotidal deposits in northwest Europe can be divided into those formed in river estuaries (the Severn, the Thames, and the Humber estuaries, and the smaller estuaries in England, Wales, and Scotland, and those of the Seine, Loire, and Gironde in France), those behind barrier islands (as along the northern Norfolk coast), those along open coasts (rare) like the Essex coast along the outer Thames estuary, those formed in macrotidal rias (mainly in Bretagne), and those formed in macrotidal bays and embayments (both in England and France). The Wash, the Bay du Mont-St. Michel and the Bassin d'Arcachon have been presented in the previous sections. In this section, attention goes to the smaller macrotidal areas.

Intertidal deposits in estuaries mostly consist of relatively narrow strips alongside the estuary, on the landward side usually bordered by high ground or a dike with a fringe of salt marsh, and from there grading into bare tidal flats down to low tide level (Allen et al. 1977; Avoine 1981; Harris and Collins 1985; Verger 1988; Dupont et al. 1994). Relatively wide tidal flats occur along the Severn estuary (up to 4.5 km) and smaller flats along the Humber estuary (up to 3.5 km), the Tays estuary (up to 3 km), and along the Gironde (up to 2.2 km), with narrower flats along the many smaller estuaries. In the Gironde, which is the largest estuary in western Europe with a length of more than 100 km, there are elongated median banks inside the estuary which are rather stable. Other, smaller, elongated banks move gradually shoreward and become attached to the intertidal deposits there. In the Seine estuary, where the mud flats have a high carbonate content (~20%), about 130 years of landfill, marsh reclamation, construction of jetties, and dredging have resulted in the creation of one deep channel, bordered by jetties, with flood channels outside the jetties; originally there were ebb and flood tidal channels and broad tidal flats. In addition to the Seine, the Loire, Humber, and Severn estuaries have been influenced to some extent by human activity.

The **Essex coast** (Dengie Flats, Foulness) at 51° 40' to 51° 55' has a spring tide range over 4.5 m and extensive sand- and mud flats with salt marshes, exposed to the southeast but protected from storm winds coming from western, northern, and northwestern directions. During storms the storm waves along the salt marsh edge can reach 3.4 m (Pethick 1992). This results in flooding and vertical lowering of the salt marsh, as well as retreat of the vegetation edge, and to accretion of the mud flats in front. After exceptionally severe storms it may take five years to restore the marshes and flats to their original condition. A similar response to storms, but less strong, occurs in the nearby, more protected Blackwater estuary.

The open coast marshes have a dissected edge with little or no ephemeral (transitional) vegetation. At the edge is a series of rills and grooves in the salt marsh that may develop into a branching network. They are formed by wave action and eroded in the lower parts of the salt marsh with interfluves at salt marsh level. Sometimes a low cliff marks the limit of the salt marsh, and may have been formed in several ways: in response to a rising sea level, by erosion of extruding sediment, by storm wave erosion, or by wave erosion and undercutting during the lower high tides, which continues during the higher high tides. The creeks in the salt marshes have wide mouths (~50 m) which rapidly decrease in width to about 5 m about 100 m farther inward. The creeks are relatively straight with few tributaries, and at their head they form a branching network of small gullies.

The **Norfolk flats** are located east of the Wash between Hunstanton and Wey-bourne at 53° N and cover a distance of 40 km, partly behind sand and gravel spits (Scolt Head Island and Blakeney Point) and dunes, partly along the open North Sea coast. The mean spring tide range is 6.6 m at Hunstanton and decreases eastward to 4.7 m at Cromer. The tides, where entering shallow water, become very asymmetric and irregular with strong flood currents and lower ebb currents of longer duration than the flood. High tide levels can vary considerably over short distances. The total intertidal area consists of about 30 km^2 of salt marsh and 60 km^2 of sand- and mud flats, while 15 km^2 has been reclaimed since about 1650 (Figure 2.9; Bird

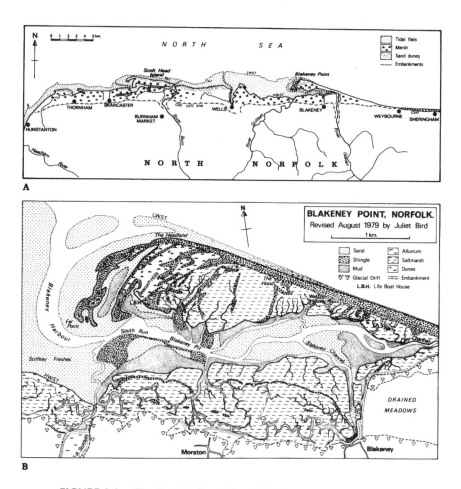

FIGURE 2.9   The Norfolk flats. (From Bird 1985a. With permission.)

1985; Pye 1992). The flats are divided into separate sections by four large sinuous channels and several smaller ones that drain parts of the interconnected flats and salt marshes. The salt marshes consist of younger lower marshes and older higher marshes. In the young marshes the drainage is poorly defined: there is a broad main channel with relatively few tributaries, probably inherited from the previous bare intertidal flat that existed before it became overgrown. Large areas (in the order of 40% of the total) are not flooded and drained through creeks but by sheet flow. In the older marshes a very sinuous network of creeks has developed, with headward erosion and usually interconnections between adjacent networks (Figure 2.10). Levees are formed along the main channels, but away from these, depressions develop where water remains standing, the vegetation decays, and saltpans are formed. Saltpans are circular or oval, poorly drained, and sometimes elongated. In the latter case they have been formed out of abandoned creeks that have not been filled in and did not become covered with vegetation.

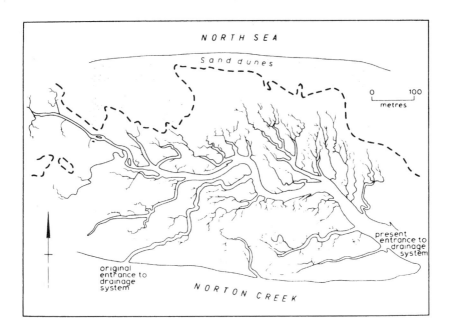

FIGURE 2.10   Channels and creeks on the Norfolk flats. (From Pethick 1980. With permission.)

The formation of intertidal deposits started about 6600 years ago at the end of the postglacial sea level rise on top of freshwater peats. After that, subsidence continued and still continues at a present rate of 1 to 2 mm · y⁻¹ because of subsidence of the southern North Sea floor. The deposition of intertidal flat sediments was interrupted around 4800 to 4400 BP and at about 3500 BP by periods of peat growth, either because of a lower rate of subsidence or, more probably, because of a lower storm frequency in the southern North Sea. Blakeney Point and Scolt Head Island (formerly a spit) are built up with sand supplied from the open North Sea and gravel from eroding Pleistocene glacial deposits along the Norfolk coast farther east. They have been in existence for at least a few thousand years and recently have expanded landward and westward over the flats and salt marshes. Small gravel deposits and ridges are present on the flats and in the marshes.

Salt marsh growth started before about 2000 BP with *Salicornia* (occasionally also *Spartina* and *Aster*; Pethick 1980), and three periods of marsh growth can be distinguished: pre-Roman marshes from before about 2000 BP, medieval marshes from around 400 AD, and the recent marshes formed after 1850. These periods are separated by transgressive periods: the Roman-British transgression, the 12th to 14th century transgression, and the recent transgression. Sediment deposition rates on the marshes range from 1 to 14 mm · y⁻¹ and are related to the elevation of the marshes (average 9 mm · y⁻¹ at low elevation, 1 to 2 mm · y⁻¹ at high elevation, the difference being 80 to 90 cm), and to the opportunities for mobilization and transport of sediment within the marshes (Steers 1938; Stoddart et al. 1989).

The **Dee estuary** at 53° 20′ N is 7.5 km wide and 18 km long with an additional 11 km of reclaimed marshes (Figure 2.11; Marker 1967; Doody 1992). It has a

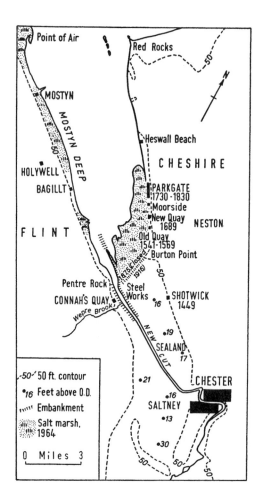

FIGURE 2.11   The Dee estuary. (From Marker 1967. With permission.)

spring tidal range of 7 m where it opens to Liverpool Bay, but in the estuary the tidal volume has been considerably reduced by the reclamations in the upper estuary. Except at spring tide, the estuary falls dry for more than 60% with sandbars and salt marshes exposed. Siltation of the inner estuary started early in the 11th century after the construction of a weir. Rapid siltation continued into the present decades with a vertical accretion of the salt marshes of up to several cm · y^{-1} in 1947 to 1963. The rate of accretion varied with the type of vegetation: the highest rates were measured in the outer parts with *Spartina* and *Salicornia*, lower rates in the outer areas of the *Puccinellia* marshes, and the lowest rates in the inner *Puccinellia* marshes. On older marshes, the rates were even lower and became negative because of compaction (as the rates were measured in cm · y^{-1} and not in g · cm^{-2} · y^{-1}). Accretion on the *Spartina* marshes farther inward, however, is also less than on the outer *Spartina* marshes, although *Spartina* is known to trap mud efficiently and to induce or accelerate mud deposition. In addition to vegetation types, accretion is

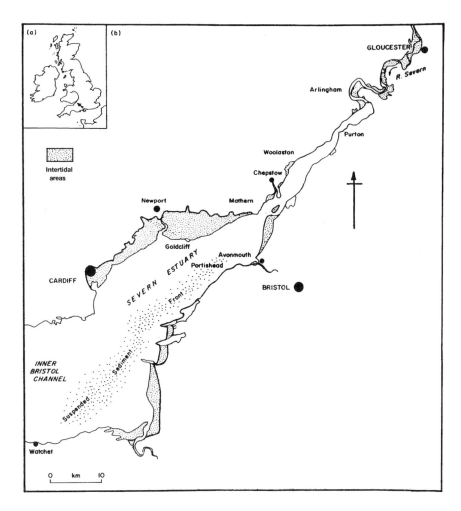

FIGURE 2.12   The Severn estuary. (Adapted from the *Times Atlas*, Kirby and Parker 1982.)

related to the frequency and duration of submergence, which for older, higher marshes is less than for the areas with younger vegetation on the lower marshes.

In the **Bristol Channel/Severn estuary,** at 51° 27′ N (Figure 2.12), ocean swell comes from the southwest, and the maximum spring tide range is 14.5 m, increasing inward from 8.6 m at Swansea to 12.3 m at Avonmouth (where the neap tide range is 6.5 m). Larger and smaller areas of tidal flats and salt marshes occur on both sides. The lower flats are bare, and the marshes are located around high tide level.

Sedimentation on the flats and marshes is determined by the rate of suspended matter supply, the organic sedimentation, the relative sea level change, and sediment compaction. The sediment supply decreases very strongly with increasing mud flat and marsh elevation, so that flats and marshes are built up at an increasingly lower rate. At first the flat is built up rapidly, and then the rate slows down within a short time. Erosion alternates with deposition, with erosion dominating in the winter and deposition in spring, early summer, and early autumn, which is related to the degree of erosional action of waves (winds).

Overall deposition on the high flats is in the order of 10 to 20 cm $\cdot$ y^{-1}. Organic deposition, which includes roots and litter, is in the order of 1 mm $\cdot$ y^{-1}, depending on the complexities of the marsh vegetation. The relative sea level rise is in the order of several mm $\cdot$ y^{-1}. The marsh sediments in the middle and upper estuary show a layering that consists of groups of sandy laminae and groups of mud laminae: the sandy laminae are considered to represent winter–spring deposition; the silt/mud laminae represent deposition in summer and autumn. Upward in the sediment the layers become thinner. The marsh edge is often terraced, and the terraces can form steps going down to low tide level. The marsh surfaces on the terraces are of different age, and the terrace elevation is lower when the marsh is younger, while the difference in elevation is related to the absolute ages of the marshes (Kirby and Parker 1983; Allen 1990).

The **Solway Firth** at about 54° 50′ N is a broad bay with extensive sand and mud flats, fringed by salt marshes (Figure 2.13). The mean spring tide range is 7 to 8 m. Large channels are present in the sandy flats; on the mud flats and mixed silt/mud flats, there are meandering gullies, 0.5 to 2.5 m deep and 0.4 to 25 m wide. Strong meandering dominates (Figure 2.14; Marshall 1964; Bridges and Leeder 1976). The gullies develop from small pools that become increasingly linear and coalescent in a seaward direction. Straight gullies are formed of 10- to 30-m length, then some sharp (almost 90°) bends occur, followed by meandering with undercutting of the walls, with the gullies increasing in depth and width downstream. The point bars within the meanders are of two kinds: with a sigmoidal cross section across the gully-axis, and with a prominent flat lower platform, 2 to 3 m wide. The first type occurs in nearly all small meanders; the second at all sharp bends in the medium- to large-sized gullies (that also have levees on the cut-bank side). Rotational slumping occurs mainly in association with the second type of point bar and can result in the formation of steep terraces. The bedding on the point bars is usually interlaminated silt/mud with mud laminae of 1 to 15 mm and silt in elongated lenses of 2- to 10-mm thickness. In large gullies the mud is present in flasers; in small gullies as persistent laminae. Large point bars have runoff rills, mud-filled scour features, burrows of *Corophium, Nephtys, Macoma,* and *Pygospio* with algal stromatolite-type formations combined with *Pygospio* tubes. Most of the deposition and erosion takes place during the late ebb at spring tide, as well as during rainfall and exposure at low tide.

The salt marshes are flooded only occasionally and are mainly wet from rain. Their edge can be accretionary (with *Spartina* and *Salicornia* and, more inward, *Puccinellia*) or erosional with a cliff of normally 1 to 2 m, but up to 3 to 4 m high. Erosion of the salt marsh is fluvial (as at the Burgh marshes in the inner estuary) or marine by waves or the tides (as at the exposed Cearlaverock marshes in the outer bay). Accreting marshes occur in sheltered areas.

Well-developed, widely spaced dendritic creek systems are present in the larger marshes. On sandy flats they have a wide funnel-shaped mouth. On mud flats the mouth is narrower. Tidal overflow over the marsh edge and the creek borders produces levees. Vertical accretion and some bed erosion produce deep creeks on the higher marshes while lateral erosion makes them wider, and headward erosion extends them inward. Lateral erosion leads to undercutting and slumping of the

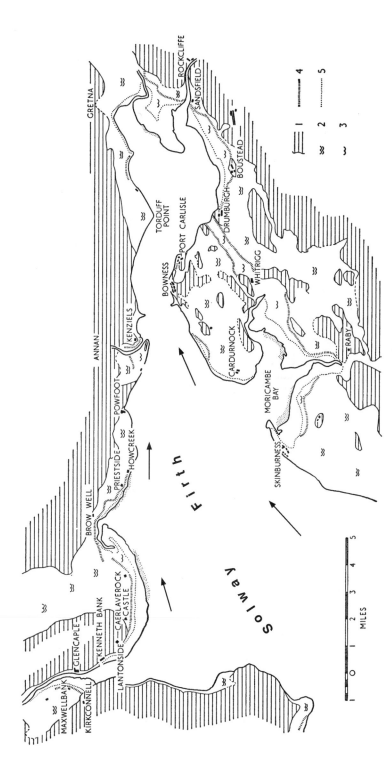

FIGURE 2.13 The Solway Firth. 1 = raised beaches (inner edge); 2 = upper terrace; 3 = lower terrace; 4 = terrace banks; 5 = inner edge recent alluvial deposits (salt marshes). Arrows indicate main directions of sediment transport.

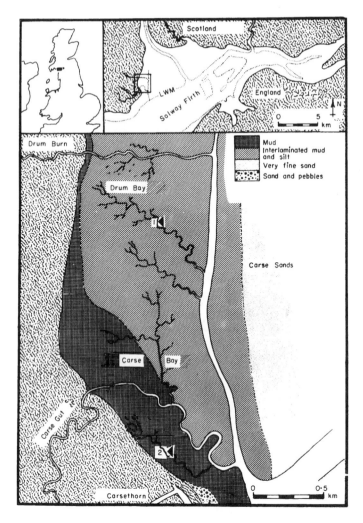

FIGURE 2.14   Tidal flat channels in the Solway Firth. (1 and 2 indicate discharge measuring stations). (From Bridges and Leeder 1976. With permission.)

walls, which is enhanced by heavy rainfall and, in the lower reaches, by waves. The creeks are usually of the same age as the surrounding marsh, but new creeks can form on old marshes after damage to the vegetation. The tidal water is then drained through the damaged part, and it was observed that full-sized creeks were formed within nine months. Saltpans are common on the marshes and can be of four types: 1) developed during the formation of the marsh as low areas between the creeks, 2) formed by erosion of the surrounding marsh, 3) formed by the blocking of creeks that then are only partially filled in, and 4) as residual lows after marginal colonization of large pans by vegetation. Strong damage to the marshes is caused by turf-stripping (e.g., by grazing cattle), by burial with overwash sand, and by turf-cutting, whereby 7 to 9 cm of top soil with marsh vegetation is removed.

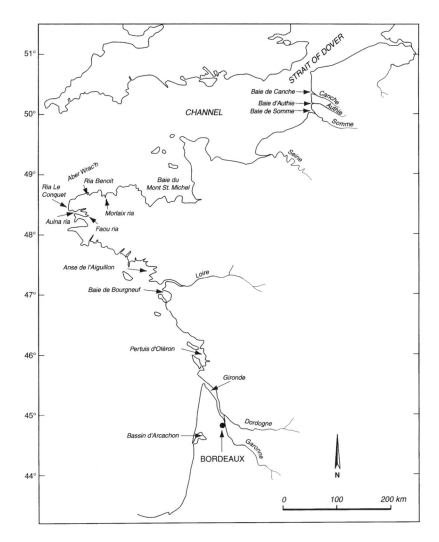

FIGURE 2.15   The Atlantic coast of France.

Macrotidal estuaries like the Solway Firth and the Severn estuary, but also the smaller ones, are usually funnel-shaped and are curved to some extent or almost meandering. Barriers are virtually absent. Jago (1980) described a dune barrier with a tidal flat behind it that shelters the Taf estuary toward the open water of the Bristol Channel, but this estuary is a tributary to the Towy estuary, which is not sheltered and has the usual funnel shape.

The intertidal deposits in most **rias in Bretagne, France** (Figure 2.15; rias are characterized by a shallow inward part and a deep outward part) are more sandy on the outer side and finer grained (pelitic) on the inward side (Figure 2.16; Guilcher and Berthois 1957; Guilcher et al. 1982; Castaing and Guilcher 1995). In some rias the sediment in the entire ria is sandy, but there is still an inward decrease in grain size. In two exceptional rias, where waves can enter more easily from the open sea, marine sediment is transported inward and a mid-bay bar or spit is formed with coarse pebbles, whose transport has been facilitated by the buoyancy of attached kelp. In the Le Conquet ria, this bar has been removed (for the construction of concrete; Figure 2.16A), but in L'Auberlac'h ria, it is still preserved (Figure 2.16D. Rias usually are not broad or large so that the intertidal deposits in rias cover only a small area (Figure 2.16). The origins of the sediment are Pleistocene (periglacial) deposits on the slopes eroded by waves during spring tide; the eroded sediment is then deposited mainly on the adjacent salt marshes. The salt marshes in almost all rias have a low erosional scarp, produced by local waves and in some places by scour from a creek (where it can become several meters high). Simultaneously, fine sediment is being deposited on the salt marsh that is eroded, and the vertical accretion almost compensates for the erosional loss of sediment. The finest sediment is usually deposited on the salt marshes. In addition to salt marshes virtually without channels or creeks, salt marshes with several types of drainage occur (as described for the Le Conquet ria (Figure 2.16A; Guilcher and Berthois 1957): a) a network of small creeks, differences in elevation of 10 to 20 cm, (probably reflecting the presence of older, partly filled-in creeks), and elongated pools without vegetation that fall dry or (on the high marshes) remain flooded; b) relatively large creeks or channels that separate large salt marsh areas with networks of small creeks; and c) numerous isolated patches of salt marsh (round or elongated, <1 m²) separated by bare mud flats, while large areas of patchy salt marsh are separated by large channels. At high levels in some rias, bare flats can also be present that are less regularly flooded. A genetic relationship has been postulated going from undissected salt marsh without creeks, to increasing dissection by creeks and channels, to a stage of a large number of separate patches of salt marsh, and, finally, to the complete disappearance of the salt marsh, after which a bare high flat is left. The fringe of high flats near the salt marshes is usually overgrown with *Salicornia*. The sediment on the bare flats along the creeks and channels can be very soft mud. The sediment in the creeks is more sandy than the sediment on the flats and may contain large amounts of mollusk shells (*Cerastoderma edule*) and shell fragments. In the Morlaix ria, it has been demonstrated that at low river discharge, fine sediment is transported inward and deposited, whereas during high discharge the fine sediment is being resuspended and transported downstream, with erosion on the lower mud flats in particular (L'Yavanc and Bassoulet 1991).

The intertidal deposits in the bays and embayments along the Atlantic coast of France have been described by Verger (1988) and include the Perthuis d'Oléron, l'Anse de l'Aiguillon, the Baie de Bourgneuf, the Baie de Somme, and the Baie d'Authie (Figures 2.17 to 2.22).

The **Perthuis d'Oléron** (Figure 2.17), between the Ile d'Oléron and the mainland coast between the Charante and Seudre rivers at about 46° N, is a wide intertidal area about 20 km long. A large north-south directed sinuous channel with several elongated islands connects an inlet in the south between the island of Oléron and the coast with a large bay in the north (Figure 2.17). On both sides of this channel are intertidal flats up to 4.5 km wide dissected by channels and numerous creeks, which mostly follow an elongated dendritic pattern. They are subparallel and oriented

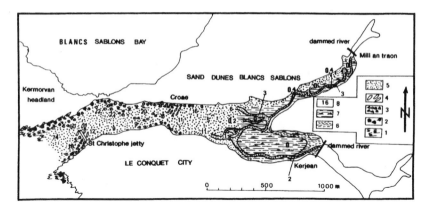

FIGURE 2.16   Sediments and channels in rias in Bretagne. A) Le Conquet Ria. 1 = rocky flats; 2 = large blocks; 3 = pebbles; 4 = former pebble spit (now destroyed by man); 5 = sand; 6 = low mud flat; 7 = salt marsh; 8 = CaCO₃ content (%). (From Castaing and Guilcher 1995. With permission.)

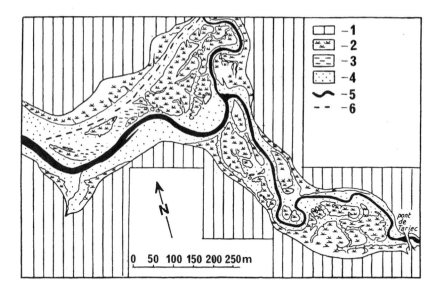

FIGURE 2.16   B) Tariec estuary. 1 = supratidal; 2 = salt marsh; 3 = high mud flat; 4 = mud flat; 5 = channel at low tide; 6 = main channel during high tide. (From Guilcher and Berthois 1957. With permission.)

at right angels to the outward edge of the flats. Salt marshes are present along some of the landward fringes and large seagrass (*Zostera*) fields on the flats along Oléron Island (Guillaumont 1991); rocky islands of Jurassic limestone are scattered through the flats. The main channel is flood-dominated with the flood entering both from the south and the north, while the ebb flows off mainly over the flats. The channels and creeks on the flats are shaped during the late ebb. Sandy flats occur on the higher

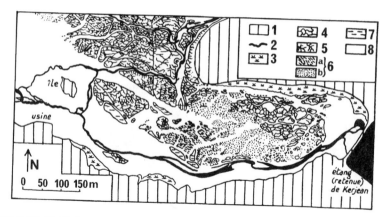

FIGURE 2.16 C) Kerjean branch at Le Conquet estuary. 1 = supratidal; 2 = main creeks; 3 = salt marsh; 4 = marsh with creeks filled with vegetation and with saltpans; 5 = marsh with numerous small creeks without vegetation; 6 = much dissected marsh with mounds (a: long, b: round); 7 = high tidal flats; 8 = (low) tidal flats. (From Guilcher and Berthois 1957. With permission.)

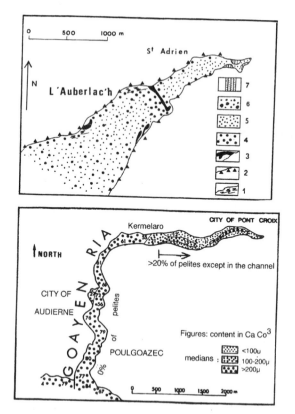

FIGURE 2.16   D) L'Auberlac'h Ria and Goayen Ria. 1 = rocky flats; 2 = cliffs; 3 = spits; 4 = coarse sediment; 5 = low mud flats; 6 = flats with mixed sediment; 7 = salt marsh. (From Castaing and Guilcher 1995. With permission.)

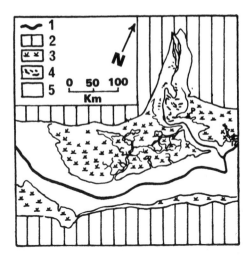

FIGURE 2.16   E) Kerroule estuary. 1 = main creeks; 2 = supratidal; 3 = salt marsh; 4 = bare mounds on mud flats; 5 = mud flats. (From Guilcher and Berthois 1957. With permission.)

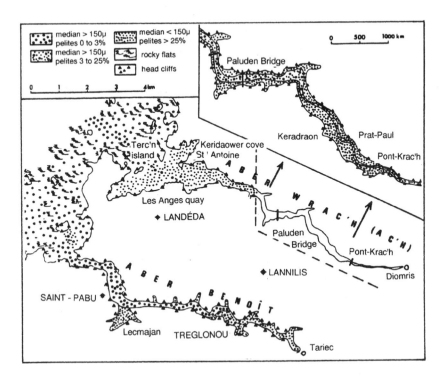

FIGURE 2.16   F) Aber Wrac'h (Ac'n) and Aber Benoit Rias. (Adapted from Castaing and Guilcher 1995.)

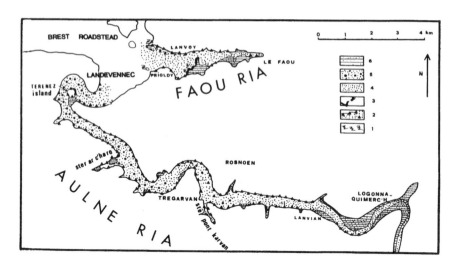

FIGURE 2.16   G) Faou and Aulna Rias. 1 = rocky flats (Faou Ria); 2 = low cliffs; 3 = spit (Faou Ria); 4 = sand and mud flats; 5 = same as 4 but with coarse shell fragments; 6 = salt marsh. (Adapted from Castaing and Guilcher 1995.)

more exposed parts; mud flats in the more sheltered areas. Gravel occurs only near the rocky outcrops and on the floor of the main channel. In the central part near the main channel, megaripple fields are present. A tidal watershed might be expected in this area where the flood from both sides meets, but instead of a tidal watershed, a connection is formed by a channel of 6 to 8 m deep which is part of the main channel. This is probably induced by the narrowing of the passage between the southern and northern flats, which results in strong currents where the tides are forced through. Salt marshes occupy small areas and are only extensive in the estuary of the Seudre river. Large seagrass (*Zostera*) fields occur on the flats east of the Ile d'Oléron.

The **Anse de l'Aiguillon** (Figure 2.18). Around the funnel-shaped estuary of the Marans river at about 46° 20′ N, intertidal flats up to 5 km wide have been formed in an almost circular basin (Figure 2.15). The lower flats are dissected by numerous creeks and small channels in a pattern similar to that in the Perthuis d'Oléron. The higher flats are almost without creeks or channels, and the inward fringe is covered with salt marshes up to 900 m wide. The large channels that cross both the high and the low flats are mostly continuations of channels that drain the reclaimed older areas surrounding the basin. They are several meters deep and 20 to 40 m wide, increasing in size downstream. They are straight or sinuous and form low levees. Where the channels and creeks follow a largely dendritic pattern, the tributary creeks tend to be oriented at right angles to the larger channel. The channels move laterally, with erosion of the flats on one side and deposition on the other, in combination with vertical accretion. Because vertical accretion takes time, the older eroding flat is usually higher than the new flat formed by lateral accretion. The levees indicate that the vertical accretion is related to the supply of sediment from the channel.

The tides cover the flats more rapidly than they fill the channels, so during the flood a flow develops from the flats to the channels and small subparallel gullies are

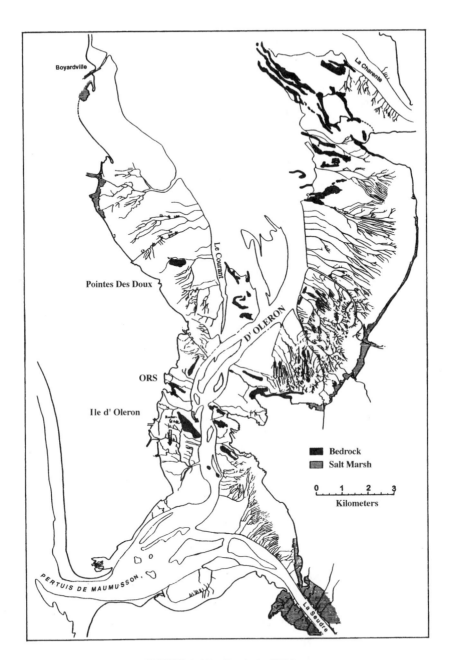

FIGURE 2.17 Perthuis d'Oléron

formed at approximately right angles to the channel. On the mud flats, regular patterns of low ridges or undulations are formed whose orientation corresponds to the current directions during submergence. Through the gullies developed during the flood, the flats are also drained during emergence. The limit between the high

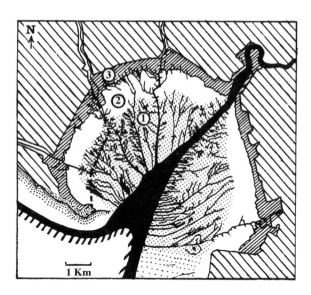

FIGURE 2.18   Anse d'Aiguillon. Black area = Marans river; 1 = central area with network of channels and creeks; 2 = first outer zone (with few channels and creeks); 3 = second outer zone (with salt marsh).

flats with few channels and creeks and the low flats is rather sharp and coincides with the neap high tide level. When between two neaps, the high tides come higher, and small initial gullies are formed that are filled in again when the flood reaches progressively lower levels. As deposition on the flats continues, the boundary between the high flats and the low flats gradually moves downward toward the central channel. Aerial photographs taken between 1923 and 1955 show this reduction in the length of the creeks from the higher flats downward.

The relatively few creeks that are present on the high flats are strongly meandering and not straight or sinuous as on the lower flats. They are almost exclusively formed by the ebb, in particular during the late ebb, when there is a rather large difference in elevation between the flats and the channel. During the flood, which comes in rather late, the creek is gradually filled with water without a strong current. On the high flats that are flooded at least 70% of the time, the larger channels are often bordered by flat levees, 8 to 27 cm high and up to several meters wide. They consist of sediment of the same size as the surrounding flats (in contrast to fluvial levees). They do not have any vegetation and are probably formed out of more consolidated mud flakes that are eroded from the channel banks during flood. They are deposited near the channels, while the less consolidated material is more disaggregated and deposited farther away on the flats. Such levees are most prominent along meanders where erosion is stronger than along the more rectilinear parts of channels and creeks. Sometimes two levees are observed that correspond to different tides. Deposition on the levees probably takes place in very quiet water so that the more consolidated and larger flocs can be segregated from the more fragile and smaller flakes and flocs. The levees are eroded when, at low tide, the sediment dries out and cracks develop that result in flakes. These are then detached by waves during stormy weather.

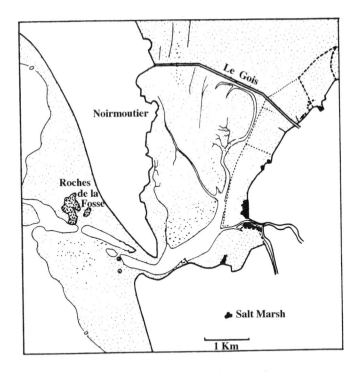

FIGURE 2.19   Baie de Bourgneuf.

Outside the few channels and creeks, the high flats are wide flat areas with only weak undulations up to 4 cm high and about 1 m wave length. Deposition of the high flats is only vertical, but there may be some transport of desiccated mud flakes and entrainment of sediment by floating algae. Bioturbation by bivalves (*Scrobicularia*) and a gastropod (*Hydrobia*) is evident, while crustaceae (*Corophium*) are more numerous on the lower tidal flats. The upper parts of the high flats are colonized by *Spartina* and *Salicornia*, the highest parts since 1950 also by *Puccinellia*, and the levees by *Aster.* Where this vegetation forms a closed front, this front tends to move forward by up to 10 m per year, as the vegetation enhances sedimentation. Between the vegetation and the bare high flats there is often a low scarp or slope of 1 to 2%. The salt marshes are flooded by 30% of the high tides at most, usually less, and are flat with relatively short creeks that begin near the dikes which form the boundary of the l'Anse de l'Aiguillon and end for the most part at the limit of the vegetation. Some, however, cross the high flats and continue as channels in the low flats down to the Marans river channel. The margins have been used regularly for sheep grazing.

The **Baie de Bourgneuf.** This bay, at about 47° N, is a tectonic basin, probably dating from the late Mesozoic, which is open to the north and has a channel and a tidal inlet in the south that form the connection with the open sea (Figure 2.19). Off the inlet, which has a maximum depth of 15 m, an ebb tidal delta has been formed. The tide comes into the bay from both the south and the north. The two tidal waves meet at a tidal watershed which is relatively near the inlet. Here, at the end of the 18th century, a dam was constructed (called Le Gois) which lies approximately at mean sea level and is dry for five hours during spring tide (four hours during neap tide). Flow over the dam goes from north to south, because of the larger flood volume in the north, and deposition on the northern side is higher. The intertidal deposits north of the dam are sandy and flat, and are reworked by numerous worms (*Arenicola*). They are without large ripples or undulations, and the sediment becomes somewhat finer toward the sides. The salt marsh along the fringes is very small. South of the dam, sedimentation is enhanced by a stone wall erected in 1850 and enlarged in 1895. Behind the wall up to 1 m of alternating layers of sand and mud have been deposited. These are bioturbated by *Arenicola* and have megaripples at the surface. Salt marshes are present in small sheltered areas and in a small river estuary. They are separated from the bare flats by a slope. Deposition on the marshes can be up to 2 cm per year.

The **Baie de Somme.** This bay is the westward extension of the broad Somme river valley at about 50° 15′ N. The intertidal deposits cover a bay that extends 14 km. Its entrance is bordered by a pebble spit (the Pointe de Hourdel) in the south and a beach ridge of pebbles and sand (the Pointe de St. Quentin) in the north (Figure 2.20). A system of ebb and flood channels has been formed which grades into the Somme river estuary. The channels are very shallow in the bay inward from the Pointe de Hourdel and the Pointe St. Quentin, and are less than 60 cm deep at low tide. Megaripples are common on the outer flats but increasingly less frequent farther inside where the flats become finer grained. Deposition during the past 85 years, mainly along the inner fringes, has been up to 7 cm · y⁻¹. Salt marshes with numerous duck ponds occur along the southern and eastern borders of the bay.

The **Baie d'Authie.** Intertidal deposits cover a much smaller area in this small bay (at 50° 25′ N) than in the Baie de Somme, but the general configuration is similar. The outward channel of the Authie river is sinuous and has been moving laterally over the tidal flat area. Between 1865 and 1955 almost the entire intertidal area was covered (Figures 2.21 and 2.22). To reduce such shifts, a dike has been constructed in the inner bay so that the land reclamations to the north would not be endangered. Flat salt marshes have been formed behind the dike to the north, as well as along the southern border of the inner bay.

A smaller bay, similar in form, is present farther north (the Baie de Canche, at about 50° 35′ N) where the channel of the Canche river is completely fixed between two submerged dikes (Figure 2.23).

2.5. MOZAMBIQUE

The coast of west and south Africa is micro- or mesotidal. Macrotidal coasts are located only along east Africa: around Beira at 19° S, the spring tidal range is 6.4 m and the neap tidal range 4.1 m, while farther north at the Zambesi river delta, the spring tidal range has decreased to 4.1 m and the neap tidal range to 2.9 m (Tinley 1985; Figure 2.24). Between the mouths of the Save and the Buzi rivers, where the tidal range is largest, extensive bare intertidal mud and sand flats have been formed. The absence of mangrove here, which is common elsewhere along the Mozambique coast, is probably caused by the strong sedimentation in combination with the high tidal range which results in regular and strong reworking of the intertidal sediments.

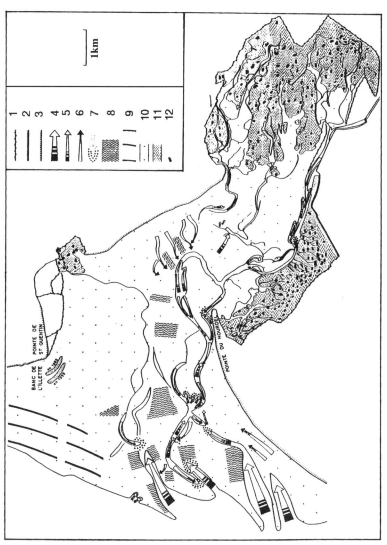

FIGURE 2.20 Baie de Somme. 1 = dike, 2 = cliff, 3 = sand and pebbles, 4 = flood inflow, 5 = flood channel, 6 = ebb channel, 7 = tidal chute (flood, ebb), 8 = megaripple field, 9 = surf ridges, 10 = flats, 11 = salt marsh, 12 = duck pond.

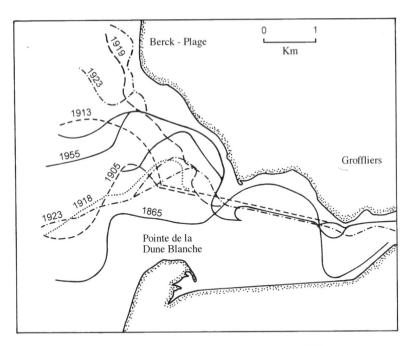

FIGURE 2.21   Position of channels in the Baie d'Authie since 1865.

Farther north at the Zambesi river mouth, where the tidal range is lower, beach ridges and cheniers form the outer edge of the delta with mangrove swamps behind them.

2.6. MADAGASCAR, WEST COAST

Along the Madagascar west coast between 12° and 25° S the tidal range generally is above 4 m. Most of the coast is protected by reef barriers, and intertidal deposits only occur where the coast is low (Figure 2.25). They are invariably covered with mangrove, which shows a clear zonation: *Ceriops* and *Avicennia* around high tide level, *Bruguiera* about 20 to 60 cm below high tide level, and *Sonneratia* and *Rhizophora* at 70 to 110 cm below the high tide level in the intertidal zone (Salomon 1978).

2.7. GULF OF KHAMBHAT (CAMBAY) AND THE GULF OF KACHCHH (KUTCH)

Between about 21° and 22° N in northwestern India, the Gulf of Khambhat (Cambay) extends in a roughly north-south direction, narrowing in width from about 60 km at the entrance in the south to a few km in the north. Here two large rivers, the Sabarmati and the Mahi, enter the Gulf from the north; two other large rivers enter from the east — the Narmada and Tapti rivers (Figure 2.26). The Gulf of Kachchh (Kutch) is situated to the west of the Gulf of Khambhat at about 22° 30′ N and

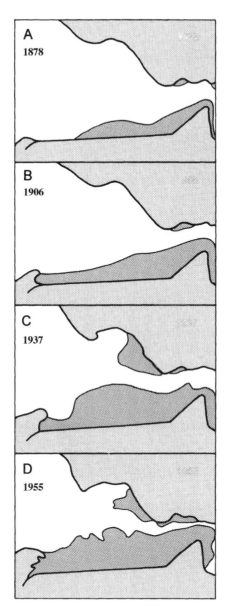

FIGURE 2.22 Growth of salt marshes (schorre) between 1878 and 1955 in the Baie d'Authie.

continues to the northeast in the form of extensive salt flats — the Little Rann — that are connected with the Rann of Kachchh and the salt flats north and northwest of the Gulf of Khambhat. These flats are probably recently filled-in former extensions of both gulfs. The salt flats to the northwest of the Gulf of Khambhat were raised during the earthquake of 1819 (Rogers 1870); Lake Nal, a brackish water lake of

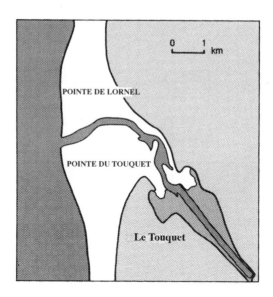

FIGURE 2.23   Baie de Canche.

35 by 6 km, is located between the northwestern end of the Gulf of Khambhat and the Little Rann, and is probably a remnant of a former connection. The mostly bare salt flats of the Rann of Kachchh and the Little Rann are flooded with 0.3 to 0.6 m of water during the southwest monsoon and spring tide. The tides are semidiurnal and reach, in the Gulf of Khambhat, a spring tide range of 9 m, and in the Gulf of Kachchh, 6.5 m (*West Coast of India Pilot* 1975). A tidal bore develops in the Sabarmati and Mahi river mouths that reaches a height of 2.4 m during spring tide in the Mahi river. Tropical storms and swells come from the south to southwest.

At the entrance of the Gulf of Khambhat, four systems of large elongated sand banks have been formed (the Malacca Banks) that are partly intertidal, while the channels between them reach water depths of 20 to 40 m (*West Coast of India Pilot* 1975). The Gulf is fringed by sand- and mud flats with larger and smaller delta flats at the river mouths. At the northern end, where the Sabarmati and Mahi rivers enter the Gulf, a large intertidal sand bank — Mal Bank — is bordered by channels with more than 22 m water depth. Inward into the Gulf, the flats are more muddy, and northward of the Bhanvnegar and Narmada river estuaries, mangrove fringes the otherwise bare flats on their inward side. In the Gulf of Kachchh, sandy and muddy intertidal flats are located at the entrance on the north side, while the south side consists of numerous small islands surrounded by coral reefs, but east of Okha the reefs are mostly dead and covered by mud banks and mangrove. On the north side, large mangrove swamps occur from Medwah Point to the east, partly behind sand ridges. The inner part of the Gulf of Kachchh is bordered by extensive mud flats and mangrove often with sand flats or sandy ridges on the seaward side. The Little Rann river flows out into the Gulf with a small delta, but is dry from November until February during the northeast monsoon. The area is much influenced by the recent port development at Kandla.

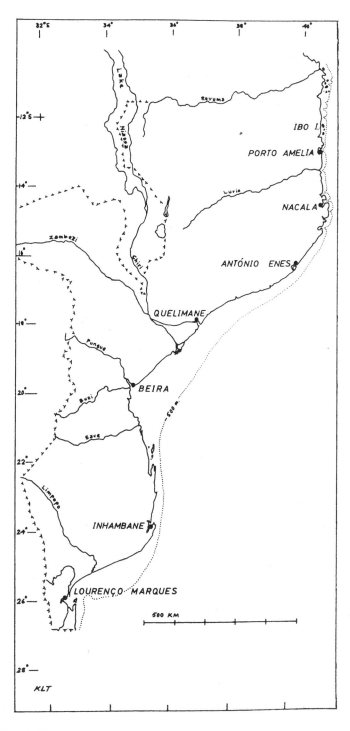

FIGURE 2.24 The coast of Mozambique. (From Tinley 1985. With permission.)

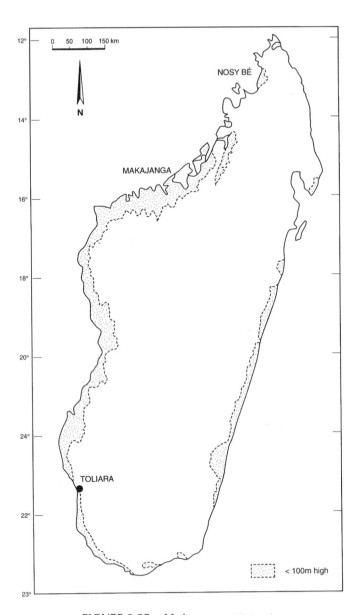

FIGURE 2.25   Madagascar west coast.

2.8.   GANGES–BRAHMAPUTRA (MEGHNA) RIVER MOUTHS

The Ganges–Brahmaputra river mouths form a broad front of more than 380 km, from the Hooghly river in the west to the Teti river in the east along the northern end of the Bay of Bengal at about 22° N. The river outflow is concentrated in the eastern part, whereas in the western part, the tides dominate. Here tidal channels and intertidal flats have formed a network of interconnecting channels and numerous

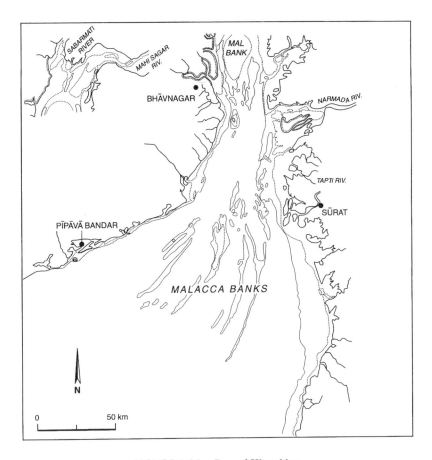

FIGURE 2.26   Bay of Khambhat.

islands with shallow, saucer-like interfluves between the creeks (Ahmad and Khan 1959, Figures 2.27 and 2.28). The eastern flats are bare of vegetation, but those more to the west are covered by mangrove forest over an area of 6000 km² (the Sundarbans), which mainly consists of *Heritiera* and *Excoecaria* trees, with a dominance of *Nypa* palms in the freshwater parts. The tides are semidiurnal and reach a range of 5.6 m at the Meghna mouth, but during monsoon storms sea level may reach 7 m above mean tide level (Islam 1978; Snead 1985). The tide-dominated channels in the western part receive freshwater from the reclaimed flat lands more inland, which are drained by numerous canals (partly former natural channels) and small sluices that open during low tide. The older channels after reclamation at first retained their original function as flow channels, but then formed levees, and deposition rose higher on the bed, so that finally the channels were much higher than the surrounding land and remained mostly dry. Farther to the south, tidal creeks and large channels were enclosed by embankments, while former flats were reclaimed down to spring tide level. During the 18th and 19th centuries lower flats were also reclaimed with dikes at low tide level (Mukherjee 1969), but because of the high cost of maintenance,

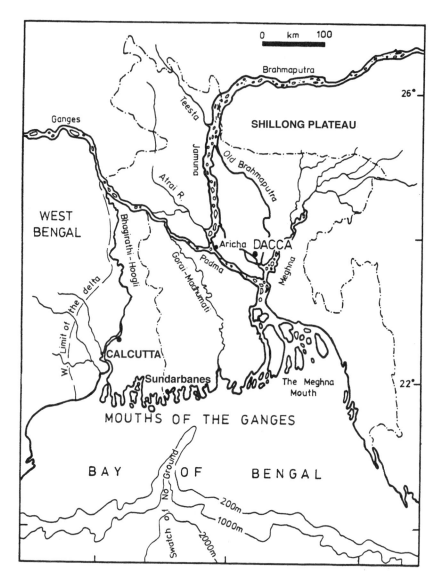

FIGURE 2.27   Ganges-Brahmaputra (Meghna) mouth.

these reclaimed areas were neglected and gradually given up. When silt could not be deposited any more on the lower flats, it was deposited on the channel bed and, with time, the dikes had to be raised to a higher level repeatedly. After the areas were abandoned, the forest returned, but in recent times it has been cleared again, so that now only the southern part of the Khulna district in Bangladesh and adjacent areas in India remain unaffected. Locally, reclamation extends as far as the shoreline (Mukherjee 1969), in particular on the western side where along the coast only scattered, partly cut mangrove forests remain (Snead 1985).

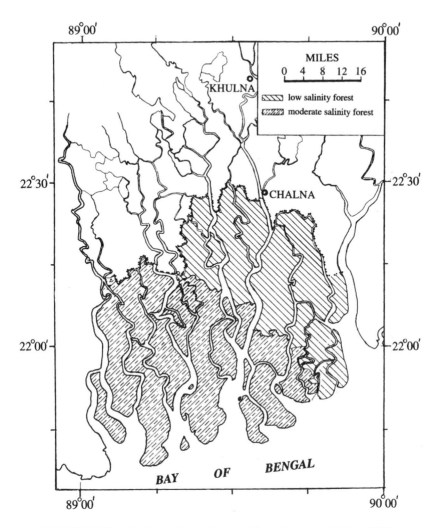

FIGURE 2.28   The Sundarbans. (Adapted from Ahmad and Khan 1959.)

The forests in the Sundarbans cover an older deltaic plain formed when the Ganges followed a more westerly course before it shifted gradually to the east, probably because of subsidence along a north-northwest axis at the present river mouth. Although very large amounts of sediment are supplied by the rivers (~500 × 10^6 tons per year by the Ganges and ~700 × 10^6 tons per year by the Brahmaputra), it is not clear whether the coast is accreting. The presence of old beach ridges and low dunes along the coast of the Sundarbans points to some accretion since their formation, but a comparison of recent surveys with maps from 1770 to the present shows little difference. Also no delta has been formed into the Bay of Bengal. The rate of subsidence is probably more or less equivalent to the sediment deposition, plus there is a loss of sediment through the Ganges canyon to the deeper parts of the Bay of Bengal where a large submarine fan has been formed (Curray and Moore 1971; Kuehl et al. 1989). Local deposition and erosion, however, can be very intensive in the eastern part. Although human interference has led to a reduction of the mangrove forest farther west, there is no indication that the present bare and quickly shifting flats in the river mouth area have ever been covered with vegetation. In the Sundarbans, fine-grained sediment is deposited during and after high tide when the area is flooded, but also here, in spite of only little evidence of erosion, the coast does not seem to be accreting, or is accreting only at a very slow rate.

2.9. IRRAWADDY RIVER DELTA

The delta of the Irrawaddy river forms a broad front at about 16° N along the Gulf of Martaban and the adjacent Bay of Bengal. Nine large distributary channels divide the river discharge (about 423×10^9 m$^3 \cdot$ y^{-1}) over a distance of about 300 km from west to east. The suspended matter brought yearly into the coastal sea is about 367 $\times 10^6$ tons per year during the rainy season, while 273×106 t $\cdot$ y^{-1} is returned by the tides, so in the order of 94×10^6 t $\cdot$ y^{-1} is transported into the coastal sea (Volker 1965; Gibb and NEDECO 1976; Figure 2.29). Most of the sand load is deposited near shore and contributes to the growth of a subaereal delta (Rodolfo 1975). The tides along the river mouth are semidiurnal, with a neap tide range of 4.0 m and a spring tide range of 5.8 m. There is a strong ebb with offshore transport of an unknown quantity of suspended material toward the east and to deeper water. Along the coast, an area of about 5300 km^2 is below spring tide level and regularly flooded. About the same area is up to 30 cm higher and can occasionally be flooded during very high tides (spring tide plus simultaneous storm surge). Marine flooding is enhanced during September/October when river discharge is low and strong winds blow from the southwest. The intertidal area consists of tidal flats covered with mangrove (*Heritiera* forest), but the tides penetrate much farther inward through the channels. Within the area flooded during spring tide are sandy ridges (former beaches, sand banks, or cheniers) where agriculture is possible. The coast advances 50 to 60 m per year at present but on an average of just 25 m $\cdot$ y^{-1} during the last 100 years. While at present about 10 km^2 is added yearly to the intertidal zone on the seaward side, this is probably compensated for by an equivalent area that rises above spring tide range because of sedimentation from freshwater floods. Above spring tide level a large area is not protected by dikes or levees and can become flooded during periods of high monsoon rainfall. During these floods, the water that floods most of the delta comes from the surrounding hills and covers the river banks before the high river flood reaches this area. The river water then flows directly to the seashore through the flood waters and deposits hardly any sediment outside the river bed. The present (1940) coast consists of mangrove flats and some sand dunes of about 100 m wide and 1 m high.

2.10. WEST COAST KOREA

Along the Korean west coast, intertidal flats occur in the southwest between about 34° and 35° 30′ N, along the central part between about 36° 40′ and 38° N, and in

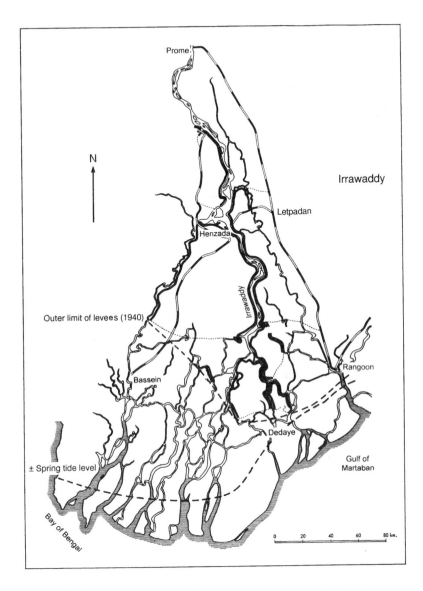

FIGURE 2.29   The Irrawaddy delta (Adapted from Volker 1965.)

the north between about 38° and 39° 50′ N (Figure 2.30). The flats are up to 10 km wide and exposed toward the Yellow Sea without any barriers. Their inward boundary is usually a dike or a seawall. Salt marshes are rare and mostly only locally present, largely the result of reclamation. The sediment is mainly silt or fine sand, with more sand on the lower flats and more fine sediment on the higher flats. The upper part of the flats can be covered with periglacial debris (angular pebbles and a fine matrix) which has moved downward by mass flow from the adjacent slopes. The entire west coast is macrotidal with, in the central part around Inchon, a maximum spring tidal

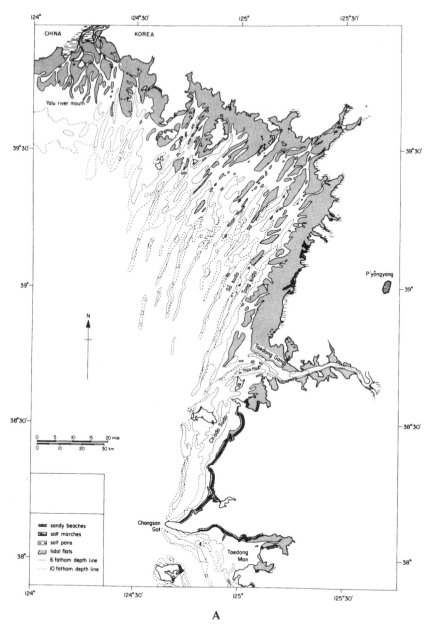

A

FIGURE 2.30 Korea coast. A. North; B. Center-West; C. Southwest. (B and C on following pages.) (From Eisma 1985. With permission.)

range near 9 m and a mean neap tidal range of over 3 m. There is a large diurnal inequality of 1.41 m, a monthly inequality of about 1 m, and a yearly inequality of 35 cm which is related to the variations in sea level in the Yellow Sea. There is probably also an 18-year cycle with a range of about 45 cm (Hahn 1980).

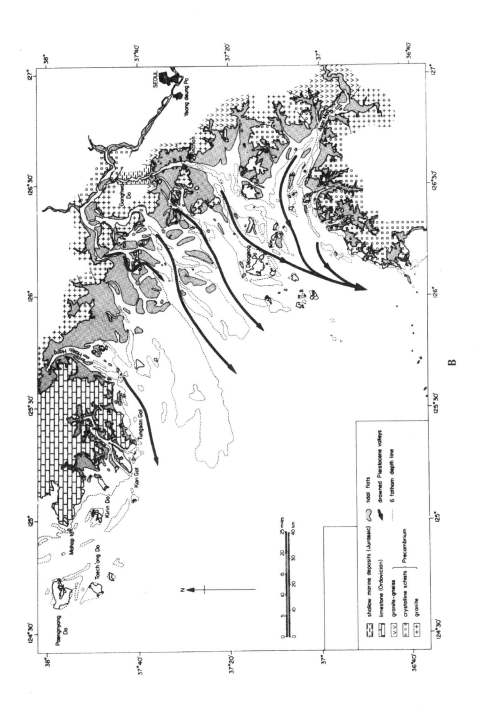

B

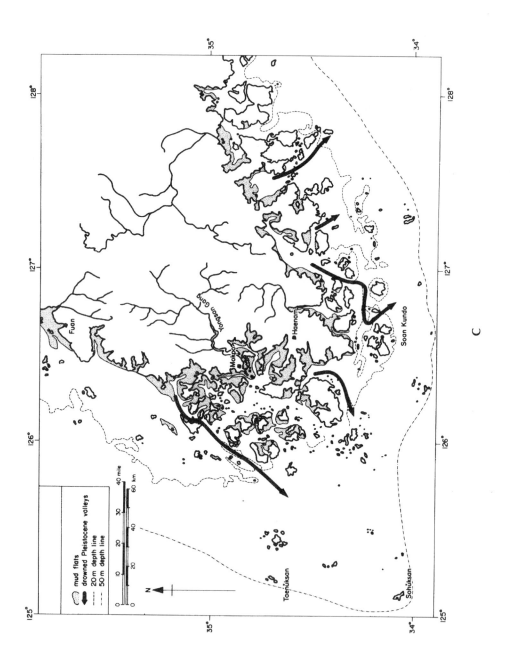

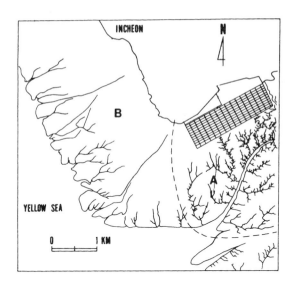

FIGURE 2.31   Tidal flats Kyeonggi Bay. (From Lee et al. 1992. With permission.)

Detailed studies of intertidal deposits have been made in Kyeonggi Bay, Changmu Bay, and Namyang Bay (Figures 2.31 and 2.32; Wells et al. 1984–85; Lee et al. 1985; Kim and Park 1985; Frey et al. 1989; Wells et al. 1990; Alexander et al. 1991; Lee et al. 1992). In general the tidal flat sediments are finer in a landward direction and, on the higher flats, strongly or completely bioturbated by crabs, polychaetes, and bivalves. Where sediment structures are present that are not due to bioturbation, a parallel to slightly wave lamination dominates. In more sandy sediment, flat current ripple lamination may be found with both flood- and ebb-foresets. Parallel lamination is thicker in sandy sediment (laminae of 1 to 5 mm) and thinner in silts (laminae < 1 mm). The sediment surface can be highly variable, ranging from watery soft to firm and from smooth to hummocky, or irregular with small depressions and mounds, usually formed by organisms, and with scour features both from waves and currents. Some parts of the flats are covered with up to 5 cm of fluid mud, and the outer tidal flats regularly consist of up to 1 m of unconsolidated soft mud. On the Namyang Bay flats there are marked differences in surface elevation with steps up to several meters high. Each step is delineated by a major channel with the higher elevation on the northern side. Deposition rates (determined with ^{210}Pb) are 5 to 9 mm·y^{-1} in the central part of the flats (where measured) and decrease both landward and seaward to 1 to 2 mm·y^{-1}.

The flats near Inchon have hardly any channels (Frey et al. 1989): only small shallow channels, a few dm deep, were present during the summer for a few months in the middle part of the flats, starting on the higher flats, while a larger, more permanent channel was present on the lower flat. The middle part of the flats had an irregular erosional surface. South of Inchon, on the flats of Kyeonggi Bay, a more or less parallel pattern of channels is present on the open flats that are more exposed to storms from the northwest, and a more dendritic pattern is present in the more sheltered parts (Figure 2.27), which, in addition to exposure, may also be related to differences in sediment grain size and water content. The sediment surface in this area rises or

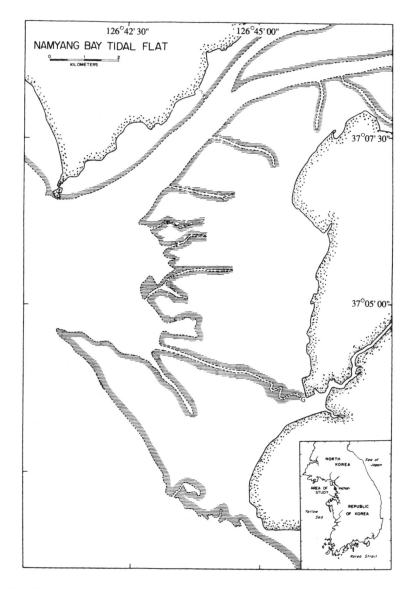

FIGURE 2.32   Namyang Bay tidal flat. (From Wells et al. 1990. With permission.)

falls 10 to 30 cm during the year because of erosion and (re)deposition. Reclamation has resulted in an increase of fine sediment on the remaining flats (Lee et al. 1992).

The intertidal flats in Namyang Bay show a locally dense network of channels (Figure 2.32). The northern part of the flats has a generally high elevation and a higher mud content than the flats in the south, with relatively little variation in texture and composition, but there is no difference in channel geometry. The channel floors tend to be flat (Figure 2.33); mid-channel shoals are usually sand or shells. The larger channels occur in the south, but only those in the northern part, which is more

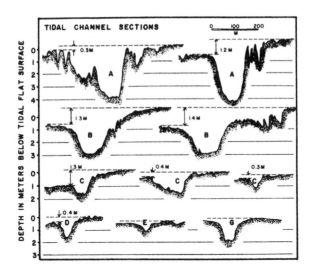

FIGURE 2.33 Tidal channel sections. (From Wells et al. 1990. With permission.)

sheltered from storm winds, have networks of small tributary channels. These are 0.5 to 1.5 m deep with usually a V-shaped profile and steep slopes (30° to 45°). Slumping along the flanks of the channels is common. Some tributary channels are strongly meandering.

The main channels are very stable, as can be ascertained from aerial photographs taken up to 35 years apart. They are relatively straight, 1 to 4.5 m deep, and oriented at about 90° to the shoreline. They are probably contemporaneous with the tidal flat sediments and ebb-dominated. The flood comes as a sheet over the flats and in the channels is only slightly earlier. When the flood is in the main channels, there is still ebb in the tributaries, so that at the tributary mouth a temporary slack water is formed and suspended mud is deposited there in usually large amounts.

A small chenier has been described from the intertidal flats in Gomso Bay, south of the Keum river, at about 32° 30' N. It consists of sand, gravel, and shells and lies on sandy intertidal flats. It started to form on the lower flats and in more than 30 years moved upward onto the high flats going from northwest to southeast (in the direction of the major storms). In 1989 it was located in front of a seawall (Chang et al. 1993). Storms, in general, lower the flats and enhance the ebb flow. Where the sediment is soft mud, the erosion can be 5 to 10 cm of sediment during one storm, and scour depressions are formed.

2.11. CHINA: JIANGSU COAST

The tidal flats along the coast of the province of Jiangsu, China, extend in a northwest-southeast direction from approximately 35° N to 32° N along the Yellow Sea. The tides are semidiurnal and come both from the southeast (the Pacific) and the northeast (the Yellow Sea, where the tides are rotating counterclockwise (Wang 1983, Ren ed. 1986; Figure 2.34). The two tidal waves meet at the coast near

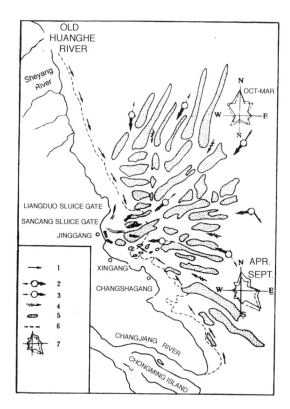

FIGURE 2.34   Jiangsu coast. 1 = longshore current; 2 = rotary tidal current; 3 = reversing current; 4 = principal wave directions; 5 = sand banks; 6 = mainland; 7 = low tide line; 8 = measured section; 9 = dike; 10 = frequency of wind directions (solid line); frequency of wave directions (dotted line).

Jianggang, which results in a large tidal range in that area. While elsewhere along the Jiangsu coast the mean tidal range is 2.5 to 4 m, it increases here to more than 4 m and up to 7 m. Also there is regularly a tidal bore of about 1 m high, and current velocities may reach 4.5 m · s^{-1} with larger flood velocities than ebb velocities.

The sediment along the Jiangsu coast has been supplied by the Yellow river which for almost 700 years up to 1855 had its mouth on the Jiangsu coast at Feihuanghekou (Figure 2.35). After this mouth was abandoned for the present one in Bohai Bay, erosion of the former river mouth produced the sediment that has been, and still is, deposited on the mud flats south of the Sheyang river. These are prograding at present. Only in the extreme south of the Jiangsu coast is there some supply from the Changjiang river.

The flats are receding around the abandoned former river mouth and also along approximately 30 km of the coast around Lusi in the south. Most of the Jiangsu coast south of the Sheyang river is protected against the large waves from offshore by tidal sand ridges in front of the coast that are present over a length of 200 km (Figure 2.34). Some ridges are partly exposed at low tide. Large typhoons, however,

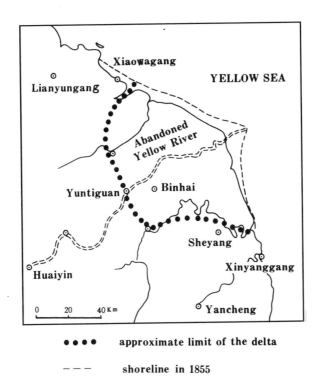

● ● ● ●   **approximate limit of the delta**

– – –   **shoreline in 1855**

FIGURE 2.35 Old Yellow River mouth, abandoned in 1855. (From Ren 1992. With permission.)

affect the entire Jiangsu coast. Against normal waves, only the old Yellow river mouth and the coast around Lusi are not protected by the banks and are directly exposed to waves from the northeast. In both areas the coast as a whole is retreating; the flats are relatively narrow (0.3 to 1 km mostly around the old Yellow river mouth; about 3 km around Lusi) and steep (with a gradient of about 4‰ (Figure 2.37). On the mud flat surface numerous tidal creeks are present, often more than 1 m deep, and at high tide a discontinuous chenier of 0.5 to 1 m has been formed that consists of mollusk shells. The chenier is not formed where the coastal retreat is rapid, probably because shells do not become available in sufficient numbers where the sediment is intensively reworked and mollusks have difficulty living. The retreat is strongest at the abandoned river mouth (100 to 110 m · y⁻¹) but is much less (15 to 30 m · y⁻¹) at a little distance and around Lusi. Simultaneous with retreating, the mud flat surface is lowered by up to 3 m in a decade. The lowered flats are covered with a thin veneer of fine sand and coarse silt.

The prograding mud flats can be up to more than 10 km wide (Figure 2.36). Around Wanggang, accretion is about 50 m · y⁻¹; protection of the coast by submarine ridges is not very pronounced here and the tide ranges 3 to 4 m. The tides are asymmetric with a stronger flood, but with an ebb that lasts 1 to 2 hours longer. The tidal currents generally follow the coast but on the flats are approximately perpendicular to the coast. The suspended matter concentrations are higher during the flood,

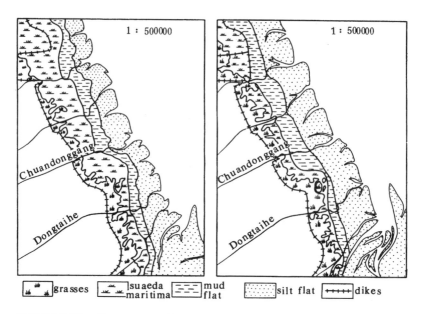

FIGURE 2.36   Jiangsu coast tidal flats. (From Ren, ed., 1986. With permission.)

as can be expected, and slack tide takes 1 to 1.5 hours. The intertidal deposits (silt and clay) are very flat with a smooth, bare surface (Figure 2.37). Large waves may affect the entire intertidal zone and storm waves often break around high tide level. The upper tidal zone (>4 km wide) has somewhat finer sediment than the lower tidal zone: muddy layers are thicker (2 to 4 cm) higher in the upper tidal zone and decrease to about 1 cm in the lower zone toward lower level. Also some rippled beds are present there. Burrowing organisms are abundant in the higher tidal zone, predominantly crabs (10 to 50 burrows per m²), and recently some *Spartina* has been planted near the high tide level to enhance sedimentation. The lower intertidal zone (more than 5 km wide) consists of silt layers of 2 to 5 cm thickness separated by mud layers less than 2 cm thick.

Subparallel sinuous channels cross the tidal flats in a direction almost perpendicular to the shoreline at 4 to 6 km intervals (Figures 2.38 and 2.39). They begin in the upper flats as creeks, and in the lower flats they become broad channels that tend to be funnel-shaped. Some channels are a continuation of inland canals that are connected with the flats by a sluice. These channels are already broad in the upper tidal zone. Each has a network of creeks and smaller channels that is separated by low interfluves from similar networks of adjacent channels. The principal channels are 3 to 12 m wide and 1 to 1.2 m deep. Not all channel networks are equally large, and while some have well-developed meanders, others are only sinuous. The sediment in the channel beds is slightly coarser than on the adjacent flats. Point bars show fining-upwards and cross-bedding. Wave ripples are common and, in the tidal creeks, herringbone-crossbedding. The channels shift frequently so that channel-type sediments are commonly found within tidal flat sediments.

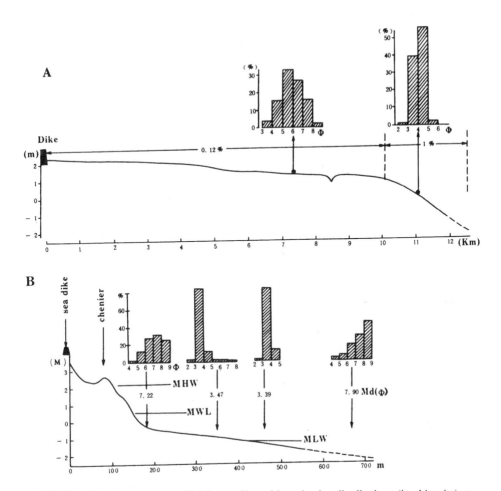

FIGURE 2.37 Jiangsu coast tidal flat profiles with grain size distributions (in phi-units) at selected points. A. deposition dominant (% indicates slope); B. erosion dominant (MHW = mean high tide level; MWL = mean water level; MLW = mean low tide level. Md = median diameter in phi-units). (From Ren 1992. With permission.)

The supratidal flat, where it is not reclaimed (dikes normally form the inner boundary), is covered with vegetation. This is more dense in the upper parts, which are hardly ever flooded, and only during exceptionally high spring tides or typhoons. At a lower level, the vegetation becomes sparser and at high tide level there is usually no vegetation, apart from *Spartina*, and the sediment is usually somewhat coarser because large waves tend to break here. In prograding flats, no cheniers have been described. Storms can raise sea level appreciably (in the order of 3 to 4 m) and large storms tend to destroy the cheniers along the retreating flats. From these flats the upper parts are strongly eroded, with deposition on the lower part. During one typhoon, more can be eroded than during an entire season of normal wave erosion.

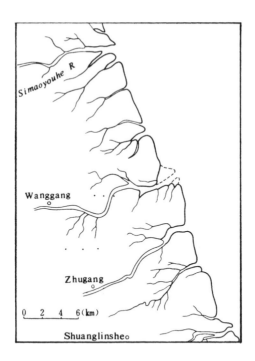

FIGURE 2.38   Channels Jiangsu tidal flats (From Ren ed. 1986. With permission.)

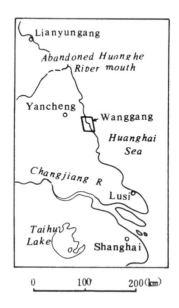

FIGURE 2.39   Location of the channels in Figure 2.38. (From Ren, ed., 1986. With permission.)

The amount of erosion depends on the elevation and width of the flats and on the direction and exposure of the coast. Animal burrows, coated with iron compounds and therefore more resistant to erosion, may remain, and after a storm they stand out as small (2 to 8 cm high) projections at the sediment surface. Depressions, found after storms and roughly parallel to the coast, are probably scoured by waves (surf) during the period of raised sea level. On the prograding flats during a typhoon the upper part is aggraded and the lower part eroded, while the finer material is winnowed out. The upper parts of the channels are filled in; the lower parts are eroded and widened (Ren et al. 1985). Erosion during a typhoon is followed by deposition. Typhoon depressions, separated from the underlying sediment by a sharp scour surface, are generally filled with sediment that is coarser than the normally deposited sediment, and the infill shows cross-bedding, wave ripples, and horizontal bedding (Ren et al. 1986). The sediment normally deposited on the flats has no bed structure or lamination that can be related to the daily tidal cycle, but the spring–neap tidal cycle can be seen in the grain size and the color of the sediment: spring tide deposits are somewhat coarser (5 to 6 f) and yellow, neap tide deposits are finer (6 to 7 f) and brownish. This is most evident in the central parts of the intertidal flats. There is also a year-cycle. Over the year there is a marked variation in sea level along the Jiangsu coast: in September sea level is 60 to 70 cm higher than in January. Because the intertidal flats have a very low gradient, there is yearly shift of about 2700 m in the position of the high-tide line. The increase in sea level results in erosion of the flats, with the formation of irregular scour depressions and mud balls, followed by resaturation of the flats to their earlier flat surface as sea level falls.

The mud flats are very young: in the retreating flats the older deposits, covered by a veneer of recent intertidal sediment, date from the period not long before the former bed was abandoned by the Yellow river. At the prograding flats, the higher flats are situated on top of a vertical sequence of earlier intertidal flat sediments that reflect the development from lower intertidal flat to supratidal flat which, at a progradation rate of 50 m · y^{-1} was not long before the present flats originated. Such sequences are exposed in recently reclaimed areas. They contain a large proportion (~30 to 40%) of storm deposits. This suggests that during storms a relatively large portion of the sediment is being removed, and subsequently returns. Recent measurements indicate an erosion of several decimeters of sediment during a storm, which is in the order of 5% of the total thickness of the vertical sequence.

The flats near Jianggang are located at the convergence of the two tidal waves. The mean tidal range increases here to 4.08 m, and the maximum spring tidal range to 6.78 m (Figure 2.40). Large amounts of sediment are deposited as shoals with interconnecting tidal channels. The shoals lie partly below mean sea level, but most of the larger shoals have become intertidal, and the highest parts, supratidal. In the channels, the ebb velocity is usually greater than the flood velocity, but on the shoals the flood velocity is greater. Waves are weak because the area is sheltered by the offshore ridges. The tidal channels tend to be parallel to the coastline, opposite the channels along most of the Jiangsu coast which are generally perpendicular to the coastline (Figure 2.38).

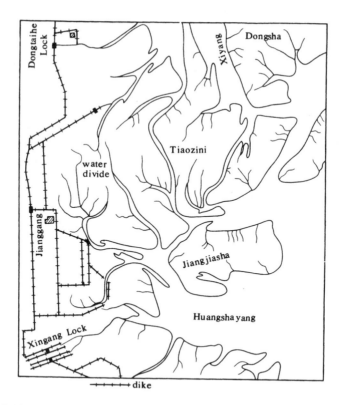

FIGURE 2.40   Tidal flats at Jianggang, Jiangsu province. (From Ren, ed., 1986. With permission.)

The supratidal flats, landward limited by a dike, are mostly covered with grass and reeds and consist of very fine-grained sediment. Their lower parts are covered with *Suaeda maritima* and there the sediment is somewhat less fine. *Suaeda* also covers the supratidal part of the shoals. The supratidal sediments have a wave bedding and are mottled. The upper tidal flat sediments have numerous animal burrows in their upper parts with thin laminations between the burrows, and cross-bedded silts deposited during typhoons. In the lower tidal flats, which contain more silt than the upper, as well as some fine sand, silt-mud laminations occur with horizontal bedding and scour surfaces on top of the finer, muddy layers, with at the transition to the subtidal zone, small cross-beddings and herringbone cross-bedding (Zhang Guodong et al. 1984).

2.12. CHINA: THE CHANG JIANG RIVER MOUTH AND HANGZHOU BAY

Intertidal flats occur in the Chang Jiang river mouth and to the south around Hangzhou Bay at about 31° 30′ N to 30° N. The maximum tidal range in the Chang Jiang river mouth is 6.0 to 6.8 m, and in Hangzhou Bay, 4.9 m at the entrance,

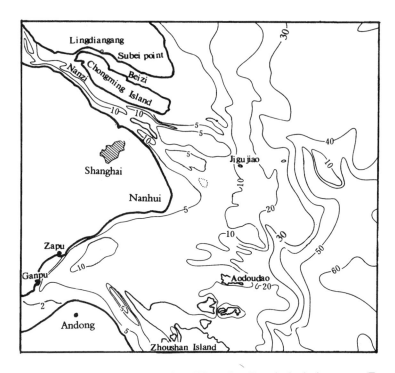

FIGURE 2.41 Chang Jiang river mouth and Hangzhou Bay; isobaths in meters. (From Ren, ed., 1986. With permission.)

increasing to 8.9 m inside. Both in the Chang Jiang river mouth and Hangzhou Bay, the sediment comes from the Chang Jiang river; only in the innermost part of Hangzhou Bay can sediment from the Qiantangjiang, the largest river that flows into Hangzhou Bay, be recognized. Considerable areas of intertidal flats are present at the Chang Jiang estuary along Chongming Island and the islands Changxin and Hengsa, around the Nanhui coast (a headland between the southern part of the Chang Jiang estuary and the north shore of Hangzhou Bay), and along the north and south shores of Hangzhou Bay (Figure 2.41). The flats in the Chang Jiang river mouth are prograding outward (eastward) with strong variations in erosion and deposition over a period of several years, as well as seasonally and during storm events. Similar variations occur on the flats along the north shore of Hangzhou Bay. Storm waves give the largest erosion; buildup of the flats is predominantly by the tides. Erosion of the flats during a storm results in the silting up of the channel entrances (Yun 1983; Zhang et al. 1993).

The intertidal flats along the eastern side of Chongming Island are up to 7 km wide and have a dense network of generally stable creeks and channels, extending radially away from the island (Figure 2.42). Some of the channel systems are interconnected. In their upper parts, the channels show a slow lateral migration (Xu 1985). The sediment is coarser on the lower flats and finer toward the upper parts. The supratidal flats have recently been reclaimed so that some channels are now

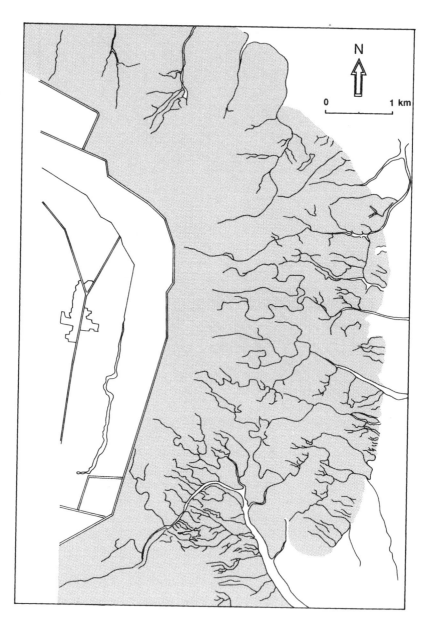

FIGURE 2.42   Tidal flats at Chongming Island. (Adapted from topographical map.)

directed along the dike and former channels were cut off. The higher flats are overgrown with reeds and, at a slightly lower level, with salt marsh vegetation; below mean high tide level the flats are bare. Sedimentation is rapid and the mud deposits with open or without vegetation tend to be very soft. Where sufficient vegetation is present, the mud is held together by the roots. Lower on the flats the sediment becomes firm. The lower flats are laminated or rippled; the higher flats have an abundance of ripples, a wavy bedding, flaser bedding, lenticular bedding, and climbing ripple laminae, as well as abundant animal burrows in the higher parts. In the near-supratidal flats, there is a horizontal layering with lamination and mud cracks (Xu 1985). Typhoons can create serious erosion: up to more than 20 cm of sediment during one storm, as well as deposition when the storm subsides. The typhoon deposits are generally coarser than the normal tidal flat deposits and show a hummocky cross-stratification at a level above normal wave base (Xu et al. 1990).

The intertidal flats along the islands of Changxin and Hengsa in the Chang Jiang river mouth at low tide level consist of fine sand that grades into mud toward high tide level and in the supratidal zone. The bare flats on the east side of Hengsa are sand and silt in thick bands with flaser bedding and some crossbedding, while on the salt marshes thick bands of sand and silt alternate. On the bare flats at the west side of Changxin island, alternating thick beds of fine sand and silt grade into thin beds alternating with wave beds, where *Scirpus* grows, and horizontal laminae in the reed zone. On the bare flats along the north and south coasts, beddings are wavy with alternating thin beds of silt and muddy silt and horizontal mud laminae on the marshes. No layering is present in the supratidal marshes, where the original layering is usually destroyed by the plant roots. Because the flats are regularly reworked, there is little benthos so that on the bare flats the sediment is only slightly disturbed and shells are scarce (Yang and Xu 1994).

A large intertidal shoal southwest of Hengsa Island is completely flooded during high tide with, on the higher parts, a grassy marsh vegetation that reflects the low salinities in the river mouth. On the north side, where the shoal is eroding, the lower, bare flats are separated from the vegetated higher flats by a low (10 to 40 cm) cliff. Deposition occurs on the southern side so that the shoal is (slowly) moving toward the southeast. The lower, bare flats are inhabited by numerous small crabs; the vegetated part is dissected by small creeks. A fine-sandy storm ridge has been deposited on the northern side near high tide level.

The Nanhui intertidal flats are located where the ebb currents from the Chang Jiang estuary and from Hangzhou Bay converge, and where the flood diverges in the opposite direction. Current velocities in this area are usually low. The flats extend seaward over a maximum distance of about 20 km and have a very low gradient (Figure 2.43). There is a marked difference in level between the Chang Jiang side and the Hangzhou Bay side, and this results in sediment transport across the flats. The net transport is landward. The high parts of the flats, which are on the landward side bordered by a dike, are overgrown with reeds and salt marsh. Storm waves, which come mostly from southeastern or northeastern directions, erode the lower flats on the seaward side: the sea level is raised, which enhances wave action (erosion) on the flats as well as the shoreward transport of sediment.

Along the Chang Jiang estuary, as well as along the Jiangsu coast, deposition is also enhanced by higher suspended matter concentrations in the flood water. When there is no storm, wind velocity and direction as well as degree of persistence determine sediment transport over the flats, more so than the tides (Yang 1991). The vegetation on the higher flats (mostly reeds and sedges) reduces the current velocities

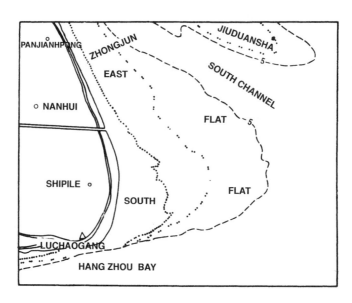

FIGURE 2.43   Nanhui tidal flats.

in the water, and also most of the wave energy is absorbed. Erosion takes place at the lower boundary of the vegetation; the vegetation enhances sediment deposition, which can be up to 9 times the deposition on the adjacent bare flats. The sediment at the marsh boundary is usually coarser than on the bare flats or on the marsh because wave action is concentrated there (Yang and Chen 1994).

The Nanhui area as well as the city of Shanghai are located east of a chenier belt 4 to 8 km wide which dates from 6800 to 3200 BP (Figure 2.44), and which indicates that between 7000 and 3000 BP, there was relatively little sediment supply. This is related to the shifting of the main channel in the Chang Jiang estuary over the last 7000 years (Liu et al. 1987; Liu and Walker 1989). After about 3000 BP, the sediment supply increased and large shoals were formed that are known from historical records (Chen et al. 1982). In particular during the past 2000 years, sediment supply from the Chang Jiang river increased because of the intensification of agriculture in the river basin.

Hangzhou Bay was formed after about 7000 BP during the Holocene transgression when the sea water reached inland of Hangzhou (Chen et al. 1990), and Hangzhou Bay was a drowned valley. Around 3000 BP, because of human activities modifying the Chang Jiang drainage area, the sediment load of the principal river (the Qiantangjiang) increased, and a delta advanced rapidly so that by 400 AD the coastline had moved 15 to 20 km eastward. Nanhui moved to the northeast, widening the bay, which resulted in an increased deformation of the tidal wave and an increase in flood current velocities. By 1200 AD the bay had approximately its present funnel-shape. After that time, some erosion on the north side and deposition on the south side occurred. The present meander bends developed, and during a combination of spring tide and high river discharge, the development of a tidal bore began that

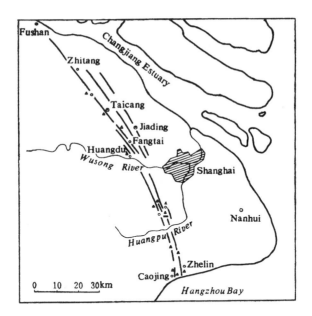

FIGURE 2.44 Cheniers (black lines) near Shanghai. (Adapted from Liu et al. 1987.)

reaches, at present, a maximum height of over 3 m. This was enhanced by the formation of a large silt bar and the reclamation of shallow flats in the inner bay, whereby this part of the bay became much narrower, and the tidal range now reaches a maximum of 8.93 m at Ganpu (Figures 2.45 and 2.46).

Along the north shore of Hangzhou Bay the flats are rather narrow: 350 m wide at the eastern end, increasing to 2300 m around Zhulin, and then decreasing to 300 to 700 m farther inward in the bay (Cao et al. 1989; Figure 2.45). Most of the flats have been prograding since 1958 (about 1000 m near Zhulin) but at the eastern end and inward in the bay, the flats are retrograding: at Lu Chao Gang from a width of 1500 m in 1958 to the present width of 200 m, and from 350 m to 270 m and from 800 m to 690 m at Quangongting and Haiyan, respectively. The retreating flats are steeper than the prograding flats; the latter also usually have a knickpoint around mid-tide level. Everywhere the flats have a dike as their landward boundary. The higher flats, above mean neap high tide level, are covered with a sometimes thin vegetation, while the remainder of the flats are bare down to mean spring low tide level. The sediment is silt and clay, which is easily resuspended in large amounts when the flood comes up over the flats and a strong turbulent frontal wave develops (Wang and Eisma 1990). This occurs in particular on the lower parts of the flats; on the higher parts, the sediment tends to be deposited during high tide and is not easily resuspended during the next flood because of consolidation and drying out. Even mud cracks can be formed during exposure when the tide is low. The sediment is mostly laminated, with finer sediment higher on the flats. Storm waves (typhoons) result in erosion of the flats with excavations of hollows (puddle structures) that are subsequently filled in, with cross-bedded, fine sediment and mud balls. Where large

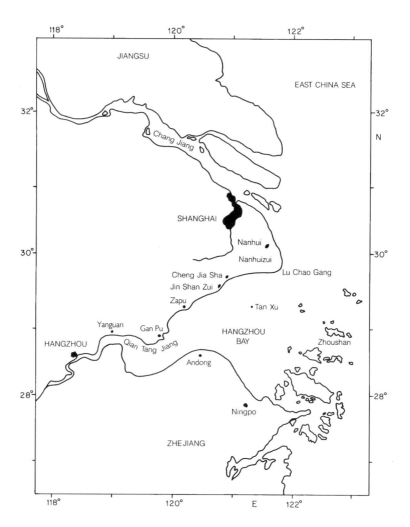

FIGURE 2.45   Hangzhou Bay. (Adapted from Ren, ed., 1986.)

waves break, the sediment consists or irregular layers of silt with clay laminae following the silt surface. Mostly typhoon structures are preserved together with erosion surfaces and lag deposits. Seasonal variations are reflected in the sediment texture: average clay content is, in the winter, 11% on the high flats and 7.2% on the low, while in summer (which is the typhoon season) these percentages are 6.7% and 4.7%, respectively. Because of the seasonal alteration of erosion and deposition, the flats tend to be regularly reworked down to 50 cm depth in the sediment over the year. Creeks or channels are rare; there are no networks of any size, and the direction of the main creek or channel, where present, is about perpendicular to the shoreline. Also animal burrows are relatively rare because of the regular reworking of the top decimeters of sediment by waves and a paucity of deep-burrowing species

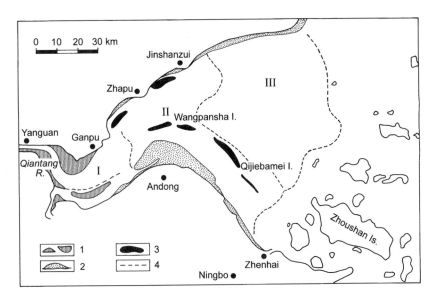

FIGURE 2.46   Tidal flats Hangzhou Bay. 1 = river mouth shoal (Qiantangjiang river); 2 = mud banks; 3 = tidal ridges; 4 = boundary of morphodynamic units. (Adapted from Feng et al. 1993.)

that rework the subsurface sediment. Crabs occur mainly in the supratidal zone or around high tide level. Burrowing worms and bivalves are found mainly lower on the flats (Cadée et al. 1994).

Along the south coast of Hangzhou Bay lies Andong shoal bordered by the channel that goes from the Qiantangjiang river mouth first to the northeast and then bends toward the southeast. The coast follows the same bend. The present point bar-like flats extend for about 12 km seaward from a dike constructed in 1977 (Figure 2.46). The dike encloses a recent tidal flat accretion plain, the Yao Bei plain, where sections of the flats have been reclaimed since 1047 (Figure 2.47). The isoline of mean tide level has been moving outward since the first measurements in 1930, and deposition probably increased progressively. Between 1959 and 1975 the outward displacement was on the average about 150 m per year, whereas between 1975 and 1984, it was on the average about 400 m per year, and between 1982 and 1984, about 1 km. The shoals consist of soft mud with mud-silt laminae and on the west side a large ebb channel that shallows in the direction of the ebb. It follows the coastline at a little distance. Small channel systems with short strait creeks occur along the entire western margin of the shoal, while on the east side are several small channels (Figure 2.48). Here sediments with crossbedding, flaser bedding, and convolute structures occur. These alternate vertically with storm (typhoon) deposits with scour features and mud balls. Scour and fill structures are more common in the western part where the channels and creeks are concentrated.

The soft muds of the upper flats are a zone of upward accretion. The channels and creeks are mainly concentrated on the middle flats, which are a zone where

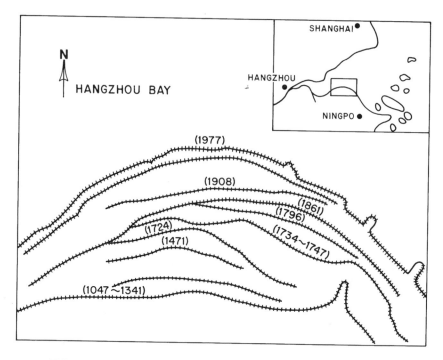

FIGURE 2.47   Dikes near Andong. (Adapted from Li and Xie 1993b.)

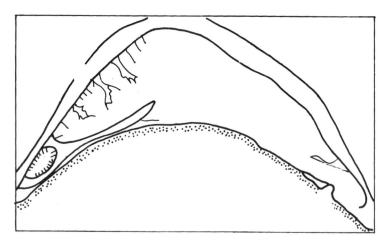

FIGURE 2.48   Channels in tidal flats near Andong. (Adapted from Li and Xie 1993a.)

sediment is predominantly being reworked, and the lower tidal flats are a zone of strong seaward accretion. Deposition rates are in the order of 0.1 to 2 cm · y^{-1}, under normal conditions but may reach 4 to 64 cm in one tidal cycle during abnormal events. Long-term deposition rates (over more than 10 years) are 2 to 5 cm · y^{-1} in the zone of upward accretion and in the zone of reworking, 1 to 10 cm · y^{-1} in the channel area, and over 50 cm · y^{-1} in the zone of seaward accretion, but there are large fluctuations. The yearly deposition is about 6×10^7 tonnes (or about seven times the supply from the Qiantangjiang, which is 7.87×10^6 tonnes per year). Eighty-five percent of this amount is being deposited in the zone of seaward accretion (Li and Xie 1993a, 1993b, Lie et al. 1993, Feng et al. 1993). Suspended sediment from the Chang Jiang estuary can be transported directly to the south shore of Hangzhou Bay along a northeast-southwest front that exists the year round. Through this front, plumes from the Chang Jiang and the Qiantangjiang merge (Su et al. 1993).

2.13. CHINA, SOUTHEAST COAST: INTERTIDAL DEPOSITS IN EMBAYMENTS

The coast of southeast China consists of numerous smaller and larger bays and river estuaries where fine-grained sediment is transported inward and accumulated. Sediment comes from the Chang Jiang estuary from which the river supply goes mainly along the coast to the south, and from the Zhu Jiang (Pearl river), from which it goes to the west. Southward transport from the Chang Jiang takes place the year round but is stronger during the winter when the southward-going coastal current (the Jiangsu current) is stronger (Ren ed. 1986). The supply from the Chang Jiang decreases to the south and ends near Fuzhou (the estuary of the Min Jiang river). From Fuzhou farther to the southwest, local supply becomes (relatively) more important up to the Zhu Jiang estuary where again a large amount of sediment is supplied to the coastal waters (in the order of about 83×10^6 tonnes per year). The bays of southeast China penetrate inland over a considerable distance and are efficient mud traps. In Zhejiang there is about 1000 km of coast with mud flats, and in Fujian about 500 km, which continues in Guangdong. In the north of China a similar kind of flat is present in the province of Liaoning at the northern end of Bohai Bay and the adjacent coast of the Yellow Sea (Figure 2.49).

Intertidal sediments in two bays — Taizhou Bay at about 28° 40′ N and Wenzhou Bay at about 27° 40′ to 28° N — have been studied in detail (Ren et al. 1986; Wang and Eisma 1988). In Taizhou Bay (Figure 2.50) the average tidal range is over 4 m and the spring tidal range more than 6 m. Suspended sediment supply from the Jiaojiang river is about 1.3×10^6 tonnes per year (Li et al. 1993), but most of the sediment in the bay is probably supplied from the Chang Jiang. The bay itself dates from the late Mesozoic and has, in large part, been filled in. The Holocene marine sediments reach a maximum thickness of about 80 m. The recently formed Taizhou bay plain is very flat and located slightly above present high spring tide level. It consists predominantly of silts (Md 6 to 15 μm). A chenier of 15.5 km is located about halfway up the plain in an approximate north-south direction and has a maximum height of 3.4 m (maximum sediment thickness is 5.8 m). It consists of sand with a large amount of mollusk shells. In general, its height, width, sediment thickness, and grain size decrease from north to south. The chenier dates from before the 12th century and marks a period of little or no progradation of the plain and

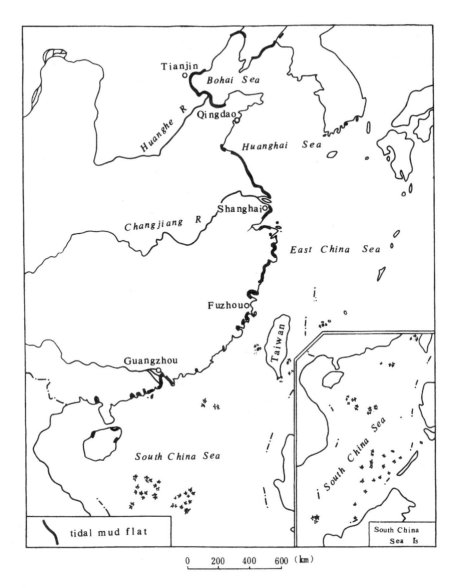

FIGURE 2.49   Distribution of mud flats along the Chinese coast. (Adapted from Ren, ed., 1986.)

little or no sediment supply. A series of sea dikes were constructed after the formation of the chenier (mostly after the 15th century) in the same direction as the chenier (Zhu 1986).

The present intertidal flats are located east (seaward) of a sea dike that was constructed in 1977. Directly outside the dike is a grassy salt marsh, 300 to 500 m wide, located above mean spring high tide level with laminated silts and mud pebbles.

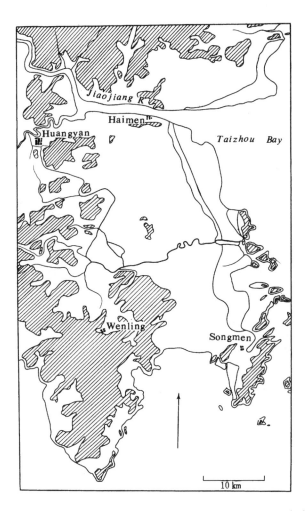

FIGURE 2.50   Taizhou Bay. (From Ren, ed., 1986. With permission.)

The intertidal flats are 5000 to 5550 m wide down to spring low tide level, and form a belt approximately parallel to the shore line. The high intertidal flats between mean high spring tide level and mean high neap tide level are sparsely vegetated and sometimes show mud cracks. Their width is about 1000 m, and the surface is generally smooth with incipient tidal gullies and occasional ripple marks. The sediment (silt and clayey silt) is thinly laminated with some flaser bedding in the upper layers. The middle tidal flats between mean high neap tide level and mean low neap tide level have a width of about 2400 m and are daily submerged. The surface is characterized by a large number of parallel gullies and ridges that run nearly perpendicular to the shoreline, and by abundant ripple marks (Figure 2.51). In the upper part of the middle tidal flats, the tidal gullies are 20 to 50 cm wide and 15 to 20 cm deep, but in the central part their width increases to 180 to 240 cm and

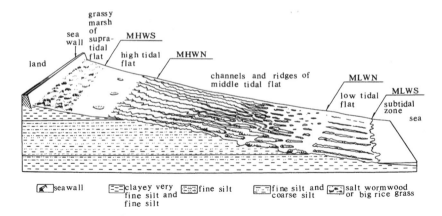

FIGURE 2.51   A tidal flat at Taizhou Bay. (From Ren, ed., 1986. With permission.)

their depth to 60 to 70 cm. The gully walls are steep, and some gullies are sinuous. In the lowest part the gully depth decreases to 30 to 40 cm, and many gullies are interconnected. The sediment in the gullies is coarser (~40 mm) than on the adjacent flats (~10 mm). The lowest tidal flats between mean low neap tide level and mean low spring tide level have few gullies or channels.

The sediment is supplied to the flats from the coastal waters. Generally the flood carries more sediment than the ebb. In the bay, the flood and ebb velocities are about the same, but on the flats a tidal asymmetry develops with a stronger flood. Deposition occurs during slack tide and where the water is shallowest, which is on the highest flats that are flooded only around spring tide and on the upper parts of the middle tidal flats, which are flooded almost every day. The returning ebb starts to scour the flat surface where the velocity becomes sufficiently strong. From there gullies are formed or maintained. Between spring and neap tide, deposition and scour zones move vertically over the flats. During the neap tide period, the sediment deposited during spring tide has time to consolidate, as it is not flooded for some days, which makes it more resistant to resuspension. In this way the sediment higher up the flats tends to stay where it was deposited, and the flats prograde as long as sufficient sediment is supplied and large storm waves (typhoons) do not occur. Large waves, because of the shallow water, usually break between the middle part of the tidal flats and the lower part of the higher tidal flats, but typhoon waves may affect the entire intertidal flats because sea level is temporarily raised.

In Wenzhou Bay (Figure 2.52) the mean tidal range is 4 to 5 m and the maximum spring tide range 7.2 m. On the mud flats at about mid-tide level, flood velocities are up to 50 cm · s⁻¹ and the ebb velocities up to 30 cm · s⁻¹, and at low tide level 80 cm · s⁻¹ and 60 cm · s⁻¹, respectively. Local river supply is about 3.4×10^6 tons of suspended sediment per year, but the sediment composition (clay minerals) of the intertidal flats indicates that the sediment along the Zhejiang coast comes predominantly from the Chang Jiang (Wang and Eisma 1988). Wenzhou Bay during the Pleistocene was a coastal valley that during the Holocene was filled with 40 to 60 m

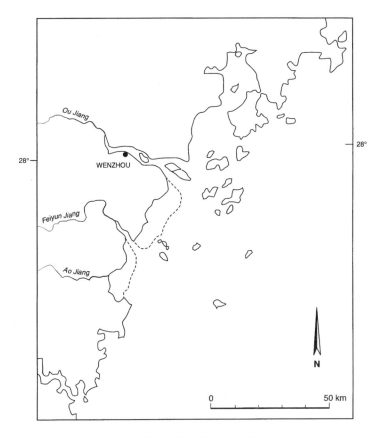

FIGURE 2.52   Wenzhou Bay.

of estuarine, deltaic, and marine deposits on top of late Pleistocene sands and gravels. The present intertidal flats form a belt parallel to the shoreline, up to 1 km wide and with a dike at the inner margin. The sediments are similar to those on the Taizhou Bay flats, only somewhat finer. Similar parallel gullies, also perpendicular to the shoreline, are present on the middle tidal flats, separated by low ridges. They are 1 to 1.5 m wide and 30 to 40 cm deep. The late ebb flows only through these gullies and deepens them. Lower on the flats, a considerable part of the sediment is deposited in the gullies so that they become shallower than they are higher up on the flats. Storms, in particular those related to typhoons, give strong erosion and a strong return flow seaward so that much sediment is transported away from the flats to deeper water. After the storm, the flat surface is rough with erosion puddles that are rapidly filled in. An unconformity remains and the infill shows oblique and horizontal laminations. In spite of the regular occurrence of storms, particularly during the summer, the average progression of the coast is in the order of ten meters per year.

From Fuzhou to the south the intertidal flats for the most part are covered with mangrove, which is present mostly on fine sediment of 10 to 50 μm median diameter.

The lower intertidal flats are usually bare up to mean tide level; the mangrove covers the intertidal flats between mean tide level and high tide level. The surface sediment in the mangrove often consists of up to several decimeters of fluid mud with a high organic content, in the order of three times higher than in the intertidal flat sediments without mangrove. In the supratidal zone the mangrove is replaced by shrubs and salt-tolerant grasses (Ren et al. 1986).

2.14. TAIWAN, WEST COAST

Along the west coast of Taiwan between 23° N and 25° 30′ N bare tidal flats form the coast over a distance of 240 km interrupted by river estuaries. They are usually up to 4 km wide but become much wider between the Tatu and Tsengwen river mouths, reaching a width of 10 km off the Chiayi coast and 15 to 30 km at the Hsinhu river mouth (Hsu 1962, 1965; Chang and Tang, 1970; Reineck and Cheng, 1978; Figure 2.53). The tidal range along the tidal flats is mostly more than 4 m, but south of the Hsinhu river mouth the range becomes between 2 and 4 m (Hsu 1962; Yin 1984). North of the Hsinhu, the flats are without barriers and exposed to the open sea, but to the south, at the lower tidal range, small sandy barriers and offshore bars are formed. The seaward extension of the flats is variable. On the Chiayi coast there was a general reduction in width between 1947 and 1961, which was partly induced by the diversion of river water on the coastal plain. Elsewhere, as at the Hsinhu river mouth, there has been an increase in width. There is a general southward shift of the flats and bars, induced by the southward transport of sediment along the coast during the winter under influence of the northeast monsoon. The west coast is sheltered toward that direction and directly exposed to the less strong winds from the south and southwest that blow during the summer monsoon, but during the winter diffracted waves from the north dominate on the west coast. Typhoons occur between June and October. They come mostly from westerly directions and do not directly influence the west coast, but they result in strong rainfall and increased discharge and sediment supply from the rivers. Also sea level may be raised by up to 3.5 m along the west coast during a typhoon, but typhoons last only a day or so, so that the monsoon winds, especially the strong winds from the north during the winter, are more influential.

The sediment deposited on the tidal flats is supplied by the Taiwanese rivers coming from the central mountains, in particular by the largest rivers, the Chosui and the Tsengwen. Both sand and mud are supplied. The outbuilding of the coast, which locally can be very rapid, is enhanced by the recent tectonic uplift of the Taiwan coast, which is in the order of 17 cm · 100 y^{-1}. North of the Tatu river mouth, where the sediment is supplied only by local (smaller) rivers, the flats are at most 3 km wide, whereas they are up to 10 km wide farther south. The estuarine and tidal flat sands along the west coast have a high content of rock fragments (approximately 35%, mostly slate fragments); the remainder is mainly quartz. In the marine sand (which has the same origin as the estuarine and tidal flat sands), the content of slate fragments may be as high as 50%, probably because of the smaller grain size of the marine sands.

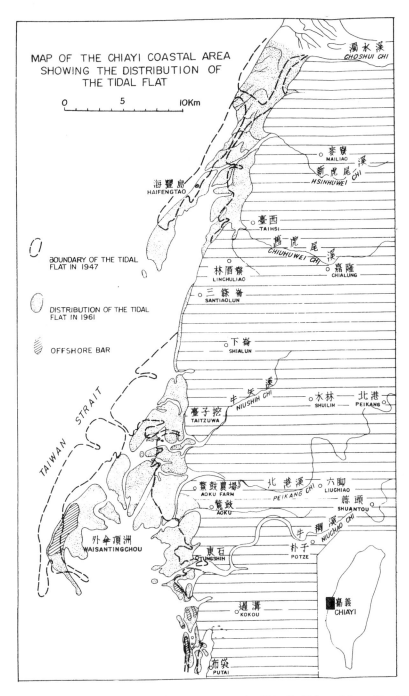

MAP OF THE CHIAYI COASTAL AREA
SHOWING THE DISTRIBUTION OF
THE TIDAL FLAT

FIGURE 2.53   Tidal flats on the Taiwan west coast (at Chiayi). (Adapted from Hsu 1965.)

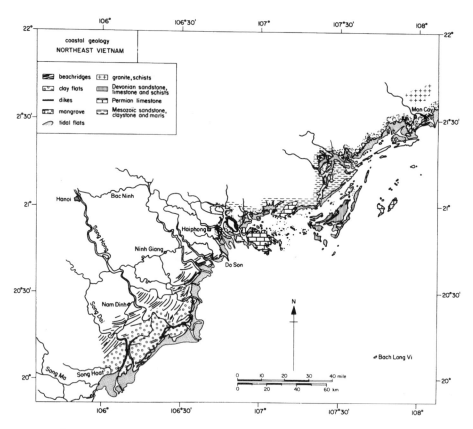

FIGURE 2.54   Red river mouths. (From Eisma 1985. With permission.)

2.15. RED RIVER MOUTHS

Along the mouths of the Red river, between about 20° and 21° N, bare intertidal flats up to 10 km wide are exposed to the open waters of the Gulf of Tonkin (Eisma 1985; Figure 2.54). The tidal range is above 4 m. The flats are bare, probably because they are regularly reworked: mangrove is only present in a few inward areas where they are well protected against waves. The intertidal deposits are mainly silts and clay; beach ridges (or cheniers) and dikes form part of the coast behind the flats; farther inland older beach ridges or chenier complexes, dating from mid-Holocene almost to the present, extend as far as 50 km inland from the present coastline (Huy and Thanh 1994). Between the beach ridges and the coast, and crossed by branches of the Red river, are extensive clay flats that are flooded only during very high tides, insofar as they are not closed off from the sea by a dike. The delta progrades seaward with about 30 to 50 m per year (maximum 120 m · y^{-1}; Thanh 1995): deposition takes place in the areas with a convex coastline, while erosion occurs (up to 10 to 15 m · y^{-1}) where the coastline is concave. Tidal channels

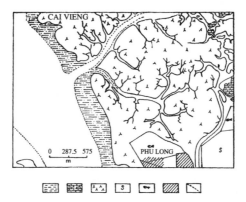

FIGURE 2.55   Tidal channels at the Red river mouth. (Adapted from Thanh et al. 1991.)

and creeks in the Haiphong–Quangyen area form a dendritic pattern (Figure 2.55; Thanh et al. 1991). The smaller creeks have their floor at mean to low tide level; the base of the larger tidal channels is below low tide level and down to 15 to 20 m water depth near the coast. Elsewhere along the Red river mouth similar channel–creek systems are present.

2.16. KING SOUND, CAMBRIDGE GULF (NORTHWEST AUSTRALIA)

Along the Kimberley coast and the Bonaparte Gulf in northwest Australia, coastal embayments are fringed by intertidal deposits. Those along the eastern shore of King Sound, the largest embayment at about 17° 15′ S and open to the north, have been studied by Semeniuk (1980a, 1980b, 1981; Figures 2.56 and 2.57). Along the outward side of the Sound, the intertidal deposits form a small fringe, but more inward, in the southern part, there are broad tidal flats with well-developed mangrove along the outer margins. The maximum spring tide range in this area is 11.5 m, the mean spring tide range 9.4 m, the neap tide range 4 to 5 m. Erosion and deposition occur simultaneously at different parts of the embayment or are locally alternating, but generally erosion dominates and the coast is retrograding, which has been assessed from aerial photographs covering the period 1949 to 1977. Around mean spring tide level, salt flats occur that are almost horizontal, and farther inland these grade into supratidal grasslands. At the boundary between the salt flats and the grassland there is a salt marsh with laminated and vesicular mud that is flooded only during the highest spring tides. Also between mean spring tide level and high neap tide level, at the seaward limit of the salt flats, usually lies a salt marsh. Below this, down to mean sea level, mangrove grows on mud that is bioturbated by an abundant fauna of crustaceae, worms, and fish. The low tidal flats, below mean sea level, are bare and consist of megarippled sand and shelley sand. They become exposed during mean low spring tide. The megaripples have amplitudes of 2 to 15 m and wave lengths of 20 to 90 m. Many are crossbedded with steeply dipping foresets (20° to

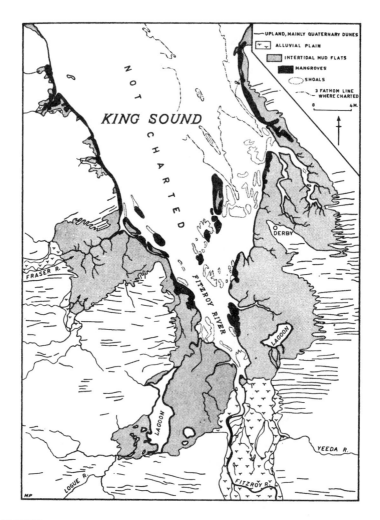

FIGURE 2.56   King Sound. (From Jennings and Bird 1967. With permission.)

35°). They lie in parallel series and occur both in the flood and the ebb direction. Small ripples of 10 to 20 cm wave length cover the megaripples in the more sheltered areas where current velocities are lower (Gellatly 1970). Between the mangrove and the low tidal flats is usually a bare slope with laminated mud/sand deposits.

Tidal creeks and channels occur at all levels above low neap tide level. There are meandering and ramifying channels that can be up to 1000 m wide and several meters deep. Banks are covered with salt marsh or mangrove, or are bare. Because of erosion, zones may be missing, and almost vertical cliffs up to 6 m high may be formed. The low tidal flats with megaripples consist mainly of clean sands. The slope above it has an interlayering of sand and mud which grades upward into laminated muds. The mangrove sediment is muddy, sandy mud, or mud with shells

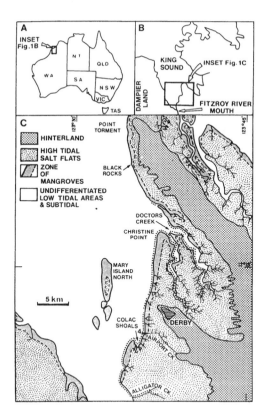

FIGURE 2.57   Detailed map of King Sound flats. (From Semeniuk 1980. With permission.)

and is thoroughly mixed by the abundant fauna. The salt flats, which have a high salt content because of flooding and evaporation, are vesiculated or laminated muds with halite, gypsum, and other evaporite minerals on the higher parts.

The recent intertidal deposits, except the sands at low tide level, form a thin veneer — less than 1 m thick, mostly less than 0.3 m — over older Holocene intertidal deposits that were formed when sea level was approximately at its present position, which means that they were formed after about 6000 BP. The oldest of the Holocene intertidal deposits (belonging to the Doctors Creek Formation) have been dated (with ^{14}C) as 7850 to 5840 BP. Sands deposited later around high tide level (the Point Torment Sands) were dated as 1190 to 500 BP. The Doctors Creek Formation consists of a sand and shell facies, a sand-mud laminated facies, a bioturbated mud facies, and a laminated mud facies, which are analogous to the present-day tidal flat zones. The Point Torment Sands are cheniers which lie around high tide level on top of earlier mangrove and salt flat deposits with mangrove deposits of 1200 BP in front and of 500 BP in the back. They are 1 to 2 m high, 100 to 1000 m long, and 10 to 20 m wide and are generally covered with low, open vegetation, but some have no vegetation at all. They consist of rather coarse, badly sorted sand that contains biogenic material, rock fragments, and mud balls. The entire belt of cheniers is about 15 km long. They were formed while retreating. Topslope beds and backslope beds with internal parallel lamination are common, but foreslope beds are virtually absent (Jennings and Coventry 1973). Mangrove with trees up to 10 m high extends from the cheniers seaward as a belt of variable width down to a little above mean sea level. The low tidal flats, that extend from the mangrove to low tide level, consist here of soft silty clays. Landward of the cheniers are high bare flats that are hypersaline and grade into a high tidal marsh that is only occasionally flooded during exceptionally high tides and is dominated by grasses and some halophytic shrubs. The cheniers end a little to the west of Point Torment (Figure 2.58). From there along the May river estuary, bare tidal flats are present with dendritic systems of channels and creeks bordered by mangrove.

Erosion occurs at all levels on the flats as sheet erosion, cliff erosion, or tidal channel/creek erosion. Sheet erosion is most important on the salt flats and lowers them to mangrove level. This is enhanced by the cavities in the mud that are formed by gas production during neap tides when the salt flats remain dry. Erosion is also enhanced by animal burrows, shrinkage cracks, and by the formation of salt crystals. During flooding at spring tide the cavities are filled with water and the crystals dissolve so that the mud collapses and a thick suspension is formed. This layer is removed with the ebb. Sheet erosion in the mangrove bares and undercuts the root systems.

Cliff erosion is largely caused by tidal scour, waves, mass slumping, and under-cutting of the mangrove root systems. Tidal creek erosion results in the development of larger creeks and channels. They progressively deepen, widen, and extend their headwaters landward. Meandering, ramifying, and bifurcating creeks are entrenched and do not migrate laterally. Small creeks are V-shaped in cross-section; larger channels have a broad U-shaped profile. The rate of sheet erosion is 1 to 3 cm $\cdot$ y^{-1}; cliff erosion varies from 0.3 m to a maximum of 90 m $\cdot$ y^{-1} (normal is about 2 m $\cdot$ y^{-1}); channels deepen 0.06 to 0.6 m $\cdot$ y^{-1}, and widen 0.3 to 4 m $\cdot$ y^{-1}. Because of erosion on the seaward side, while the sediment surface on the landward side is lowered, the mangrove migrates landward, usually keeping pace with erosion. The erosion is partly compensated for by accretion (locally as a recent continuation of the Doctors Creek Formation), but on the whole, net erosion is rather rapid in most areas. Although the ages of the deposits are not very clear, it seems that the post-glacial infill is mainly represented by the Doctors Creek Formation, which continues to be deposited in some areas, but at present is chiefly being eroded. It is not known when this erosion started nor what is causing it.

Farther to the northeast at about 15° S similar intertidal deposits, but without erosion, occur on the east side of the Cambridge Gulf at the mouth of the Ord river (Thom et al. 1975; Figure 2.59). The mean spring tide range here is 5 to 6.6 m, and the maximum spring tide range reaches about 9 m. Besides the Ord river some smaller rivers flow out into the Gulf. A combination of high spring tide, onshore storms, and high river discharge occasionally gives very high floods that cover vast areas, including parts of the supratidal zone. Because of this flooding, and probably also because of the high temperatures, halophytes grow in this zone, covering low hummocky dunes of sand blown in from the lower salt flats. The high tidal flats are

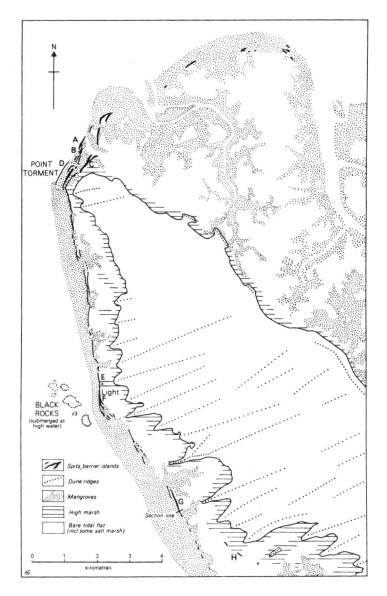

FIGURE 2.58   Flats at Point Torment, King Sound. (From Jennings and Coventry 1973. With permission.)

flooded during spring tide and dry out during neap tides when the tidal range is 2 to 3 m. They are similar to the high tidal flats in King Sound. Mangrove covers the mid-tidal zone; a lower mangrove zone is flooded during normal high tides, an upper zone only during spring tides. Both zones are separated often by a low step. The low tidal flats are only exposed during low tide and are bare, apart from some

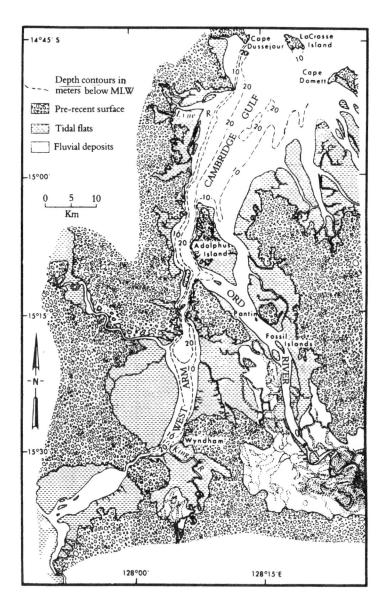

FIGURE 2.59   Cambridge Gulf, Ord river. (From Thom et al. 1975. With permission.)

mangrove seedlings. In the inner parts of the estuaries the sediment consists of soft dark-brown to black clays and silts with a fauna of mudskippers (*Periophthalmus*), an almost amphibious fish, that makes shallow burrows (see Chapter 9). In the less estuarine outer parts, the sediment is generally sandy with a variety of bedforms (ripples, megaripples).

Between the Ord and Victoria rivers, over a distance of about 120 km to the east, extensive sandy tidal flats occur that extend for up to 1000 m seaward from narrow barriers, barrier spits, and beach ridges of fine sand. They have been formed within the high tidal flats mainly between 2020 BP and 1210 BP (Lees 1992); at least one chenier is more recent. They reflect periods of reduced mud supply. Parts of the low flats are covered with mangrove (*Avicennia*) that can withstand the waves that pass through them. Wave ripples are formed within the mangrove and high-tide swashlines farther landward. Where the mangrove grows, the flats are slightly higher and reach mid-tide level.

During the Holocene sea level rise, mud flat formation started here around 8000 BP with at least a partial plant cover. Around 7000 BP the extent of the tidal flats was only slightly smaller than at present, except in the delta area where river-supplied sediment was subsequently deposited. A tidal drainage system developed that has continued more or less the same up to now. Around 6000 BP sea level remained for some time at –3.0 to –2.5 m, and a mangrove cover developed that resulted in an organic layer that has been preserved and contains fragments of crustaceae (*Thalassina*) and gastropod shells (*Cerithidae*). During the following 5000 years up to now, relative sea level remained stable within ±1 m of the present level. Vast areas of tidal flats reached mean high tide level during this period, and hypersaline sediments with mud cracks developed on the highest, least flooded parts. In the Ord river channel there is at present a rapid formation of islands with mangrove, as well as progradation of the tidal flats in the outer estuary. Where the tidal flats reach a level of ±3 m above low spring tide level, fine sediments (muds) dominate that become covered with mangrove (*Avicennia*). Waves generally create erosion, destruction of the outer mangrove fringe, and the formation of beach ridges, but this occurs only during severe storms. Between the storms, accretion of the wave-affected flats occurs both vertically and laterally. In the inner estuary of the Ord river, where wave action is small or negligible, vertical accretion dominates, mostly resulting in hypersaline salt flats, with mangrove only along the channel margins.

2.17. NORTH AUSTRALIA

The north coast of Australia from the Bonaparte Gulf at 129° E to the Gulf of Carpentaria at 137° E (latitudes 11° to 15° S; Figure 2.60) is macro- to mesotidal with semidiurnal tides that grade into mixed tides with a strong semidiurnal component to the east. The mean spring tidal range is 5 to 5.5 m southwest of Darwin. It decreases to 2.5 to 3.5 m in the Van Diemen Gulf and increases again to 3.5 to 4.5 m along the northeast coast (Arnhem Land). The winds are southeasterly during the summer and westerly and northerly during the winter, but strong winds are associated with tropical cyclones that pass several times a year (Galloway 1985). Intertidal and supratidal flats occur all along the Australian north coast, but extensive flats are present in particular along the south coast of Bonaparte Gulf at the estuaries of the larger rivers (Ord, Keep, and Victoria rivers), along the Van Diemen Gulf, Boucaut Bay, and along the west side of Carpentaria Bay mainly along the mouth of the Roper river. The tidal flats, creeks, and estuaries are mostly fringed by mangrove, but going eastward the intertidal flats gradually become bare saline mud flats and saltpans. Along the Bonaparte Gulf and along the central Arnhem Land

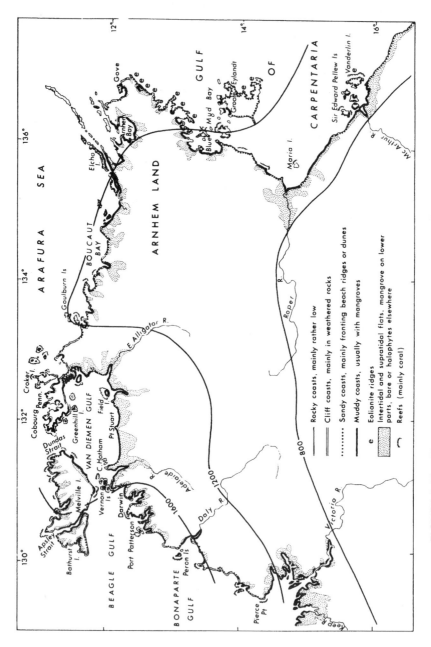

FIGURE 2.60   The north coast of Australia. (From Galloway 1985. With permission.)

coast, swampy alluvial plains are present that are probably former tidal flats. The supratidal flats along the Bonaparte Gulf and the Gulf of Carpentaria become increasingly bare in a southward direction because of the decrease in rainfall.

The sediment from the rivers that flow into Bonaparte Gulf goes in large part to the north. The sediment supply from the Keep and Victoria rivers is seasonally strong and mainly goes into the buildup of intertidal mud islands and supratidal flats that are dissected by creeks. The flats are mostly covered with mangrove. The supratidal flats are bare and only occasionally flooded during the highest tides. Along the open coasts with a long fetch, the mangrove is often damaged by storm waves. Inward along the rivers, at low salinities, the low flats are also occasionally flooded; they are covered with stunted mangrove. Around Darwin, extensive mangroves occur along drowned valleys, and wide fringes of mangrove are also present along the south coast of the Van Diemen Gulf with bare clay flats that grade into plains covered with sedges and grasses on the landward side. Similar mangrove fringes with bare supratidal flats on the landward side occur along the more sheltered parts of Melville Island, Bathurst Island, and the Cobourg Peninsula. Along some very sheltered parts, mangrove is also present on the open coast in front of beaches; bare supratidal flats are present where the coast is protected by fringing reefs

Along the central part of Arnhem Land extensive flats and supratidal flats occur with a halophyte vegetation and are dissected by creeks fringed with mangrove in sheltered parts. On the exposed parts of the coast, the mangrove fringe, tidal flats, and supratidal flats, where present, are narrow. In the Gulf of Carpentaria the muddy tidal flats and sparsely vegetated supratidal flats along the west side, generally become more silty and sandy toward the south and east.

2.18. BROAD SOUND (QUEENSLAND)

Broad Sound is located along the Australian northeast coast at about 22° S, open to the north, and ending on the south side in a funnel-shaped river estuary (Cook and Mayo 1977). Three other estuaries are on the west side of the bay, and some large tidal channels appear on the east side (Cook and Mayo 1977; Figure 2.61). The tidal range is 9 to 10 m, reaching 11 m during spring tide and 12 m at high tide during onshore storms. This large tidal range occurs only for a relatively short distance along the coast; 200 to 250 km to the south and north, the tidal range decreases to less than 4 m (Figure 2.62). The tides are semidiurnal and mostly asymmetric, with the ebb being of longer duration than the flood. The degree of asymmetry depends on the tidal range of the moment, on the prevailing winds, and on the local topography (it is stronger over sand banks). Because of the tropical climate, there is an excess of evaporation over the year, so that during the dry winter period (August/September) the ebb water has a salinity of 42 to 45‰ S and is more saline than the flood, which has a salinity of 36 to 37‰ S).

The tidal deposits consist of bare intertidal sediments (sands, muds, mollusk shells), mangrove swamps (mud) and channels (mud and sands), and supratidal flats with little or no vegetation (muds). In the subtidal area shallow marine deposits (gravel, sand, mud) occur, partly in the form of sand ridges and sand banks. Above

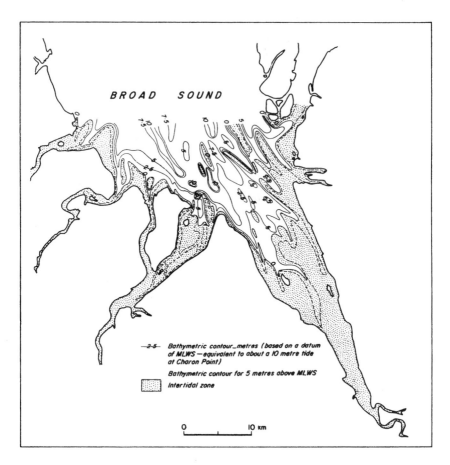

FIGURE 2.61   Broad Sound. (*Source:* Cook and Mayo 1977.)

the supratidal zone, which is only flooded during high spring tides, are extratidal coastal grasslands that are flooded only occasionally for a short time at high tide during storms. Schematically the relation of tidal zones and depositional environments is shown in Figure 2.63. All channels, except those in the river estuaries, are tidal channels and may extend from the subtidal area far into the mangrove swamp and in some places into the supratidal mud flats and coastal grasslands. There is little relief except for a low escarpment up to 1 m high that is widespread and forms the boundary between the intertidal and the supratidal zones. It is formed by wave erosion and slumping after repeated wetting and drying, but is usually lacking at the mouth of the four estuaries, where depositional features dominate and a mangrove pioneer vegetation is present. The flats as well as the subtidal sand ridges are prograding in a seaward direction. The sediment is at present coming mainly from the upstream regions of the rivers, where Holocene sediments are being eroded.

The intertidal sediments are predominantly sand. Gravels are rare and relatively abundant only where pre-Holocene gravels, local Paleozoic bedrock, or former beach

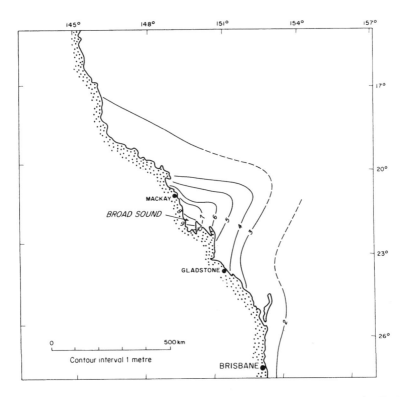

FIGURE 2.62   Spring tidal range along the coast around Broad Sound. (*Source*: Cook and Mayo 1977.)

deposits are reworked. The sands are generally well-sorted, and up to 30% consist of carbonate (remains of bivalves, echinoderms, gastropods, or reworked carbonate nodules formed during the Quaternary by subaereal weathering). Mud occurs in laminae or as mud balls. There are few living organisms in the intertidal sands: crabs (hermit and fiddler crabs) are relatively common, causing bioturbation and producing fecal pellets; foraminifera are abundant in some protected pools which result in locally calcareous sediment. Some plant debris is present which comes from the mangrove swamps. Generally the sand is brown and oxidized; reduction and grey colors occur only at more than 5 cm below the sediment surface. The sands show a variety of current-induced bedforms: current ripples, oscillation ripples, cross bedding, megaripples. The last are up to 1 m high with a wave length of 3 to 15 m.

Cheniers have been formed near mid-tide level in several areas around Broad Sound after sea level reached approximately its present level (about 6000 years ago). They may reach a length of 50 km, a width of 5 km, and a height of 2 m and are mainly composed of coarse mollusk shell debris. The youngest date from about 700 BP (radiocarbon dating), the oldest from about 5000 BP (Cook and Polach 1973). They may have been formed in response to changes in local depositional conditions, probably related to fluctuations in sediment supply. The greatest concentration of cheniers is on the southwest side where the wave fetch is largest.

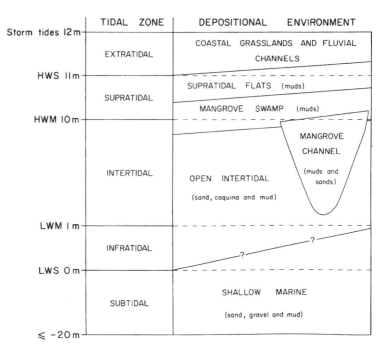

FIGURE 2.63   Zonation of intertidal deposits at Broad Sound. (*Source*: Cook and Mayo 1977.)

The intertidal muds in the upper estuaries range from sandy mud to mud and are predominantly calcareous. They form a narrow zone 100 to 250 m wide seaward of the mangrove swamps. The seaward boundary is sharp and sometimes has a low step. Where there is active progradation, some scattered mangrove trees are present, together with mangrove fragments (from leaves and twigs to complete trunks), and there is usually a thin veneer of benthic algae. Generally the mud has a depositional dip of 4° to 10° in a seaward direction and is laminated where not bioturbated; mud deposition is considered to occur mainly during spring tides, but some lamination may have been formed by benthic algae. Mud clasts are relatively common on the outer edge; in soft prograding muds there is also some load casting. Mud is supplied to the upper reaches of the estuaries (where mud is predominantly present) mostly during the summer wet season (January and February), when river discharge is at its maximum. During the dry winter, when river discharge is low, mud is mainly winnowed out of the sands. Transport of both sand and mud then tends to be inward. The fauna in the mud deposits consists chiefly of crabs and mudskippers, which bioturbate the sediment.

Mangrove channel deposits may occur in the entire range from the subtidal zone outside the mangrove to the supratidal zone, but they are mostly intertidal. They extend to 5 km inland; the maximum width of the zone is about 500 m. The wider channels have a regular arcuate form; many smaller ones are straight with sharp directional changes. All channels are bordered by mangrove trees that extend down the channel walls. The channels are both ebb and flood channels. At low tide there may be only a few centimeters of water left (up to about 1 m at the channel mouth). The channels are V-shaped, as can be seen at low tide, with slope angles up to 30°. The muddy walls are unstable and are common, ranging from small slumps to large-scale ones that are several meters across and include trees. The channel walls become more stable at slopes <10°. There the mud is laminated with a seaward depositional dip. In the channel floors the sediment is usually unstratified and sometimes sandy where the mud has been winnowed out. Some channels are half-circular to almost circular with open ends at both sides.

Mangrove swamps occur in the upper part of the intertidal zone and in the supratidal zone; usually the swamps form an irregular belt locally 5 km wide along the mangrove channels. The boundary with the bare intertidal sediments is the low escarpment mentioned above, but it is gradual where the mangrove is colonizing, i.e., where the sediment is prograding. The seaward limit of the mangrove is determined by water depth, the landward limit probably by the high salinity of the sediment and lack of sufficient moisture. The landward boundary with the supratidal flats is gradual over a few meters. Within the mangrove there is usually a change from low bushy mangrove on the seaward side to tall mangrove trees (>10 m) more inland. *Rhizophora* dominates, with *Aegiceras* at the seaward edge, *Avicennia* along the channels, and *Ceriops* at the (upper) boundary with the supratidal flats. The dense mangrove vegetation modifies the rather high-energy environment into a low-energy environment between the trees, so that suspended matter settles out. The muds are fine with silt/sand-sized carbonate components (fragments of bivalves and gastropods, and whole gastropod shells). At the seaward edge the mud is soft and grey; in the swamps the mud is usually mottled dark yellowish brown, and on the landward side brownish-grey. There are few sedimentary structures because of large-scale bioturbation by crabs, worms, and some gastropods, and by roots. The sediment is well aerated, and wood or leaf fragments are strongly oxidized. In a landward direction the sediment dries out more easily and mud cracks and algal films become more common. The mangrove swamps are considered dynamically stable; the eroding parts provide sediment for progradation in other parts.

Supratidal flats are extensive low-relief mud flats up to 5 km wide, located landward of the mangrove swamps and seaward of the coastal grasslands and uplands. There are locally narrow supratidal channels, with a maximum width of 50 m, that extend for several kilometers into the coastal grasslands and merge with fluvial channels. Some channels have been dammed to regulate the water level for the maintenance of pastures. The landward limit of the supratidal flats is often a low escarpment of 10 to 60 cm high. Also isolated remnants of coastal grassland are separated from the supratidal flats by such an (erosional) step. The scarp is probably caused by sheet-flooding of the supratidal flats when they are flooded for 1 to 2 hours during high spring tides. Pools of brackish water may persist for several weeks. During the summer there is probably an extensive cover of freshwater.

The mud on the supratidal flats (which also often contain some fine terrigenous gravel) is less calcareous than in the mangrove and on the intertidal flats. The top

5 to 10 cm of sediment is dry and hard (and only damp after flooding) with some wood fragments. The surface sediment and the interstitial water are very saline (2 to 3 times the salinity of sea water), and evaporites are formed (gypsum mainly and some halite). Crabs and large gastropods are common, and these are small halophyte plants, but most of the flats are bare with a thin film of filamentous algae. The mud is usually finely laminated and has a rubbery texture, probably because of the algae. When the flats are flooded, there is extensive algal growth. After desiccation, a blistered, cracked, and flaked surface remains with "rolls" which are formed after cracking. Plant debris is blown in, and twigs, carried by the wind, give groove casts. Tracks of birds, cattle, and kangaroo are common. In the shallow drainage channels are striations, ripple marks, small mud balls, cracks, and mud breccias. These channels decrease in depth toward the extratidal zone where they are only several centimeters deep.

The Holocene onshore deposits are entirely late-Holocene and date from 6000 BP or later. Almost everywhere 1 to 3 m grey sand or muddy sand, former open-intertidal deposits, are overlain by up to 2 m of dark grey to black mud, with abundant mangrove fragments (mangrove swamp deposits), with 1 to 2 m of olive-grey mud that is almost devoid of wood fragments (supratidal flat deposits) on top of this. Cheniers with shell debris occur in some areas. Elsewhere fluvial muds overlie the olive-grey mud on which a thin soil has been formed. The mangrove swamps grow progressively laterally as well as upward, until they become so high that hypersaline conditions prevail. Then the mangrove dies and supratidal flats are formed. ^{14}C dating of mangrove deposits and cheniers has shown that the vertical rate of accretion of the late-Holocene deposits averages 0.5 to 1.5 m per 1000 years. The lateral rate of progradation is in the order of 1.2 to 1.7 km per 1000 years. Since 6000 BP, sea level has been stabilized at approximately the present level. There has been an increase of about 1 m in the tidal range, and there may have been a period of increased storm activity between 4000 BP and 2500 BP, but in general sea level has not varied more than about ±1 m around present spring high tide level. East of the central part of Broad Sound (on the Torilla Plains), there may have been some tectonic uplift of up to 1 m per 1000 years. A schematic representation of the Holocene history is given in Figure 2.64. The estuaries have become progressively more V-shaped, which has led to a slight increase in tidal velocity and resulted in the present-day conditions of erosion being almost balanced by deposition, with sediment redistribution as the principal process (Cook and Mayo 1977). The average thickness of the supratidal deposits is 0.6 to 0.8 m, of the mangrove deposits 1.2 to 2.0 m, of the open intertidal deposits 1.2 to 2.3 m, and of the Holocene 3.1 to 4.7 m. A large part (about 40%) of the intertidal sediments is locally formed marine carbonate, decreasing to about 4% in the supratidal deposits; the remainder comes from the rivers. The sediment dispersal is schematically indicated in Figure 2.65.

2.19. COOK INLET, GLACIER BAY, ALASKA

Cook Inlet is a partially submerged tectonic basin along the Gulf of Alaska at about 60° N with a spring tidal range of about 6 to 9 m and a mean tidal range of 4 to 6

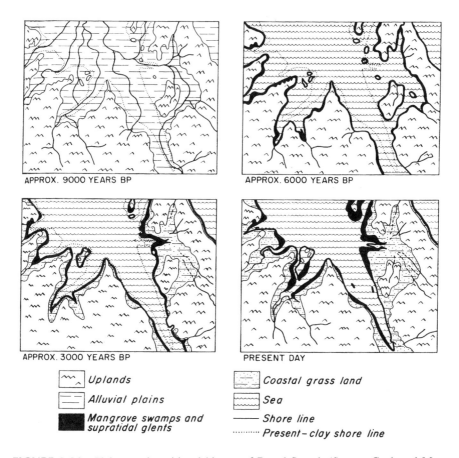

APPROX. 9000 YEARS BP

APPROX. 6000 YEARS BP

APPROX. 3000 YEARS BP

PRESENT DAY

⌄⌄ Uplands	Coastal grass land
Alluvial plains	Sea
Mangrove swamps and supratidal glents	—— Shore line
	·········· Present – clay shore line

FIGURE 2.64 Holocene depositional history of Broad Sound. (*Source*: Cook and Mayo 1977.)

m. Tectonics are highly active, which episodically results in earthquakes, and there is active volcanism along its shores. The inlet is generally sheltered, but large storm waves irregularly penetrate far into it. About 1% of the shoreline consists of intertidal deposits in small bayheads along Kamishak Bay on the west side and Kachimak Bay on the east side (both in Lower Cook Inlet) and in the inner parts of the inlet. In the latter area, in Turnagain Arm, the tidal range reaches a maximum of 9 m. The deposits consist of intertidal shoals at stream mouths, and mud flats and salt marshes located at the head of several bays (Ovenshine et al. 1976; Hayes and Michel 1982; Figure 2.66). Intertidal deposits inside Turnagain Arm were formed following the 1964 earthquake: 18 km^2 of silt deposits with a thickness of 1.5 m at the seaward side and 0.9 m at the landward side, consisting of sediment eroded from intertidal banks in Turnagain Arm. Sedimentation was probably induced by subsidence during the earthquake and was still continuing in 1973. Tidal stream channels developed as a continuation of freshwater streams with strong flow during ebb tide, and these have levees along their upper reaches (Figure 2.67). During high tides and storms, levees are locally

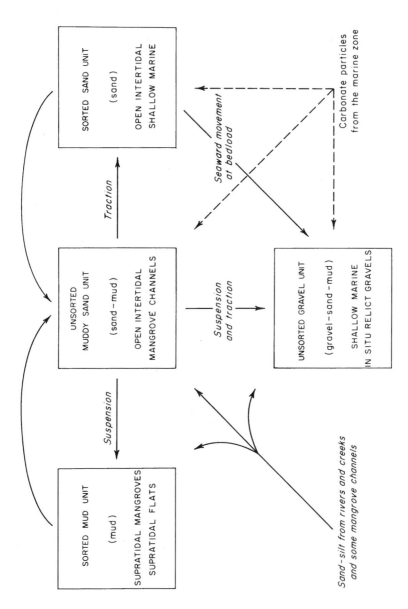

FIGURE 2.65   Sediment dispersal in Broad Sound. (*Source:* Cook and Mayo 1977.)

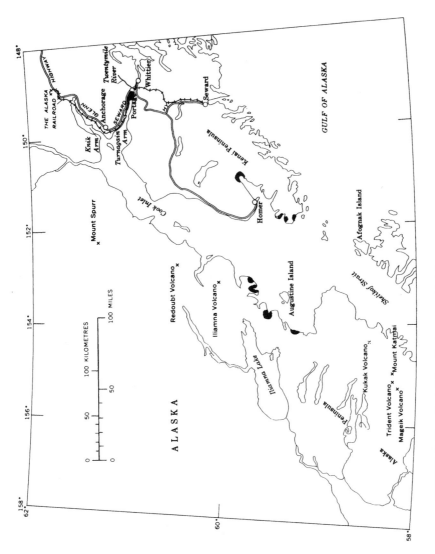

FIGURE 2.66 Cook Inlet. Dark areas: tidal flats. (Adapted from Ovenshine et al. 1976.)

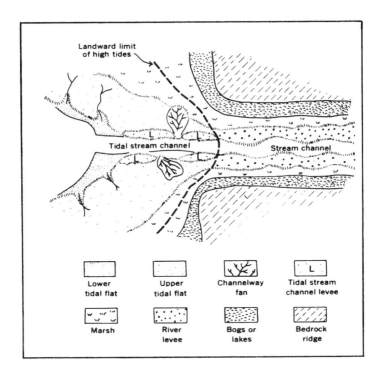

FIGURE 2.67   Schematic representation of a tidal channel in the Placer river area. (From Ovenshine et al. 1976. With permission.)

breached and overflow fans are formed that extend into and prograde over the adjacent saltwater marshes. They are flooded during about half the high tides and covered by a large number of closely spaced, shallow creeks or channels 0.5 to 1.5 m wide and 0.3 to 0.5 m deep. They are a depositional feature. The lamination of bank sediments dips toward the channels parallel to the bank face. Between the creeks and channels, the sediment is sparsely overgrown with salt grasses.

The fans grow and grade into the upper tidal flats, which are nearly horizontal and usually separated from the lower tidal flats by a 1- to 2-m-high erosional scarp. Their sediment is laminated and contains small voids where air has been trapped during rapid flooding. The landward part has a sparse flora. The salt marshes are covered with a dense vegetation of salt-tolerating grasses. Laminations, where present, are on a millimeter-scale. On the salt marshes, less than 30 cm of sediment has been deposited since 1964, as compared to 0.9 to 1.5 m on the higher and lower intertidal flats. The lower tidal flats consist of sediment that is almost saturated with water, usually rippled but unstable: it easily turns into quicksand.

The deposits formed after 1964 lie on top of an older, more consolidated silty deposit, and below this are at least 300 m of silt. There probably have been repeated earthquakes with subsidence followed by sedimentation. Earthquakes result in inundations, enlargement, and extension of tidal channels and headward erosion. Surrounding

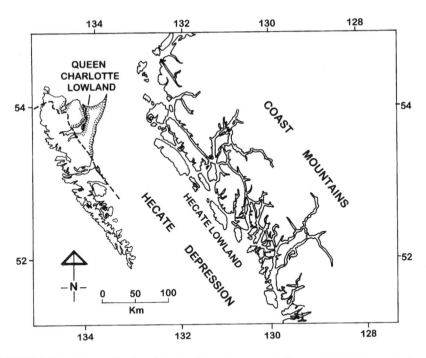

FIGURE 2.68   Queen Charlotte Islands. (From Owens and Harper 1985. With permission.)

low-lying areas become active intertidal areas, and intertidal deposition progrades over former freshwater marshes. Because of the rapid deposition, the original situation is more or less restored before the next earthquake. In general, tectonic activity at Cook Inlet results in an episodic strong supply of sediment that cannot be removed by the tides in spite of their large range, but is gradually redistributed by waves as well as by the tides. This results in wide tidal flats (and wave-cut platforms) in the lower intertidal areas and accumulation of sediment in the subsiding areas.

In Glacier Bay at about 59° N with a tidal range of 7 m at spring tide and 2 m at neap tide, clay/silt/mud flats form an up to 1.5-m-thick veneer over fluviatile/deltaic sands and gravel and are dissected by fluvial channels. They have a width of 300 to 1400 m and are flatter where they are wider. Sediment deposition occurs during neap tides when the flats remain partly submerged, and no sediment escapes from the flats; erosion occurs during spring tides when the flats are completely exposed around low tide (Smith et al. 1990).

2.20. QUEEN CHARLOTTE ISLANDS (CANADA)

Along the Canadian west coast at about 54° N, the Queen Charlotte Islands lie offshore in the Pacific Ocean separated from the mainland coast by Hecate Strait (Clague and Bornhold 1980; Figure 2.68). The tides in the northern part of Hecate Strait are mixed to diurnal and have a range up to 8 m (at Prince Rupert). The west side of the islands is exposed to ocean waves, consists of fjords, and has a narrow shelf with small deltas inside the fjords. The east side along Hecate Strait is more sheltered and waves are lower. Here are beaches and tidal flats, particularly along the largest and most northern island (Graham Island). Amos et al. (1995) describe a barricade of intertwined tree trunks on the wide intertidal area along eastern Graham Island, which is formed by trapping the logs between northward driven storm flow and the southward brach of a tidal gyre. Sea ice is of little importance. On the opposite side of Hecate Strait the coast is more indented with deltas in elongated bays.

2.21. COLORADO RIVER DELTA AREA

Extending for about 50 km along the Colorado river mouth and farther south for about 60 km along the east coast of Baja California, at about 31° 30′ N in the northern Gulf of California, wide tidal flats cover an area of about 2000 km^2 (Thompson 1968, 1975; Meckel 1975; Figure 2.69). The flats are exposed to the south and, in particular, to the larger waves which come from the southeast. This results in a northerly longshore sediment transport. The tides have a range of about 8 m at spring tide with a maximum of 10 m, and are semidiurnal to mixed with a sequence of a higher high water, a higher low water, a lower high water, and a lower low water.

The intertidal flats consist of nearly horizontal high flats around extreme high tide level, which are flooded (partly) only during extremely high spring tides, and are partly supratidal and not flooded at all. Between spring higher high tide level and spring lower low tide level an upper zone occurs with multiple beach ridges (perhaps cheniers) and mud flats with a slightly seaward dip. North of Estero Beach (Figure 2.70) the high flats are bordered by beach ridges, with a lower intertidal zone seaward that continues below the lowest tidal level as a subtidal zone down to about 11 m water depth.

The sediments on the high flats are 80 to 90% fine silt and clay with gypsum and halite where the flats fall dry. The climate is arid, and evaporation exceeds precipitation at all times. The fine sediment and the evaporites form a porous mixture without lamination and with hummocky relief. In depressions (the Arroyo Diablo and Las Salinas), large evaporite deposits occur. In Las Salinas there is more than 1 m of evaporites over an area of 15 km^2. The deposits in the Arroyo Diablo are thinner (up to 80 cm) and less extensive.

The beach ridges form semicontinuous belts of subparallel ridges. They have the character of cheniers, but this has not been ascertained. They are located between mean sea level and maximum spring tide level and have low dunes of 0.6 to 1.0 m high. The beach ridges are interrupted by inlets or gaps with channels. Usually the channels are small and occur only in the immediate vicinity of the gaps, but at Estero Beach, where four major inlets are present over a distance of 17 km, a channel drainage network has been formed that extends 3 to 6 km inward. The channel floors near the inlets are approximately at mean sea level, but because of local scour may be 1 to 2 m deeper. The series of beach ridges near mean sea level was formed

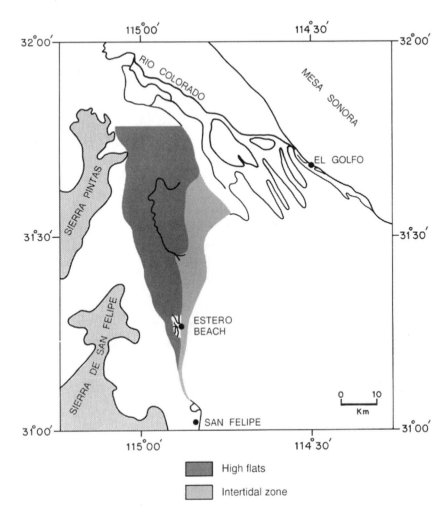

High flats

Intertidal zone

FIGURE 2.69 General map of the tidal flats at the Colorado river mouth. (Adapted from Thompson 1968.)

between 1675 and 690 BP, but because dating was done on mollusk shells that may have been reworked, datings made at the same locality may vary by 600 years. An older series of beach ridges at approximately spring tide level dates from 3000 to 2200 BP. The beach ridges are storm deposits and have been formed during periods of low mud supply from the Colorado river. During the past 90 years this has been caused by diversion of the river to the Salton Sea basin farther north, and by the construction of the Hoover Dam. It is likely that also in earlier times the river, when the bed was raised by sediment deposition, was diverted to the Salton Sea basin, which is approximately 80 m below sea level.

In the upper tidal flats, between mean low and mean high tide level, brown laminated muds dominate. The laminae are <1 to 8 mm thick (average 2 to 3 mm);

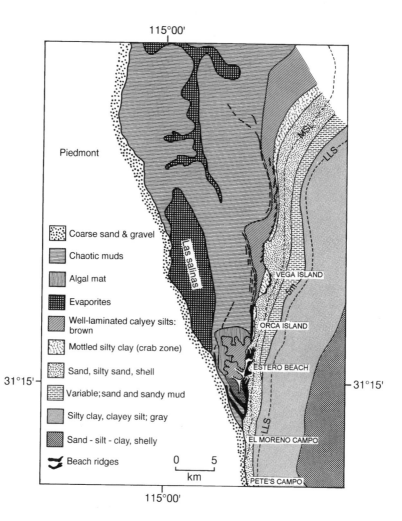

FIGURE 2.70   Sediments on the tidal flats near the Colorado river mouth. (Adapted from Thompson 1968.)

also wavy laminae and silt ripples occur. In some areas the laminated muds are overlain by fine sediment with evaporites, capped by an algal mat. The laminae, which reflect the paucity of benthic fauna, are related to the spring tidal cycle. Every spring tide, one layer is deposited and reworked during the rising tide, so that relatively coarse laminae are formed. Desiccation during neap tide results in evaporite formation and shrinkage cracks. Angular fragments are loosened where and when currents become intense.

The lower intertidal sediments, between mean low tide and spring low tide levels, are sandy and characterized by megaripples of 0.5 to 1 m high with a wave length of 150 to 200 m. They are extended roughly parallel to the shoreline, their shoreward side is steep, and they are dissected by ebb tidal gullies. A transgressive veneer of coarse sand has been formed, following the reduction in sediment supply from the river, and has been produced by reworking of the existing deposits. Benthic mollusks are common in older muds (which are partly reworked into mud balls armored with shell material), but there are few in the recent deposits, presumably because of the increased reworking of sediment. Crabs (mainly fiddler crabs, *Uca sp.*) are common above neap low tide. They cause sediment to become mottled, produce mud pellets, and mix sediment and mollusk fragments. They are concentrated in the muddy banks of channels and in the lower intertidal zone.

The flats were formed during the Holocene sea level rise. Low intertidal flat deposits occur at about 13 m depth, followed upward by higher intertidal deposits with the highest tidal flats in their present position at the top, where they partly overlap the Pleistocene piedmont deposits. Because an older relief has been buried, the sequence is not everywhere completely present.

2.22. BAY OF FUNDY

The Bay of Fundy is located on the east coast of Canada at about 45° N and is open to the southwest. It has an approximately rectangular shape with a depth of about 50 m at the entrance, and it ends in the northeast in two smaller bays with less than 12 m water depth: Chignecto Bay in the north and the Minas Basin with Cobequid Bay in the south (Dalrymple et al. 1975; Figure 2.71). The tides are semidiurnal

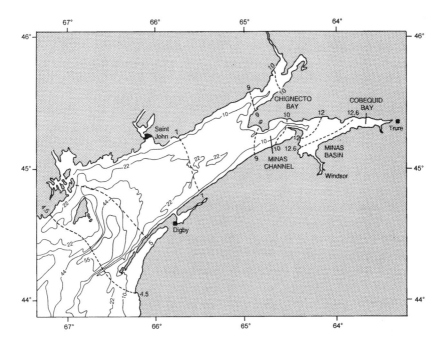

FIGURE 2.71   Bay of Fundy; broken lines indicate equal tidal range. (Adapted from Dalrymple et al. 1975.)

with a mean tidal range of 3.5 m at the entrance of the bay; they increase in range to more than 10 m in Chignecto Bay, and to more than 12 m in Cobequid Bay with a maximum of 16.3 m. The diurnal inequality during spring tide is about 1.5 m. The large tidal range develops in two stages: amplification of the oceanic tide as it moves over the shelf, and further amplification within the bay. There is a resonance effect at the bay and the adjacent shelf, and within the bay there is a strong effect of inward shallowing and narrowing. The tidal range has increased substantially during the past 4000 years because of the changes in the dimensions of the bay during the rise of sea level.

Intertidal deposits have been formed as a thin veneer on wave-cut benches, as estuarine clay flats, as accumulations in the lee of bedrock islands, as partly subtidal sand banks, and as salt marshes (de Vries Klein 1963, 1964; Knight 1980). Wave-cut benches, formed by the undercutting of coastal cliffs, are widespread along the bay: approximately 75% (by area) of the intertidal deposits have been formed on such benches. They have a thickness of more than 60 cm, which increases at the mouths of the tributary streams that extend across the intertidal sediments. The lithology of the local rock to a large degree determines the sediment properties of the veneer sediments (color, composition, granularity). Wave action during rising tide is the main process that influences sedimentation and sediment reworking. Fine material is winnowed out and deposited during slack water at high tide when the entire flats are flooded. Some of this deposit remains behind during ebb tide. Thus, the sediment near high tide level is poorly sorted but increasingly better sorted toward low tide level. Only few streams cross the intertidal zone: they have a braided stream pattern with channels up to 5 cm deep that can shift rapidly, particularly near the low tide mark. Ripples are common on the intertidal bench flats (both oscillation and current ripples), as well as rhomboid ripples, current lineations, scour pits, and animal tracks. Ripples often are flat-topped by current scouring. Sediment structures formed by stream flow through the tributary channels (ripples, lineations, flutes, grooves, lenticular bedding, and imbricated bedding of pebbles) are mostly destroyed by the waves during the tidal cycle. Characteristic are megaripples of coarse sand up to 1 m high and 3.5 m long that are often covered with interference ripples and current ripples oriented at right angles to the megaripple crests. The megaripples are formed by high velocity flow; the superimposed small ripples probably by ebb flow parallel to the megaripple crests.

At the mouths of the larger rivers entering the Bay of Fundy, such as at the Cornwallis and Avon river mouths in the Minas Basin (Figures 2.72 and 2.73), where wave-cut cliffs and benches are absent, the intertidal deposits largely consist of river-supplied sediment. A thin deposit of sand forms the upper limit of the flats where they border bedrock glacial deposits or salt marshes. The clay flat sediments are brown at the surface but black below 5 cm from the surface (Amos and Long 1980). Bottom fauna — gastropods, worms, and burrowing bivalves — are abundant. The flats are bare with meandering creeks that migrate laterally. The channel bottoms contain only few mollusk shells or shell debris; the sediment is unstratified and lacks primary structures.

In the lee of bedrock islands, such as occur along the north and south sides of the Minas Basin, a high sandy flat with fine gravel develops which is drained by

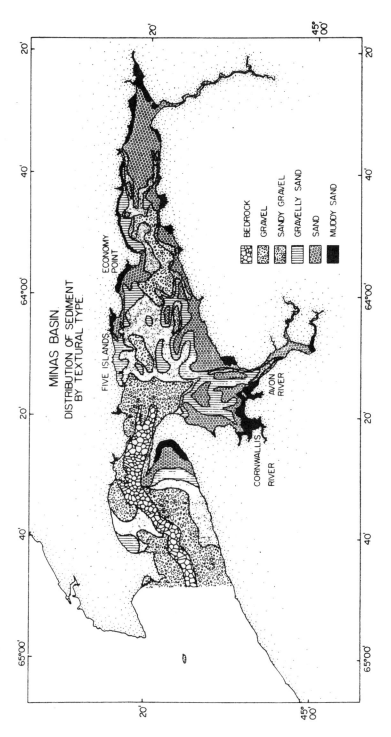

MINAS BASIN.
DISTRIBUTION OF SEDIMENT
BY TEXTURAL TYPE.

BEDROCK
GRAVEL
SANDY GRAVEL
GRAVELLY SAND
SAND
MUDDY SAND

ECONOMY POINT

FIVE ISLANDS.

AVON RIVER

CORNWALLIS RIVER

FIGURE 2.72 Sediments in the Minas Basin, Bay of Fundy. (Adapted from Greenberg and Amos 1983.)

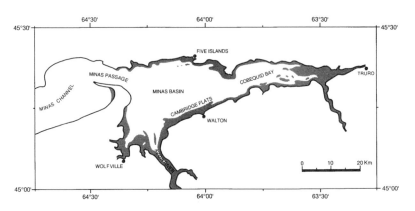

FIGURE 2.73   Distribution of intertidal flats in the Minas Basin, Bay of Fundy. (Adapted from Dalrymple et al. 1975.)

braided streams. A lower flat (which at Five Islands along the north of the Minas Basin accounts for 95% of the tidal flat area) consists of poorly sorted silt, clay, gravel, and sand with bivalve shells. At low tide this is drained by laterally meandering creeks with coarse material concentrated on the channel floors. Point bar deposits cover this. During rising tide the flood follows the creeks before flooding the flats. The mud flats are laminated with laminae of 2- to 6-mm thickness: those with only silt are thinner; those with silt and sand, thicker.

In Cobequid Bay sand bars are both intertidal and subtidal. Waves in this area, because of the short fetch, are low (<50 cm high mostly). The bars during the winter are covered by ice when the intertidal flats are also covered with an ice-sediment crust and the shoreline is protected by an ice-foot. The sand flats have only few large bed forms, mostly braided channels. The elongated sand bars are separated by relatively deep channels and are mainly formed by the tides with waves and river flow being secondary. During the ebb the water is concentrated in the channels, while the flood flows over the banks and the sand moves counterclockwise around the bank (Dalrymple et al. 1975, Knight 1980). The mud flats (brownish silt and clay) receive little sediment from the rivers in this area, but mainly from elsewhere (in suspension) and from nearby salt marsh erosion. The sediments are laminated (1 to 2 mm laminae) and rippled with some bioturbation by mollusks and worms. Within the mud laminae small lenses of silt and fine sand are sometimes present. The mud flats are drained by a system of small channels and creeks up to 2.5 m deep (Knight and Dalrymple 1975). Sandbars are present in the middle to lower intertidal zone and are generally linear in an east-west direction.

Intertidal transverse sedimentary bedforms include, in addition to small-scale ripples <5 cm high and megaripples 5 to 50 cm high, large megaripples (sand dunes) up to 70 cm high with a shorter wave length and a greater steepness than the megaripples, and sand waves up to 3.4 m high with megaripples superimposed on them (Dalrymple et al. 1978). All these bed forms occur in sands; the sand waves occur in sand coarser than 308 mm. The sand dunes and sand waves are associated with higher current velocities than the megaripples and the small ripples.

Salt marshes occur widely in Chignecto Bay and the Minas Basin at high tide level adjacent to estuarine clay flats. In some areas they have been reclaimed during the past centuries. The sediment is almost exclusively grey or blue-grey silt and clay covered with a thin layer (10 to 15 cm) of plant material with trapped (brown) clay. No stratification or vegetation structures have been observed.

Deposition started in the Bay of Fundy during the Holocene sea level rise. In the Minas Basin, the first marine deposits date from around 8500 BP and were formed in the deepest parts. They consisted of nontidal marine muds deposited when currents were very weak (Amos 1978). The first tidal influence dates from 6300 BP when sands were deposited, followed by poorly sorted gravels with sand, shell debris, and intact shells, probably transported by ice rafting. With a rising sea level, gradually the present intertidal deposits were formed along the margins.

2.23. ST. LAWRENCE ESTUARY

In northeastern Canada, the inner part of the St. Lawrence estuary inward of Capucins has a semidiurnal tide with a maximum range of 6 to 8 m near Québec and currents up to 1.5 m · s⁻¹ (Figure 2.74). From December to April the shore is covered with 0.5 to 1.5 m of ice that extends 4 to 5 km into the estuary. During winter, the wind comes mostly from the west and pushes the ice toward the opposite side of the estuary. Salt marshes occur on the south shore mostly between Baie de Belle Chasse near Québec, and Métis, and along the north shore between Beauport (Québec) and Beaupré. Farther to the north small salt marshes occur inside protected bays. Dionne (1972) distinguished the following types of salt marshes: pitted salt marshes, boulder-strewn salt marshes, rocky salt marshes, and muddy salt marshes covered with a continuous vegetation. The pitted salt marshes have numerous depressions of 10 to 1000 m length formed by ice erosion. Most of the depressions (and the largest ones as well) occur around mid-tide level. The depressions can be up to 5 m wide and up to 50 cm deep (exceptionally up to 80 cm). They have straight, steep walls and a flat to slightly concave bottom. The vegetation along the edges is torn. Isolated pits in the upper salt marsh are occasionally flooded during storm surges and high tides, and often contain decomposing plants and algae, as well as anthropogenous debris (pieces of paper, textile, metal, wood, plastic, and bottles). During the summer they dry out. Many pits can grow together into large elongated depressions. The pits and depressions are formed by ice containing frozen sediment and vegetation. In spring the ice breaks apart and single slabs are lifted and moved by the tides, while the vegetation is ripped out and vegetation and sediment remain frozen in the ice. After the spring melting, randomly distributed pieces of salt marsh and sediment 30 to 60 cm thick can be seen that have been transported in this way. They can be displaced over hundreds of m. Where the tidal flats are bare, grooves and tracks as well as a polygonal pattern may by formed by ice that is moving over the sediment surface and by mud being pressed into ice-cracks by the weight of the ice on the (soft) mud (Dionne 1971, 1972). Small mud mounds of rounded conical shape, with steep slopes of 40° to 65° and 5 to 25 cm high, can be formed during the break up of the ice when air, and possibly also water, trapped in soft mud is expelled by load pressure (Dionne 1973).

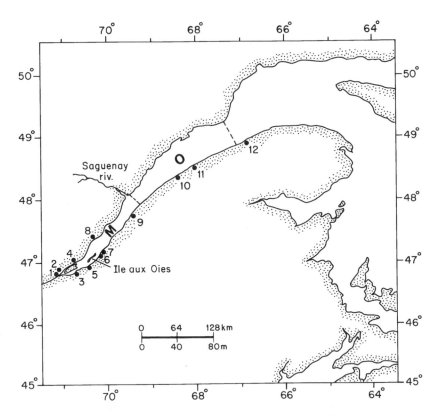

FIGURE 2.74   The St. Lawrence estuary. 0 = outer estuary; M = middle estuary; 1 = Québec; 2 = Beauport; 3 = Bale de Bellechasse; 4 = Beaupré; 5 = Montmagny; 6 = L'Islet; 7 = St. Jean-Port-Joli; 8 = Baie St. Paul; 9 = Rivière du Loup; 10 = Rimouski; 11 = Métis; 12 = Capucins. (Adapted from Dionne 1972.)

Some depressions may have been formed in other ways, such as in other salt marshes that are not covered by ice for some time: by differences in sedimentation, by alteration of old channel beds, by irregular extension and growth of the salt marsh, by wave erosion when flooded, and by the formation of pools of salt water that destroy the vegetation. The pitted salt marshes along the St. Lawrence estuary occur mainly in the central part of the estuary between Rimouski and Rivière-du-Loup on the south shore and at Bay St. Paul on the north shore.

Boulder-strewn salt marshes along the St. Lawrence estuary include virtually all salt marshes in the estuary where the tidal range is above 2 m. In some salt marshes, stones may cover nearly 60% of the entire marsh surface. Small boulder salt marshes have also been observed elsewhere in northern Canada up to the east coast of Baffin Island and Prince Charles Island (72° N). The sediment between the blocks, which may be up to 1.5 m in diameter, is a mixture of clay, sand, and small gravel colonized by halophytes, in particular *Spartina*. Blocks may be on top of the sediment or almost completely buried, with all intermediate possibilities. They frequently are pushed by moving ice, can be pushed over several meters' distance, and may even be lifted and deposited somewhere else. Usually algae (*Fucus, Ascophyllum*) grow on the larger ones.

Salt marshes with a rocky substrate occur only in a limited area on the south shore of the St. Lawrence estuary at L'Islet and St. Jean-Port-Joli and on the island of Ile aux Oies, where flatted rock is covered by a 10- to 30-cm-thin layer of mud. This layer is more or less continuous and is covered with a halophyte vegetation adapted to a low salinity. Muddy salt marshes with a continuous vegetation occur in the inner estuary around Montmagny and the Baie de Bellechasse, where the tidal range is largest and the water is almost fresh. The deposits of mud reach a thickness of up to 2 m. The influence of ice is relatively small and the effects of ice erosion are quickly obliterated by new deposition of mud. Recent salt marsh erosion in the Montmagny area is probably related to the presence of numerous geese which make 10-cm-deep holes in the vegetation in search of food (Dionne 1985).

2.24. HUDSON STRAIT–UNGAVA BAY–BAFFIN ISLAND

In northern Canada, a large macrotidal coastal area stretches from Labrador along the Hudson Strait to the northern Hudson Bay. Like most of northern Canada this region was covered during the Wisconsin glacial period with an ice sheet of approximately 3000-m thickness. This lasted up to about 8000 BP. When the ice melted down, the sea entered through Hudson Strait, and the land mass gradually rose 120 to 150 m in isostatic recovery. Along Ungava Bay, where the spring tidal range is above 9.5 m and may reach 15 to 16 m, the rate of emergence rapidly declined after 6000 BP. In the Leaf Bay area, which was studied by Lauriol and Gray (1980; Figures 2.75 and 2.76), the area submerged by the sea was gradually reduced between 7000 and 1000 BP, which was accompanied by a strong decline in the size of the waves because of the reduction of the fetch. The land emergence was probably not a continuous process, but interrupted by periods of relative sea level stability.

The present intertidal areas are characterized by the presence of large boulders and boulder fields and by boulder barricades. Ice covers the flats from December to early June, but ice movement occurs mostly from mid-May to early June when the ice cover is breaking up and melting. On the flats, ice cracks develop (mostly in concentric zones about 30 m across) and meltwater channels are formed. During this period of alternating melting and freezing the boulders move slowly down slope even on very flat slopes because of frost heaving and gravity. The large concentration of the boulders on the mud flats can be explained by this slow downward movement; it presumably started at the upper limit of submergence during the postglacial period and kept pace with the displacement of the intertidal zone downward to its present position. Also in spring, ice with boulders in it may move down slope, which can take place even on very flat slopes and is enhanced during ebb tide. Actual rafting of the boulders is unlikely because of their size of up to several meters across (although boulders up to 1.5 m in diameter are known to be ice-rafted; Dionne 1972). There is only a vague relationship between the boulder fields and the presence of moraines as a source of the boulders.

Boulder barriers are huge walls of boulders that approximately follow the contour lines and occur near river mouths, mostly in bayheads. They are also found

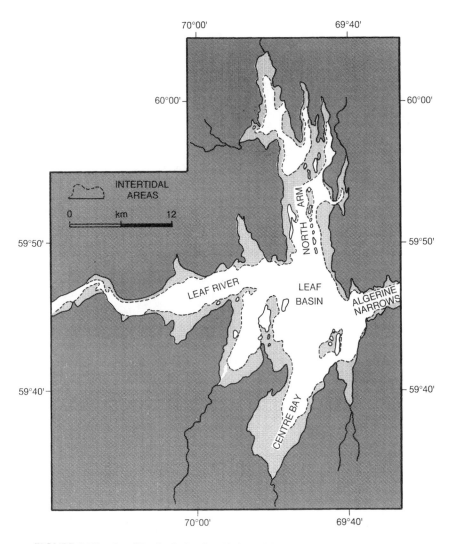

FIGURE 2.75   Leaf Basin, Labrador. (Adapted from Lauriol and Gray 1980.)

above the present intertidal zone at approximately 25, 40, and 60 m above sea level. The barriers are probably formed by the interaction of the strong tidal currents with river currents during the spring melting period when the ice cover breaks up. The more stable ice in the intertidal zone then acts as a barrier where moving ice with boulders is accumulated; opposing currents during flood then enhance the building up of ice and boulder dams. Ice rafting of boulders is not necessary to produce the boulder barriers because boulders that are pushed over the tidal flats downward will also accumulate on the barrier dams. Similar boulder barriers occur farther east along the Labrador coast where the tidal range is much smaller (1.4 m), but where the surface waves reach heights of more than 7 m and the wave swash zone is more than 6 m high (Rosen 1980). Here waves interact with the river outflow and boulders

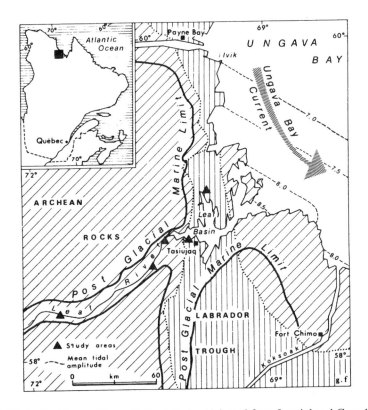

FIGURE 2.76   Ungava Bay with Leaf Basin. (Adapted from Lauriol and Gray 1980.)

are accumulated in boulder ridges. The boulder barriers at higher levels around the Leaf Basin are thought to represent former intertidal zones which developed during the pauses in the land emergence at 5300, 5600, and 6000 BP. Apart from the boulders, the tidal deposits are generally muddy without much structure but usually with ice cracks. As in the adjacent areas with a similar emerging coast (Hudson Bay, Labrador), the tidal deposits are thin (<50 cm): soil formation is slow in the subarctic climate, so only small amounts of fine sediment (clay, sand) have been available, and the sediment that accumulated during the past 3000 to 5000 years has remained only a thin layer. No flora (salt marsh) or bottom fauna has been reported from this area.

Along southeast Baffin Island between 61° N and 67° N the tidal range is almost 12 m at Lake Harbour, Hudson Strait, and in the inner parts of Frobisher Bay, and 4.5 to 8 m at Resolution Island and Cumberland Sound. The ice starts to break up in June, beginning at the head of the bays, which within 10 days results in a close cover of pack ice that gradually moves away with the westerly winds. Freezing begins in late October or early November, so that the coast is ice-free for about four months at best, but pack ice may return to the bays with easterly winds during these months. Intertidal deposits have been studied in Koojesse Inlet (Frobisher Bay) and in Pangnirtung Fjord (Cumberland Sound) by McCann et al. (1981). In Koojesse

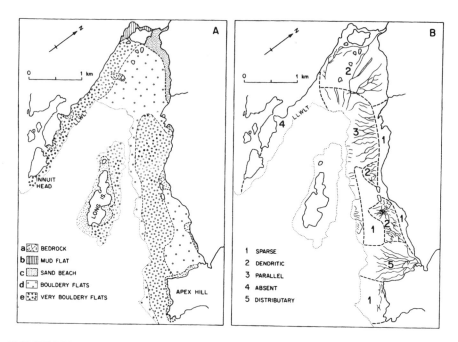

FIGURE 2.77 Frobisher Bay with sediments and channel types. (From McCann et al. 1981. With permission.)

Inlet mud flats with salt marshes around high tide level up to 1 km wide occur in sheltered areas with part of the intertidal profile consisting of bedrock. Cobbles and boulders occur everywhere on the flats on the surface or are embedded in the sediment. On the upper parts of the flats they are less numerous. The sediment on the flats ranges from clay to boulders but mostly is coarse sand with a veneer of fine sediment covering the surface of the flats, mud being present only on the upper flats. Boulders are common, and also largest, in a 250- to 350-m-wide shore-parallel zone where boulders and boulder mounds have been deposited by grounded iceflows at 3 to 6 m above the lowest sea level. Reworking of the sediment is usually limited to the upper 10 cm but goes deeper during the episodic occurrence of large waves. Burrowing is done by polychaetes and mollusks, polychaetes being the most common. Channels on the flats are very shallow without lateral displacement and can best be seen on aerial photographs. On some flats they are sparse or absent, but on the northeast side, where the flats are broader, they occur as dendritic, parallel, or distributary systems. (Figure 2. 77). There are no associated sediments on the flats, and the channels are therefore considered to be erosional.

In Pangnirtung Fjord (tidal range 4.8 m) intertidal flats occur mainly on the southeastern side and are 100 to 600 m wide with a nearly continuous asymmetric boulder barricade on the outer fringe, 12 to 15 m wide and up to 1.7 m high with a steeper landward slope. The size of the boulders is mostly between 0.8 and 1.5 m but can be up to 3.5 m. The boulder barrier is very stable and has been unchanged for up to 15 years. The flats are strewn with boulders that decline in number shoreward. The sediment distribution on the flats (mud, sand, gravel) is influenced by fluvial input during the summer. In spite of the tides, scour features and other superficial effects of ice action are still visible here.

On Baffin Island no evidence for ice rafting was seen, but material deposited on the ice surface may be entrained. Intertidal ice may rise and fall with the tides; around large boulders this results in a cone-shaped deformation structure. Down slope movement during a long period may have also resulted in the accumulation of boulders here at low tidal levels as in Leaf Basin, while rafting or bulldozering will have resulted in the formation of a boulder barrier. The flats at Koojesse Inlet probably have a sediment deficit so that erosion by shallow surface channels is visible, but the flats along Pangnirtung Fjord, where such channels are absent, receive recent fluvial supply. The present coast has a tendency to submerge. The uplift after deglaciation probably started around 11,000 BP and ended around 3000 BP.

2.25. SAN ANTONIO BAY (ARGENTINA)

This sheltered bay along the coast of Argentina at about 40° 50′ S has a mean spring tidal range of about 8 m, a maximum spring tidal range of 9 m, and a neap tidal range of 5 m. The intertidal deposits consist of two types: a more inward salt marsh without freshwater inflow and with a dendritic pattern of small channels and creeks and sandy intertidal shoals at the outer part of the bay (Schnack 1985; Figure 2.78).

2.26. SAN SEBASTIAN BAY (TIERRA DEL FUEGO)

San Sebastian Bay is located in the eastern (Argentinian) part of Tierra del Fuego at about 53° 10′ S. It is open to the east and exposed to waves from that direction, but access from the Atlantic Ocean is restricted by a gravel spit (the Paramo Spit; Isla et al. 1991; Figure 2.79). The tides are semidiurnal with a maximum range of 10 m. The flood enters along the south of the bay, and the ebb returns from the north along the inside of the spit. Because of the dominance and persistence of westerly winds in this region, the ebb currents dominate. The spit is formed by oblique waves from the north which induce a southward longshore transport, while waves from the east are reduced by the westerly winds. Because of bottom friction, higher waves (>50 cm) are common during high tide, lower waves (<20 cm) during low tide. One river, the San Martin, enters the bay from the west where several short gravel ridges are present. Longshore currents in the bay are toward the west along the south shore and to the north along the inside of the Paramo spit.

Within the bay, *Salicornia* mud flats have been formed around and above mean spring tide level. The strong winds from the west, with velocities over 60 km/h during 200 days of the year, deflate the mud surface which is rarely flooded. Mud clumps 1 to 2 m wide and covered with *Salicornia* are scattered over the flats, leaving large deflation ponds which are enlarged by further deflation. On the westside of the clumps, the *Salicornia* is progressively buried, and shrubs and grasses take over. During winter, the flats are covered with ice and snow.

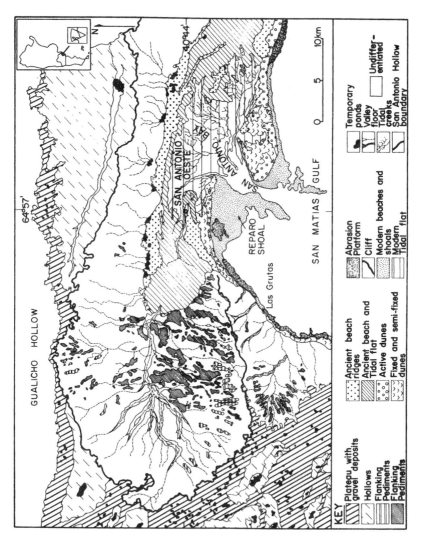

FIGURE 2.78   San Antonio Bay. (From Schnack 1985. With permission.)

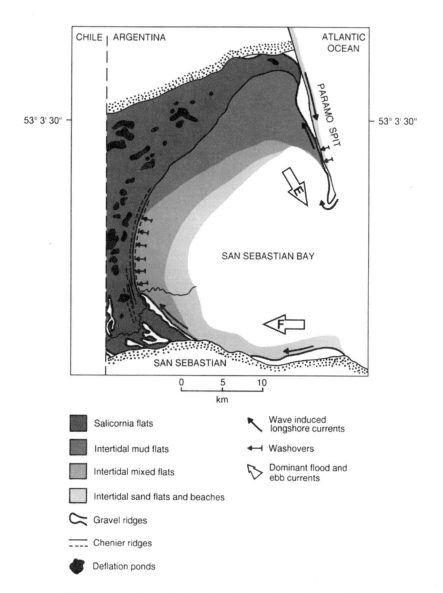

FIGURE 2.79   San Sebastian Bay. (Adapted from Isla et al. 1991.)

In the south of the bay, where waves from the east as well as the flood tide reach the eastern margin, pebbles are transported westward and have been deposited in ridges, leaving a few gaps where the San Martin river passes. The pebbles can be as large as 20 cm in diameter and locally are overlain by sand dunes. Farther north along the western margin of the bay, cheniers have been formed on the mud flat, consisting of reworked shells and sand. The oldest series of cheniers is continuous; the younger ones are interrupted by washover channels. The most recent cheniers are 0.8 to 1.0 m high with washover channels every 400 m. Here the cheniers are recurved and washover fans of mud and shells are formed. The cheniers are storm deposits and their formation is considered to be favored by a lowering of relative sea level during the Holocene (Isla et al. 1991).

The sediment building up the cheniers and the intertidal flats comes from erosion of Miocene and glacial deposits and is abundantly available. The intertidal flats cover a wide area along the inner margin of the bay with mud flats that begin where the cheniers end, i.e., where mud deposition becomes possible and storm waves are not able to form cheniers any more. Seaward of the mud flats, which are up to 10 km wide, are mixed flats with flaser bedding as well as wavy and lenticular bedding, and farther seaward there is a sand flat often covered with ripples. The mud flats can de divided into an upper zone between spring high tide level and neap high tide level, and a lower zone down to low tide level. The upper zone is 3 km wide, very flat, and with a uniform surface. The sediment consists of 3 m of silt and clay in flat beds with abundant mud cracks and dominated by wind action. The lower zone has an upper part, 3.6 km wide, and a lower part up to 2.5 km wide. The upper part consists of 3 m of laminated clay interlayered with salt and interrupted by flaser and wavy bedding. Meandering tidal channels are formed as well as shallow ponds that are elongated in the direction of the dominant wind and are about 1 m wide. Within the channel deposits, point bar deposits and clay chips, imbricated in the direction of the ebb flow and covered by mesoscale crossbeds, are common, as well as rotational slumps and solifluction phenomena. The lower part consists of more sandy deposits with a thickness of about 3 m and with silt laminae, flaser and lenticular bedding, and bioturbation. In the northernmost area, channels are straight and oriented normal to the direction of the prevailing winds. Migration of the channels, influenced by the wind, results in elongated sand banks on the channel floor. More to the south the channels are meandering and may be up to 3 m deep and 50 m wide. Sediment transport through the channels is highest during spring tide and during the winter, when the water is cold and ice rafting takes place.

The gravel spit (El Paramo) consists of a sandy matrix with gravel that is mostly smaller than 5 cm in diameter. Sand becomes dominant along the beach around low tide level. In the northern corner of the bay, where mud is deposited directly onto the spit, water flows out of the spit over the flats during low tide. Rills with a spacing of about 1.5 m are formed in the sediment surface, going down slope. The spit consists of several dozen beach ridges at its northern end, differentiated according to size and shape of the pebbles. The largest are present on the storm berms, which reach up to 9 m above low tide level. In the central part of the spit, there are beaches and storm berms on the Atlantic side as well as on the bay side with only 50 m distance between the two berms. Cobbles up to 40 cm in diameter occur on the storm berms and the narrowest part is cut every 100 m (approximately) by washover channels. Waves pass over the spit during episodic very high tides (in the order of twice a year). After the overflow from the sea, the washover sediment is reworked by waves generated in the bay, which are more persistent (and much smaller) than those on the Atlantic side. In the southern part of the spit, the beach at the end extends in three different directions, forming a triangle. The modern beach along the spit is generally being eroded on the Atlantic side.

The sedimentary history of San Sebastian Bay is not well known. The Holocene (Flandrian) transgression ended about 7000 BP, after which sea level has dropped 3.5 m since 5500 BP. Marshes progressively covered mud flats on the eastern side of the bay, and the southward growth of the Paramo spit progressively increased the tidal effects (and the southward extension of the intertidal flats) as well as restricted the effects of ocean waves within the bay.

2.27. COLOMBIA–ECUADOR

Along the coast of Colombia, south of Cabo Corrientes at about 5° 60′ N down to about 1° 30′ N, south of Tumaco along the north coast of Ecuador, dense mangrove occurs, partially fringed by sandy beaches and interrupted by short stretches of cliffs (West 1956; Schwartz 1985; Figure 2.80). Farther south smaller mangrove areas occur at Ancon de Sardinas Bay and at Cojimiés (both with salt marshes) at 1° 30′ N and 0° 30′ N, respectively, as well as a larger mangrove and salt marsh area, also with mud flats, in the Gulf of Guayaquil at about 3° S. The tides are semidiurnal with a 4 to 6 m range at spring tide along the Colombia coast. Southward this range decreases to about 3 m at the ocean coast of Guayaquil, where it increases inward to more than 4.5 m. The mean tidal range here is 1.8 m along the ocean, increasing to 3.3 m inward. Waves and ocean swells come from the west and southwest.

The mangrove in Colombia and northern Ecuador forms a tidal fringe along a narrow coastal plain. Numerous small streams from the nearby mountains form estuaries and coalescent deltas with tidal channels. West of Buenaventura Bay this is interrupted over a distance of 40 km by cliffs of Tertiary rock; other interruptions occur over a short distance just south of Buenaventura, and over a distance of only about 24 km at Las Tortugas on the Ensenada de Tumaco. The intertidal sediments, which occur for a distance of more than 600 km, consist of a belt of shoals and mud flats immediately off the coast, followed inland by a series of discontinuous sand beaches (frequently absent), interrupted by tidal inlets, estuaries, and wide mud flats, with a zone of mangrove forest about 1 to 4.5 km wide at the back of this. Farther inland the mangrove grades into a belt of freshwater tidal swamps, where the higher grounds are covered by equatorial rainforest. About half of the intertidal area is fringed on the seaward side by sandy beaches; where they are absent, the coast is fringed by extensive mud flats and, in quiet bays and sheltered estuaries, by mangrove. The mangrove belt, which normally is only a few km wide, can become very wide near river mouths, where the mangrove reaches a maximum width of 23 km. Along the entire coast the mangrove (*Rhizophora* mainly) reaches heights of more than 30 m. In addition to *Rhizophora, Avicennia,* and *Euterpe sp.*, a freshwater palm, cover wide areas. Where the mangrove is disturbed more inland (either by man or by strong wind), a brackish water fern (*Acrostichum*) is common. The patches of high, drier, and firmer ground, which probably have become higher through decay of the mangrove vegetation and the formation of peat, are covered with palms and equatorial forest.

Tidal channels are usually short and meandering. They form complex systems with wide channels on the seaward side that rapidly narrow inward and become small creeks. At high tides the forest is flooded; at low tide the smaller creeks fall

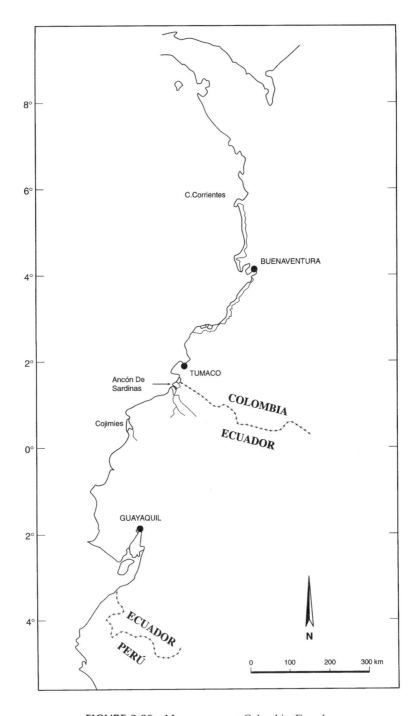

FIGURE 2.80   Mangrove coast Colombia–Ecuador.

dry, while the larger, deeper channels become small creeks. Where beaches are present, the seaward section of the mangrove, as well as the tidal channels and lagoons, are arranged parallel to the coast and form interconnections between the estuaries. In the mangrove, soft mud is deposited, which is colonized by crabs that dig holes in it and form small depressions at the mud surface. Sandy areas, present in the mangrove, are probably remnants of former beach ridges or cheniers, or may be former levees formed along (former) stream channels. Erosion of mangrove by waves is common on their seaward side. An earthquake in 1979 with its center near Tumaco resulted in a tsunami wave, liquefaction of Holocene sediments, and subsidence of the coast by 1.6 m over a distance of 200 km north of the Ecuador border. This resulted in flooding, erosion, and reworking of the sediments (Herd et al. 1981).

At the Ancon de Sardinas Bay, an area of intertidal deposits of 500 km^2 consists of extensive shoals and flats covered with mangrove and salt marsh and drained by tidal channels. A similar but smaller intertidal area is present at Cojimiés behind a coastal barrier. Extensive intertidal areas with mangrove and salt marshes are present again in the Gulf of Guayaquil, where a large estuary and delta are formed by the Guayas river. Along the east side of the Gulf, small deltas with mangrove have been formed by small rivers coming from the east. Strongly meandering channels and creeks occur in the salt marshes and mangrove of the Isla de Puna and in those extending farther inward up to Guayaquil, as well as along the east and south coast of the Gulf. The salt marshes, both at Cojimiés and in the Gulf of Guayaquil, have been extensively excavated to make shrimp ponds. The mangrove mostly forms a fringe around the salt marsh.

2.28. MACROTIDAL DEPOSITS

When discussing macrotidal deposits, it should be taken into account that those in Africa and South Asia have been little studied, as can be seen from the previous sections, and that also relatively little is known about those in South America. The most intensively studied are those in Europe, East Asia, Australia, and North America (although little is also known about the intertidal deposits along the Queen Charlotte Islands). On the basis of the well-studied macrotidal deposits, we can make several observations.

Sandy barriers, barriers islands, and spits are virtually absent, which agrees with the conclusions of Hayes (1975), but spits and barriers of pebbles or gravel are present, where pebbles or gravel, are available, as is the case in macrotidal areas along south and east England, northern France, and Tierra del Fuego. Sandy spits are rare, even where sand is available, because sand can be moved more easily than pebbles or gravel by the tides and therefore is more easily dispersed by tides with a large range. Nevertheless, there are sandy barriers at the entrance of the Bassin d'Arcachon and at the Baie de Somme, where there is strong longshore transport of sand and the barriers are part of a larger system of sand transport along the coast. In both cases, the entrance to the basin behind the barrier has remained open — although the inlet to the Bassin d'Arcachon has been deflected to the south, while along the Les Landes coast north of Arcachon there is evidence that other (former) stream mouths have been closed, although these estuaries must have also been macrotidal. Perhaps the fact that the tidal range is not large here (spring tidal range 4.9 m) is also of influence. It follows that it is not merely the presence of macrotides but also the intensity of longshore transport of sand that determines whether a sandy barrier or spit is present in a macrotidal area. In the same way the barrier at the Taf river estuary may be explained by strong longshore sand transport. For the small beach ridges described along the Irrawaddy river delta and the Sundarbans, it is not clear whether they are cheniers marking areas that are not prograding at present, or whether they are (old?) beach ridges.

Cheniers are present around the Zambesi river mouth, along the Chinese coast (Jiangsu), at the Red river mouth, King Sound and the coast between the Ord and Victoria rivers, at Broad Sound, near the Colorado river delta, and at San Sebastian bay. They are recent or subrecent; older (Holocene) cheniers are present at the Chang Jiang river mouth, along Taizhou Bay, and at Wenzhou Bay. None of the cheniers can be related to varying sand supply from a nearby river. Cheniers consist of reworked mollusk shells and sand. They occur between mid-tide level and spring high tide level (most are located around high tide level). They mark periods of nonaggradation or recession of the coast and an absence of fine-grained sediment supply. When this supply is resumed and the coast prograades again, the chenier remains as a low ridge in the fine-grained deposits. Present formation of cheniers at a macrotidal coast can be seen around the former Yellow river mouth along the Jiangsu coast. Shell ridges in the Baie de Mont-St. Michel and a small chenier along the West Korean coast started to form on the lower tidal flats and moved upward and landward following the direction of the principal storm winds and waves. Off Korea a small chenier moved from the low flats to the high flats near a dike in more than 30 years.

Channels and creeks occur in a wide variety on macrotidal flats:

Straight: in the Wash, along the Essex coast, at the Anse d'Aiguillon, the Solway Firth, Broad Sound, and San Francisco Bay.

Sinuous: in the Baie du Mont-St. Michel, the Norfolk marshes, the Anse d'Aiguillon, the Baie d'Authie, the Ganges and Irrawaddy deltas, Namyang Bay, Jiangsu, the Chang Jiang estuary, Taizhou Bay, the Red river mouth, King Sound, Ord river, Broad Sound.

Meandering: in the Wash, the Baie du Mont-St. Michel, the Solway Firth, the Anse d'Aiguillon, the Irrawaddy delta, Namyang Bay, the Red river mouth, King Sound, Bay of Fundy, San Sebastian Bay, Colombia-Ecuador.

Parallel channels: on the Korea coast, in Jiangsu, Taizhou Bay, Wenzhou Bay, Koojesse Inlet.

Dendritic channel systems: in the Bassin d'Arcachon, along the Essex coast, the Norfolk marshes, the Solway Firth, the Irrawaddy delta, Kyeonggi Bay, Jiangsu, the Red river mouth, King Sound, Koojesse Inlet, San Sebastian Bay.

Elongated dendritic channel systems: in the Perthuis d'Oleron, Anse d'Aiguillon.

Distributary channels: in King Sound, Koojesse Inlet.

Braided channels: in the Baie de Somme, Meghna estuary, Bay of Fundy.

Interconnecting channels: in the Sundarbans, the Bassin d'Arcachon, the Irrawaddy river delta, the Chang Jiang river estuary.

Few or no channels: in the Baie de Bourgneuf, Inchon Bay, north Hangzhou Bay, Nanhui (Chang Jiang river mouth), the Colorado river delta, the St. Lawrence estuary, Koojesse Inlet.

Lateral shifting channels: in the Baie d'Authie, Anse d'Aiguillon, Jiangsu.

There is no type of channel, creek, or channel system typical for macrotidal areas. Elongated dendritic channel systems, distributary, and laterally shifting channels do not occur often. Meanders and sinuous channels that do shift laterally mostly follow relatively small belts and do not rework wide areas, except possibly along the Jiangsu coast. Virtually all channels and creeks on macrotidal areas are incised and erosional. Only in the Baie de Bourgneuf is there a tidal watershed (fixed by a low dike).

Sediment structures also occur in a wide variety. Predominant and occurring in all macrotidal areas are approximately horizontal parallel lamination and sandy silt/silty clay or sand/silt parallel layering. Also common are small ripples a few centimeters high (both current and oscillatory ripples) and ripple structures in the sediment. Megaripples up to 70 cm high are present in sandy sediment in the Wash, the Baie de Bourgneuf, the Perthuis d'Oleron, the Baie de Somme, King Sound, the Ord river estuary, Broad Sound, the Colorado river delta, and the Bay of Fundy. Sand waves up to 3.5 m high occur in the Wash and the Bay of Fundy. Scour marks occur on the flats of China and Korea (where they are related to the regular passage of typhoons as well as to winter storms from the north), and in the Wash, the Solway Firth, King Sound, the Bay of Fundy, and the St. Lawrence estuary. Mud balls resulting from erosion of more or less consolidated mud deposit (often after slumping along a channel or creek wall) have been observed in the same areas, as well as in Broad Sound. Mud cracks occur on the higher flats that regularly fall dry for some time; they are formed mainly in areas with a dry warm climate or season (the Korean, Chinese, and Australian coasts, and the Colorado river delta) but also in the Wash, the Baie du Mont-St. Michel, and San Sebastian Bay. An erosional scarp between salt marsh or mangrove and the bare flats below is described from the Bristol Channel, the Wash, and the macrotidal bays along the French Atlantic coast, from Broad Sound, and from King Sound. Lenticular and flaser bedding are described from the Baie du Mont-St. Michel, the Bassin d'Arcachon, the Solway Firth, the Chinese coast, the Bay of Fundy, and San Sebastian Bay. Ice structures — boulders and boulder ridges on the flats, pits, and other scour features — are known from the St. Lawrence estuary and northern Canada as well as, to a much smaller extent, from the northern part of the Bay of Fundy and even the Baie du Mont-St. Michel. Load structures and convolute bedding occur where muddy deposits are soft, watery, and almost fluid in the Bay du Mont-St. Michel and along the southern shore of Hangzhou Bay. Intertidal fluid muds are described from the Baie du Mont-St. Michel, Korea, and the Chinese coast. Other sedimentary features have been observed only in one or perhaps up to three macrotidal areas: intertidal quicksand in Cook Inlet; mud drapes in the Wash; graded beds in the Baie du Mont-St. Michel; surface furrows, grooves, striations, and current lineation in the Wash, Broad Sound, and the Bay of Fundy; storm bars in the Baie du Mont-St. Michel; vesicular mud or sand in King Sound, Broad Sound, and Cook Inlet; channel levees in the Baie du Mont-St. Michel, the Anse de l'Aiguillon, the Norfolk marshes, and the Solway Firth; surface undulations (up to 4 cm high) in the Anse de l'Aiguillon; saucer-shaped interfluves in the Sundarbans; crossbedding along the Chinese coast and in San Sebastian Bay; channel deposits along Jiangsu; deflation ponds in San Sebastian Bay.

The listing here of sediment structures is by no means exhaustive. Sediment features have not been studied to the same extent even in all macrotidal areas that have been studied relatively well. The virtual absence of channel deposits reflects the paucity of channels that rework a wide area: the intertidal channels and creeks are almost all erosional, and they produce few sedimentary structures. Many surface features are easily destroyed by the tides and surface waves, which leave mainly parallel layering or lamination and ripple marks behind. Benthic fauna, which may (partly) destroy sedimentary structures, is abundant in many areas but virtually absent in other areas. Usually some of the original sedimentary structure remains where bioturbation is strong; complete reworking by bioturbation is rare.

Macrotidal river estuaries are almost invariably funnel shaped or approaching it; they may be somewhat curved (Gironde, St. Lawrence) or sinuous with a few large bends almost like meanders (the Severn estuary, Solway Firth, Hangzhou Bay). The mechanism for this funnel shape has been discussed by Wright et al. (1973) and was found to be related primarily to the tidal wave length, not to the tidal range.

The influence of **large-scale climate** is prominent. Climate determines the extent of ice-dependent features, the presence of salt marshes or mangrove, of (bare) salt flats and of flats in general, but this is not specific for macrotidal deposits.

3 Mesotidal Deposits

Mesotidal deposits are formed where the tidal range is between 2 and 4 m, but, as with the macrotidal deposits, the limits are not sharply defined. They may vary because of diurnal or longer-term inequalities in the tidal range, short-term variations related to the wind regime, and variations in tidal range within the same intertidal area. Nevertheless, mesotidal deposits, grouped within the loose limits used here, have some common characteristics and are usually associated with sandy barriers, barrier islands, spits, tidal deltas, and inlets (Hayes 1975). Like macrotidal deposits, mesotidal deposits are only found where there is shallow coastal water, an open coast or an open connection with the sea so that the tides can come in, and granular material available for building up sediment. In some large, mainly mesotidal areas, such as the Wadden Sea and the north coast of South America between the Amazon mouth and the Orinoco delta, the tidal range varies from microtidal to macrotidal, but the mesotidal deposits dominate, so they are grouped in this chapter.

3.1. THE WADDEN SEA

Along the southeast coast of the North Sea between about 53° N and 55° 30' N the Wadden Sea, with a length of 450 km, extends mostly behind a series of 21 barrier islands in The Netherlands, Germany, and Denmark (van Straaten 1951, 1954; Postma 1954, 1957, 1961, 1967; Reineck 1967, 1970, 1975; van Straaten and Kuenen 1957, 1958; Oost and de Boer 1994; Oost 1995; Figure 3.1). Smaller islands and sand flats also occur on the seaward margin of the Wadden Sea. The total number of islands is about 50. Several rivers and streams flow into or through the Wadden Sea (the Elbe, Weser, Ems, Eider, and Varde Å being the largest) and, together with the outflow from canals and from Lake Ÿssel (which is filled with freshwater from the Rhine), they give a lower salinity along the inner margins. The Wadden Sea is 15 to 25 km wide in its westernmost part, decreasing to 3 to 4 km in Germany, then increasing again to 10 to 12 km in the German Bight and 17 km at Süderoog, and then decreasing again to a width of 5 km at Fanø in Denmark and 2 km at Skallingen. The tidal range is 1.2 m at the western end near Den Helder in The Netherlands. It becomes 4.1 m in the German Bight and decreases from there to 1.3 m at Skallingen in Denmark. The tidal wave comes from the south and west and follows the coast eastward to the German Bight, where it turns to the north along the German–Danish coast. The tides enter the Wadden Sea twice a day through the inlets. Between Den Helder and the German Bight, they enter the inlet which is situated more to the west

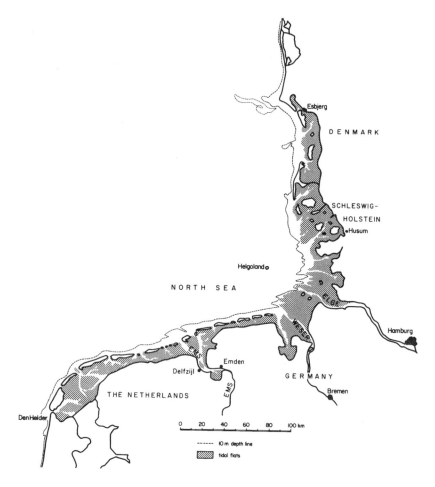

FIGURE 3.1   The Wadden Sea.

earlier than in the inlet more to the east. The tidal wave that has entered the more western inlet therefore has already traveled for some time through the Wadden Sea when the tides enter the next inlet to the east. Where the tides from two successive inlets meet inside the Wadden Sea, a tidal watershed is formed where sediment with a high mud content is deposited and the Wadden Sea is relatively shallow. Because of the difference in distance travelled by the tidal waves from the two inlets, the tidal watershed is not located halfway between the two inlets but more toward the east. Currents are generally strongest in the inlets and remain strong in the channels, but on the flats they become weaker in a landward direction as well as toward the watersheds because of the friction of the inclining tidal flat surface. On the tidal watersheds, large gyres develop where the currents from both sides meet (Zimmerman 1974). There is little or no residual transport over the tidal watersheds, but during strong westerly winds, there is some transport to the east, and there is some transport through a few shallow channels that cross the watersheds.

From Texel to the Jade, the barrier islands become gradually smaller. They are absent between the Jade and Eiderstedt, and north of Eiderstedt they are smaller than west of the Jade. Here they are also more circular in shape and not so well aligned along the coast. Large well-aligned island barriers with tidal watersheds behind them are present again from the island of Sylt to the north. From the German Bight to the north along the German coast up to the island of Sylt, the tidal watersheds are less well developed than between Den Helder and the Jade.

The tidal watersheds divide the Wadden Sea into a number of tidal basins, each connected with a tidal inlet and with an ebb tidal delta on the North Sea side. Some basins are interconnected, as are the basins of the Texel and Vlie inlets, but usually there is little flow from one basin to another. At the estuaries of the larger rivers — the Ems, Weser, and Elbe — one or two inlets are directly linked to the river. Sometimes two tidal basins share one inlet, as happens with the inlets between the islands of Ameland and Schiermonnikoog (Figure 3.2) and between Juist and Norderney. These inlets are divided into separate sections by a shallow area that extends through the inlet up to the North Sea coast. From Den Helder to the Jade and from Sylt to the north, the basins have the same pattern: two large channels diverge from the inlet inward with a large shoal in between. Often secondary channels follow the Wadden Sea coast of the adjacent barrier islands. Between the Jade and Sylt the tidal channels extend directly from the North Sea landward into the flats and form small extended tidal basins with watersheds that have become partly supratidal with flat sand islands. From Eiderstedt to Sylt there is a transition zone where the islands,

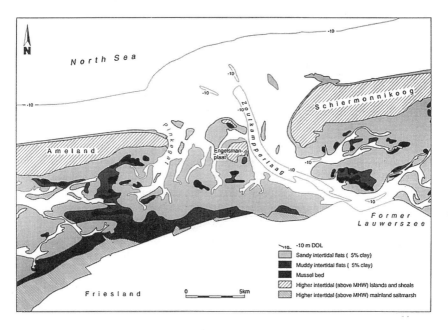

FIGURE 3.2 Flats around the inlet between the islands of Ameland and Schiermonnikoog. (Adapted from Dijkema 1989; From Oost 1995. With permission.)

tidal basins, and channels are larger. The estuaries of the Elbe and Weser are funnel-shaped: this, and the absence of barrier islands in the German Bight up to Sylt, is probably related to the tidal range which here reaches to above 4 m and to the change in the orientation of the coast toward the predominant westerly winds. Two large embayments, the Dollard and the Jade, and several smaller ones (Helmsand, Jord-sand, and the Ho-Bukt) extend from the Wadden Sea inland, with salt marshes on their inner borders. Several others have been reclaimed, such as the Ÿssellake (former Zuiderzee) in 1932, the Middelzee between 1000 and 1505 AD (Van der Spek 1994), and the Lauwerszee in 1967, while the Dollard has been much reduced in size by reclamations since 1520 AD, when it had its greatest extension (Oost 1995).

In most of the Wadden Sea the tidal deposits are bare, with the inner margins covered by salt marshes. Also parts of the outer (ebb-tidal) deltas at the North Sea side of the inlets are intertidal, while in the Texel and Vlie inlets flat sandy islands are present that are submerged during high floods and winter storms. The pattern of the delta shoals is related to the interaction of the offshore tidal currents, the tidal currents through the inlet, and waves. Waves and longshore drift tend to force the ebb-tidal delta into a down-drift asymmetry, while shore-parallel tidal currents tend to force it into an up-drift asymmetry (Sha and de Boer 1991). Large areas of (former) salt marshes and high flats have been reclaimed so that most of the Wadden Sea is now bordered by dikes on the landward side, as well as on the Wadden Sea side of the barrier islands. Dikes (dams) also connect the islands of Nordstrand, Nordstrandisch-moor, Oland, Langeness, and Sylt with the mainland. The barrier islands, which are mostly covered with dunes and on the North Sea side have a sandy beach or a sand flat, have been changed by the construction of sand dikes, which on Terschelling and Schiermonnikoog has resulted in the formation of large salt marshes (behind sand dikes dating from 1930 and from 1950 to 1978, respectively). Several islands (Texel, Fohr, Amrum, and Sylt) have been formed around a core of Pleistocene glacial deposits.

The channels extending from the inlets, with their smaller channels and gullies, form a more or less dendritic pattern (Figure 3.3). The main channels are straight or sinuous and may be up to 20 m deep near the inlet. Inward they decrease in depth to less than 1 m (at low tide) and generally reach to below low tide level. Their width also decreases inward. The main channels as well as the smaller ones may be ebb- or flood-dominated channels, or ebb and flood may follow the same channel with neither dominating the channel entirely but with ebb dominating mainly one side, and flood, the other side. In channel bends, tidal chutes (or tidal wedges; van Straaten 1954) may be formed, whereby the channel is cut off or shortened. The chutes rapidly become shallower in the direction of the main flow (Figure 3.4). Along the sides and the head of the channels, creeks or gullies are formed on the flats. They are at most 1 to 2 m deep and fall dry during low tide: during flood they are gradually filled in, but during the ebb, and in particular during the late ebb, water flows through them toward the channels with currents up to 80 cm · s⁻¹ (Figure 3.5). They are usually strongly meandering and are better developed in sheltered areas, where the sediment is muddy and more or less consolidated, than in sandy deposits where the sand is mobile and regularly reworked by waves (Figure 3.6). Deeply incised gully systems with a tendency to very strong meandering develop particularly

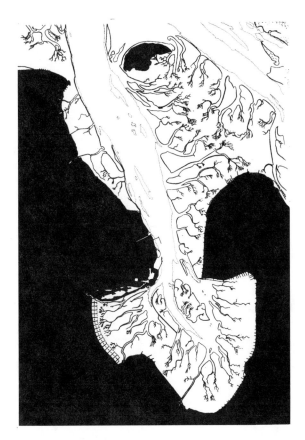

FIGURE 3.3 Channels and creeks in the Jade (German part of the Wadden Sea). (From Reineck, ed., 1982. With permission.)

in the lower mud flats. At exceptionally low tides, deep gullies may also be formed in sand flats; these disappear, however, during the next flood. The floor of the major gullies can be much lower than the adjacent flats. They seldom fall dry and are rich in benthic fauna. Locally, very shallow sinuous or meandering gullies may be formed which become braided where the slope is flatter or where more sand is being transported. On the highest flats there are no distinct water courses, probably because currents there are too weak and the flats are covered with only a thin cover of water during high tide.

The channels move laterally with erosion on one side and deposition on the other. In some channels lateral erosion is reduced or even stopped by resistant sediment (consolidated muds, glacial till). Recent lateral displacement and lateral as well as vertical erosion and/or deposition can be estimated from comparing the series of sea charts that are available for parts of the Wadden Sea from the 16th century on, and from the soundings that have been made since the early 19th century. There is a residual influx of sediment into the Wadden Sea, sand as well as suspended matter, with deposition mainly on the watersheds, in the embayments, along the

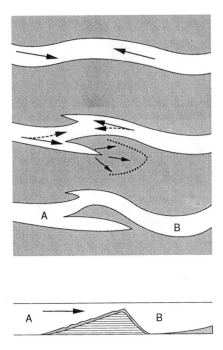

FIGURE 3.4   Chutes in tidal channels. (Adapted from Van Straaten 1954; From Oost 1995. With permission.)

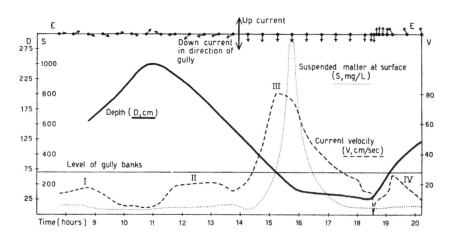

FIGURE 3.5   Currents, water depth, and suspended-matter concentration in a tidal gully. (Adapted from Van Straaten 1954; From Oost 1995. With permission.)

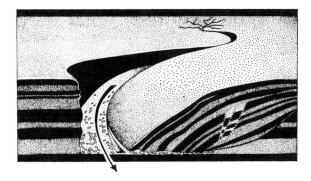

FIGURE 3.6 Schematic view of a meandering gully. (Adapted from Reineck 1967.)

inner margins, and along the dikes that connect former islands with the mainland (Figure 3.7). The recent lateral shifts in the channels, as indicated by the charts and soundings as well as by information from bottom sediments (cores), indicate that the intertidal (and subtidal) deposits are primarily channel deposits. Lateral deposits, including bands of fecal pellets and mud balls, are formed during the lateral displacement of gullies and channels. Lateral gully displacement up to 15 cm per day for 12 days has been observed (van Straaten 1951). Lateral migration of channels may be 100 m per year (Oost and de Boer 1992) but is restricted by the formation of ebb- and flood-dominated tidal chutes. In dynamically stable areas (such as the Dollard), channels and gullies, except very small ones, may not change their position for decades (Zhang Yong 1988).

The tidal flats are generally subdivided into salt marshes (at high tide level), mud flats (usually present near high tide level), sand flats (near low tide level), and mixed flats in a broad transition zone between mud flats and sand flats. Wave action dominates on the flats, tidal currents in the channels. The tidal flat surface is generally flat or undulating and usually dissected by channels and gullies. On the salt marshes the surface is often irregular, and on the sandy flats there are large fields of wave and current ripples. In wide, deep channels, sand dunes up to several meters high and hundreds of meters long may be formed whose dimensions increase with the depth of the channel. Megaripples and ripple fields occur along the larger channels, where the tidal currents are strong. Along the edges of the flats, where the water is deeper and where tidal currents are weaker than in the channels, fine sand and locally muddy sands occur. Mud or muddy sand are not present on the windward side of a channel margin, because it is either not deposited or because it is winnowed out by the waves after deposition during a period of calm weather. Mud flats are generally found in areas that are sheltered against waves (and wind), as can be seen in the distribution of mud in the Jade (Figure 3.8).

On the sand flats, a wavy bedding is common as well as ripples, but laminated sands, flaser bedding, and lenticular bedding are more common on the mixed flats. Tidal lamination, with sand deposited during ebb and flood and mud deposited as fecal pellets or from suspension during slack tide, have an overall thickness of 6 to 7 mm, reflecting a full tidal cycle. When the level of the tidal flat surface is reached,

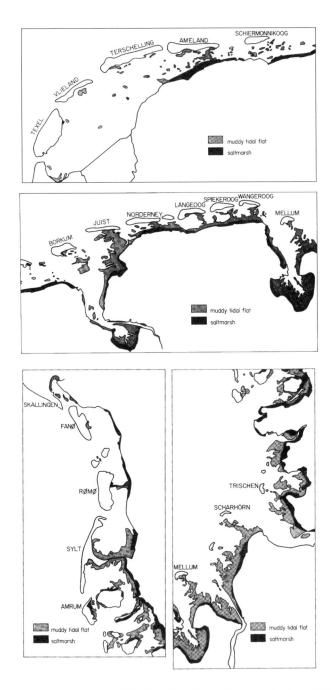

FIGURE 3.7   Mud deposits in the Wadden Sea. (From Eisma and Irion 1988. With permission.)

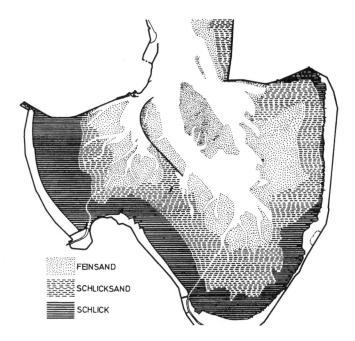

FEINSAND

SCHLICKSAND

SCHLICK

FIGURE 3.8   Sediment grain size in the Jade. Feinsand = fine sand; Schlicksand = muddy sand; Schlick = mud. (From Reineck, ed., 1982. With permission.)

deposition slows down, there is more episodic wave erosion and bioturbation increases. Where gullies develop, there may be rapid lateral and vertical erosion, as well as lateral deposition. Usually the floors of gullies and creeks consist of relatively coarse sand and finely laminated clay/sand with laminae of restricted lateral extension. Sandy laminae are rippled, and clay laminae consist mostly of fecal pellets, material eroded from older clay deposits, and peat debris. Burrowing benthos is found mainly in the upper part. On point bars, the sediment layers are slightly curved and dip toward the gully bed (longitudinal cross bedding). Interlayered sand/mud bedding is common in these deposits. Climbing ripples occur only at the mouths of gullies. Channel floors consist almost entirely of sand with ripples and cross-lamination, sometimes with herringbone patterns where the current direction has alternated. Between the sandy ebb- and flood-laminae, mud laminae may occur, probably deposited during slack tide. These mud laminae can be relatively thick and may contain more mud than deposits formed out of suspension during slack tide, because of the deposition of fecal pellets in the laminae. The mud laminae often follow the surface of the rippled sand (mud-drapes). Domination of either ebb or flood in a channel or in part of a channel can be distinguished by the domination of flood or ebb sand laminae. Fecal pellets, mud-drapes, slump structures, layers of relatively coarse sand, mud (clay) balls, and shell beds are common in channels. Benthos and bioturbation are rare, except in abandoned channels that are being filled in with often muddy sediment. This occurs regularly near new embankments, such as Ÿssellake dike, that cut off (former) channels.

Some sand may also be present within the mud layers as thin laminae. Mud on the flats is usually finely laminated when not bioturbated and occurs mainly on flats that are sheltered against large waves. Reworking of the tidal flat surface occurs mainly by waves during the winter: an original surface layer of up to 20-cm thickness may be mixed or removed. Generally the bedding on the flats is horizontal except near laterally migrating channels or gullies, where the channel infill has a longitudinal cross bedding. Ice occurs yearly on the flats in the German Bight, and more episodically in the Dutch and Danish parts, which are less enclosed by land. In the German Bight, ice up to 3 m thick can be formed (Breitner 1953); where the flats are frozen, ice-rafting of sediment is common. Moving ice disturbs the sediment and leaves tracks on the tidal flat surface, as do rain drops and the feet of many birds who feed on the flats. Also bird droppings, rill marks, deflation marks around shells or other obstacles, gas pits, foam marks, and flood level marks often can be found on the flats. Mud cracks are rather rare because of the temperate climate and the limited time the flats fall dry.

Benthic organisms form tracks, as well as feces, small mounds, and small depressions on the tidal flat surface; worms produce tubes that may stick out (e.g., *Lanice sp.*). On the higher mud flat, *Corophium*, a burrowing crustaceae, forms U-shaped tubes and reworks the top sediment down to a depth of about 5 cm. Bioturbation, not only by *Corophium* tubes, but also by worms and mollusks, is strong in these flats. The sediment of the *Corophium* flats is muddy but somewhat coarser than salt marsh deposits; it contains many fecal pellets and thin layers of peat detritus, foraminifera, ostracods, and echinoderm spines. On sand flats with no permanent gully systems, *Arenicola*, a burrowing worm that reworks the top 15 cm of the sediment, can be very numerous: it forms a feeding cone depression on one side of each (curved) tube and a heap of feces on the other. The tubes reach a depth of 15 cm in the sediment. Sediment structures are rare on such flats because of the strong bioturbation and the reworking of sediment by waves. Around mid-tide level mussel banks (*Mytilus edulis*) are often present and partly cultivated (Figure 3.9). The sediment is predominantly muddy: mussels concentrate suspended matter in pellets that are deposited between the shells. As the mussels cannot live in mud, they work themselves upward through the sediment so that a bank is formed, which gradually becomes eroded on the windward side, with new mussels attaching themselves on the lee side. This can result in the displacement of the bank by several meters per year. The mussel banks show a variation of distributed mud and undisturbed laminae. When the mussels die and no longer protect the sediment against wave action, the mud is usually rapidly reworked. The mollusk shell beds occur locally on the flats as well as on the floors of gullies and channels, and mollusk shell fragments as well as whole shells are a common constituent of the sediment in sand as well as in mud. Shell fragments are mainly produced by predators who crush the shells (ducks, fish, crabs), but some, like eiderducks, actually prefer the hard-shelled *Cerastoderma edule* as food. *Salicornia* grows around high tide level in front of the salt marshes, and in the eastern Wadden Sea, seagrass and green algae cover the flats from high tide level to about 1 m below high tide level, in particular along gullies. Nesting birds are common on the flat sand islands above high tide level that are scattered through the Wadden Sea, mainly on the seaward side.

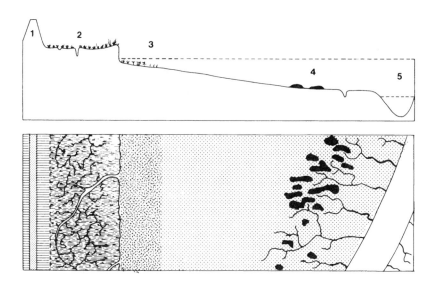

FIGURE 3.9   Schematic view of (1) a dike; (2) salt marsh with small channels and creeks; (3) intertidal flat with *Salicornia;* with (4) mussel banks (*Mytilus*); and (5) gullies and small channels as well as large channels. (From Van Straaten 1954. With permission.)

The sediment on the salt marshes, flooded during exceptionally high tides and storm floods, usually shows an alteration of sandy and muddy laminae that can extend laterally over several hundreds of meters. The sand laminae decrease in thickness upwards: from up to 5 mm in the lower parts to about 2 mm in the upper parts. The laminae are often wavy, reflecting the uneven marsh surface. Plant roots traverse them vertically and layers of mollusk shells deposited by storm waves form horizontal intercalations. Where former cuts by meandering creeks have been refilled, mollusk shells and mud balls are often present at the basis. Marsh creek sediments show sand laminae, ripple structures, and herringbone lamination. The seaward limit of a salt marsh is often a low erosional scarp, up to 1 or 2 m high, formed by waves during normal high tides when the salt marsh is not flooded. The growth of almost all present salt marshes in the Wadden Sea has been induced by human activity, but now the salt marsh area no longer increases. The marshes are mostly eroded, but some sedimentation continues, especially in the lower marshes. The recent erosion coincides with a rise in mean high tide level during the last 25 years (Bakker et al. 1993).

The Wadden Sea is a remnant of a much larger intertidal area that existed south of Dogger Bank in the North Sea between about 9000 and 6000 BP. This is at present a relatively shallow area, where tidal flats started to be formed during the post-glacial sea level rise (Eisma et al. 1981). During the Holocene, with rising sea level, the intertidal areas shifted in the direction of the present coast, while new areas were regularly flooded. When the sea level rise, caused by melting ice, ended around 6000 BP, relative sea level continued to rise at a lower rate (10 to 20 cm per 100 years) because of subsidence of the southern North Sea. Erosion of the coast and flooding alternated with periods of peat formation (Figure 3.10). The large embayments along

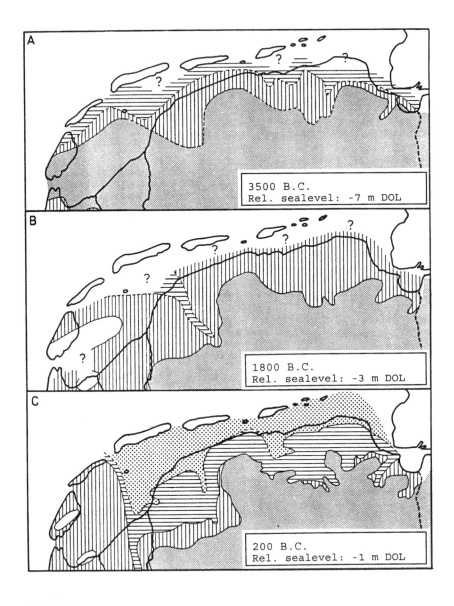

Peat
Clay
Sand
Higher areas

FIGURE 3.10 Changes in the distribution of peat, clay, and sand in the Dutch part of the Wadden Sea since 3500 BC (DOL = Dutch Ordnance Level). (Adapted from Mazure et al. 1974; From Oost 1995. With permission.)

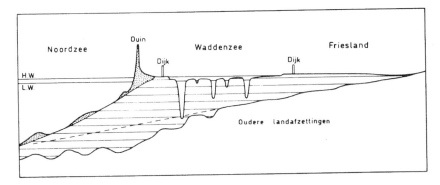

FIGURE 3.11   General profile through the Wadden Sea. (Adapted from Van Straaten 1954.)

the Dutch–German Wadden Sea date from the Middle Ages, but when dike construction and land reclamation started around 1400 AD, the Wadden Sea became smaller and the present profile developed (Figure 3.11). Gradually almost all land lost during the previous centuries was recovered with the largest reclamations in 1932 and 1967, when the former Zuiderzee was closed off by a large barrier dam, and the Lauwerszee with a smaller dam. Nevertheless, until recently large areas along the coast were occasionally flooded during storm surges, in particular during the 16th and 17th centuries.

Present deposition in the Wadden Sea includes slow vertical deposition that keeps pace with subsidence, which is in the order of 1 to 2 mm · y^{-1}. Deposition rates, determined with ^{210}Pb in laminated, undisturbed, fine-grained sediment, are also in the order of 1 to 2 mm · y^{-1}, in salt marshes somewhat higher (Pheiffer-Madsen 1981; Bartholdy and Pheiffer-Madsen 1985; Eisma et al. 1989). Because of resuspension by waves, mud deposited on the flats during calm weather can form a more permanent deposit in sheltered areas. During storms, however, these areas can be reached by large waves, so that only some mud may remain behind that has consolidated during the calm periods. In the Dollard area, about 1 to 2% of the mud deposited during the summer remains behind after erosion by winter storms (De Haas and Eisma 1993). During calm weather, a mud film may also be deposited on a sand flat, which is resuspended during the next period of windy weather (as also happens on the Belgian–Dutch North Sea beaches; Depuydt 1972). Mud deposition is enhanced by diatoms and other organisms which glue mud particles together with mucus or form fecal or pseudofecal pellets, the latter consisting of material not ingested by the organism. Also mud is eroded from older, more consolidated mud deposits, and reworked into mud balls or mud pebbles and deposited as such within sandy deposits. Mud drapes, deposited during slack tides and mainly consisting of fecal pellets, are easily buried by sand.

The sandy sediment in the Wadden Sea consists mainly of quartz grains with some feldspars and up to more than 30% biogenic calcium carbonate (mainly mollusk shell fragments; Figure 3.12). The muddy deposits consist of clay minerals and silt particles (quartz, feldspars, heavy minerals) with up to 10% of organic matter (fine plant detritus, diatoms, and other small organisms) and up to 20% calcium

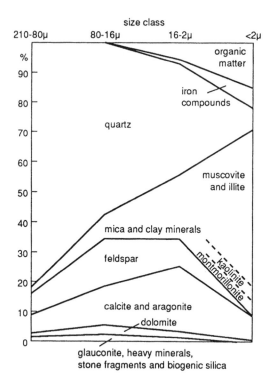

FIGURE 3.12 Composition of Wadden Sea sediment (Dutch Wadden Sea) in relation to grain size. (From Van Straaten 1964, in Oost 1995. With permission.)

carbonate (detrital and biogenic, such as shell fragments and sea urchin spines). The sand comes from erosion of the North Sea coast along the barrier islands, from adjacent parts of the mainland coast, and from the nearby sea floor. Very little sand is supplied by erosion of older (Pleistocene) deposits. Sand entering the Wadden Sea from the North Sea coast and floor is first temporarily deposited on the outer banks of the ebb-tidal deltas in the inlets and then transported inward. Because of the interaction of tidal current and waves, large amounts of sand can be transported even at low current velocities. Silt deposits are rare: the silt is transported in suspension, mostly flocculated together with the clay and organic matter, and deposited as "mud." In addition to sand, mud also comes in from the North Sea (in suspension). It is supplied from the Flemish Banks off Belgium and southern Holland and from the French Channel coast and is transported with the residual northward current along the Belgian–Dutch coast up to the German Bight and farther along the German–Danish coast. Smaller amounts of mud come from the rivers and other freshwater outflow and from miscellaneous sources such as waste discharges, atmospheric deposition, and net primary production (Eisma 1981, 1994; Bartholdy and Pheiffer-Madsen 1985).

In the Wadden Sea, there is a residual transport of suspended material inward with deposition of mud predominantly on the tidal watersheds, the inner margins, and in the embayments. This is related to a tidal asymmetry with a long period of low current velocities around high tide and a short period around low tide. This type of asymmetry favors deposition on the flats during high tide when water depths are small, over deposition in the channels during low tide, when water depths are comparatively large. Another type of tidal asymmetry with a strong flood of shorter duration and a slower ebb of longer duration favors transport during the flood as the amount transported is related to the second or third power of the current velocity. This asymmetry is found in the German Bight. Also, the current velocity needed to resuspend deposited mud is higher than the velocity at which it was deposited, so that some mud remains behind when the ebb begins to flow. This depends, however, on the degree of consolidation of the mud, and organisms may also influence deposition and resuspension by fixing mud, as seen above. Storm waves, however, can resuspend all mud even where it has dried out superficially during low tide, except where it is very much consolidated.

3.2. MESOTIDAL RIVER ESTUARIES AND RIAS IN WESTERN EUROPE

Most of the coasts of the United Kingdom and France are macrotidal with the exception of parts of Scotland and East Anglia, which are mesotidal, and small areas in Wales and along the Channel coast, which are microtidal. Other mesotidal areas in western and northern Europe, besides the Wadden Sea, include the Schelde estuary in the southern Netherlands and Belgium and the estuaries of the Tejo and the Sado in Portugal, the rias in northern Spain, and most of the White Sea.

The mesotidal Scottish coasts, on the west side of Scotland as well as on the east, contain numerous small salt marsh areas in sheltered bays and fjords. Much larger intertidal areas are present in the **Schelde estuary** around 51° 30' N, which consists of the Wester Schelde (the present river estuary that connects Antwerp with the North Sea) and, north of it, the Ooster Schelde. The Ooster Schelde was formerly the main river estuary, but since early medieval times it has gradually lost this position. Since then it has increased in size because of coastal erosion; in 1867 it was completely cut off from the Schelde river by a dike. It continues as a separate embayment with, until recently, some influx of freshwater from the Rhine. In 1980 it was partly closed by a dam with gates that can be lowered when a storm surge is expected. When the gates are open, which is most of the time, the tides enter the Ooster Schelde, albeit somewhat reduced because of the reduction in the size of the inlet. The tides are semidiurnal with a spring tide range at the North Sea side of 3.29 m and a neap tide range of 2.31 m. Inward, the range increases at spring tide to 3.79 m and at neap tide to 2.91 m (at the inner margin). In the Wester Schelde the spring tide range (at Vlissingen) is 4.47 m, and the neap tide range 3.02 m. Inward the tidal range increases to 5.45 m at spring tide and to 3.97 m at neap tide (at Bath near Antwerp).

The Ooster Schelde has three major sinuous channels, which are up to more than 40 m deep, with secondary channels and smaller branches toward the inland margins (Figure 3.13; Nio et al. 1980). The Wester Schelde has one large sinuous main channel, where the water depth is maintained by dredging to ensure the passage of ships to Antwerp. There are secondary channels, and in the bends tidal chutes or

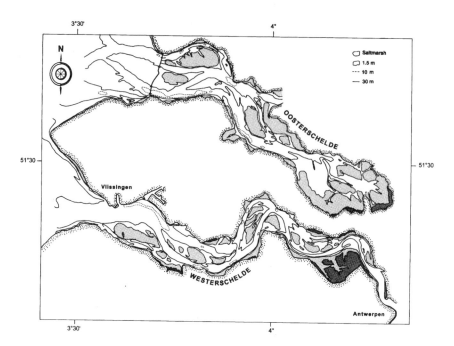

FIGURE 3.13   The Ooster and Wester Schelde. (Adapted from hydrographic charts.)

wedges occur, which in the Ooster Schelde are fewer in number. Both Ooster and Wester Schelde have an ebb-tidal delta on the North Sea side. Large sandy shoals, some with small sand flats extending above the high tide level occur in the Wester Schelde from the estuary mouth to the river, or approximately to the border between The Netherlands and Belgium. The shoals are situated within the bends and bordered by the main channel and a secondary channel. Narrow shoals with salt marshes are scattered at high tide level along the dikes on the inner fringes. A large intertidal area with extensive salt marshes is present in the inner estuary ("the drowned land of Saaftinge"). In the Ooster Schelde, two large shoals with intershoal channels occur in the outer half of the embayment with narrow strips of intertidal shoals along the bordering dikes. Two other large shoals are present about halfway, and smaller shoals, dissected by secondary channels, fill the innermost part. Here also salt marshes fringe the margins of the embayment at about high tide level, but not as extensively as in the inner Wester Schelde.

The large intertidal shoals in the outer part of the Ooster Schelde are covered with patches of flood- and ebb-oriented megaripples. In the intershoal channels ebb-dominated bedforms dominate, and on the channel margins there are flood-dominated bedforms, with transitional bedforms in-between. This pattern is strongly influenced by the wind, and on the shoals there is a lateral change from large-scale ripples in the western part to slightly sinusoidal sand dunes on the eastern side.

Sand waves are formed mainly during strong winds and tides when current velocities are above 50 cm · s^{-1}. A levee along the border of an intershoal channel, however, does not have large-scale bed forms but only small (wave) ripples. Strong westerly (onshore) storms cause strong sediment transport over the shoals as well as deposition in the channels to the east of them; easterly winds are weaker and less frequent so that there is much less transport in a westerly direction. Intershoals tend to shift laterally in an easterly direction, and the shoals with them. Sets of ripple laminae, crosslamination, and truncation of older sets of lamination are common features. Within the sand layers, mud drapes occur. Also backflow ripples and well-developed bottom sets, indicating a relatively high flow of sand in suspension, have been observed. Large-scale bedforms are abundant at the surface of the shoals, but mostly horizontal to low-angle lamination and small-scale crossbedding are pre-served in the sediment. Convolute lamination can develop (and be preserved) where air is trapped in the intertidal sands at the margin of a sandy shoal (de Boer 1979). Benthic microorganisms (algae) stabilize bottom sands and, to some extent, influence sand transport (de Boer 1979, 1981).

Shoals in the Wester Schelde have the same characteristics as those in the Ooster Schelde but generally are somewhat smaller. Large-scale cross stratification formed during the dominant tide (ebb or flood, depending on time and place in the estuary) alternates with some erosion and/or deposition during the subordinate tide (Boersma 1969, Boersma and Terwindt 1981): large sets of crosslaminated sand are separated by erosional unconformities and solitary trains of small-scale sets of laminae going in the opposite direction. Migration of bedforms occurs mainly during spring tide and is virtually absent during neap tide. Accordingly, surface bedforms are much more pronounced and show more variation during spring tide than during neap tide: sharp-crested straight sand waves and lunate dunes are present during spring tide, whereas during neap tide the shoals (those that were studied) had an essentially flat surface.

The salt marshes (Van Straaten 1954; Figure 3.14) have been formed mainly by vertical deposition during flooding at high tide: a laminated sediment is formed, traversed by plant roots, and with levees along the creeks and the salt marsh edge. The levees are formed when the water during flood starts to move over the edges and sediment is trapped by the vegetation. The edge of the salt marsh is often a low scarp, which is the result of deposition of sediment on the marsh during spring tides, followed by erosion during normal high tides. Erosion and deposition may be simultaneous. The material eroded at the scarp can be deposited on the marsh behind it. Generally on salt marshes, where plants enhance sedimentation, the vertical accretion is higher than on the adjacent bare flats where the accretion rate remains the same. A gradual transition can be found where sedimentation strongly dominates erosion. Where the marsh edge is exposed to the prevailing winds, there is usually a scarp. Where the edge is more sheltered, a gradual transition is more likely to be found. Ridges along the marsh edge cliff are formed by material thrown up by the waves or as natural levees formed during high floods.

Creeks in the marshes are numerous and strongly meandering (Figure 3.14A) with only major creeks and small channels going toward the adjacent open water. Creeks usually are of the same age as the marshes, and their depth is related to the vertical accretion of the marsh surface. Measurements during 14 tides in a creek in the Wester Schelde marshes (Saaftinge) have shown that, most of the time, the creek was exporting mainly plant material (macrophytes), but that the total amount

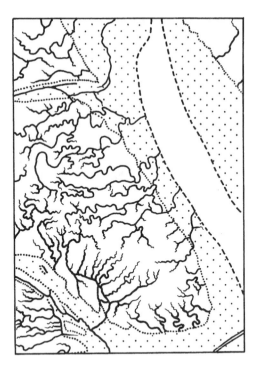

FIGURE 3.14 Salt marsh channels Ooster Schelde. (From van Straaten 1954. With permission.)

involved (about 0.2% of the above-ground plant biomass) was too small to influence the accretion of the marsh significantly (Hemminga et al. 1996). The creeks are formed on a hummocky surface and increase in length by headward erosion. They are regularly anastomosing because of the original surface morphology, because of capture of adjacent creeks, or because of shortcuts. Different parts of the creeks may be active during different stages of the tides: flow, starting through the lower parts, can be taken over by flow in another direction through the higher parts. Raised banks may close off large depressions, normally filled with vegetation. Without vegetation they become saltpans where the salinity may increase considerably because of evaporation. Rainwater strongly reduces the salinity. Strong salinity fluctuations may inhibit the development of a salt marsh vegetation.

The **Tejo (Tagus) river mouth**, with a large intertidal area of about 45 km length, is situated on the coast of Portugal at about 38° 40′ N (Figure 3.15). It is connected to the coastal sea by a broad channel of about 12 km length and has an ebb-tidal delta on the seaward side. The river flows into the intertidal area from the north and follows an anastomosing pattern through the flats toward the exit channel. Several small streams flow into the tidal area from the south. From the entrance (or exit) channel inward, two flood channels extend into the tidal area on both sides of the Tejo channels. The tides are semidiurnal with a range of 1 m during neap tide

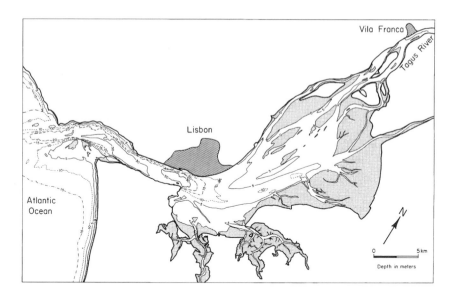

FIGURE 3.15   The Tejo estuary. (Adapted from Vale and Sundby 1987.)

and 4 m during spring tide (at Lisbon). Large intertidal areas have been used for oyster culture, and large parts have been reclaimed for industry in the south.

Directly south of the Tejo river mouth is the **Sado river estuary.** Inside the estuary on the north side is a large intertidal area of about 15 km length, and on the south side are three smaller intertidal areas (Figure 3.16; Moreira 1992). The estuary is largely bordered by dikes and separated from the sea by a sandy spit with dunes. The tides are semidiurnal with a range of 3.7 m at spring tide, increasing to 4 m inside the estuary. Waves up to 1 m high are formed locally by northerly winds, but large waves from the ocean do not pass through the inlet. The intertidal area on the north has a large central channel (the Marateca channel) with a system of interconnected smaller channels and intertidal flats with numerous larger and smaller patches of salt marsh. Salt marshes fringe the margins. The smaller intertidal areas south of the channel have a similar character on a smaller scale; the most western intertidal area has smaller flats but a very large main channel. A very small intertidal area, about 2.5 km long, is located on the inside of the spit near its tip. It also has a small channel with flats and fringing salt marshes.

The salt marshes are flooded only during spring tide and are separated from the bare lower flats (that are flooded twice daily) by a scarp of 0.5 to 2.0 m. The flats are sandy near the channels and muddy at a greater distance with sandy levees and local spits. The outer edges of the flats are locally covered with small oyster reefs where organic detritus and fine sediment are also trapped. The inner parts of the flats are covered with *Spartina*. The salt marshes are dissected by networks of meandering creeks and saltpans. The salt marsh scarp is retreating inward almost everywhere with only a few exceptions; undercutting by flood currents results in the collapse of the scarp. The retreat of the scarp can be accelerated by anthropogenic

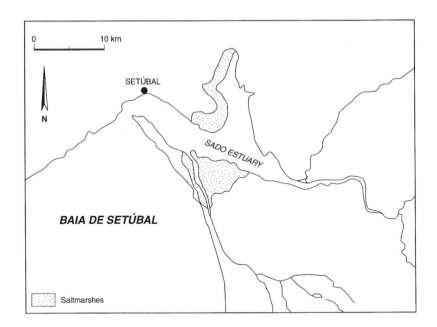

FIGURE 3.16   The Sado estuary.

destruction of the vegetation. Where vegetation is scarce or absent, deep desiccation cracks may occur, and saltpans and holes may become numerous. Rotational sliding of large blocks of sediment takes place, and erosion by the tides may result in a rapid retreat of the scarp. Or large blocks of peat may collapse and be eroded by tidal currents or storm waves. Meanwhile the sand flats prograde inward over the mud flats throughout the estuary, which is probably in response to a relative rise in sea level. On the salt marshes an average vertical accretion rate of $1.9 \text{ mm} \cdot \text{y}^{-1}$, varying from 0.8 to $3.1 \text{ mm} \cdot \text{y}^{-1}$, keeps pace with this rise. Higher rates have been measured on sandy channel levees and in saltpans, and the highest rates are found in well-vegetated areas. Erosion is limited on the flats and is considered to result more in a redistribution of sediment than in a net loss of sediment from the area.

The **rias in northern Spain** are mesotidal with intertidal deposits in the sheltered parts that are less exposed to the large waves from the North Atlantic Ocean and the Bay of Biscay. The **Bahia de Baiona** at the entrance of the **Ria de Vigo** at 42° 20′ N contains a small intertidal area (about 1 km²) behind a spit with a complex pattern of channels which end in the estuary of the Rio Minor. (Figure 3.17; Alejo et al. 1990). In the estuary, intertidal shoals are covered with megaripples and small ripples. On the north side of the estuary is an elongated intertidal sand flat with relict channels that remains dry most of the time. It is reworked by a worm, *Arenicola*. The flats south of the estuary consist of salt marshes and mixed flats with *Arenicola*. Two small streams enter this intertidal area from the southeast. The salt marshes consist of an upper part that is only flooded during spring tide, and a lower part, covered with *Spartina*, that is flooded during every high tide (Figure 3.18). The sediment is muddy with some sandy levees that are inhabited by worms (*Nereis*), a

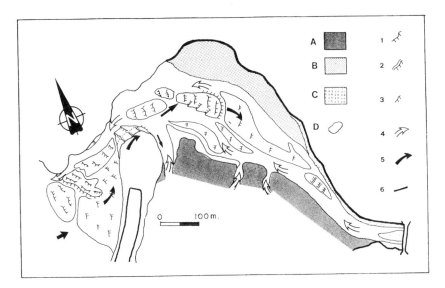

FIGURE 3.17 Baia de Baiona. A = intertidal zone; B = tidal flat; C = blocks and stones; D = sand banks. 1 = megaripples; 2 = alignments of algae along megaripples; 3 = ripples; 4 = ebb currents and fluvial discharge; 5 = flood currents; 6 = artificial structures. (From Alejo et al. 1990. With permission.)

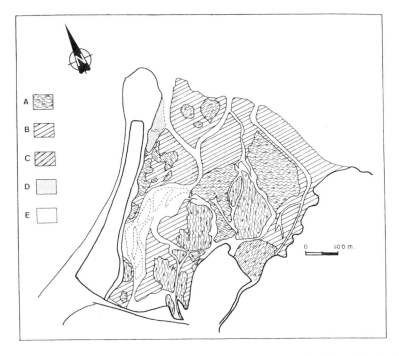

FIGURE 3.18 Intertidal area Baia de Baiona. A = salt marsh; B = mixed flats; C = *Arenicola* flats; D = inner *Arenicola* flats; E = channels. (From Alejo et al. 1990. With permission.)

crustaceae (*Corophium*), and a small gastropod (*Hydrobia*). Mixed sand/mud flats occur in the channel meander bends and are often covered with green algae (*Enteromorpha*) and mollusk shell accumulations (mainly *Cerastoderma edule*). There is strong bioturbation by worms (*Nereis*, and locally some *Arenicola*) and mollusks (*Venerupis, Cerastoderma*). The channel slopes, where they have been eroded, are often muddy and covered with *Zostera*. Near the spit that separates the flats from the open water to the west, sandy flats are present, inhabited by large numbers of *Arenicola*. Since 1956 the area has been considerably eroded, in particular the spit and the stream channels. This has resulted in a general decrease in size of the intertidal area. The eroded material is mainly deposited in the deeper parts. The erosion is related to a general decrease in sediment supply, which is the result of the construction of walls and roads, and of port development.

Inside the Ria de Vigo is a small tidal flat with two main channels fed by small streams, by numerous small, sinuous channels, and by (often) meandering creeks. The channels are ebb channels; the flood enters over the flats. A river channel (from the Oitaben river) traverses the flats along the east side (Figure 3.19; Vilas Martin

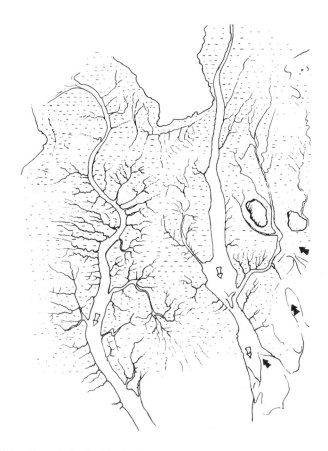

FIGURE 3.19 Channels in the Ria de Vigo. Black arrows = flood direction; white arrows = ebb direction; horizontal broken lines = tidal flats. (From Vilas Martin 1983. With permission.)

1983). Erosional channels and creeks are present in the mud flats, mainly where the sediment is more cohesive and consolidated. On the mixed flats the creeks still show ramifications that are not present on the sand flats. Most of the sediment is bioturbated near high tide by roots of salt marsh vegetation, and at lower levels by mollusks and worms (*Arenicola, Lanice*). The flats are covered by small wave and current ripples, whereas in the channel floors that are more sandy crossbedding is common.

The **Ria Formosa** near Faro on the south coast of Portugal has a tidal range that varies from 1 to 3.5 m with considerable semidiurnal and fortnightly variations. The ria is a shallow lagoon with muddy flats and salt marshes behind sandy barriers separated by several inlets. The tidal area is dissected by a network of interconnected channels of varying width.

Farther north, along the coast of Cantabria at about 7° 15′ W, a small intertidal area is present in the **Ria de Foz** (Figure 3.20; Castaing and Guilcher 1995). The spring tide range at the entrance of the ria reaches above 4 m. The highest parts, above high tide level, are covered by salt marsh. The bare intertidal flats can be subdivided into mud flats, mixed sand/mud flats, and sand flats. The channel floors are generally sandy, the channel slopes clay/silt. Adjacent to the salt marshes, large tidal flat areas are muddy and covered with green algae (*Enteromorpha*) with some gastropods. At a lower level, the mud surface tends to become fluid mud and is inhabited by mollusks (*Scrobicularia, Venerupis*) and worms (*Nereis*) and their predators (crabs). There is strong bioturbation. Because of the lateral migration of small channels and creeks, muddy slopes are formed which become steep where blocks of consolidated mud have slumped down. On the more sandy channel slopes, rill marks, small ripples, and megaripples occur while mollusks (*Lutraria, Solen, Cerastoderma*) and worms (*Arenicola*) rework the sediment. In the channel floors, which are sandy, cross stratification capped by small ripple systems with little bioturbation is common. In the outer part of the ria intertidal sandy shoals are present with considerable sand movement during submergence, which results in fields of straight and lingoid megaripples covered with small current and oscillation ripples.

3.3. GUINEA–BISSAU–GUINEA

The coast of Guinea–Bissau and Guinea between about 12° 20′ and 9° N consists of several large and numerous small river mouths and channels with extensive mangrove swamps covering intertidal deposits that can be up to 50 km wide (Figure 3.21; Diop 1985). In the north of Guinea–Bissau this mangrove is a continuation of the mangrove swamps along the Casamance river mouth in Senegal. Regionally, and in particular north of the Geba river at Bissau, the actual coast is formed by sandy ridges and small cliffs. Seaward of the ridges, cliffs, and mangrove are bare sandy or muddy intertidal shoals that may extend up to 20 km from the shore. The spring tide range is up to 4.6 m in the north at Bubaque and 4.2 m at Conakry, but the neap tide ranges are 0.21 and 1.1 m, respectively. Channels in the mangrove are interconnected and meandering with frequent cutoffs. Adjacent to the mangrove on the landward side, high flats occur ("tannes") that are mostly covered with a halophyte vegetation and are locally bare. The high soil salinity is related to occasional flooding and evaporation during a long dry period between November and May. The

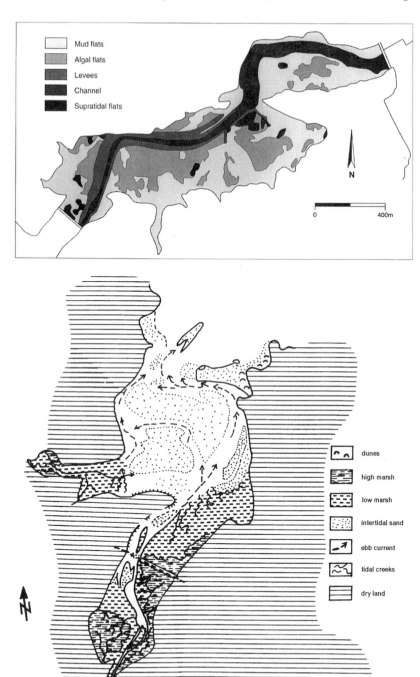

FIGURE 3.20   Top. Ria de S. Vicente. (From Gonzalez Lastra et al. 1984. With permission.); Bottom. Ria de Foz. (From Castaing and Guilcher 1995. With permission.)

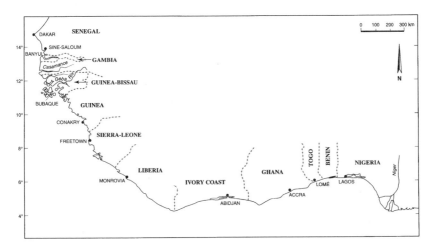

FIGURE 3.21   The coast from Guinea Bissau to Guinea.

present coastal deposits were formed after the Holocene sea level rise that ended at about 6000 BP with a maximum level of +2 m. The high flats and ridges are considered to mark stages of coastal retreat after this maximum (Diop 1985).

3.4. CAMEROON AND EQUATORIAL GUINEA

Between 5° N and 1° N the coast consists of two large embayments and a large river estuary with extensive areas of mangrove swamp with spits and sandy beaches, with cliffs in the north at about 4° N, and with coral more to the south (Figure 3.22). The tides are semidiurnal with a spring tide range of 2 to 4 m and a mean tidal

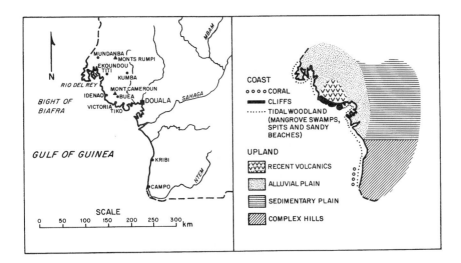

FIGURE 3.22   Cameroon coast. (From Schwartz 1985. With permission.)

range that drops to below 2 m in the south. Large swell comes from the west from June to September.

3.5. TANZANIA AND KENYA

Along the Tanzania and Kenya coast, river estuaries are small except where the rivers flow into shallow coastal waters: here small deltas have been formed (as at the Rufiji and Wami river mouths) with intertidal margins (Figure 3.23). Mangrove is abundant on these margins, and between Pangani and Kenya, where the coast is protected against large waves, mangrove occurs along the open coast. The tides are semidiurnal; the tidal range is 2 to 3 m. Along the Kenya coast large embayments with channels and islands are present around Lamu between about 2° S and 2° 30' S, with smaller embayments around Mombasa at 4° S and near the border with Tanzania at about 4° 40' S (Figure 3.24). The bays are fringed and partly filled in with mangrove that grades inland into papyrus swamp. The spring tide range (at Mombasa) reaches a maximum of 4 m in November; the average range varies between 2.5 and 3.6 m.

3.6. NORTHWEST PERSIAN GULF

Extensive tidal flats, with marshes, lagoons, spits, several large islands (Bubiyan, Failaba), and two large inlets (Umm Qasr, Bandar Dahpur) are present in the northwestern Persian or Arabian Gulf at about 29° 30' N (Figure 3.25). They are mostly part of the combined delta of the Euphrates, Tigris, and Karun rivers with the Shatt el Arab as the principal channel; at its entrance the tidal range is more than 3 m. In Kuwait Bay neap tides reach a range of about 3.5 m and spring tides 4.3 m (Khalaf and Ala 1980). Waves are low; waves larger than 50 cm are rare. Strong winds come mostly from the northwest and raise dust and sand. The intertidal deposits are mostly muddy with a relatively high content of carbonate. Detrital minerals supplied by the principal rivers dominate in the area north of Kuwait, while carbonates (low- and high-Mg calcite, aragonite, dolomite) dominate in the south, farther away from the Shatt el Arab. Dolomite north of Kuwait is probably supplied as aeolian dust, with probably also a considerable part of the detrital minerals.

3.7. INDUS RIVER DELTA AND RANN OF KUTCH

Besides small intertidal salt flats along the Makran coast (the largest is at the Dasht river mouth), where the mean spring tide range is about 1.8 m, and along Mani Lagoon, where the mean spring tide range is 2.1 m, large salt flats lie along the Indus river delta behind about 15 short sandy barrier islands between Karachi and the Indus river mouth (Figure 3.26). Farther east salt flats extend into the Rann of Kachchh (Kutch). Between Karachi and Kori Creek in the east, the spring tide range reaches 2.6 m (Snead 1985).

The barrier islands are mostly barren and flat with locally low dunes. They are separated by tidal inlets that vary in width from a few hundred meters to more than

(A)

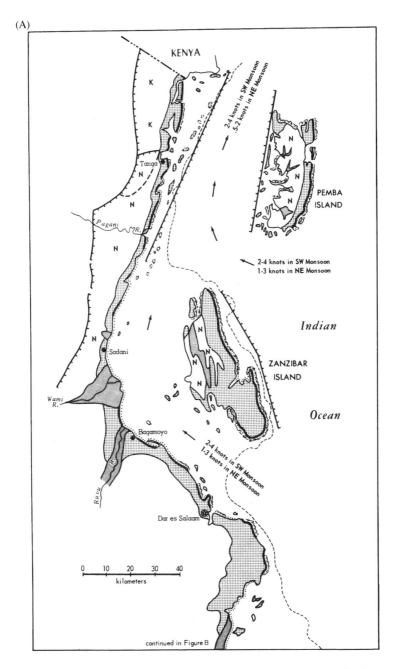

FIGURE 3.23 Tanzania coast (continued on following page). (From Alexander 1985. With permission.)

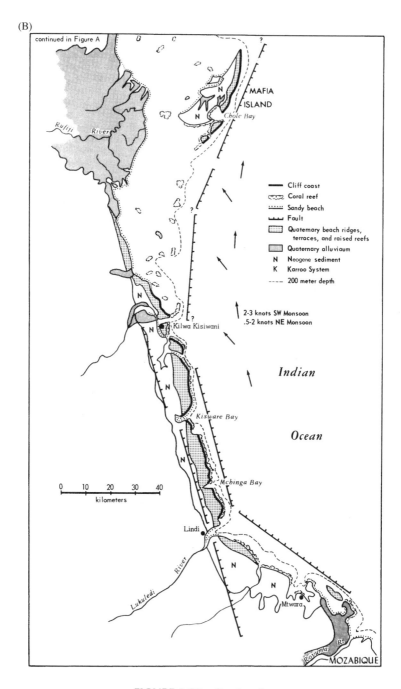

FIGURE 3.23 *Continued.*

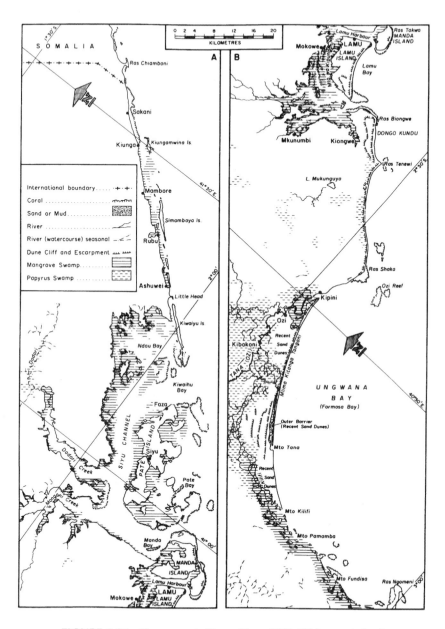

FIGURE 3.24 Kenya coast. (From Ojany 1985. With permission.)

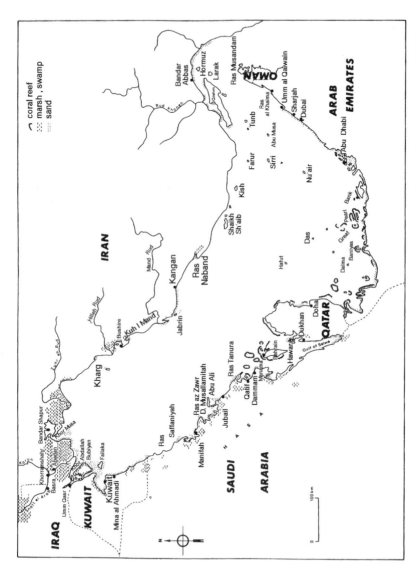

FIGURE 3.25   The Persian Gulf. (From Sanlaville 1985. With permission.)

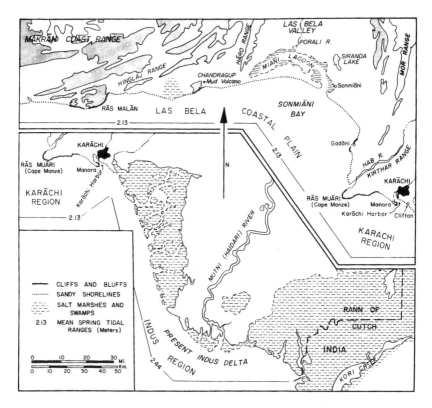

FIGURE 3.26 The Indus river mouth. (From Snead 1985. With permission.)

1.5 km. East of the Indus river mouth, mud flats extend to the shoreline, and there are few barriers. At high tide the flats are normally flooded as far as 5 to 6.5 km inland, but during the monsoon rains, when the river flood coincides with high tide, flooding may occur up to 32 km inland. Also in summer during storm surges from the southwest, vast areas can be flooded. The lower delta, because of the repeated flooding, has many tidal creeks with small levees and overbank deposits, which are fringed by low mangrove. Where the flats fall dry, salt covers the sediment and dry salt is also present below mud that has been deposited on top of it. Older beds of the Indus river or of its tributaries dating back probably as far as Roman times are still present. Because of the flooding, the strong evaporation that results in dry saline muds, and the strong mud deposition, mangrove is rather rare.

The mud flats in the delta are up to 40 km wide. In the river mouth the tides reach inward as far as Tott. Mangrove fringes the seaward part of the mouth and grades inward into *Tamarix* bushes. The eastern part of the delta was still open sea in the 8th century but had been filled in by 1600 and a salt flat had been formed. During the early 19th century most river branches in the delta were abandoned, and in 1835 only the present main channel remained (Wilhelmy 1968). During the past 5000 years, the delta has progressed by approximately 30 m per year (about 1 km in 25 to 40 years during the past 230 years; Wilhelmy 1968), but due to dam construction and diversion of river water for irrigation, the amount of sediment that reaches the sea through the river mouth has been reduced to less than 25% (Wells and Coleman 1984). Mud transport along the coast is toward the south and into the Bay of Kutch. The eastern part of the delta and the Rann of Kutch are subsiding but, nevertheless, were filled in between 800 and 1600 AD.

3.8. NORTHERN GULF OF THAILAND

In the northern Gulf of Thailand at about 13° 20′ N, mud flats and mangrove extend on both sides of the Chao Phraya river mouth near Bangkok up to Phetchaburi in the west and Chonburi in the east (Pitman 1985; Figure 3.27). The spring tide range is 3.6 m, the mean tidal range 2.1 m, decreasing to the west to 3 m and 1.8 m, respectively. Bare intertidal mud flats extend up to 7 km seaward at the Chao Phraya river mouth and up to 4 km at the Mae Khlong river mouth as well as along the Phetchaburi and Chonburi coasts. Along the remaining parts of the coast the tidal flats are less than 1 to 2 km wide. Landward of the flats is mangrove and, more inward, freshwater *Nypa* swamp, but along most of the coast the mangrove has been cut and turned into rice fields with fish ponds. Locally at the back of the flats are

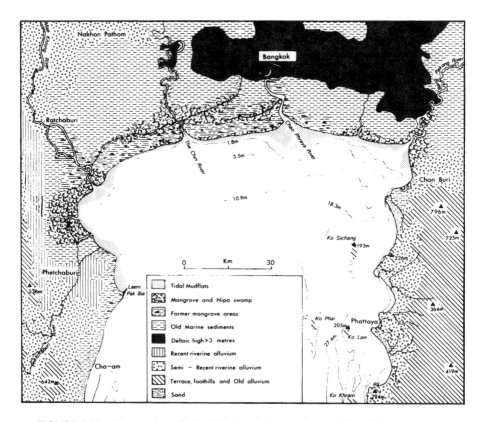

FIGURE 3.27   The northern Gulf of Thailand. (From Pitman 1985. With permission.)

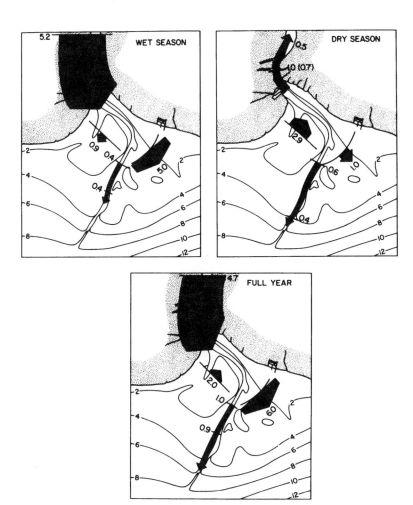

FIGURE 3.28 Sediment flow at the Chao Phraya river mouth over the year. Numbers indicate 10^6 tons (total during the indicated period). (From NEDECO 1965. With permission.)

supratidal salt flats. The entire plain is a large delta plain formed by deposition from five rivers: the Chao Phraya, the Mae Khlong, the Phetchaburi, the Tha Chin, and the Bang Pakong. The Chao Phraya is by far the largest. Based on ^{14}C dating of peats in the delta, the delta front has advanced about 4 m per year on the average during the past 5000 years. At present a zone about 5 km wide along the coast is regularly flooded during high tides. Large parts of the flat area more inland up to north of Chanat is flooded every year with freshwater (Ohya 1975).

The mud on the flats is mainly supplied in July/August at the beginning of the wet season. At that time winds are weaker and waves smaller so that the flats progress outward. During the dry season, when the river discharge is low, the tides go farther inward, and waves transport mud from the flats eastward when the winds come from southerly directions (Figure 3.28; NEDECO 1965). At the outer limit of the mud flats off the Chao Phraya river mouth lies a large intertidal sandbar with megaripple fields and small-sized wave- and current-ripples.

3.9. WEST COAST OF BURMA (MYANMAR), THAILAND, AND PENINSULAR MALAYSIA (ANDAMAN SEA COAST)

The coast of Burma south of the Irrawaddy, the west coast of Thailand, and the southwest coast of Peninsular Malaysia between about 16° N and the Equator are, particularly in Burma and Thailand, much indented with numerous islands and bays (Figure 3.29). The tides are semidiurnal and the coast is mesotidal, with a tidal range between 1.9 and 2.7 m along the west coast of Thailand and between 2.4 and 4.1 m along Peninsular Malaysia (Pitman 1985; Teh Tiong Sa 1985). Not much is known about the Burma coast. Along the Thailand west coast bare mud flats and mangrove dominate, interrupted by sandy beaches and dunes, and by rocky headlands and cliffs mainly around Phuket and some of the islands. Because of recent submergence, the coast is often steep, and intertidal deposits are restricted to river estuaries, bays, and sheltered areas behind islands and bay mouth barriers, where they are less exposed to waves and swells from the west and south. The largest mangrove swamps, extending inland for up to 10 km along estuaries and tidal creeks, occur between Phuket and Satun, grading inland into freshwater swamps. The tidal channels are sinuous and interconnecting with steep banks; the tidal deposits are 3 to 5 m thick and lie on top of gravels and earlier (river or mangrove) deposits. Along Peninsular Malaysia, mangrove swamps also dominate, fringing almost the entire coast and alternating with some sandy beaches and spits. The mangrove on the average has a width of 12 km, but its inner margins are often cleared for agriculture. Bare mud flats occur in front of the mangrove and can be up to several km wide — in exceptional cases, as at Pantai Remis, they reach 6 km. The mangrove is probably rather recent. At Pantai Remis, older Holocene mangrove was drowned during the Holocene sea level rise, and a sand barrier was formed in front of the former mangrove deposits (Figure 3.30). The recent mangrove was formed in front of the sandy barrier and at present advances at a rate of 12.5 m per year (between 1881 and 1961; Koopmans 1964).

3.10. MEKONG RIVER MOUTH

Off the Mekong river mouth between 8° 30′ and 10° 30′ N occur bare tidal flats up to 6 km wide around the main distributary mouths (the Bassoc, Mekong, Vaico, and Dong Nai; Figure 3.31). Smaller flats occur along the coast farther to the southwest, which reflects the transport of sediment from the river mouth in that direction by waves and a coastal current. Behind the flats are often extensive mangrove swamps and numerous cheniers up to 50 km inland. The tides are diurnal with a strong semidiurnal component during the period of the equinoxes. The range is 3.5 to 4 m (Kolb and Dornbusch 1973; Eisma 1985).

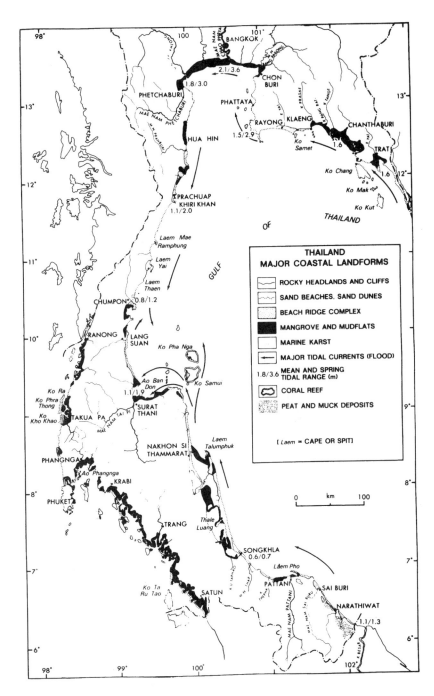

FIGURE 3.29 The coast of Thailand along the Andaman Sea and the Gulf of Thailand. (From Pitman 1985. With permission.)

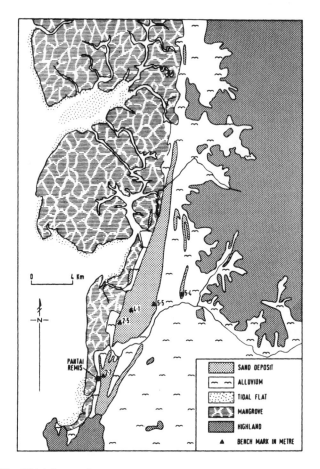

FIGURE 3.30   Tidal flats and mangrove at Pantai Remis, west coast Peninsular Malaysia. (From Teh Tiong Sa, 1985. With permission.)

3.11. THE PHILIPPINES

The Philippines consist of numerous smaller and larger islands which are mostly volcanic. The numerous coasts present a variety of beaches, beach ridges, backswamps, small river deltas, mud flats, low cliffs, and coral reefs. Siltation along the coast is often heavy because of the loose volcanic soils and deforestation in the river basins. Intertidal mud flats are usually overgrown with abundant mangrove, which occurs everywhere in sheltered areas but often has been cleared for the construction of fish ponds and for rice cultivation. The tides are mixed; the tidal range reaches more than 2 m in the southern Philippines and decreases between the northern islands to less than 2 m. Tropical storms regularly pass over the coasts along the east side of the Philippines.

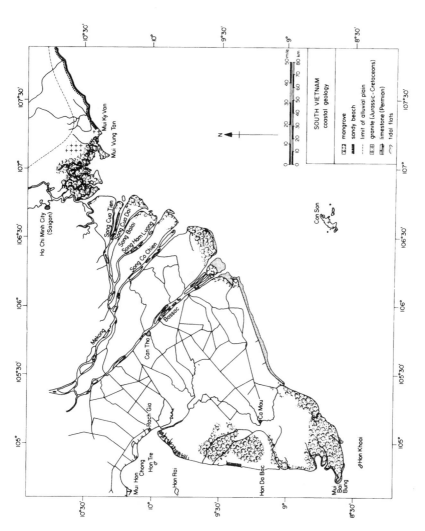

FIGURE 3.31   The Mekong river mouths. (From Eisma 1985. With permission.)

3.12. INDONESIA

In Indonesia large intertidal areas, mostly covered with mangrove, occur on the larger islands (Sumatra, Kalimantan, Irian Jaya), less on Java and Sulawesi, and on the numerous smaller islands. Mesotidal areas occur along the east coast of Sumatra (where the maximum spring tide range is 2.0 to 3.8 m), East Kalimantan (maximum spring tide range 2.1 to 2.8 m), southern Flores (maximum spring tide range 2.4 to 2.8 m), and southern Irian Jaya (maximum spring tide range 2.5 to 4.1 m, but locally 6 m). The remainder of the Indonesian coast is microtidal (Figure 3.32; Bird 1985b). The tides enter the Indonesian waters from the Pacific Ocean through the East and South China Sea and the Philippine waters, from the Indian Ocean through the Timor and Arafura Seas, and they have a complex mixed or diurnal character. Waves are generally low; tropical storms do not reach Indonesia, which is located around the Equator, while swell reaches Indonesia mainly from the south. The principal intertidal areas are along the mesotidal coasts with some exceptions that are located along microtidal coasts and are included here.

Mangrove and bare tidal flats form an often broad fringe with a maximum width of about 20 km along the southeast coast of Sumatra, in particular along the broad embayments where the large rivers discharge (Verstappen 1973). The rivers on Sumatra flow from the central mountains in a northeastern direction and form broad estuaries and delta plains where deposition is rapid. The port of Palembang, now 75 km inland, was close to the shore in the 15th century, which gives an average yearly accretion of 150 to 200 m (Figure 3.33). The Djambi delta progressed 7.5 km during the last century or 75 m per year. This rapid sedimentation is related to the relative nearness of the high mountains, the mostly volcanic and deeply weathered soils, the strong monsoon rains, and the shallow depth of the coastal sea. The freshly deposited outer part of the mud flats, below mean tide level and regularly reworked by waves, are bare but inhabited by numerous burrowing crabs and mudskippers.

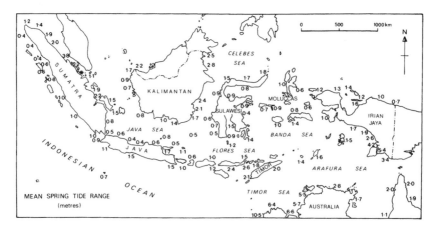

FIGURE 3.32   Indonesia; the numbers indicate spring tide range (in meters). (From Bird 1985b. With permission.)

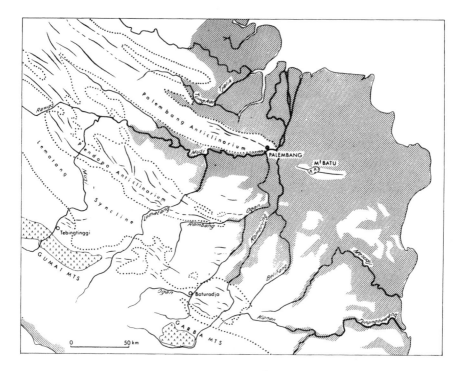

FIGURE 3.33 Alluvial plains, mostly covered with mangrove, near Palembang. (From Verstappen 1964. With permission.)

Extensive mangrove forests are also present along western, and in particularly southern, Kalimantan, as well as along the south coast of Irian Jaya, where broad low swamps are present with wide muddy river estuaries. The lobate Mahakam river delta on the east coast of Kalimantan (between 0° 20′ S and 0° 55′ S) is about 70 km long and 40 km wide. From the head of the delta a fluvial delta plain, cut by three river distributaries and mostly covered with *Nypa*, extends for 10 to 20 km (Figure 3.34; Magnier et al. 1975; Allen et al. 1979). The main delta plain is marine and dissected by sinuous or meandering tidal channels and creeks which are connected to about 15 tidal inlets, 10 of which are connected to the three distributary channels. The marine plain is largely covered with mangrove, and inward also with *Nypa*. The channel floors, banks (levees), and point bars consist of coarse-to-fine sand with mud flats between the channels that consist of silt, clay, and organic matter. On the seaward side of the mangrove are bare mud flats that may extend seaward up to 7 km.

The small river deltas in Indonesia often have an outer fringe of mangrove up to 2 km wide (sometimes even 10 to 12 km wide, as along the Manukwam delta in northern Irian Jaya) or had so before the mangrove was cut. This was done mostly along the island of Java, which is densely populated. Along the delta fronts are usually some beach ridges or cheniers, as well as mud flats, in particular where sedimentation and coastal accretion are rapid as along the Solo river mouth

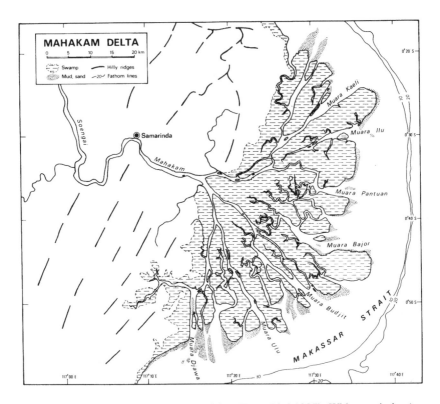

FIGURE 3.34   The Mahakam river delta. (From Bird 1985b. With permission.)

(Verstappen 1964; Hoekstra 1989). Along the south coast of Java near Cilacap an extensive mangrove swamp with tidal channels and creeks borders a shallow estuarine lagoon.

3.13. GULF OF PAPUA

Along the south coast of Papua, New Guinea, at the Gulf of Papua along the Torres Strait, rivers form a series of broad estuaries with inlets and tidal flats with mangrove, sandy shoals, and sandy beaches. Most of the intertidal areas are intersected by a network of sinuous and meandering channels that widen seaward, where the scour becomes stronger. The estuaries tend to be funnel-shaped. The tides are semidiurnal with a mean tidal range of 2.4 to 2.8 m, but with a spring tide range of 4 m at the Fly river mouth (neap tide range 1 m) and above 3 m at the Purari river mouth (Bird 1985; Wolanski et al. 1995). The Fly river estuary is wide and funnel-shaped, with three main channels and numerous intertidal shoals. The tidal currents are asymmetric with a strong flood dominance. A tidal bore develops in the inner estuary upstream of Lewada (Figure 3. 35). Under normal conditions most of the freshwater discharge goes through the south channel, but during strong trade winds, which

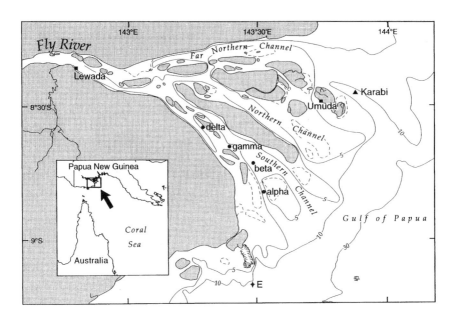

FIGURE 3.35   The Fly river mouth. (From Wolanski and Gibbs, 1995. With permission.)

come from the southeast, the discharge follows the north channel. A turbidity maximum is present during spring tides but not during neap tides.

The Purari river mouth is divided into three main channels and numerous sinuous secondary channels as well as sinuous and meandering creeks that are interconnected and partly flow into the Pie river estuary (Figure 3.36; Thom and Wright l983). Sandy beaches are present between the channel mouths; sediment transport along the shore is generally from east to west, except where interrupted by a channel. Large waves come from the southeast from May through September when average wave height is 1.3 m, but they reach only an average height of 30 cm during the rest of the year. In the channels and river mouths, vegetated intertidal shoals are common. The upper delta plain is a freshwater swamp covered with rain forest, where the main river channels dominate with levees up to 2 m high. Between the main channels, smaller, interconnected, sinuous channels are present. Between these channels, open inter-distributary swamps with traces of abandoned sinuous channels can be observed on aerial photographs. Here, presumably, peat is being formed. The lower plain is dominated by brackish water and mangrove; the mangrove zonation patterns (*Excoecaria* with *Sonneratia* and *Nypa* grading into *Rhizophora* and *Bruguiera*) are associated with salinity gradients. The active river mouths include those of the three distributary channels: the Purari to the east, the Ivo-Urika in the center, and the Varoi-Wame to the west. The central (Ivo-Urika) distributary has a sandy bed with downstream migrating megaripples of 0.5 to 1.0 m height; the water becomes brackish only near the mouth where the channel bifurcates. Wave effects are conspicuous here in the form of longshore bars backed by swales. The Purari distributary has two large

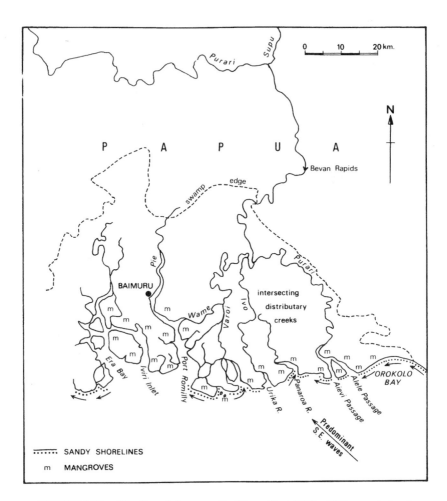

FIGURE 3.36   The Purari river mouth. (From Bird 1985c. With permission.)

outlets with broad arcuate bars that are partly intertidal and have a ridge- and runnel-topography. At the Wame distributary mouth, tidal flow is more important, as is indicated by the funnel shape where it discharges into the Pie river estuary. The distributary channels are the main source of sand along the coast. The suspended matter remains trapped in a broad near-shore brackish-water zone.

To the east of the Purari river delta up to the Fly river estuary is an extensive complex of tide-dominated, funnel-shaped estuaries with relatively small rivers extending inland, the Kikori river being the largest. Mangrove covers the tidal flats, where accretion is mainly vertical. Tidal currents dominate; in the main direction of the tides, linear subtidal ridges are formed that extend offshore. In the tide-dominated estuaries, the channel beds consist mostly of mud; flow velocities are weaker; and channel depths are deeper because of bidirectional scour. The intertidal deposits in the Purari delta (levees, backplains, channels) are largely muddy with an organic matter content above 10%.

3.14. GULF OF CARPENTARIA AND NORTHERN QUEENSLAND

In the Gulf of Carpentaria (northern Australia) between 3° N and 18° S, the tidal range is 2 to 3 m along the eastern shores, up to 4 m near Karumba in the south, and 1 to 3 m along the western shores. Along the Pacific coast of Queensland, between Torres Strait and Townsville, the tidal range is 3 m, but around Broad Sound it increases to more than 10 m (see Section 2.17). The tides, both in the Gulf of Carpentaria and along eastern Queensland, are mixed, but toward Brisbane they become semidiurnal (Galloway 1985; Hopley 1985).

The west side of the Gulf of Carpentaria (facing east) is a low plain with embayments (Figure 3.37). In the more sheltered parts intertidal and supratidal flats have been formed that have their largest extension around the Roper and McArthur rivers. The more exposed parts of the coast consist of beaches, spits, and sand ridges with low dunes. The intertidal flats are mostly covered with mangrove; the supratidal flats are bare or have some halophytes. The intertidal flats in the Roper river area are dissected by sinuous creeks, fringed by mangrove.

Along the south and east side of the Gulf, facing north and west, similar flats are present with more salt marsh and cheniers (forming chenier plains) and numerous meandering tidal creeks fringed with mangrove (Figure 3.38; Rhodes 1982). This coast is much influenced by storm surges from the north, which result in damming the rivers and in river floods, which is enhanced by the shallow depth of the coastal waters. The muds at low tide level are light to dark brown and gently undulating. They show an interlayering of sand and mud, with mud layers of 2 to 6 cm thickness and sand layers of 1 to 4 cm. The surface is covered with small ripples of 1 to 3 cm height. Accretion is vertical as well as horizontal until the flat reaches a level where either cheniers or high tide mud flats are formed. Between the low mud flats and the cheniers or high tide mud flats are intertidal muds, which are rarely more than 30 cm thick, 100 to 300 m wide, and covered with mangrove, which thus forms a narrow zone at the seaward edge of the chenier plain and along the channels that cut through it. This deposit is continuous over a distance of 20 km. The flats around high tide level are bare and deeply cracked with salt flats located between the cheniers. During the wet (monsoon) season they are covered with blue-green filamentous algae. During the dry winter the flats and algae dry out, salt effloresces in the cracks, and salt wedges are formed. This destroys the lamination that was formed of algal mats (up to 1 cm thick) and silt/clay layers, so that below a few centimeters' depth in the sediment no lamination is found. Deflation of the high mud flats during the winter results in dunes up to 1.5 m high that consist of clay held together by salt and organic matter. They can become overgrown (and fixed) by grasses. The high tide muds lie on top of intertidal or low tide muds, are approximately 2 m thick, and continue horizontally over distances up to 50 km. The chenier plain is cut by meandering tidal channels. The channel floors are sandy; point bars and overbank deposits are muddy. The channels are usually fringed by mangrove. The chenier plain was formed about 5800 BP, and the cheniers indicate periods with low fine sediment supply, which are probably related to periods of low rainfall.

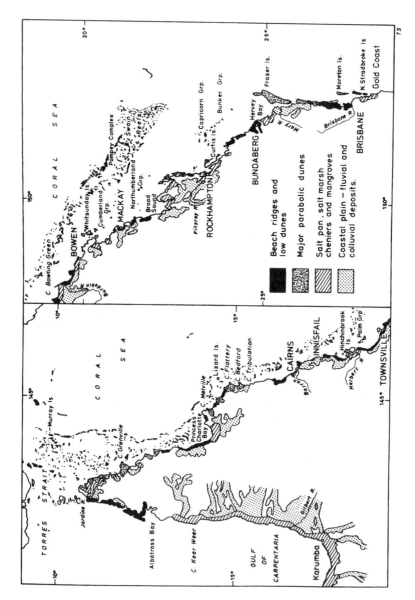

FIGURE 3.37   The eastern Gulf of Carpentaria and Queensland coast. (From Hopley 1985. With permission.)

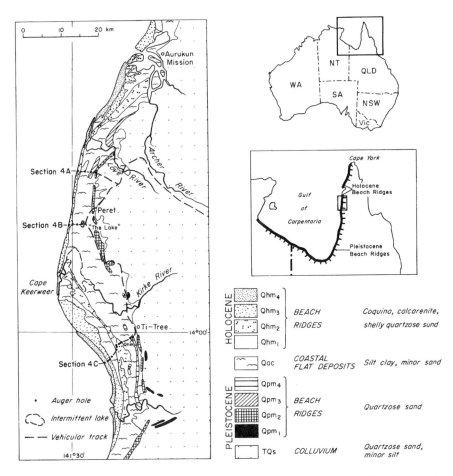

FIGURE 3.38 Beach ridges and coastal flats along the east coast of the Gulf of Carpentaria. (From Smart 1976. With permission.)

From Albatross Bay to the north, no intertidal deposits are present; the coast consists of beach ridges and low dunes. The east coast of Queensland is more irregular with smaller and larger bays and embayments and limited intertidal areas with salt flats, salt marsh, cheniers, and mangrove (Figure 3.37). The largest areas are located in Newcastle Bay, Princess Charlotte Bay, and in the Hinchinbrook area. In Princess Charlotte Bay, at about 14° 30′ S, a chenier plain 25 km wide has been formed with an outer fringe of mangrove and up to 5 km of bare saline flats with isolated cheniers (Figure 3.39). At the back of this are extensive clay plains (Chappel and Grindrod 1984). The lower intertidal mud flats near low tide level are dark grey with some muddy sand and sandy mud. They are often rippled at the surface with small-sized ripples. From below mean tide level to near high tide level a 300-m-wide mangrove zone is present with dark grey muds and sheets of fine-to-medium sized shell gravels of 5 to 40 cm thickness. Mangrove also fringes the tidal creeks.

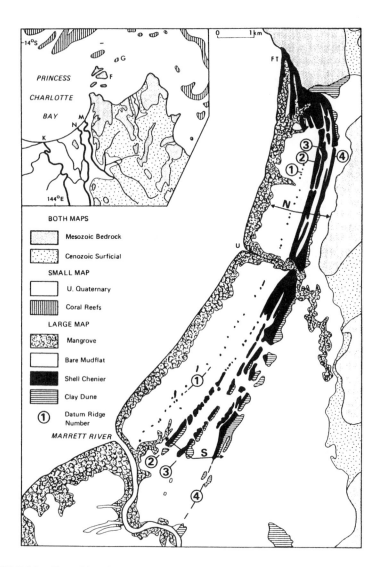

FIGURE 3.39   East side of Charlotte Bay. M = Marrett river; N = Normanby river; K = Kennedy river; F = Flinders Island; G = King Island. (From Chappell and Grindrod 1984. With permission.)

Behind the mangrove and between the cheniers are the upper intertidal muds that reach to high tide level. They consist of grey saline muds with algal mats that desiccate during the dry season when deep mud cracks are also formed. Above the high tide flats, a layer of up to 10 cm of supratidal flat clay lies at the surface between the cheniers. This layer is deflated by strong winds, resulting in clay dunes up to 4 m high that can become covered with grasses. The cheniers consist of shelly gravel. They carry some vegetation and reach 1 to 2 m above high spring tide level. The chenier plain started to form about 6000 years ago.

While the chenier plain at Princess Charlotte Bay is very similar to the chenier plain along the south coast of the Gulf of Carpentaria, the intertidal deposits in Missionary Bay (Hinchinbrook Island) and between Hinchinbrook Island and the mainland are very different. They lie at about 18° S and are formed in a wet tropical climate. They have a tidal range of about 3.2 to 3.5 m (Figure 3.40A and B). Between Hinchinbrook Island and the mainland the tidal flats are covered with mangrove with bare mud banks on the seaward side. Channels between the bare mud banks are shallow (1 to 2 m), whereas between the mangrove flats they are much deeper (4 to 6 m, with scour holes to 20 m). The mangrove channels have steep banks and

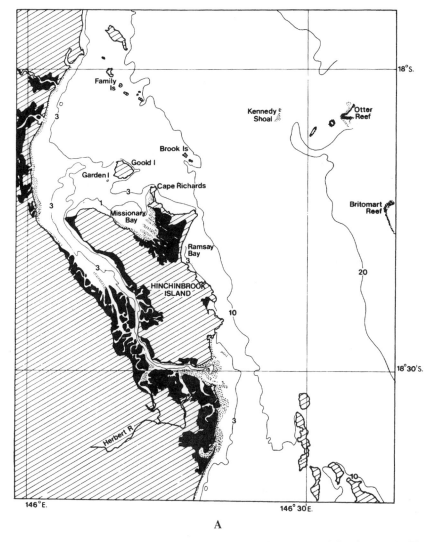

A

FIGURE 3.40   Hinchinbrook Bay (A) and Missionary Bay (B) (on following page). (From Wolanski et al. 1980. With permission; and Grindrod and Rhodes 1984. With permission.)

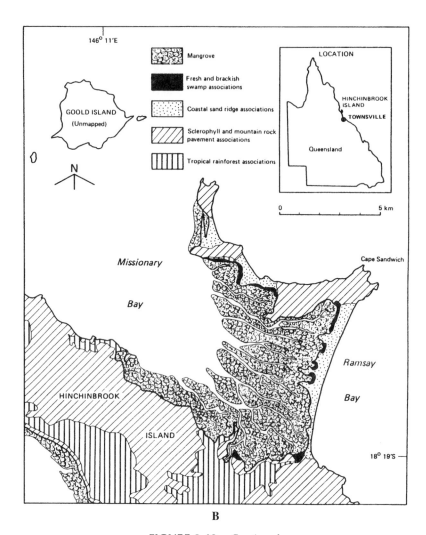

B

FIGURE 3.40   *Continued.*

a relatively flat bottom, with the mangrove growing on the banks at both sides of the creeks (Wolanski et al. 1980). In Missionary Bay, at a tidal range of about 4 m, 50 km² of mangrove is formed in a rectangular bay of 8 by 10 km. It is closed to the southeast and northeast by sand ridges connecting former islands (Figure 3.40B; Grindrod and Rhodes 1984). Muds at low tide level are brown to black with a homogeneous top of 1 to 1.5 m thickness, which is strongly bioturbated by crustaceae. On top of the low tide muds, which may be up to 2 m thick, are intertidal organic muds on which the mangrove has settled. They contain very little shell debris and up to 10% of organic carbon, and they have a thickness of about 2.5 m. The mangrove is drained by sinuous, sometimes meandering channels that generally go to the northwest. The channel floors have different types of sediment: 1) a layer of coarse angular to subangular sand 1 to 3 m thick, that is formed where the channel is an extension of a gully in the adjacent hills that supplies most of the sediment; ebb-oriented megaripples of 20 to 40 cm height occur in these sands; 2) an organic mud of 0.5 to 1.25 m thickness, dark brown to black, with an average 4.5% organic carbon and up to 10% of fine shell fragments; 3) a grey-olive mud with up to 60% calcium carbonate, mostly consisting of broken shell assemblages from estuarine mollusks; this deposit is mostly subtidal but becomes intertidal in some channels; 4) a homogeneous brown-to-black mud on intertidal point bars in meandering channels or creeks; they become more sandy down slope on the point bar and contain less than 10% calcium carbonate and more than 5% organic carbon.

Also around Cairns (about 16° 30' S) at a mean spring tide range of 2.1 m, intertidal sand- and mud flats with mangrove swamps are present, alternating with sandy areas that include sandy barriers and spits (Figure 3.41). They were formed at the same mean sea level that exists at present, which has experienced only small variations during the past 4000 to 6000 years. The sediment of the intertidal deposits has mainly been supplied by the Barron river.

Large flat deltas have been formed farther south at the mouths of the Burdekin river (19° 40' S) and the Fitzroy river (23° S). There are extensive saltpan areas and sinuous tidal creeks (Burdekin estuary) or strongly meandering creeks (Fitzroy river estuary), fringed with mangrove and up to 2 km wide (Figure 3.42; Jennings and Bird 1967).

3.15.   FIRTH OF THAMES, NEW ZEALAND

On the northeast coast of North Island, New Zealand, at about 37° S, the Firth of Thames is an elongated bay, open to the north with three small rivers flowing into it in the south (Woodroffe et al. 1983). The tides are semidiurnal with a spring tide range of 2.8 m and a neap tide range of 2 m. Waves are mostly low local waves, as the ocean swell hardly penetrates the Firth; the largest waves, which come from the northwest during storms, are approximately 1.5 m high. The residual flow along the east side of the Firth goes to the south. Along the west side, it goes to the north. The inner margin of the bay is fringed with tidal deposits and with mangrove up to 300 m wide along the south shore. The muds along the south shore are strongly bioturbated and contain skeletal carbonate, mostly from mollusks. Deposition mainly takes place during the flood, which deflects the outflow of the Waihou river toward the flats. On the southwest side between Miranda and Kaiaua is a chenier plain 8.5 km long and up to 2 km wide (Figure 3.43), with its ridge systems roughly parallel to the long axis of the Firth. The southern part of the ridges consists predominantly of bivalve shells, while the northern part is of well-rounded gravel. The landward limit of the plain is a Pleistocene marine cliff. Sediment transport in general is from north to south along the chenier plain. The cheniers are situated within interridge alluvium that consists of former intertidal deposits and salt marsh. At present most of the plain is drained and cultivated. It is partly dissected by several large channels that have straight and sinuous parts, while the largest one (Miranda stream) has a few meanders. The chenier plain was formed within the past 4000 years. The modern cheniers consist of bivalve shell ridges, the lower part of which lies below high spring tide level and consists of sand and fine mollusk shell fragments

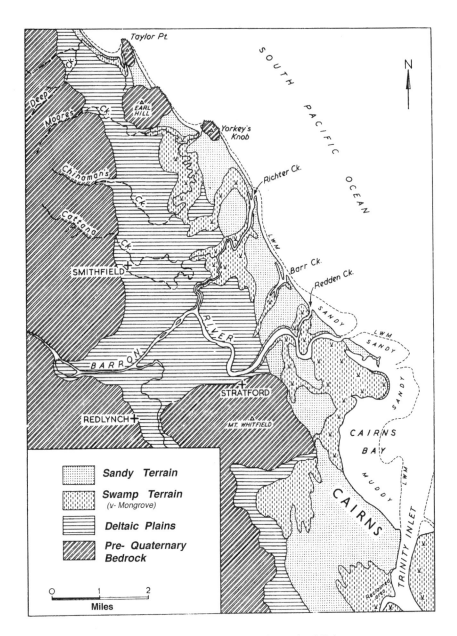

FIGURE 3.41   The coastal plain north of Cairns.

(hash). The upper part is a storm ridge consisting mainly of entire bivalve shells. Often there is a bench at high spring tide level on the seaward side and washover fans on the landward side. Cheniers move landward by as much as 9 m in seven months 10 m in 7 days during a storm). Tidal flat deposits and remnants (stumps)

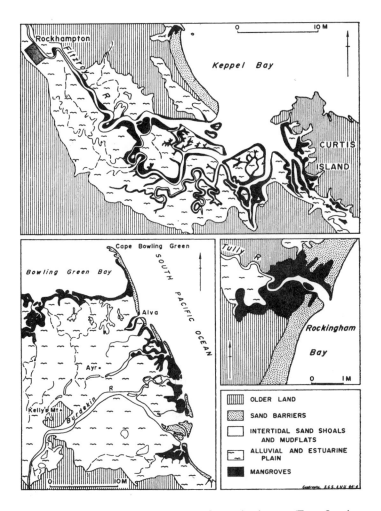

FIGURE 3.42   Principal river mouths along the Queensland coast. (From Jennings and Bird 1967. With permission.)

of mangrove appear on the seaward side of the moving chenier. The landward movement of a chenier stops when a new chenier is formed in front.

On the lee of the ridges are embayed tidal flats, which may be bare or covered with mangrove or salt marsh. The mud is black, anaerobic, without shell fragments, and with less than 10% sand and silt. These muds lie 0.57 to 1.66 m above mean sea level (measured at Auckland) and are inhabited by a burrowing bivalve and by a snail that lives on the sediment surface. The higher flats are mostly bare and show desiccation cracks. Mangrove up to 4 m tall lives on the flats up to about 1.4 m above mean sea level. Salt marsh (*Salicornia, Suaeda*) occurs on the regularly flooded parts, mostly between 1.6 and 1.8 m above mean sea level. The sediment below the mangrove and the salt marsh is 90% or more mud and generally less than

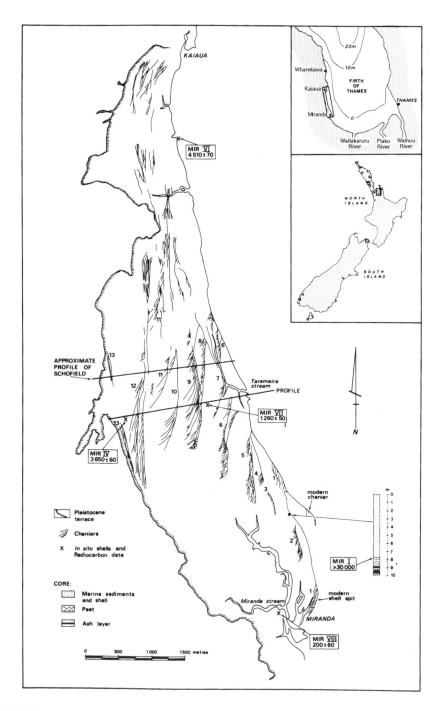

FIGURE 3.43   Chenier plain along the western side of the Firth of Thames. (From Woodroffe et al. 1983. With permission.)

5% carbonate. Mangrove starts to grow where the mud is at 1.20 m above mean sea level, and sedimentation continues below the mangrove. There is no evidence that salt marshes, which grow at higher elevations, replace mangrove. At the same elevations bare mud flats may also be present.

3.16. BRISTOL BAY, ALASKA

Bristol Bay at about 58° N is an elongated bay with a tidal range of 3.75 m and lined by mud flats and marshes, as well as by sandy beaches and numerous lagoons (Walker 1985).

3.17. FRASER RIVER DELTA

Along the delta front of the Fraser river delta at about 49° N at a mean tidal range of about 3 m, intertidal flats up to 6 km wide have been formed. On the landward side around high tide level these are bordered by up to 1 km of salt marsh, mostly in front of a dike (Figure 3.44; Milliman 1980). During the past century the delta

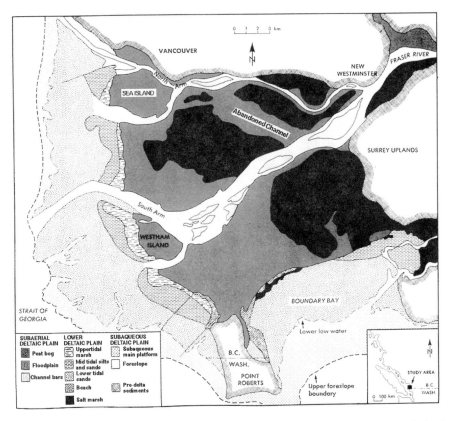

FIGURE 3.44   The Fraser river delta. (From Williams and Roberts 1989. With permission.)

has been growing seaward at a rate of about 9 m per year. The flats are dissected by several river distributaries and by tidal channels. About 80 to 85% of the river discharge goes through the main channel (Main Arm); the remainder through two other distributaries. Two channels, including Main Arm, are kept navigable for oceangoing ships by jetties (Mathews and Shepard 1962; Owens and Harper 1985; Williams and Hamilton 1995).

The tidal channel floors consist of fine sand with silt in the inner parts. Bars are sandy, sometimes with layers of sandy silt on top. The river channel floors consist of coarse sand with pebbles or gravel and are scoured down to 22 m below low tide level (or 24 m below the river water surface). The river channels are rather stable: only one abandoned distributary has been observed. The intertidal flats are sandy with current and wave ripples and at some places large networks of interconnecting gullies or small channels. Dendritic systems of sinuous or meandering channels start with incised beds on the high flats and continue down to low tide level. Vegetation is sparse on the flats (mainly *Ulva*, patches of eel grass, and benthic diatoms); it is mainly present on the muddiest patches. Mud patches and pools are formed on the upper part of the flats, in local depressions, and where waves and currents are weak. The dendritic drainage systems develop in these pools where they are not very permeable. Fauna is also rather sparse; only *Callianassa* (a shrimp) and *Macoma* (a bivalve) are common. During high river floods wide sand sheets are deposited on top of finer sand that was probably deposited during the waning stage of the previous flood. A seasonal layering is generally formed of annual layers containing vegetation fragments and silt/fine sand laminae, which can have a minimum thickness of 1.5 cm. Because of erosion, the layers are always incomplete. Low (50 cm high) sand swales of 50 to 100 m wave length have developed on several parts of the flats by waves and wave refraction.

In the central section of the flats off Lulu Island, the lower flats are sandy with virtually no silt or organic matter. Bedding is rare, and horizontal and graded, if any. Wood fragments are also rare. The silty top sediment is evenly and very thinly laminated with 15 to 20 laminae over 2.5 cm. The middle tidal flats form the seaward edge of the upper flats, which are covered with a continuous salt marsh vegetation up to supratidal level. The sediment on the middle flats is mainly organic-rich silt with a horizontal bedding. The seasonal layering occurs with an optimal development of mud pools and tidal channel systems. The maximum thickness of this deposit is at least 4 m, but on the average its thickness is 1 to 2 m. The salt marsh sediment on the upper flats consists of organic-rich clayey silt to silty fine sand which becomes generally finer and with more fine plant debris in a landward direction. This deposit has a thickness of 1 to 2 m with its basis approximately at mid-tide level and toward the basis a little more sand than at the top. The lower part lies between 280 and 365 cm above mean low low tide and is exposed without interruption for over 18 hours during low tide; the middle part lies between 365 and 415 cm above mean low low tide and has an uninterrupted exposure for at least 63 hours. The high part between 415 and 440 cm above mean low low tide has an uninterrupted exposure for over 700 hours (Hutchinson 1982). The deposition rate on the marshes in 1964 was largest on the middle part (8.5 mm · y^{-1}), lower on the highest part (6.1, 6.3 mm · y^{-1}), and lowest on the lower marshes (2.6, 3.7 mm · y^{-1}). The deposition rate based on ^{14}C data is lowest (2.2 mm · y^{-1}), which probably reflects the greater compaction of the older sediments, but also the rate between 1964 and 1991 (3 to 9 mm · y^{-1}) was lower than between 1954 and 1961, when the rate was between 6 and 20 mm · y^{-1}. The present marsh surface shows signs of erosion: 10- to 20-cm-high scarps are formed and erosion goes to below the plant roots so that islands or bastions of marsh deposits are left with vegetation on top. The marsh vegetation tends to trap sediment through attenuation of waves and currents, which is probably the reason the sedimentation rate is highest on the middle part of the marsh, where the vegetation is more dense than on the lower marsh and submergence is of longer duration than on the high part of the marsh. The erosion after 1964 is probably related to increased dredging of sand at the mouth of Main Arm, which results in a lower sediment supply to the flats and marshes. The marsh vegetation consists mainly of *Scirpus* (two species), *Carex*, and *Triglochin*. On the high parts above high tide level, the salt marshes, where not backed by a dike or disturbed by causeways or other artificial structures, grade into freshwater peat bogs that lie 3 to 5 m above sea level.

The Fraser river delta is less than 11,000 years old but more than 7300 years old. In spring and summer it receives large quantities of sediment which is mostly deposited on the delta front. The sediment on the silty flood plain is about 4 m thick and lies on top of sands deposited at or below low tide (Johnston 1921; Luternauer and Murray 1973; Luternauer 1980; Williams and Roberts 1988).

3.18. YAQUINA BAY, SAN FRANCISCO BAY

Along the Oregon coast at about 44° 35' N, the Yaquina river has formed a wide estuarine bay at the mouth (Figure 3.45). Intertidal deposits up to 3 km wide border this bay up to the Yaquina river and range from fine/medium sands to sandy silts and clays. The tides along the Oregon coast are mixed semidiurnal with maximum spring tide ranges of about 4 m and a neap tide range of about 1.8 m (Komar 1985). Waves are generally high along the coast: about 2 m in summer, 3 to 4 m during the winter, and up to 7 m or more during winter storms. The chief sediment source is the Yaquina river; locally reworked Pleistocene sands and coastal sands transported by littoral drift come in from the sea with the tides, while sand from the coastal dunes is blown inward by the wind. Sediment deposition on the flats is largely seasonal with maximum deposition in winter and early spring, when runoff is highest and strong southerly winds give a strong littoral drift along the coast and inward transport of dune sands into the estuary. In summer there is little deposition.

Marine sediments supplied from the coast, extend about 2.5 km inward from the estuary mouth; sediments of mixed marine/fluvial origin occur up to about 10 km inward. The tidal flats consist of fine sand along the channel banks; on the south side (on the Sallys Bank Flats) the tidal flat sediments grade laterally into fine sandy silts with a progressively poorer sorting. Based on sediment composition, the southern flats have more material of marine origin, the northern flats of more fluvial origin. The tidal flat sediments contain less than 1.2% of organic carbon and less than 5% calcium carbonate (Kulm and Byrne 1966, 1967).

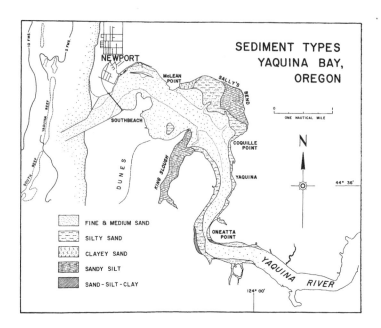

FIGURE 3.45   Yaquina Bay. (From Kulm and Byrne 1967. With permission.)

San Francisco Bay is an elongated, almost north-south directed bay connected to the Pacific Ocean by a narrow, more than 20-m-deep entrance (the Golden Gate; Figure 3.46). The bay consists of a southern part — South Bay — and a northern part — North Bay — which are connected by a narrower central part. In the northeast, the Sacramento and San Joaquin rivers enter the bay, and several small streams enter in the southeast. The bay is fringed by about 170 km of mud flats and salt marshes, which are concentrated along the north and south shores and have a maximum width of 2 to 3 km. The mean tidal range is about 1.80 m, increasing in the bay to about 2.70 m. Waves are only low local waves that can become more than 1 m high during storms (Pestrong 1972). Salt marshes in the bay, generally located around high tide level, are dominated by *Salicornia* and *Spartina*. *Zostera* is poorly developed or absent. *Distichlis*, also a salt grass, is distributed in a zone between *Salicornia* and *Spartina*. A mud-boring clam (*Geukensia*) occurs in large colonies and perforates channel banks. Shell debris in the marshes is mostly from shells of *Ostrea*, which is widespread in the bay and forms extensive beds. Saltpans on the marshes are roughly circular and bare of vegetation; they have been formed contemporaneously with the marsh, or after the damming of a creek. The creeks in the marshes are strongly meandering and grow both headward and in depth as the combined result of erosion and marsh accretion (Pestrong 1965). Ebb flow dominates in the creeks.

Most of the sedimentation takes place around *Spartina* plants, which effectively trap sediment because of flow reduction around them. The very sinuous or meandering marsh channels become less sinuous on the bare tidal flats where hydraulic flow controls the channel form and not the vegetation. Because the sediment is very

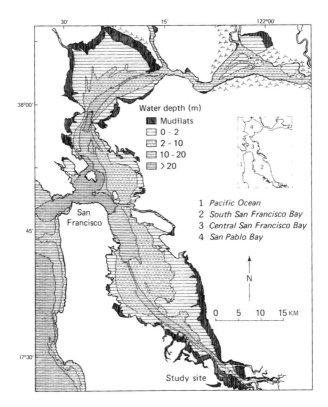

FIGURE 3.46   San Francisco Bay. (From Nichols and Thompson 1985. With permission.)

uniform, the channel floor, point bars, and levees do not differ much in sediment characteristics. The levees receive a little more silt than the flats farther away from the channels and are exposed somewhat longer during low tide and therefore become more consolidated and desiccated. The upper flats have abundant mud cracks, which develop especially when neap tides coincide with hot, dry weather. Cracks rarely extend more than 7 or 8 cm into the sediment. The channel beds contain much shell debris as well as small mud domes where gas (air) is being trapped below the sediment surface. Mud balls are formed at the edge of a receding marsh where mud is being eroded and reworked. The marsh sediment shows well-developed parallel lamination, where not disturbed by vegetation or burrowing organisms. The tidal flat sediment in the upper 5 to 10 cm is light grey-brown like the marsh sediment, but blue-black below that level because of anaerobic conditions; the boundary is sharp.

Sedimentation started in the bay during the Pleistocene with an intertidal mud deposit that was probably formed in shallow, relatively quiet water. After an intermediate period of sand deposition (upper Pleistocene/lower Holocene), the present bay mud was deposited, which reaches a thickness of about 40 m. Deposition of the bay mud started about 6000 to 10,000 years ago. The average sedimentation rate is in the order of 3 to 20 cm · 100 y^{-1}.

3.19. THE U.S. EAST COAST: GEORGIA, SOUTH CAROLINA, MAINE

Most of the US East Coast is microtidal except the coasts of Georgia and the adjacent part of South Carolina, and the coast of Maine.

The **Georgia coast** consists of a series of eight large (Holocene-Pleistocene) barrier islands separated by tidal inlets (Figure 3.47). At the back lies a very flat coastal plain with tidal bays and large tidal embayments flanked by up to 12-km-wide salt marshes and tidal flats that cover an irregular Pleistocene topography. The tides are semidiurnal with a spring tide range of 2 to 3 m, a mean range of more than 2 m, and a storm range up to 3.4 m (Edwards and Frey 1977). The extension of the marshes is related to the tidal range: where the spring tide range is less than

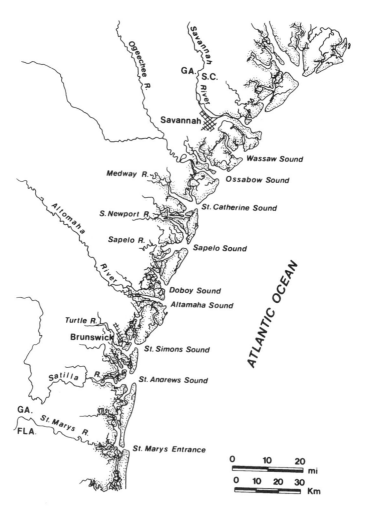

FIGURE 3.47   The Georgia coast. (From Hayes 1985. With permission.)

1 m (as in Florida), marshes are less well developed and fringe a shallow lagoon or form a narrow strip between a Holocene sandy barrier and Pleistocene mainland. Where the tidal range reaches about 2 m, as along the Georgia coast, marshes can extend inland for tens of kilometers, which is enhanced by the presence of the flat coastal plain. The marshes are almost completely tidal with each channel system geographically isolated, with very little freshwater influx and the flood discharge almost equal to the tidal ebb discharge. The tidal range also controls the barrier geometry as well as the size and the spacing of the inlets (Nummedal et al. 1977).

The inlets along the Georgia and southern South Carolina coast are tide-dominated (Hubbard et al. 1979) and grade into wave-dominated inlets in central and northern South Carolina. The almost entirely sandy intertidal deposits on the channel banks at the inlets are rippled with small current- and wave-ripples to megaripples and large sand waves. Channel-margin bars extend far into the coastal sea along the main ebb-channels and represent an equilibrium between the ebb-tidal currents and the coastal waves. When the banks emerge during falling tide, they are dominated by wave-generated horizontal lamination, small ripples and small megaripples. During the flood tide, when they are submerged, wave action can alter this completely by removing the larger ripples and filling in trenches generated by the swash during low tide. In the wave-dominated inlets, which occur in North Carolina (at a mean tidal range of 90 cm and wave heights between 0.9 and 1.1 m) as well as in Florida, ebb-tidal deltas are small, often with numerous small channels, while the flood-tidal deltas are generally larger. Wave-generated swash dominates, producing ripples and low-amplitude transverse bars on the banks, which are destroyed during the next flood. The flood-tidal delta can remain subaereal for some time and the (sandy) sediment becomes bioturbated so that well-preserved sediment structures are not common. In the transitional inlets in central and northern South Carolina, where the tidal range is 1.4 to 1.6 m, tidal point bars are formed covered by megaripples 30 to 40 cm high and 1 to 3 m long, with mud deposition in the scour troughs on the bars. A rainstorm can completely destroy the megaripples. The upper bar surfaces are dominated by landward flow and by strong interaction between tidal flow and waves, with megaripples as the dominant bed-form. The bars have a complicated internal sediment structure of megaripple deposits interbedded with large-scale horizontal or slightly dipping beds with planar lamination.

Waves during normal winter storms (that usually blow in an offshore direction) become up to 1 m high. Hurricanes occur only a few times in a decade (Hayes 1985). The Holocene component of the barrier islands diminishes toward the south: Sapelo Island (near the Florida border) consists largely of Pleistocene barrier deposits. The inlets are estuarine or purely tidal; the tidal estuaries are more numerous and dominated by sea water with some freshwater coming from rainfall on the marshes and from groundwater runoff from the islands (Imberger et al. 1988). The fluvial estuaries are fed by rivers coming from the coastal plain or the Piedmont mountains farther inland. The sediment in the tidal estuaries is more muddy than the river-estuary sediment and has more biogenic structures. The tidal estuary behind Sapelo Island (Duplin river) has a strong ebb-dominance with the flood being of longer duration (about 1 hour) than the ebb. Bed forms (sand dunes and megaripples)

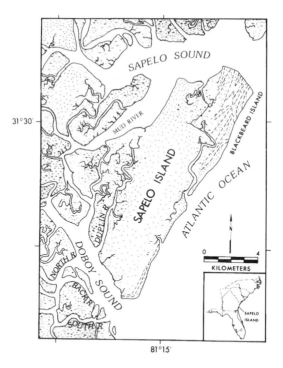

FIGURE 3.48   Sapelo Island and adjacent flats. (From Zarillo 1985. With permission.)

are ebb-oriented and in equilibrium with ebb-related bed shear values (Figures 3.48 and 3.49; Zarillo 1985).

The salt marsh around the island and behind the sandy coastal barrier can be divided into muddy low marshes and sandy higher marshes (Edwards and Frey 1977). The silt/clay ratio is approximately constant, but the sand content is variable. Mud and organic matter probably come from erosion and redistribution of local and regional sediments, whereas the sand comes from erosion of the local Pleistocene and Holocene barrier islands, including the Pleistocene sands that underlie the marshes. Carbonate comes from local mollusks (oysters) and organic matter from the marshes, but peat is rare. The creek banks and estuary margins have mixed mud and sandy sediments with wavy and lenticular bedding, occasionally flaser bedding, convoluted and load-casted laminae, slump and fault structures, and local discontinuities, together resulting in a chaotic bedding. Erosion by burrowing mud crabs (*Panopeus*) along the creek and channel banks is important. Mud cracks develop on drying out, and the irregular surface induces an increased erosion. The bank surface can then be smoothed and become stable. Slumping of large arcuate blocks of sediment can locally be common, where the banks have become over-steepened. Many structures are obliterated by bioturbation so that laminae are more common in the *Salicornia-Distichlis* (higher) marshes, where the rate of sediment deposition is probably higher. Along the terrestrial margin of the marshes sand sheets are deposited over the marsh, probably during storm floods, which results in a large-scale bedding that can subsequently be destroyed by bioturbation. Burrowing is

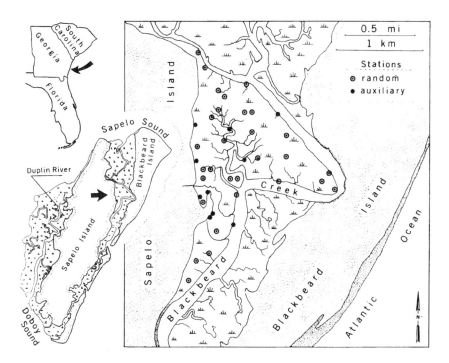

FIGURE 3.49   Creeks at Sapelo Island. (From Edwards and Frey 1977. With permission.)

predominantly done by a worm and a number of crab species, as well as by plant roots. Crabs produce burrows with openings, scrape the mud surface, and make pellets during both feeding and burrowing, which enhances erosion of the mud. An area of 0.5 m² can be completely reworked within one week by 10 crabs.

Along the adjacent coast of **South Carolina** the river estuaries become wider and the barrier islands smaller and more irregular; at Winyah Bay the large arcuate bays begin that continue northward up to Cape Hatteras with increasingly elongated narrow barrier islands. The southern part of the South Carolina coast is still mesotidal, but the spring tide range decreases to 2.4 m at Kiawah Island (Figure 3.50), and the coast becomes microtidal north of Winyah Bay. Along the South Carolina coast the ebb-tidal deltas become very large and the inlets migrate rapidly (60 to 70 m · y⁻¹ at Kiawah Island).

The salt marsh sediment is generally silt-clay except near beach ridges and barrier islands, where it becomes more sandy and has an organic content of less than 10 to 15%. The organic matter mainly consists of plant debris and roots. *Spartina* with *Salicornia* is common at higher elevations. Bare tidal flat sediment is usually fine-grained, cohesive, and with abundant oyster shells (*Crassostrea*). Point bar deposits are normally more sandy and channel or creek deposits even more so. Mollusk shells and other organogene carbonate is usually transported into marsh creeks from outside (remains of mollusks, echinoderms, and coral with bryozoa). Channels and creeks usually meander and can migrate laterally rather fast, up to 75 m in 24 years (Ward 1981). Meandering creeks or channels, where they cut

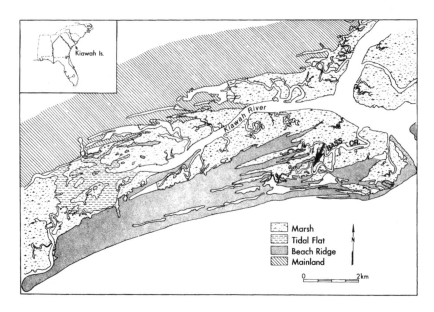

FIGURE 3.50   Kiawah Island. (From Ward 1981. With permission.)

through older, often more sandy deposits, can form steep-cut banks that may become up to 8 m high. As there is usually no, or very little, net water flux, deposits and bed forms tend to be related to the origin of the sediment supply, to the tidal asymmetry, and to waves. Waves in tidal creeks and channels are mostly locally developed and low, or are in the form of swell coming in (Barwis 1978). They affect the sediment during emergence. Oscillation ripples, usually interfering with current ripples, occur both on sandy and muddy deposits. Bank slumping occurs mostly below the plant roots (*Spartina* reaches 30 to 80 cm into the sediment), where the sediment slope is 15° to 45°. It is enhanced where the sediment structure has been weakened by burrowing organisms (in particular by crabs, *Panopeus sp.*). Slumped sediment, where incorporated in a channel bed, gives a chaotic bedding. Biotic sediment structures include the feeding tracks of birds, feeding pits of rays, alligator wallows, the surface tracks of snails (*Nassarius, Littorina*), and the burrows of crabs (up to 150 per m²), mollusks, and polychaetes. The latter are also indicators of current direction: they can be aligned parallel to the flow or oriented into the flow, or can produce aligned scour pits or current lineations (Frey and Howard 1969; Mangum et al. 1968; Meyers 1972).

The coast of **Maine**, extending northeast to the Bay of Fundy in Canada, is characterized by a large number of bays, coves, and several larger embayments. It consists mostly of cliffs with sandy pocket beaches and about 3000 salt marshes and tidal flats. Most of these cover only a small area, and fringe the more exposed parts of the coast. Only a few percent of the intertidal deposits have been formed behind barrier beaches; these deposits are usually larger (Fisher 1985; Jacobson et al. 1987).

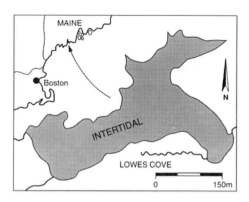

FIGURE 3.51   Lowes Cove. (From Anderson et al. 1981. With permission.)

Lowes Cove at about 4° 40′ N covers only 600 by 150 m (0.09 km²) and has a
mean tidal range of about 3 m (Figure 3.51). It is protected so that even during
storms waves do not reach more than 10 cm. The flats are muddy, bioturbated by
benthic mollusks (*Mytilus, Mya, Macoma*) and two species of crabs, and are fringed
by a 15-m-wide *Spartina* marsh. Repeated leveling has shown appreciable changes
in sediment volume, both in space and in time, during a period of one year (Anderson
et al. 1981), which indicates that erosion as well as deposition take place. Erosion
is caused by waves (resuspension) and by ice during the winter when biologically
induced sediment stability is at a minimum. Biodeposition is a major factor in
sediment accumulation and is much influenced by the population density of the
organisms and their filtration rate. Erosion and deposition may occur simultaneously
in different parts of the flats, and cyclic patterns, both spatial and temporal, may
occur. Also there is exchange with adjacent estuaries: eroded material is deposited
elsewhere, and material is deposited that comes from a neighboring intertidal area.

Wells marsh, at about 43° N, is quite different: it is relatively large (8 km²) and
situated directly at the back of a barrier beach (Figure 3.52). It is drained by a tidal
stream (the Webhannet), has a mean tide range of 2.6 m, and spring tide range of
3.0 m. The marsh peat is fibrous and sandy with a very small amount of fine material
(Jacobson 1988). The Maine coast has been (and is) much influenced by variations
in sea level rise after the postglacial sea level rise ended at about 6000 BP. After
5000 BP, the rate of sea level rise increased from 1.2 mm · y⁻¹ to 1.87 mm · y⁻¹ and
then decreased to 0.6 mm · y⁻¹ at the beginning of the 20th century. During the past
40 years the rise has been 1.1 mm · y⁻¹ at Portsmouth, but 3.2 mm · y⁻¹ in eastern
Maine, which is probably caused by differential crustal warping. Reliable maps for
the Wells area are available from 1794 onward, which allows a study of the changes
in the marsh over about 200 years. Between 1794 and 1872 the estuary and the
stream channel widened by drowning or by erosion of the salt marsh, perhaps in
response to an increased sea level rise. A similar change occurred during that period
in the smaller streams that enter the river and estuary from the north. This was
probably enhanced by the relative ease of lateral shifting of the channels and creeks

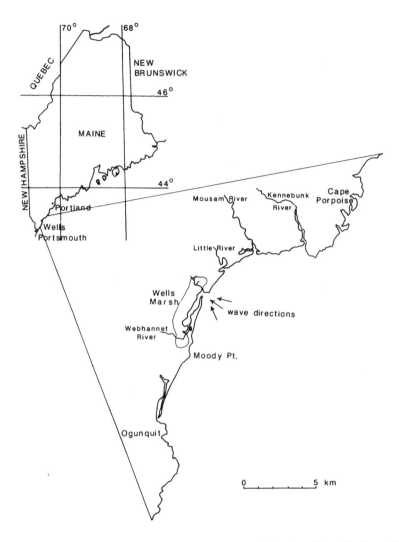

FIGURE 3.52   Wells Marsh (in 1956 with 1872 margin) and location. (From Jacobson 1988. With permission.)

because the sediment does not consist of consolidated clays and silts, but of (cohesive) peat: undercutting and rotational slumping are common features. Between 1872 and 1956 stream bank erosion was balanced by deposition on the northwest bank so that the Webhannet was displaced toward the southeast. After 1956 the estuary was influenced by the dumping of dredge spoils and by stabilization of the sandy barrier, preventing erosion and overwash. In this way the barrier has not retreated during the past 60 years. (Between New Jersey and North Carolina a retreat of 1.5 m · y^{-1} on the average is common and is compensated by overwash deposition on the landward side.)

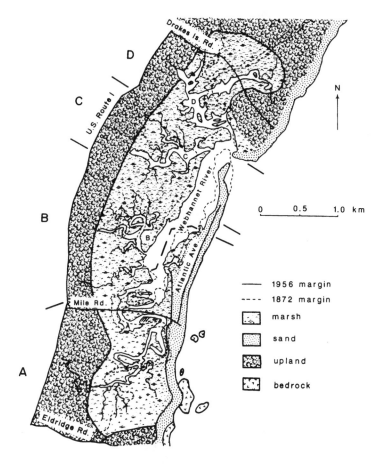

FIGURE 3.52 *Continued.*

On 26 marsh sites along the Maine coast deposition rates were estimated with brick-dust layers as a marker (Wood et al. 1989). The marshes belonged to four types: back barrier marshes (usually large with a high marsh vegetation), fluvial marshes (at upper estuary margins and variable in area), bluff-toe marshes (along the base of coastal bluffs, usually small with often a low scarp along the seaward edge), and a transitional marsh of freshwater peat intersected by a salt marsh. The highest deposition rates were found on the back barrier marshes (1 to 13 mm · y^{-1}; average 7.5 mm), lower values on the fluvial marshes (1.5 to 6 mm · y^{-1}; average 2.5 mm), and the lowest on the bluff-toe marshes (0.5 to 4 mm · y^{-1}; average 1.9 mm). On the transitional marsh a rate of 5.3 mm · y^{-1} was found. There was no relation to local sea level rise over the period 1940 to 1982 or to the tidal range. On some marshes more than 20% of the sediment was supplied by ice rafting. There is a clear relation between the position of the marsh along the coast (back barrier, inner estuary margin, bluff-toe), which suggests that the net supply of sediment to the back barrier marshes is highest where the vegetation is also highest because it probably has a higher sediment trapping efficiency than the lower vegetations on the other marsh types.

3.20. SOUTHERN HUDSON BAY, LABRADOR

The coasts along the southern Hudson Bay are mostly low energy, emerging coasts, often fjord-like, with a low tidal range, and with a maximum tide range near 3 m. The exposed open Labrador coast is a high energy coast with large waves, but within the bays and fjords the coast is sheltered. The area became deglaciated around 8200 BP with a strong isostatic rebound during the following 2000 years. In the southern Hudson Bay area (James Bay), the coast is still rising at about 1 m/100 years (70 to 120 cm/100 years). Along the western and southern James Bay the coast is very flat and poorly drained with beach ridge complexes, wide salt-brackish and fresh-water marshes, and river estuaries (Figures 3.53 and 3.54). Along eastern James Bay and Labrador the coast is more fjordlike and much indented with an irregular shoreline, and with tidal flats and salt marshes in the embayments (Figure 3.55). After the break up of the Wisconsin ice sheet, two separate ice covers remained for some time: the Hudson Bay ice sheet in the west and the Quebec ice sheet in the

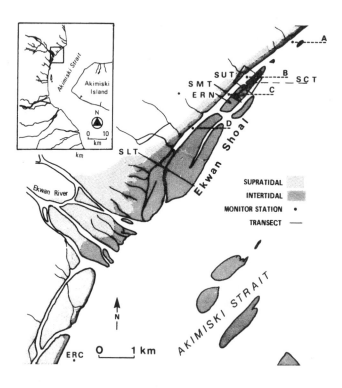

FIGURE 3.53 Ekwan Shoal in James Bay. (From Grinham and Martini 1983/1984. With permission.)

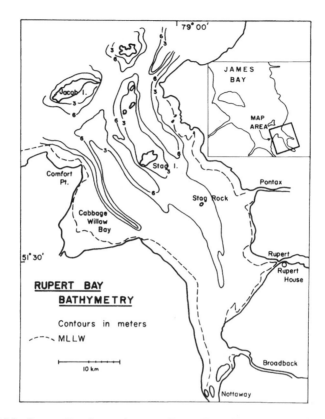

FIGURE 3.54   Rupert Bay in southeastern James Bay. (From d'Anglejean 1980. With permission.)

east. Around 7900 BP the sea entered through the Hudson Bay and Hudson Strait, and the Tyrrell Sea was formed with a maximum elevation of 200 to 300 m.

Because of continuing isostatic recovery with a residual of 120 to 150 m, this sea gradually withdrew, leaving a sediment blanket of stratified grey silts and clays around James Bay (Martini et al. 1980; Dionne 1980). In the intertidal areas the Tyrrell Sea sediment became covered with up to 40 cm of Recent clay and sand, as well as thousands of boulders that are randomly distributed among all tidal levels. Only where the flats are very wide, as along southern James Bay, do boulders occur only occasionally, which is probably related to the large distance from a source (moraines). The boulders on the flats accumulate through the action of alternating freezing and thawing in combination with gravity (Lauriol and Gray 1978) or by drift ice (Dionne 1980). The only other marked intertidal features are depressions made by the ice that covers the intertidal flats for up to 6 months with a layer 1.0 to 1.5 m thick. The ice, when it starts to break up and is moving over the flats, contains frozen sediment and/or vegetation, which is ripped out when the ice moves and is transported over some distance, leaving a depression behind. In some areas parallel boulder ridges extend roughly perpendicular to the reworked moraines. Boulder barricades have been observed only along macrotidal coasts or where waves

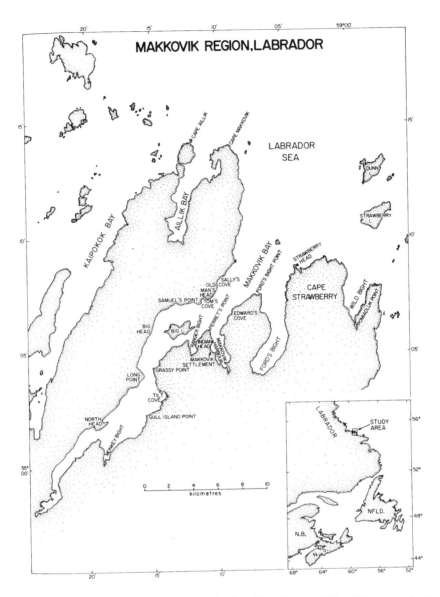

FIGURE 3.55   The Makkovik region, Labrador. (From Rosen 1979. With permission.)

are very large (see Section 2.24). Salt marshes are located around high tide level or are supratidal, and landward they may grade into freshwater marsh. They are flooded only during high water levels and consist mainly of mud with boulders. They have a dendritic drainage pattern. Where waves are large, as off Labrador, salt marshes are only found in protected areas at the head of embayments. This also occurs in fjord-like surroundings where intertidal deposits only occur at the head of the fjords.

3.21. THE SOUTH AMERICA NORTH COAST: AMAZON RIVER MOUTH–GUYANAS–ORINOCO DELTA

The coastal area between the Amazon river mouth and the Orinoco delta, a distance of over 1400 km, is directly influenced by sediment supply from the Amazon. Mud is transported westward over the inner shelf in suspension and along the coast in the form of a series of large mud banks. The mud banks end up at the Orinoco delta and only a relatively small amount of mud passes through the Gulf of Paria to the north coast of Venezuela (Eisma and Van der Marel 1971; Eisma et al. 1978; Milliman et al. 1982). At the Amazon mouth, at about 1° N, the spring tide range is up to 4 m with an average range of 2.4 to 2.8 m. Directly to the northwest of the river on the Amapà coast at 2° N, the tidal range increases to 5 to 11 m and then decreases again to less than 4 m at the Rio Cassiporé at 3° 40′ N and to 1 to 2 m (mean range) along the Guyana coast north of 4° N (Allison et al. 1995a). Along the Suriname coast the maximum spring tidal range remains around 3 m (with a mean tidal range of 1.80 m), but along the Guyana coast and the Orinoco delta it increases again to a maximum spring tide range of 4 m. Most of the north coast of South America is therefore mesotidal or almost microtidal with a macrotidal area near the Amazon river mouth. It is discussed here under one section because the South American north coast forms a (major) mud transport and deposition system.

In the Amazon estuary the main discharge of the river follows the channels north of Marajó Island, with only a small part flowing south of the island to the Para river and the Tocantins (Figure 3.56). The intertidal area includes a broad zone along the river and estuary with mangrove on the seaward side and more inland a higher

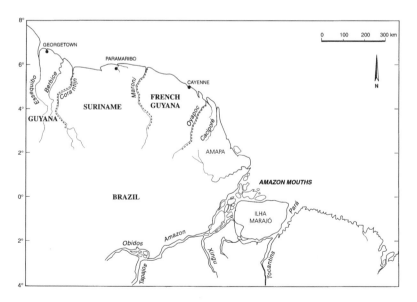

FIGURE 3.56   The Amazon river mouth.

intertidal area covered with grassland and (freshwater) flooded forest. There are no bare intertidal areas. In the intertidal estuarine zone irregular depressions occur of a few hundred square meters that are filled in with tidal marsh. South of Marajó island the Amazon river is connected with the Para river by a large number of tidal channels called "furos" with currents coming from opposite directions. There are tidal watersheds where the opposite tidal waves meet and diverge again. Marajó Island is dissected by winding channels, 6 to 40 m deep and 50 to 450 m wide. They have probably been formed during subsidence of the river mouth area with simultaneous deposition of mud between the channels. The intertidal area includes almost the entire western half of Marajó. The eastern part is flooded by the river during the rainy season only (i.e., from February to July; Sioli 1965; Murça Pires 1965).

About 10 to 15% of the suspended matter supply of the Amazon (or about 150 × 106 tons y⁻¹) goes along the Guyana coast toward the northwest in the form of large mud banks (Figure 3.57). Directly adjacent to the river mouth, the Amapà coast is eroding, and older, consolidated muds are exposed that date from more than 1000 years BP (Allison et al. 1996). Here about 1 to 40 × 10⁶ tons are added yearly to the sediment that goes westward (or only a few percent). Amazon mud is deposited here on the inner shelf, separated from the shore by relict muds as well as from the modern muds deposited on the inner shelf farther north. The northern Amapà coast near to Cayenne forms a transition zone toward the coastal mud belt that extends as far as the Orinoco mouth. Intertidal muds are deposited here at a rate of 0.5 to 4 cm · y⁻¹. On the landward side they are fringed by mangrove with seasonally flooded grassland farther inland (as in the Amazon mouth area). Observations along

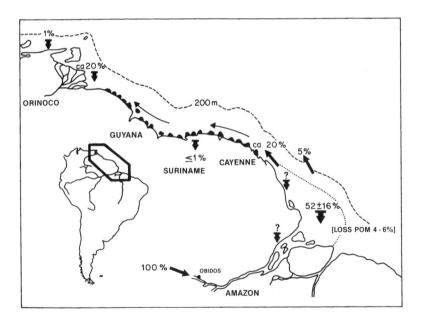

FIGURE 3.57 Dispersal of Amazon-supplied suspended matter along the Amazon–Orinoco coast and shelf. (From Eisma et al. 1991. With permission.)

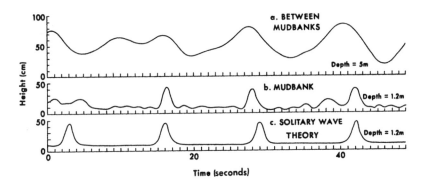

FIGURE 3.58 Waves between and on mud banks and solitary waves (From Wells and Coleman 1978. With permission.)

a small part of the coast — the results of which can probably be extrapolated along the entire northern Amapà coast where it is accreting — indicate that rapid sediment deposition (in the order of 1 cm per day) takes place from January to June when the trade winds are stronger and the suspended matter concentrations on the adjacent shelf are highest. In this way an intertidal mud deposit of about 1.5 m thickness is formed seasonally. These muds are reworked by the tidal currents and by surface gravity waves that are transformed into solitary wave-like waves because of the high suspended matter concentrations (Figure 3.58). These are up to more than 10 g · l⁻¹ and give the muds the character of fluid muds (Wells et al. 1979; Wells and Coleman 1981a; Wells and Kemp 1986). Deposition of the reworked mud results in a lens-shaped mud deposit above mean high tide level at the limit of wave influence. The deposits formed during the first half of the year are almost completely remobilized during the second half and probably partly transported along the coast to the north-west. Because of the rapid sediment deposition during the first half year, the shoreline along the northern Amapà coast yearly progrades by several meters through colonization of the fresh mud by mangrove (*Avicennia*). The lower part of the mud remains bare, probably because of the intensive reworking of the sediment by waves and tides and through remobilization later in the year.

Longshore transport from the northern Amapà coast to the northwest along the Guyana coast goes in the form of large mud banks that are 20 to 30 km wide and 50 to 60 km long (Figure 3.59). They are asymmetric, generally oriented obliquely N 24° to the shoreline, with a steeper northwest (leeward) side and a flatter southeast side which is directed toward the dominating currents. They have their largest thickness (~15 m) around the 10 to 15 m depth lines and landward as well as seaward become thinner (NEDECO 1968; Allersma 1971). They move at a rate of about 1 to 5 km per year. This is related to the angle of the coastline with the direction of the trade winds: along Suriname, where this angle is larger, migration is faster than along Cayenne and Guyana (Augustinus 1987). About 1 × 10⁸ tons moves along the Guyana coast in this way yearly (together with 1.5 × 10⁸ tons in suspension). The mud transport is enhanced by its mobility: the mud is easily resuspended by small waves and net transport by relatively weak currents can be very large (Wells 1983).

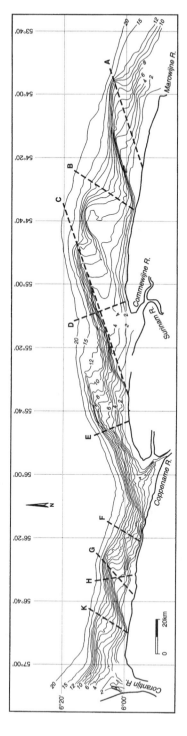

FIGURE 3.59 Depth contours of mud banks along the Suriname coast. (From Augustinus 1978. With permission.)

A roughly 30-year period of mud bank accretion alternates with a period of approximately equal duration with interbank erosion (often with the formation of a sandy beach and a chenier), which reflects the westward migration of the banks. What causes the fluctuations in sediment transport is not known, but the fluctuations as well as the migration of the mud in the form of large banks, may be related to fluctuations in trade wind direction and strength lasting several decades (Eisma et al. 1994). A relatively small amount, in the order of 1% or less of the Amazon mud supply (or about 0.1×10^8 tons $\cdot$ y^{-1}), is deposited along the Guyana coast together with a series of cheniers. The cheniers end at a river mouth where the sand, if it reaches that far, remains behind while the mud banks continue westward. Most of the cheniers begin at a river mouth where sand is available but do not continue all the way to the next river mouth when the sand supply is not sufficient (Augustinus 1980,1987; Augustinus et al. 1989; Prost 1989, Daniel 1989).

The mud banks are partly subtidal and have, at least along the Suriname coast, their lower limit at the 20 m isobath. The intertidal part is bare often with mangrove on the inner (landward) side. In the direction of the longshore transport, mud capes are formed (Allison et al. 1995a) in analogy to sand spits, which have a similar morphology. A mud deposit several meters high extends above low tide level for tens of kilometers westward from the cape or the left bank of a local river mouth and deflects the river to the west. The central part of the mud cape is higher and can be capped with mangrove or with supratidal grassland or rainforest. In Cayenne it is covered with salt marsh. At the Cassiporé river mouth (Amapà coast) a mud cape of 70 km length has been formed, which, at the present accretion rate, must have started to form 600 to 1200 years ago (Figure 3.60). Sixteen mud capes can be distinguished between Cabo Cassiporé and the Orinoco delta; they are identified by the deflection of the rivers and their spit-like shape. They are not present at the mouths of the larger rivers (the Marowijne, Corantijn, and Essequibo) probably because the discharge of these rivers is large and the tidal currents in the river mouths are too strong. It is not clear whether all present mud capes are growing: some appear to have halted at a river mouth; others seem to have been rendered inactive by the development of a new mud cape farther seaward. When a new mud cape is formed, this results in a seaward growth of the coastal plain by 10 to 40 km.

Along the Suriname coast the intertidal mud bank surface consists on the west side (where the deposition takes place) mainly of fluid mud ("slingmud"). On the eastern side, where mud is resuspended (eroded), the mud is more consolidated. The western side therefore has a flat surface of wet mud covered with expelled water, surface flow structures, and a very indistinct drainage system. The main channels are parallel to the slope and are relatively straight. Tributary channels are generally more sinuous, or may be meandering where they follow directions more perpendicular to the main channels (Augustinus 1980; Rine and Ginsburg 1985). Toward the eastern side of the bank, where the mud is increasingly consolidated, the surface becomes pitted and the channels deeper. Local erosion results in irregular shallow depressions with steep walls, that become deeper and wider through continuing erosion. An irregular pattern of low depressions is formed with flat-topped bastions. The depressions tend to merge, and straight channels develop up to 1 m deep with

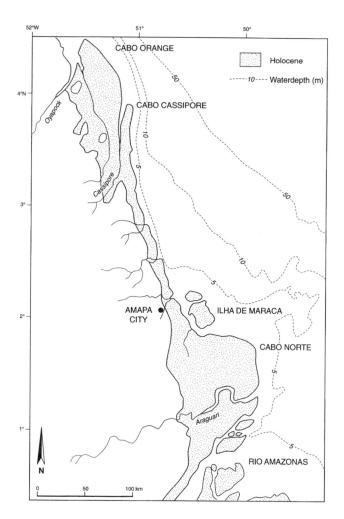

FIGURE 3.60   Coast between the Amazon river mouth and French Guyana, with mud capes at Cabo Cassiporé and Cabo Orange. (Adapted from Allison et al. 1996.)

a flat and steep bank. Abrasion (by waves) continues and around mean high tide level a low scarp, up to 1 m high, indicates the landward limit of erosion.

Along the Cayenne coast the channels on the west side of the intertidal mud bank flats are small, numerous, and interconnected (anastomosing), as could be ascertained from aerial photographs taken during low tide (Figure 3.61; Froidefond et al. 1987, 1988). On the east side the channels are less numerous, larger, and better developed, and may persist farther inland where mangrove has colonized the mud flats and the channels may run parallel to the shoreline for several kilometers before they turn seaward. Three of the six mud banks have two intertidal mud flats (which differ considerably in size and shape), one on each side of a local river mouth. Here

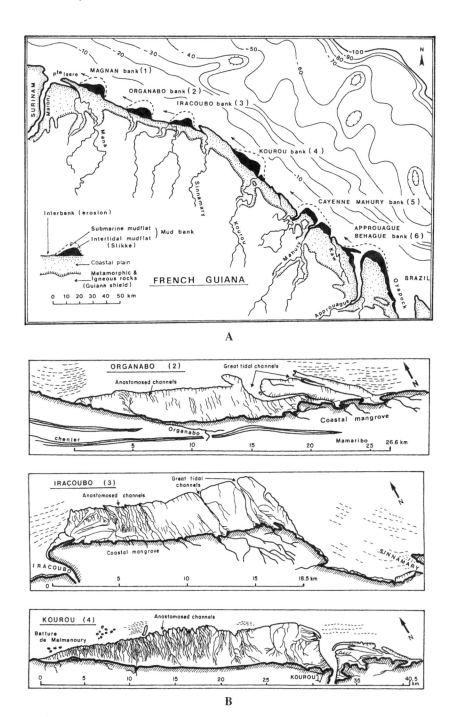

FIGURE 3.61 Mud banks along the Cayenne coast. A = distribution; B = at low tide (aerial photographs); C = erosional coastline (on following page). (From Froidefond et al. 1988. With permission.)

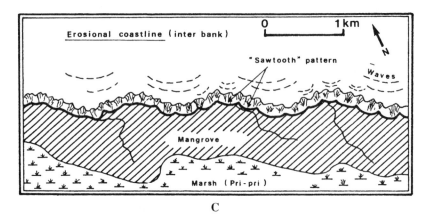

C

FIGURE 3.61 *Continued.*

(as at the Approuague and Kourou rivers) mud spits have been formed toward the river mouth which look like small versions of the mud capes described above.

Mud flat surface sediments in Suriname are laminated with thin laminae of fine sand and silt ranging in thickness from a few tens of microns to some hundreds of microns (Augustinus 1980; Rine and Ginsburg 1985). Mud "laminae," however, are massive, structureless layers, which rarely contain burrows and form approximately 80% of the deposits. They may be up to 2.2 m thick. The parallel laminations may be graded mixtures of clay, silt, and mud with sharp basal contacts. Sometimes laminations consist of slightly higher concentrations of silt and sand than the surrounding matrix. Subparallel lamination, lenticular and micro-crosslamination indicate lateral movement of the mud and are probably related to small-scale wave transport. Organic remains such as plant fragments and sponge needles may occur in the laminae and generally lie parallel to the bedding planes. Clay laminae range from about 500-micron thickness to a thickness of several centimeters. Scour-and-fill structures occur in some cores, but completely bioturbated sediment has hardly been observed, and most sediments are only little bioturbated. Small fish make holes in the mud in which they hide, and worm tubes may occur locally in large numbers. Around high tide level, many burrowing crustaceae (crabs) are present. Some laminae contain a high proportion of fecal pellets.

On the Amapà coast a larger variation of sediment structures has been observed than along the Suriname coast (Allison et al. 1995a), which is related to the formation of the seasonal surface layer formed January through June, and to the presence of more sand on the sediment. The seasonal surface layer is stratified with parallel to wavy (occasionally lenticular) silt laminae separated by clay-rich layers up to 25 cm thick, that are faintly laminated. The silt laminae have a thickness up to more than 1 cm and are laterally continuous over a distance of about 10 cm. Sometimes there is interlamination with plant fragments (wood), platy minerals, and fecal pellets. Sand laminae (<1 cm thick) and thin sand beds of 1 to 5 cm thickness, are present in particular near the Cassiporé estuary. One type of sand laminae is 1 to 3 cm thick, often with two to five individual layers, subparallel to the bedding, and with sharp basal and surface contacts. They are laterally continuous over 100 to 200 μm or less. Some are cross-laminated with alternating layers of sand, plant material, platy minerals, and mud. The other type of sand laminae is 0.5 to 5 cm thick, with a gradational upper boundary and a gradational or sharp basal contact. Clean sand with less than 10% mud is sometimes present at the basis. Hiatal surfaces (Allison et al. 1995a) are formed on truncated laminations or burrowed surfaces. The hiatal surface at the basis of the seasonal surface layer usually contains lag-deposits and burrow infill of lateritic sand grains, fecal pellets, iron-cemented burrow casts, and wood fragments. In the mud deposits a layered microfabric of platy minerals (micas and clays) is present parallel to the bedding. Solitary layers in an unoriented matrix occur, as well as groups of interlaminated layers that often are regular and are laterally continuous over more than 3 cm. Cyclic interlamination and very closely packed individual layers <0.01 mm thick occur, consisting of very small individual platy minerals. Bioturbation is rare, except above mean high tide level. In May there is a slight concentration of polychaete burrows in the upper 10 cm of the seasonal surface layer.

Based on the velocity of the mud banks and their volume in relation to the present supply from the Amazon, the mud bank system between the Amazon mouth and the Orinoco delta has existed for at least 1000 years, with the oldest mud banks near the Orinoco. Sedimentation of clay deposits with cheniers started when sea level had approximately reached its present level around 6000 years BP (Figure 3.62), and since then it has proceeded in three depositional periods separated by periods of erosion lasting a few hundred to approximately 1000 years (Brinkman and Pons 1968; Augustinus et al. 1989; Eisma et al. 1991). The last depositional period — the Comowine phase — started about 1000 years ago, is still continuing, and is the period covered by the present mud banks. There are indications that shorter fluctuations in the formation of the more permanent coastal mud deposits (as distinct from the temporary deposition in mud banks) exist and may be related to the fluctuations in the direction and force of the trade winds. Also there is probably a relationship between periods of more or less rainfall in the Amazon and Orinoco river basins and the implication that more mud was being transported and supplied to the sea during the rainier than during the drier periods.

3.22. THE SOUTH AMERICA EAST COAST

Along the South American east coast north of Bahia Blanca the tidal range is less than 4 m and only becomes larger between Fortaleza and the Amazon river mouth (Figure 3.63). The tides are mixed between Buenos Aires and Rio de Janeiro and elsewhere are semidiurnal. North of the salt marshes at Mar Chiquita, which are microtidal (see Section 4.21), extensive mud flats and salt marshes have been formed along Samborombón Bay at the south side of the Rio de la Plata. Along the coast of Uruguay, where the tidal range is less than 1 m but storm surges from the southeast can reach 2.7 m at Montevideo and 3.7 m at Colonia, small estuarine lagoons occur at the river mouths. Salt marshes are present between Montevideo and Azarati in sheltered areas but are not of large extension. At the border between Uruguay and

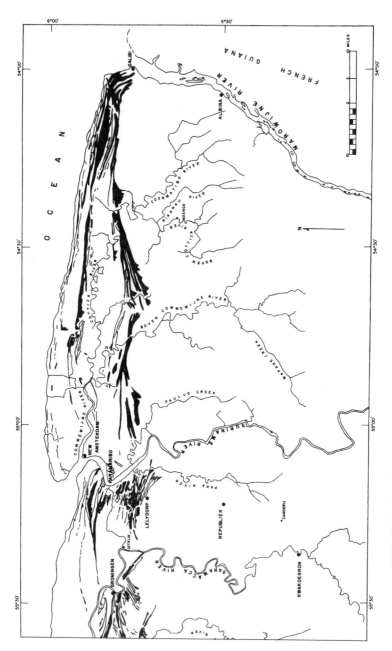

FIGURE 3.62   Cheniers along the Suriname coast. (From J. H. Vann 1959. With permission.)

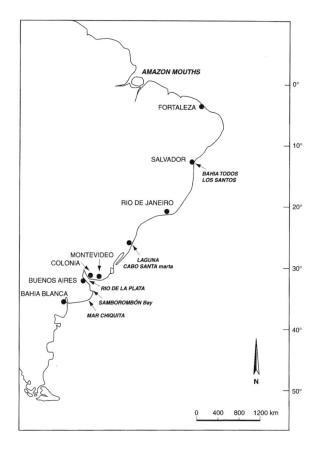

FIGURE 3.63   East coast, South America.

Brazil up to Laguna (where at Cabo Santa Marta at about 29° S mangrove has its southern limit) large lagoon streams have been formed behind a wide sandy barrier fringed by salt marshes and sedge swamps. Farther north, all along the coast up to the Amazon river mouth, there are numerous small lagoons, embayments, and small river mouths with sometimes a large extension of mangrove. South of Bahia Todos los Santos, the rivers form small deltas; farther north they flow out into usually small estuaries.

3.23. BAHIA BLANCA, ARGENTINA

Along the coast of Argentina at Bahia Blanca, around 39° S, a large estuarine bay, approximately 70 km long and widening to about 60 km on the seaward side, consists of large shoals with tidal flats, marshes, and supratidal areas, that are dissected by four large channels, on the average 10 m deep (Figure 3.64). The tides are semidi-urnal with a mean range of about 3 m, ranging from 2 m offshore to 3.5 m at Bahia. The entire area is ebb-dominated because of freshwater inflow from the Sauce Chico river. Also the net sediment transport is outward, and a large ebb-tidal delta has been

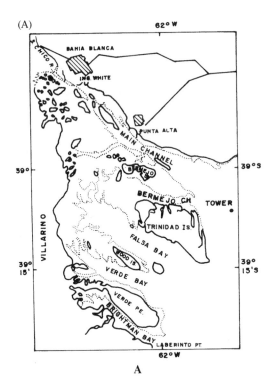

A

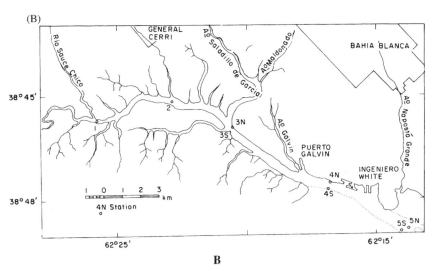

B

FIGURE 3.64   Bahia Blanca: A = the bay (From Aliotta and Farinati 1990. With permission.); B = the main channel (From Piccolo and Perillo 1990. With permission.); C = distribution of older Holocene tidal flats (on following page). (From Piccolo and Perillo 1990. With permission.)

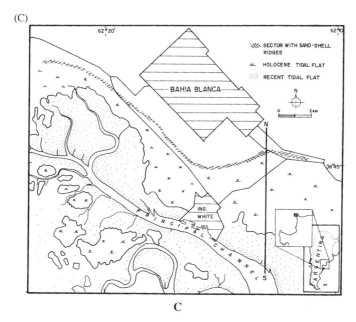

FIGURE 3.64 *Continued.*

formed with sand waves in the main channel (Aliotta and Perillo 1987; Aliotta and Farinati 1990). Winds come mainly from the northwest or north.

The marshes and tidal flats are dissected by a dense network of interconnected tidal creeks and (ebb) channels. The main channel shows erosion on the inner side of the bends and deposition on the outer side, which indicates a general shift in the channel system (Perillo and Sequeira 1989). The creeks are very sinuous or meandering and migrate laterally with widening of the channels. Low cliffs are formed along the banks followed by rotational slumping, resulting in erosional scarps. Slumping starts with the formation of crescent-shaped tension cracks, followed by fracturing, ending after slumping in a concave upward slope and a cusp-shaped hollow. On the average there are about six slumps per km where the banks are strongly affected. The largest cusps are up to 30 km wide; cusps are smaller (~15 m) where they occur in groups. Slumping is not affected by the presence or absence of vegetation on the surface sediment, but is enhanced by the presence of interconnected decapod holes, which occur around mid-tide level at an average of about 11 per meter. The holes may occur in 40% of the bank surface and are up to 50 cm deep. The slumps are the result of over-steepening of the banks, which mainly have a slope of 50°, which after slumping is reduced to 38° and finally to a stable slope of about 25°. Slumping occurs during falling tide shortly after 30 to 50 cm of sediment has emerged and may be induced by waves (even such as produced by a passing boat). The tidal flat sediment is rather stable with a parallel layering down to 70 cm depth, with sandy layers of 1 to 2 cm (Ginsberg and Perillo 1990). The salt marshes are covered mainly by *Spartina*.

The present flats probably date from the end of the postglacial sea level rise. Shell ridges on the inner margins near Bahia date from about 4600 BP. They are cheniers and lie on top of older sand/silt/clay deposits which are laminated and show flaser bedding. The main tidal channel follows the principal valley of the Late Pleistocene to Early Holocene delta complex of a former Colorado river (which now flows about 100 km farther south into the sea) and of the former Desaguadero river. The (later) southward migration of distributary streams has formed the major channels of the Falso, Verde, and Brightman bays (Perillo and Siqueiro 1989; Aliotta and Farinati 1990).

3.24. MESOTIDAL DEPOSITS

Mesotidal deposits in northwestern Europe, Australia, North America, and along the north coast of South America between the Amazon and the Orinoco river mouths have been studied rather intensively, whereas those in Africa, South and Southeast Asia, as well as in Alaska and the remainder of South America have been relatively little studied. In the latter list, however, there are only a few areas with intertidal deposits of importance. On the basis of the well-studied mesotidal deposits, the following summary can be made.

Sandy barriers, barrier islands, and spits are present in most mesotidal areas: the Wadden Sea (but not in the German Bight, which is mostly macrotidal), the Sado river estuary, the Bahia de Baiona and the Ria de Vigo in northern Spain, along Guinea, Guinea-Bissau and the Cameroon coast, at the Indus river mouth, along the west coast of Thailand and Peninsular Malaysia, in Indonesia, at the Purari river mouth, along the east coast of Queensland, at Yaquina Bay, and along Georgia, South Carolina, and Maine on the U.S. east coast. Sandy barriers and ridges, as pointed out by Hayes (1975), occur predominantly in meso- and microtidal areas, which is corroborated here for the mesotidal areas, except that in half the mesotidal areas described here, there are no barriers. The same applies to the presence of tidal deltas, which are present in mesotidal and microtidal areas and not in macrotidal areas, as indicated by Hayes (1975), but are present in only nine of the 28 mesotidal areas discussed here. As far as barriers are concerned, this can be related to the tidal range (neglecting that Hayes (1975) refers to barrier islands only): almost all mesotidal areas without barriers have a tidal range between 3 and 4 m, or a little over 4 m, and only part of the time a lower range around neap tide (Tejo estuary, Bay of San Vicente, the Kenya coast, the northern Persian Gulf, the northern Gulf of Thailand, the Mahakam delta, the Fly river estuary, the west side of the Gulf of Carpentaria, the north coast of South America [partly] and Bahia Blanca). The Firth of Thames, the Fraser river delta, San Francisco Bay, Labrador and Hudson Bay, and most of the South American north coast have a tidal range below 3 m but are protected against waves (the South American north coast by the presence of large mud banks), which suggests that the lower tidal range is only responsible for the preservation of barriers formed primarily by waves. The tidal deltas in the mesotidal areas are only ebb-tidal deltas (flood-tidal deltas have been described only from microtidal areas; see Chapter 4). In the mesotidal areas without ebb-tidal deltas the maximum ebb-tidal currents do not occur late in the tidal cycle near low water (Hayes 1975) so that no, or much less, horizontal segregation occurs between ebb and flood currents in the tidal channels. Also, the formation of an ebb-tidal delta may be prevented by strong wave action or longshore currents along the coast which carry the sand away. The fact that also no flood-tidal deltas are formed points to absence or only a limited development of a time-velocity asymmetry in the inlets or river mouths. Mesotidal boulder barriers and ridges are present only along Labrador and the Hudson Bay and, although formed under the influence of ice, are comparable to the macrotidal gravel and pebble ridges (see Chapter 2).

Cheniers are present at the Mekong river mouth, in Indonesia, along the south and west sides of the Gulf of Carpentaria, along the east coast of Queensland, along the Firth of Thames in New Zealand, along the Guyana coast of South America between Cayenne and the Orinoco river mouth, and in Bahia Blanca. As in the macrotidal areas, they occur at or above mid-tide level and consist of sand, shells, and shell fragments. Most are recent and sub-recent, some are older but Holocene. Where known, they mark periods of stabilization or regression of the coast, but a relation with an interruption of sediment supply is not everywhere clear.

Channels and creeks, as on the macrotidal flats, occur in a wide variety:

Straight: in the Wadden Sea, the Mahakam delta, the Firth of Thames, at the Amazon river mouth, and on the Guyana mud banks.

Sinuous: in the Wadden Sea, the Ooster and Wester Schelde, the Ria de Vigo, at the west coast of Thailand, in the Mahakam delta, at the Purari river mouth, in Hinchinbrook Bay (eastern Queensland), the Burdekin delta, at the Firth of Thames, the Fraser river delta, in San Francisco Bay, along the coasts of Georgia, South Carolina, and Maine, at the Amazon river mouth, and on the Guyana mud banks.

Meandering: in the Wadden Sea, Ooster and Wester Schelde, the Sado river estuary, the Bahia de Baiona, the Ria de Vigo, the Purari river mouth, along the Gulf of Carpentaria, in Hinchinbrook Bay, at the Fitzroy river mouth, along the Firth of Thames, in the Fraser river delta, in San Francisco Bay, along Georgia and South Carolina, and on the Guyana mud banks.

Dendritic channel systems occur in the Wadden Sea, Ooster and Wester Schelde, the Fraser river delta, in James Bay, and on the Guyana mud banks.

Interconnecting channels: in the Ooster and Wester Schelde, the Ria de Formosa, along the west coast of Thailand, at the Purari river mouth, the Fraser river delta, on the Guyana mud banks, and in Bahia Blanca.

Parallel, distributary, and **braided channels and elongated dendritic channel systems** were not observed in the more intensively studied mesotidal areas. In some cases no channels were mentioned at all (particularly in the less intensively studied areas), but it is likely that in some mesotidal areas, such as the large high salt flats east of the Indus river mouth and along the Gulf of Carpentaria, and along Labrador and James Bay, channels are only present in a rudimentary form or not at all, because in these areas with a very dry or a very cold climate the flow of water is limited. Lateral displacement of channels is only described in the Wadden

Sea, the Ooster and Wester Schelde, and the Ria de S. Vicente. **Tidal chutes** are only mentioned in the Wadden Sea, the Ooster and Wester Schelde; tidal watersheds are mentioned in the Wadden Sea and the Amazon river mouth (landward of Marajó island). **Channel levees** have been described in the Ooster Schelde (few), the Mahakam delta, at the Purari and Fly river mouths, and in San Francisco Bay. The Guyana coast between Cayenne and the Orinoco river delta is the only (mesotidal) area with large migrating mud banks and mud capes. Intertidal fluid mud occurs here as well as at the Amazon river mouth and in a few isolated locations in other mesotidal areas (e.g., Riade Foz).

As with macrotidal deposits, **sediment structures** in mesotidal deposits occur in a wide variety. Approximately parallel lamination or somewhat thicker parallel beds are indicated in all descriptions of sediment structures in mesotidal deposits. Also small ripples, both current and oscillatory ripples, occur in all mesotidal areas. Megaripples are less common and are described in the Wadden Sea, the Ooster and Wester Schelde, the Baie de Baiona and the Ria de S. Vicente, at the Chao Phraya river mouth (at the northern Gulf of Thailand), the Fraser river delta, and along Georgia and South Carolina. Sand waves occur in the Fraser river delta, and sand dunes occur only in deeper channels, e.g., in the Wadden Sea. Sand/silt/mud laminae or layers (often wavy) occur regularly in the Wadden Sea, the Ooster and Wester Schelde, and the Fraser river delta; sand sheets occur along the coast of Georgia and South Carolina. Cross-bedding occurs in the same areas; in the Amazon mouth–Guyana muds a micro-form exists. Longitudinal cross-bedding is observed in the Wadden Sea. Other sediment structures are described only in the Wadden Sea, the Ooster and Wester Schelde, and the Georgia–North Carolina mesotidal deposits. These are herringbone structures, climbing ripples, convolute bedding and loadcasts, deflation structures (ridges behind obstacles, flats), gas pits, rill marks, foam-line/flood level marks, erosional scarps (low cliffs, steps, banks) at the seaward edge of the salt marshes, shell beds, mud drapes, mud balls, and slump structures. This indicates that detailed observations of sediment structures have only been made in a few mesotidal areas. To this can be added that many sediment structures are destroyed by bioturbation, by the tides, and by surface waves. Nevertheless, more observations of sediment structures have been made than have been mentioned. Slump structures have also been observed (with chaotic bedding) in Bahia Blanca, mud balls in San Francisco Bay, where low gas domes are also formed. In the Amazon mouth–Guyana muds, graded lamination and small-scale scour and fill structures were described. The many sediment structures are only partly related to tidal characteristics, but mainly to basic principles of sediment transport and deposition by currents and waves, as will be discussed in Chapter 7. Mud cracks are related to climate and occur in dry areas, in particular around high tide level, but also in temperate areas (Wadden Sea, Georgia coast) during the summer. Also the salt flats at the Indus river mouth, along the Gulf of Carpentaria, and the east coast of Queensland are related to a dry climate. Features produced by ice (cracks, pits, tracks, wedges) are clearly related to cold climates but can also be formed in temperate areas (Wadden Sea) during the winter. As was already mentioned for the macrotidal areas in Chapter 2, climate, to a large extent, also determines the presence or absence of mangrove, salt marshes, and bare flats.

Mesotidal estuaries and river mouths have no typical configuration and are less generally funnel-shaped than macrotidal estuaries but can sometimes be clearly funnel-shaped, as the Fly river estuary shows, which has a spring tide range of 4 m. The only other mesotidal river mouth with a spring tide range of about 4 m is the Tejo river mouth, which is confined in a channel with rocky banks.

4 Microtidal Deposits

Tidal areas are microtidal when the tidal range is less than 2 m, but this range is not very well defined: the upper limit for the reasons given in Chapter 3, the lower limit because a zero tidal range does not occur in nature and the regions with a small microtidal range are much influenced by wind effects on the water level. The so-called nontidal coasts where wind effects dominate and sea level may be temporarily raised by up to 2 m, are discussed in Chapter 5. Hayes (1975) indicates that tidal flats and salt marshes become gradually more rare when the tides become more microtidal and are altogether absent at a tidal range below 30 to 50 cm. Usually wind effects cannot easily be separated from tidal effects at these low ranges, but a tidal influence on microtidal deposition is not absent even at very small tidal ranges. In general the same conditions apply for the formation of microtidal deposits as for the formation of other intertidal deposits: shallow water depth, an open connection with the sea, and the availability of granular material. Where one of these conditions is not met, as along steep rocky coasts and in lagoons that are closed off from the sea, no tidal deposits are present.

4.1. DYFI ESTUARY, WALES

The Dyfi (or Dovey) estuary in Wales at about 52° 30' N, with its adjacent coastal/deltaic plain, is roughly triangular in shape, open to the west toward Cardigan Bay, but partly closed by a sandy spit (Figure 4.1). With a length of about 9.5 km it covers an area of about 45 km^2 with the estuary covering about 17 km^2. The intertidal deposits consist of bare mud flats and a broad fringe of salt marsh along the south side with a network of tidal creeks. The tides are mixed-semidiurnal with a range that is normally less than 2 m with an average of 1 m, but may reach 2.1 m during extreme spring tides and almost 5 m during spring tide in combination with an onshore storm (Shi et al. 1991a, 1991b, 1995).

The tidal flats are sandy at the lower levels and in the channels and become finer (muddier) upward, reaching 90% silt/clay in a higher mud zone, which is separated from the sandy lower flats by a lower mud zone with silt/clay percentages up to 40%. In the salt marshes the amount of silt/clay reaches more than 90%. The sediment consists mainly of sand/mud couplets; the thickness of both the sand and mud layers in each couplet varies in relation to current speed and tidal range, as well as to differences between the individual neap and spring tides. The couplets mainly reflect diurnal tidal inequalities and the neap-spring tide cycle. The mud

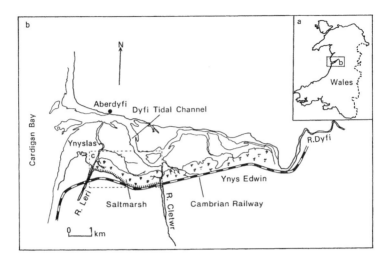

FIGURE 4.1   The Dyfi estuary. (From Shi et al. 1995. With permission.)

layer of each couplet is deposited from suspension during slack tide, the sand layer from asymmetrical bidirectional currents either during the ebb or during the flood. The recent as well as the older deposits reflect the mixed-semidiurnal tides prevailing at present. The salt marsh accretion rate, determined from marker beds consisting of red sand, brick or coal dust, and from the thickness of the individual sediment laminae, was found to vary between 8.0 and 42.0 mm · y^{-1} during the past 80 years (average of 10 mm · y^{-1} during 1977–1989) and decreasing with time (Shi 1993). Creek density increased between 1966 and 1989, ranging from 8 to 11 km^2. The creek patterns consist of one or more large sinuous creeks with numerous small creeks or gullies (Figure 4.2), forming a dendritic network. With time, the creek network became larger, generally more dendritic and complicated, with some creeks becoming narrower, most becoming longer, and some creeks shorter. There has probably been sedimentation in some creeks, while most were cut back, and some creeks, or parts of them, may have been closed by vegetation (*Spartina*).

Estuarine deposition in the Dyfi estuary and the estuarine plain started at about 10,000 BP with relatively deep water subtidal deposition under low-energy conditions of rapidly accumulating sediments in a progressively shallower environment. This changed around 6000 BP to shallow water tidal and intertidal high-energy deposition in an estuary that was considerably larger than the present estuary. Around 3500 BP, the present shallow-water low-energy estuarine salt marsh-dominated deposition began (Shi and Lamb 1991). The age of the present salt marsh is probably rather young; it may have started to grow around 2000 BP.

4.2. THE WHITE SEA, NORTHERN SIBERIA, THE GULF OF ANADYR, AND THE SEA OF OKHOTSK

Along the White Sea, tidal flats and salt marshes occur along the Kandalakhskaya Guba, where the tidal range is between 2 and 4 m, along the deltas of the Onegin

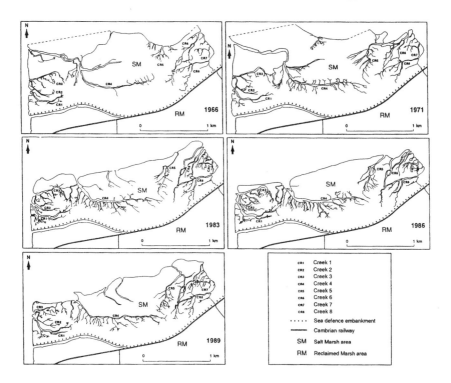

FIGURE 4.2   Development of channels and creeks in the Dyfi estuary. (From Shi et al. 1995. With permission.)

and Northern Dvina rivers, and in the Gulf of Mezen where the coast is macrotidal with a tidal range up to above 6 m and a spring tidal range up to 10 m. East of the Kanin Peninsula (Figure 4.3) wide flats are present along Cheshkaya Guba, where the tides are mesotidal. Farther east, along the north coast of Siberia, the tidal range drops to less than 2 m, with a slightly higher range at the New Siberian Islands (2.1 m) and a minimum range around 3 cm at Bear Island. Similar tidal ranges are found in the Bering Strait and southward in a large part of the Gulf of Anadyr. Along the southern coasts of the Gulf of Anadyr, the range increases to more than 2 m and increases further to more than 4 m along the north coast of the Sea of Okhotsk, where a maximum of 12 m is reached in the funnel-shaped Bay of Penzhinskaya. The tides are semidiurnal from the White Sea to the Gulf of Anadyr, where the mixed tides of the Pacific become dominant. No intertidal deposits and salt marshes occur along the Siberian coast north of about 73° N because the ice-free period is too short to allow prolonged intertidal sediment transport and the buildup of intertidal flats. Also the snow- and ice-free period with higher temperatures is too short for the growth of salt marsh vegetation. At slightly lower latitudes, intertidal deposits and salt marshes have been formed along the Pechora Gulf, the Ob river mouth, the Yenisei river estuary, and along the Lena river delta with only patchy salt marshes. The Ob, the Yenisei, and the Lena are the largest rivers in north Asia: they flow northward through Siberia and discharge into the sea, the Ob through an elongated

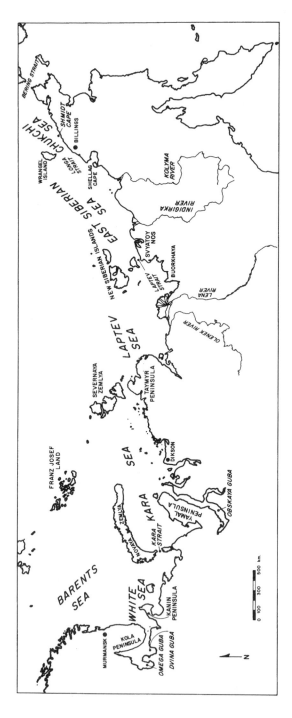

FIGURE 4.3   North Siberia and the White Sea. (From Zenkovich 1985. With permission.)

bay (a liman), the Yenisei somewhat farther to the east into an estuarine bay; the Lena has formed a large fan-shaped delta in the Laptev Sea, which is now largely relict. Tidal deposits and salt marshes east of the Lena delta occur at the deltas of the smaller Omoloi, Yana, Alazea, and Kolyma rivers, which are also rather large and, like the Lena delta, mostly subrecent. Recent deposits, including some tidal flats and salt marshes, form only a small fringe.

Along the East Siberian Sea low flat areas predominate, which during storm surges (in summer) can be flooded up to 30 km inland; during southern winds the sea bed is exposed for about the same distance. Although the tides are insignificant, the flats have the character of tidal flats (see Chapter 5). From Cape Billings to the east, the coast is formed by a series of lagoons that are subdivided by transverse bars into lakes that are up to 10 km wide. The direction and location of the transverse bars is related to the fetch of the most frequent winds. The lagoons extend to the east almost as far as the Bering Strait.

Along the Pacific coast of Russia, extensive intertidal flats are present along the Gulf of Anadyr, often behind barrier spits and in particular around the mouth of the Anadyr river, farther south along Kamchatka, in embayments along the north coast of the Sea of Okhotsk, and along Sakhalin. Large lagoons are present along the northern coast of Sakhalin. The Amur river in eastern Siberia discharges through a relatively narrow river mouth into the narrows between the mainland coast and Sakhalin (Zenkovich 1985; Kaplin 1985).

4.3. NORTHWEST AFRICA

Several lagoons are present along the Atlantic coast of Morocco: Moulay Bou Salham at about 34° 50' N, and the Oualidia and Sidi Moussa lagoons around 30° N behind sandy coastal barriers along a predominantly steep coast. Farther south, along the Saraoui coast, sebkhas border the sea, while at the Banc d'Arguin in Mauritania there is a shallow coastal sea area that is partly intertidal with shoals and sandy tidal flats that extend into the Atlantic Ocean between 21° N and 19° N. The spring tide range varies from 1 to 1.5 m at Moulay (0.15 to 1.0 m at neap tides), to 2 m at Oualidia, and 1.2 to 1.6 m (mean 0.8 to 1.1 m) at the Banc d'Arguin. The lagoon Moulay Bou Salham is about 6 km wide and 11 km long, 1 to 1.5 m deep, and fringed with bare tidal flats and salt marshes (Figure 4.4; Carruesco 1989). It follows a zone of subsidence and is connected with the sea by a narrow sinuous channel that cuts through the coastal dune barrier and is a continuation of a wadi (Oued Drader). In winter, which is the rainy season, the salinity in the lagoon is between 5 and 30‰ S (average 18‰), in summer around 35‰. The sediment is mostly mud with more than 20% clay, but along the margins it is more sandy with 10 to 20% clay. Sandy deposits are present in and along the channel that connects the lagoon with the sea. The sediment in the lagoon is supplied predominantly through the Oued Drader.

The Oualidia lagoon is an elongated basin 8.5 km long and less than 1 km wide with a relatively wide, short inlet and intertidal shoals and one main channel, a secondary channel, and a number of interconnecting channels, 0.5 to 1 m deep (Figure 4.5; Carruesco 1989; Bidet and Carruesco 1982). At the entrance the shoals are

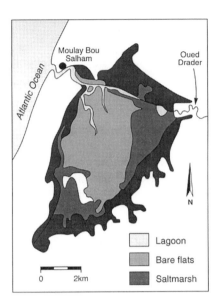

FIGURE 4.4   Moulay Bou Salham. (From Carruesco 1989. With permission.)

sandy and bare, while more inward they are muddy and covered with salt marsh; bare mud flats form a narrow fringe along the salt marshes. The salt marshes consist of *Spartina, Zostera, Suaeda,* and green algae on the lower parts of the flats, with *Salicornia* and *Obione* on the higher parts. The sandy shoals are entirely marine and contain marine mollusk shells and shell fragments. Inward the lagoon is brackish during most of the year. There is no sediment supply from inland except by the wind and probably most of the sediment comes from reworking and deflation of older deposits. The sands have a low content of organic carbon (0.1 to 0.4%); in the muddy sediment this increases to a maximum of 3.1%.

Also there is no sediment supply to the Banc d'Arguin from inland except by wind, but this is probably quite substantial because strong winds blow regularly offshore from the Sahara (Altenburg et al. 1982). The intertidal deposits consist of sebkhas (barren sand flats, that are only flooded during extremely high spring tides) that cover about 140 km², bare sandy intertidal flats (about 80 km²), muddy intertidal flats (about 220 km²), seagrass beds (covering about 196 km²), *Spartina* marshes (about 25 km²), and mangrove (about 5 km²) which has its northern limit here at the northern tip of Nidra island (Figure 4.6 and 4.7; Wolff and Smit 1990). Sandy deposits with less than 12% clay and silt dominate; seagrass beds occur on finer muddier sediment with up to 31% clay and silt. Besides seagrass and *Spartina* there are some benthic macroalgae and benthic diatoms. *Hydrobia, Bittium,* and other small gastropods live on the muddy sediments. Burrowing bivalve mollusks (*Anadara senilis, Loripes lacteus,* and *Abra tenuis*) are very common as well as the fiddler crab (*Uca tangeri*), which is particularly numerous around high tide level. *Anadara* occurs in densities up to 350 per m². A large number of polychaetes live on the flats but not in large concentrations. They are small in size and do not give much bioturbation.

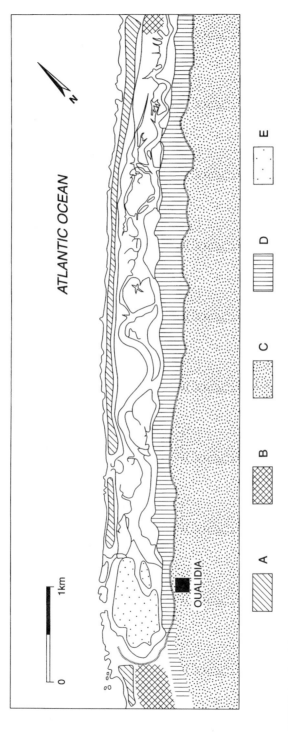

FIGURE 4.5   The Oualidia lagoon. A = aeolian soils; B = clayey alluvium; C = red Mediterranean soils; D = lithosols and regosols; E = sand banks of flood tidal delta. (From Carruesco 1989. With permission.)

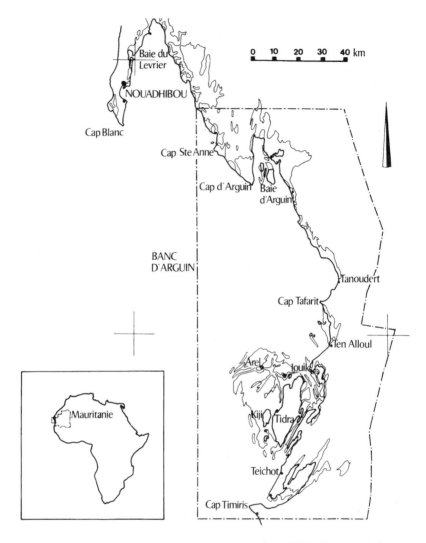

FIGURE 4.6   The Banc d'Arguin. (From Altenburg 1982. With permission.)

4.4.   SENEGAL AND GAMBIA

Along the Senegal and Gambia coasts, limited intertidal areas occur at the Senegal river mouth at about 16° N in combination with sandy ridges and small spits, and much wider tidal areas at the mouths of the Sine and Saloum rivers at about 13° 50′ N and at the Casamance river mouth at about 12° 30′ N (Figure 4.8; Guilcher 1985) The tidal range varies along this coast between 1.2 and 1.6 m during spring tides and 0.5 and 0.7 m at neap tides. Swell comes mainly from the northwest and is generated by storms at higher latitudes. Locally generated smaller waves also come from this direction. The northwest waves have resulted in longshore sediment (sand)

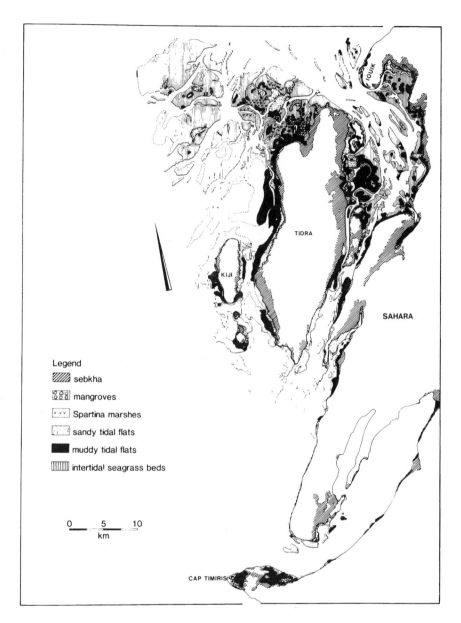

FIGURE 4.7   Tidal flats at the southern Banc d'Arguin. (From Wolff and Smit 1990. With permission.)

transport to the south and a strong deflection of the Senegal river mouth in this direction, and to a lesser extent a southward deflection of the Sine and Saloum river mouths. The intertidal areas are mostly covered with mangrove (*Avicennia, Rhizophora*), which has its northern limit a little farther to the north at the Banc d'Arguin in Mauritania (Guilcher and Nicolas 1954). The wide intertidal marshes at the

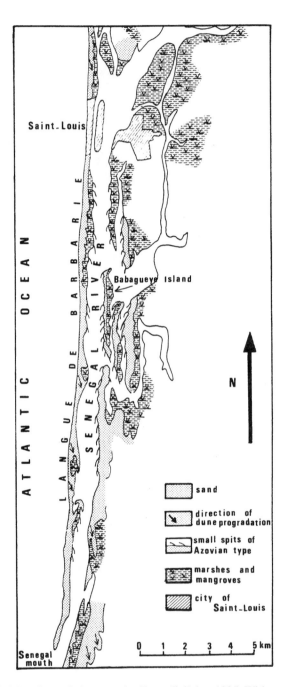

FIGURE 4.8   Senegal river mouth. (From Guilcher 1985. With permission.)

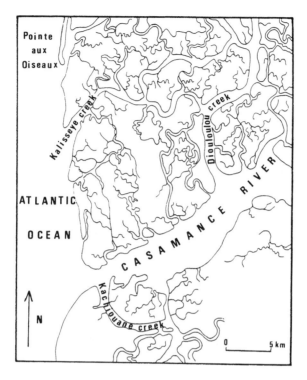

FIGURE 4.9   Casamance creeks. (From Guilcher 1985. With permission.)

Sine–Saloum and Casamance river mouths are dissected by numerous sinuous or meandering interconnected channels and creeks (Figure 4.9). The mangrove here is larger and generally in better condition than the stunted mangrove at the Senegal river mouth, as they occur more to the south along a coast with higher rainfall. Between the mangrove and the terrestrial vegetation more inland, bare muddy areas called "tanns" occur around high tide level. They are flooded during spring tide but remain dry during neap tides. Because of evaporation, the salinity in the sediment increases during the prolonged periods of exposure and becomes too high for mangrove growth. Many of these areas have now been reclaimed for rice cultivation. The Gambia river mouth does not have wide intertidal areas because it is cut into sandstone.

4.5.   SIERRA LEONE–BENIN

The coast of Sierra Leone, between 9° N and 7° 30′ N consists of cliffs and beach ridges separated by large embayments with numerous river mouths and creeks fringed by intertidal areas covered with mangrove (Figures 4.10 and 4.11). The tides are semidiurnal with a maximum range of 2.8 m at Freetown. Swell comes from the southwest. The freshly deposited mud is colonized by *Rhizophora* that can reach a height of 20 m. Sandy deposits are mostly covered with *Avicennia*; *Conocarpus*

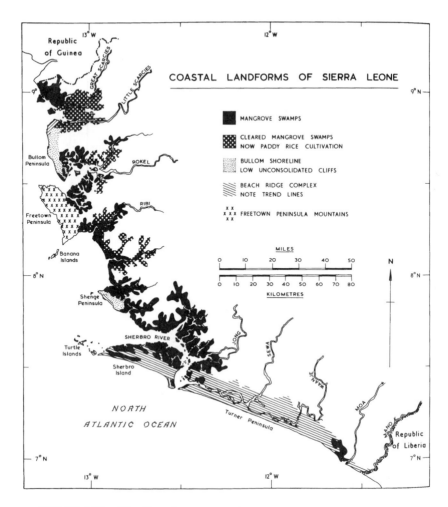

FIGURE 4.10   The Sierra Leone coast. (From Scott 1985. With permission.)

and *Laguncularia* are scattered around high tide level. Part of the mangrove has been cleared for rice cultivation. Bare mud flats occur below mean tide level.

Around the Freetown peninsula the low intertidal deposits consist of sands on intertidal channel bars and point bars with ripples, sand waves, and sand dunes in both the ebb and flood directions, and low-angle crossbedding. The periodic bedload movement inhibits burrowing bottom fauna: there are no bivalves and only locally some tube-building polychaetes. The higher intertidal deposits cover large areas and consist of muddy sands with up to 10% mollusk shell debris, fine sands with *Uca* burrows, and mud flats of silt and clay in approximately equal amounts. The muddy sands are covered with straight-crested asymmetrical ripples or are plane surfaces with current lineations. They are dissected by a dendritic network of mostly meandering channels that leave concentrations of shell debris behind on an erosion surface during lateral migration. Three species of bivalves that occur in densities of up to

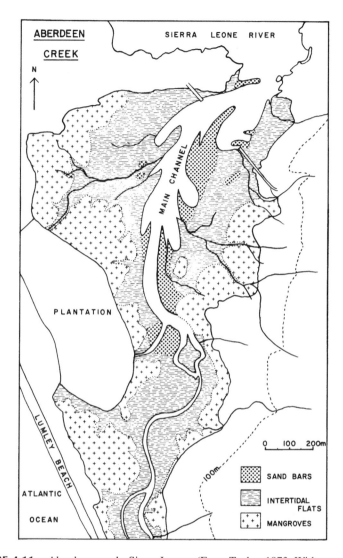

FIGURE 4.11   Aberdeen creek, Sierra Leone. (From Tucker 1973. With permission.)

20 individuals per m², make burrows down to 10 to 40 cm deep. Shallower burrows are made by another species of bivalve, by polychaetes, and crustaceae. Two gastropods live on the sediment surface in densities up to 50 individuals per 100 grams of sediment. Algal mats are formed during the dry season in the pools and channels, so that at the end of the dry season a gelatinous mat is present. Debris of higher plants can be up to 20% of the sediment. Close to the main channel oval depressions (scour pits) occur that are up to 15 cm deep and a few centimeters to one meter long. They are asymmetrical with an upcurrent side that is steeper than the downcurrent side. On the rim, sediment can be deposited that was excavated during their formation. They start to form when the flow is diverted by an inhomogeneity on the

sediment surface (a crustaceae burrow, worm tubes, a bivalve shell). They are formed mainly by the ebb tidal currents; those made by the flood currents are usually not preserved. The fine sands with *Uca* burrows occur predominantly on the higher intertidal areas where the groundwater is 10 to 40 cm below the sediment surface at low tide. *Uca* makes up to 25 burrows per m², which are curved and go 40 cm deep. The crabs scrape the surface sediment for food and produce fecal pellets of 1 to 4 mm. Larger pellets, up to 1.5 cm, are formed during burrowing. There is little structure in the sediment because of the bioturbation; laminations can be seen where the activity of the crabs is minimal. Ripples on the sediment surface are straight-crested and sinusoidal current ripples that may be rounded by waves, and oscillation ripples. Polychaetes are not common. The mud flats consist of silt and clay in about equal amounts and occur in restricted, quiet areas in the upper reaches of the estuaries. There are polychaetes, crabs, few bivalves, some mudskippers, and gastropods on the sediment surface. Few sediment structures can be seen because of bioturbation.

Mangrove (*Rhizophora, Avicennia,* and on higher grounds *Laguncularia*) is widely distributed above mid-tide level up to high tide level. The sediment is dark-grey mottled muddy silt and sandy mud, with much organic matter, and roots and rootlets of *Rhizophora.* There are burrowing crabs with oysters, barnacles, and gastropods on the roots and trees. Sediment deposition, enhanced by the mangrove, results in an elevated area commonly delineated by a bank up to 30 cm high. There are marked differences between the dry and the wet season: during the dry season, there are persistent westerly winds that result in the development of spits that may close off the estuaries from the sea, as the river discharge is low or negligible. Also because of this, salt water penetrates into the estuaries, the flood currents dominate, and much bed load is brought in from offshore and moved inward by the flood. In the channels the main bed forms are sand dunes in the flood direction. With the dry season the numbers of burrowing crabs increases and evaporation results in higher salinities and the development of algal mats, with the formation of mud cracks on the exposed areas. During the wet season there is intense wave action only during storms and the estuaries remain open while river discharge is high and salt water is driven out of the estuary. The ebb currents dominate and much suspended material and plant debris supplied by the rivers can be deposited on the flats. Ebb-tidal sand dunes dominate on the banks in the channels, while on the exposed flats bed forms are obliterated and channel systems are formed by the intense precipitation. Scour pits are commonly formed and bioturbation is reduced because of a decrease in the number of crabs and bivalves (Tucker 1973).

From Sherbro river at about 7° 30′ N to the east, beach ridges, interrupted by some river mouths with mangrove, form the coast. This continues in Liberia, there interrupted by small lagoons (the largest being Lake Piao) mostly associated with river mouths and lined with mangrove. Farther east along the Ivory Coast after about 250 km of cliffs, beach ridges and small lagoons appear again around Abidjan and continue eastward to Axim in Ghana. From there a rocky coast dominates again as far as the Volta river delta in eastern Ghana, then beach ridges and lagoons with coastal marshes continue into Togo and Benin as far as the Niger river delta. Along this stretch of coast from Sierra Leone to the Niger delta, a total distance of about 2100 km with about 500 km of cliffs, the tidal range is less than 2 m. The swell comes from the southwest to south-southwest and is generated in the South Atlantic Ocean. It can be large and result in a strong high surf. The lagoons are usually small but some are very large (Ebrie and Aby lagoons near Abidjan are 566 km^2 and 424 km^2, respectively; Keta lagoon in the Volta river delta is 250 km^2). They are shallow with about 1 m water depth and up to 4 m in channels. Some are closed off from the sea, some only temporarily, by a sandbar, which near Abidjan has been cut by a canal that is kept open by a jetty that deflects the longshore (eastward) sand transport. The Volta river delta has a sandy barrier with large lagoons (mostly freshwater lagoons) on the delta plain. Also where there is an open connection with the sea, the tides generally do not penetrate very far into the lagoons. Flooding of the marshes surrounding the lagoons occurs mostly by freshwater during the rainy season (generally October through February). The lagoons have muddy bottom sediment with more sandy deposits along the margins where waves may affect the bottom. Sandy deposits also occur in the generally muddy marshes behind the coastal ridges and are probably (mostly) the remains of former spits and beach ridges dissected by meandering marsh creeks. The marshes are covered with mangrove where the water is still somewhat brackish, but most of them are freshwater marshes covered with grasses. Salt production has led to the cutting of mangrove and excavation of numerous small ponds (Guilcher 1959, 1985; Burgis and Symoens 1987).

4.6. THE NIGER DELTA

The Niger delta at about 5° N has a classic triangular-arcuate shape with distributary channels diverging outward from the river (Figure 4.12). The coast is formed by a series of 21 barriers with beach ridges interrupted by larger and smaller inlets. The inlets are mostly tidal but at least seven are directly connected with the distributary branches, while two are connected with other, smaller rivers. The Cross river inlet east of the delta and the Lagos-Lekki lagoon in the west are separate systems not connected with the Niger. The tides are semidiurnal with a range of about 1 m near Lagos and 1.6 m at the Opoboi river. Farther east the tidal range increases to 2.6 m at the Imo river mouth east of the delta and 2.8 m at Calabar on the Cross river (Allen 1964, 1965a,b; Usoro 1985).

Behind the barriers is a belt of about 9000 km^2 of intertidal swamp, 480 km long and up to 16 km wide, consisting of intertidal deposits largely covered with mangrove and dissected by a network of channels and creeks (Figures 4.13, 4.14, 4.15, and 4.16). Inward the intertidal swamps grade into a forested river floodplain or border on a terrace formed on late Tertiary and possibly Pleistocene deposits. The contact between the flats and the floodplain is irregular: tongues of fresh levees extend into the swamps, and tidal channels cut backward into the floodplain. Overall both the floodplain and the intertidal swamps grow in a seaward direction, the latter by erosion of the back side of the beach ridge barriers.

The channel network consists of large feeding channels, often connected by a river distributary, and smaller channels that connect the feeding channels with elevated intertidal flats that fall dry every falling tide and which are dissected by small channels and creeks. The large feeding channels are up to 1 km wide and end (or start) at the tidal inlets that are up to 2 km wide. In the wider parts sandy shoals

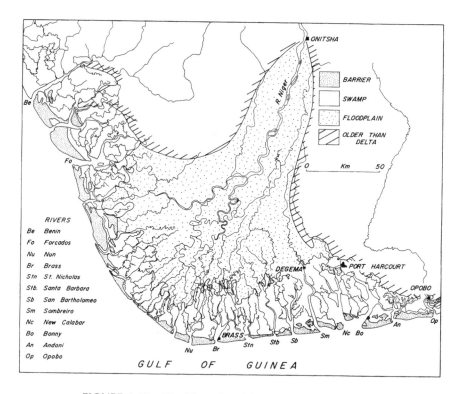

FIGURE 4.12   The Niger river delta. (*Source*: Allen 1965b.)

are formed that are partly intertidal, as are the ebb-tidal deltas on the seaward side. Mud deposition starts in the open water: initial banks grow above low tide level, followed by colonization with mangrove along the edges. Further accretion and vertical deposition continue until a sizeable intertidal flat has been formed that rises above mean sea level. When mangrove colonizes the flat by reducing the flow velocity, it traps sediment and reduces scour. Where the sediment has become firmer and freshwater dominates, the mangrove is replaced by tropical forest.

The large channels are straight or gently curved with sometimes hooked bends, 9 to 15 m deep, and generally aligned at right angles to the coastline. The smaller channels and creeks are dendritic and strongly meandering with backward extensions into the tidal flats and the river floodplain. Where they border the sandy coastal barriers, there is strong erosion; sandy remnants of eroded barriers remain as higher ground in the swamps, suitable for settlements and farms. Point bars are formed by the meandering channels and creeks, and in the widest parts of the larger channels, just inside the inlets are large complexes of islands and mud flats (intraswamp deltas). The mangrove (mainly *Rhizophora*) forms laterally, spreading roots and aerial roots that make the channel borders relatively stable: often uniform tall trees up to 30 to 45 m high are present on both sides of a channel where usually a low levee is formed. On the interchannel flats farther away from the channel, where the sediment is less sandy and soil conditions are generally less favorable, the mangrove is smaller and

FIGURE 4.13 Channels in the Niger delta marshes. A = two meanders near channel capture; B = meandering channels; C = reduction of inner-channel flat because of meander; D = strongly meandering channel with ox-bow formation; E, F = capture of open-ended channel by meander growth. (From Allen 1965a. With permission.)

usually older (second generation). On most of the interchannel flats the mangrove is sparse and only present in the form of low shrub. Numerous crabs inhabit the channel edges and have an effect opposite that of the mangrove by rendering the sediment cavernous with burrows. The result is a slow erosion by meandering channels and creeks with a meander belt advancing in the general direction of the strongest flow and gradually changing the shape of the interchannel flats.

The flats grow by lateral as well as by vertical accretion. The drainage of each flat tends to be toward the center, away from the higher channel borders, with the major tidal watershed close to the perimeter of the flat, which tends to be several decimeters higher than the center. The watershed is breached at only one or two places. The flats can be between 1.5 and 12 km across and at least the large ones are elongated in the direction of the most direct drainage, i.e., at a steep angle to the coastline. In general, lateral accretion is balanced by lateral erosion so that net-deposition is mainly vertical.

The river and its distributaries supply sand, whereas the silt and clay are mainly carried inward from the coastal sea by the flood tide. The wind has little effect inside the barriers, but the large swell from the south southwest gives considerable long-shore transport along the coast toward the east as far as the Cross river estuary. Within the swamps there is redistribution of sediment by lateral erosion and rede-position. The larger channel beds are usually sandy with a rapid lateral fining to

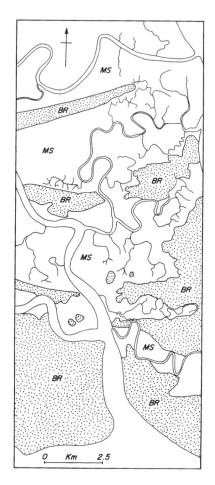

FIGURE 4.14   Channels in the Niger river delta near an inlet. MS = mangrove swamp; BR = beach ridge island. (From Allen 1965a. With permission.)

silts on point bars. Vertically there is interlayering of sand, silts, and plant debris with strong grain size contrasts. Sand beds, of a few millimeters to 50 cm in thickness, are well sorted and may consist of fine to very coarse sand with some fine gravel. On the channel banks sands are finer on top; laminated sediments are rare, but small-scale as well as large-scale cross stratification occurs, as well as bioturbation. Sandy layers usually have an erosional basis. The fine sediment (silts, clays, decaying organic matter) are poorly sorted with beds of a few centimeters to more than one meter thickness. Laminations of plant debris and mica can be up to 18 cm thick. In calm creeks, tangles of driftwood with a mud infill may accumulate. Sediments on the interchannel and intercreek flats consist of fibrous mangrove root mats or black organic-rich silty clays with crab burrows at the edges of the flats. Only the surface deposits along the swamp margins are sandy where there is some supply of river sediment or contact with eroding beach ridge barriers. Shallow tidal

FIGURE 4.15   Channels in the Niger river delta at the contact between the marshes and higher (Pleistocene) grounds. MS = mangrove swamp; AL = (freshwater) alluvium. (From Allen 1965a. With permission.)

channels and creeks are mostly eroded in silt/clays; only the deeper channels cut into river sands or elder deposits.

Formation of the delta started during the Cretaceous and since then deltaic sediments have been deposited. The volume of the sediments has led to subsidence but, in general, delta outbuilding has prevailed, and the effect of subsidence has been relatively small compared to the Pleistocene sea level changes. Deposition of the present sediments started at least several thousand years ago when sea level reached its present position. Sediment supply outpaces subsidence, as it has done mostly during the past, so that the river floodplain and the intertidal marshes are progressing outward.

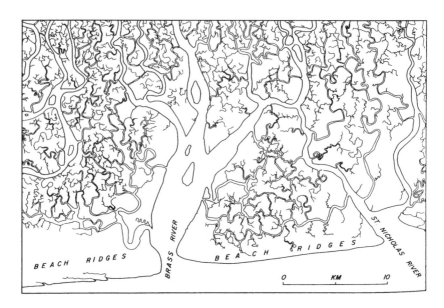

FIGURE 4.16   The Brass river inlet, (*Source:* Allen 1965b.)

On the west side of the Niger delta is the shallow Lagos-Lekki lagoon with a maximum depth of about 3.5 m and connected with the sea by a channel at Lagos. It has a variable salinity which is highest during the dry seasons November to March and late August to early September. The bottom consists of sandy mud to organic-rich mud with shelley sands at the southern margin where the beach ridge barrier that separates the lagoon from the sea is eroded on its landward side. The lagoon is fringed with mangrove where the water is brackish and with reeds or shrub where the water is mostly fresh.

On the eastern side is the estuarine complex of Calabar/Rio del Rey, which is 70 km wide on its seaward side and extends 50 km inland (Figure 4.17). Four main rivers flow out into the sea in this area (the Cross, Akpa, Lokole, and Meme). As in the Niger delta, the rivers supply sand, while silt and clay come in from the sea. Eastward longshore transport of sand stops west of Calabar so that the shoals and mangrove of the complex directly border the open sea. The mangrove swamps in the complex are up to 16 km wide with only some beach ridges on the seaward side along some exposed parts. The mangroves are cut and dissected by wide, curved channels and small meandering channels and creeks that form island flats consisting of organic-rich silt and clay. Most islands are covered by uniform stands of relatively tall trees around areas of relatively short trees. Along open water the mangrove-covered islands grade into bare mud flats that are below mean tide level. Shoals in open water are partly intertidal, up to 3 km across, and separated by channels up to 10 m deep and 2 km wide.

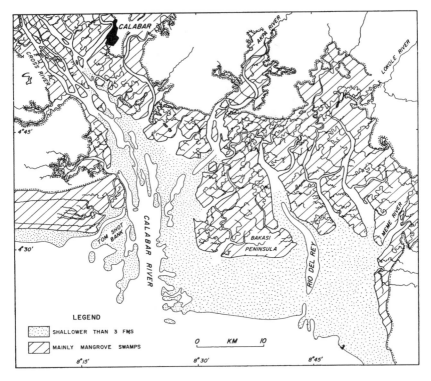

FIGURE 4.17   Calabar, Rio del Rey. (*Source*: Allen 1965b.)

4.7. ZAIRE RIVER ESTUARY

Along the west coast of Africa at about 6° 5′ S the Zaire river estuary is bordered by about 28 km of mangrove up to 10 km wide on the north side, and about 45 km of mangrove up to 15 km wide on the south side (Meulenbergh 1974; Figure 4.18). The mangrove (mainly *Avicennia*) living on consolidated grey clay is dissected on the north side by several large interconnecting channels that are up to more than 1 km wide and numerous smaller channels and creeks that are aligned in the direction of the river estuary, which is also the direction of the river outflow (Figure 4.19). Two large channel systems have their mouths on the seaward side at an embayment behind a spit that separates the northern mangrove area from the sea. The other larger channels branch out from the main river outflow channel, where the main outflow is concentrated in the center. Little freshwater from the river passes through the mangrove area, where the flow is predominantly tidal. The mangrove area on the south side is separated from the sea by a broad spit and protected by a shallow bank that separates it from the main river outflow (which turns here to the northwest). The channels in the mangrove follow a more south-north direction than on the north side, which reflects a water supply from the adjacent hills and terraces. Only toward the east, where the mangrove is near the main river outflow, are the channels and

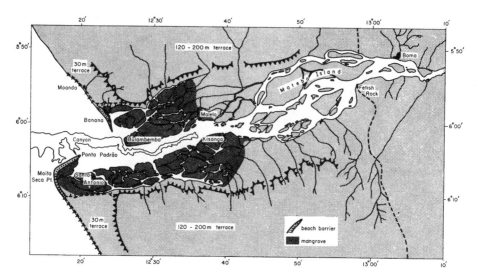

FIGURE 4.18   Zaire river mouth. (From Eisma and Van Bennekom 1978. With permission.)

creeks more aligned in the direction of the river flow. As on the north side, the channel mouths on the south side are widened because of the tides coming in. Inland, east of Malela, the mangrove grades into freshwater swamps that are partly cultivated. In the main river channel the river outflow (the second largest river discharge in the world) dominates in the upper 3 to 10 m. Elsewhere in the estuary and in the freshwater swamps, the tides dominate as far inland as upstream of Boma. They are semidiurnal with a range of 1.3 to 1.5 m at spring tides, 0.4 to 0.6 m at neap tides. Along the center of the estuary a canyon extends seaward from near Malela, where water depths drop from about 10 m to more than 90 m. It continues into the Angola Basin down to more than 4000 m and has a depth of about 450 m between the spits at the river mouth. Waves and swell from the southwest can only penetrate into the estuary between the spits and can hardly reach the intertidal areas inside. The flow through the mangrove channels and the mixing of salt- and freshwater show large seasonal variations. In the northern mangrove area the salinity may remain well below 10‰ S at low tide in January, but may reach 35‰ S in August during high tide. The channel floors, where studied (mainly in the northwest part of the estuary), are sandy and grade laterally into the grey compacted mud on the mangrove flats. Locally along the side of a channel there is brown, clayey sand with grey clay fragments that have been eroded from the flats. There is only local erosion, related to channel migration, and comparison of the recent channels with charts from 1877 indicates that the position of channels and creeks has hardly changed during the last 100 years (Meulenbergh 1974). As far as is known, no further sediment studies have been carried out in the intertidal deposits.

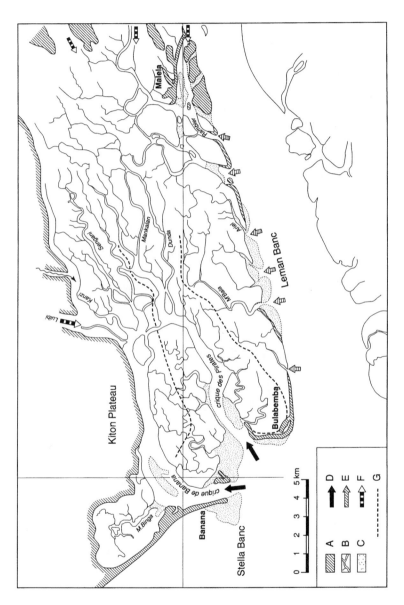

FIGURE 4.19 Channels on the north bank of the Zaire river estuary. A = mainland (supratidal); B = mangrove marsh; C = shallows; D = high salinity water; E = low salinity water (<5 in ‰ S); F = freshwater; G = marsh divides. (Adapted from Meulenbergh 1974. With permission.)

4.8.   LANGEBAAN LAGOON

At about 33° S behind a dune barrier lies Langebaan lagoon, about 17 km long and 4.5 km wide, connected with Saldanha Bay by an open channel. More than half of the lagoon is filled with bare intertidal deposits and salt marshes; the remainder consists of subtidal flats and tidal channels with a maximum depth of 8 m (Figures 4.20 and 4.21; Flemming 1977). The tides are semidiurnal with a range of approximately 1.8 m. Ebb velocities are about 20% larger than flood velocities. The time-

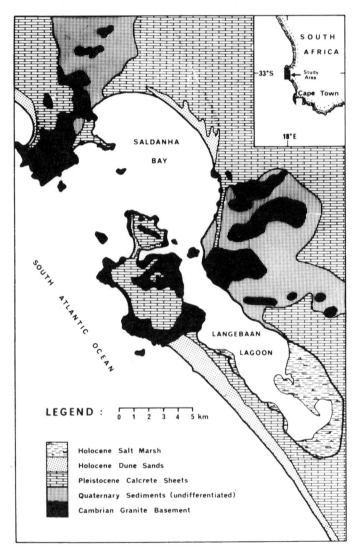

FIGURE 4.20   Location of Langebaan Lagoon. (From Heydorn and Flemming 1985. With permission.)

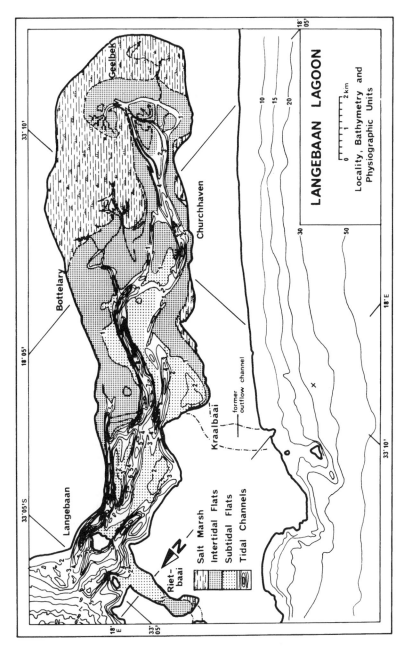

FIGURE 4.21 Langebaan Lagoon. (From Flemming 1977. With permission.)

velocity asymmetry increases inward with a short period of low current velocities around low tide and a long period around high tide, but this is absent at the entrance of the lagoon. Waves hardly penetrate inward because the lagoon is sheltered by Saldanha Bay, but winds can influence the flow in the lagoon considerably because of its shallow depth. The lagoon is divided into two longitudinal sections by a large meandering channel with secondary channels. The shoals and tidal flats as well as the channels are stabilized by fossil oyster reefs that underlie most banks and flats. Because of a time delay between the moments of high tide and current reversal (which increases inward to at least 30 minutes), major inflow and outflow paths do not coincide.

The sediment is generally sandy with about 40% calcium carbonate at the entrance. This decreases inward (southward) to less than 5%. The noncarbonate part of the sediment is predominantly quartz with rarely more than 5% other (accessory) minerals. The carbonate as well as the quartz are derived from erosion of fossil dunes. The inward decrease in carbonate is explained by selective transport (Flemming 1977). Mud content is very low as there is no river sediment supply. Only in the south in the salt marshes does the mud content increase to 5 to 10%. Bioturbation is strongest on the lower intertidal flats, especially on the west side which is sheltered from the wind. On the east side more ripples are formed. Higher on the flats runoff structures are frequent. Low (10 to 20 cm high) evenly spaced bars parallel to the main wind direction occur and extend with decreasing height into the lower flats. Near high tide level, ripples are common, with fecal pellets in the troughs. Between the upper flats and the lagoonal beaches are numerous rill marks formed by water running off *Callianassa* burrows. These rills may develop into meandering runnels ending in miniature deltas. From the burrows small sand volcanoes may be built up with extruding sediment. On the sides of the sand bars, water level marks and slump structures are common. No current-generated bedforms are present except on an intertidal sand spit in the center of the lagoon.

Erosion/reworking of salt marsh deposits results in mud pebbles, that occur both along the lagoonal beaches and in the meandering creeks or channels that undercut the marshes. Desiccation cracks are formed on exposed flats (the climate is semi-arid with hot dry summers and winter rains). Dry tidal pools extend from near high tide level to within the supratidal zone. Mud crack formation is limited by the low mud content. Salt marshes are formed by *Zostera*, which may form beds down to below mean tide level. Around mid-tide level it is replaced by *Spartina*, and toward high tide level by *Arthrocemum* and *Salicornia*, which extend well into the supratidal zone. In tidal pools, algal mats are common but not well developed, and other organisms are poorly represented because of the large salinity fluctuations (during the year up to 40‰). Bioturbation is strongest on the lower intertidal flats; its diversity increases toward the higher flats. Extensive populations of *Diapatra*, a polychaete, commonly occur along channels together with the *Zostera* beds. Upper flats are mostly characterized by surface tracks and traces of hermit crabs (*Diogenes*), small gastropods (*Assiminea, Nassarius, Littorina*), a sand shark (*Rhinobatus*), tubes of another polychaete (*Euclemene*), fecal pellets of *Callianassa*, and burrows of *Callianassa* and *Upogebia* (also a crab). *Upogebia* occurs mainly in the southern part of the lagoon. Internal sediment structures on the tidal flats include horizontal lamination in the upper 5 cm of the sediment on top of completely bioturbated sediments, but generally sediment structures are not common because of bioturbation. Besides horizontal lamination, ripple cross-lamination may occur in the upper few centimeters of the sediment, together with herringbone structures and large-scale dune crossbedding in the channel sediments.

The intertidal deposits were built up after the Flandrian transgression, which reached a peak height of about 3 m above present level, as old coves and cliffs as well as dated mollusk shells indicate. The highest level was probably reached at about 5500 BP. Since then relative sea level became gradually lower but still remained rather high until about 2000 BP. During the period of higher relative sea level most of the sediment was flushed through the lagoon, leaving a coarse lag deposit in the channels. Substantial sedimentation started around 2000 BP when the channel area merged and a former connection with the sea at Kraalbaai was filled in. Around 1700 BP sedimentation stopped, and the area became stable with only some minor fluctuations in relative sea level and local truncation of marsh deposits.

4.9. KWAZULU–NATAL

The Kwazulu–Natal coast, from about 33° S to the Mozambique border at Ponta do Ouro at about 27° S, is characterized by cliffs with small river estuaries in narrow valleys, but from Mtunzini, at about 29° S, to the north lies a northward widening coastal plain with a series of lagoons that are almost closed off from the sea by coastal barriers (Figure 4.22). The tides are semidiurnal with a mean spring tide range of about 1.80 m and a mean neap tide range of 0.5 m. Swell up to more than 7.5 m high (median 1.49 m) comes from the southeast to east, which is roughly perpendicular to the coast, while smaller wind waves come from the southwest or northeast, which is more parallel to the coast.

One of the small river estuaries, the Mtamvuna estuary at about 30° 45′ S, has been studied in detail by Cooper (1993). This river drains an area with a humid subtropical climate with episodic and seasonal discharge variations. Suspended matter concentrations in the river are generally low (2 to 41 mg · l⁻¹). The estuary is located in a gorge flanked by cliffs up to 200 m high. The adjacent coast is rocky and the shelf only 10 km wide with a thin cover of sediment. The estuary is connected with the sea by a narrow channel through a sandy barrier. Apart from the barrier and some isolated patches of sand and muddy sand, the deposits in the estuary are mud and sandy mud with up to more than 10% of organic carbon (loss on ignition). In the intertidal parts of the lower estuary mangrove is present (*Bruguiera, Avicennia*) with dense communities of crabs and mudskippers, and inward, at very low salinities, grading into reed swamps. Levees up to 4 m high are laterally restricted because of the narrowness of the river valley. Muddy deposits occur close to mean tide level and are generally bioturbated by invertebrates and plant roots. Sandy overbank deposits have thin bedding with surface trails of arthropods and small vertebrates. The barrier sands are partly intertidal or supratidal and contain up to 10% of (skeletal) carbonate with mollusk shells and abraded foraminifera with little organic carbon (less than 5%) and a graded low-angle landward dipping planar bedding. The foreshore has graded beds dipping more strongly seaward. There are a flood-tidal delta

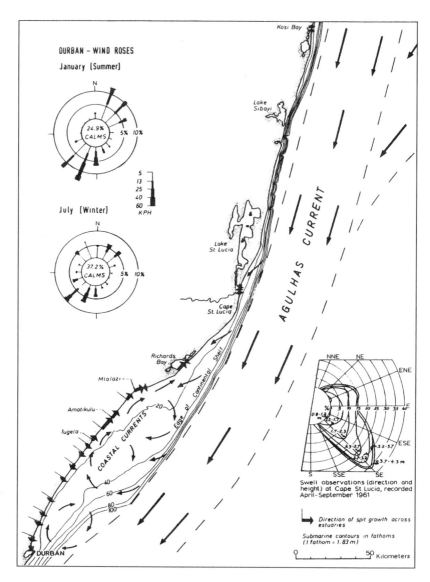

FIGURE 4.22   The coast of Kwazulu-Natal. (From Orme 1972. With permission.)

and shallow washover fans that are sparsely inhabited by crabs and prawns (*Callianassa*). The seaward side of the barrier is exposed to high waves, which has limited the formation of an ebb-tidal delta. The barrier tends to be displaced inward but was displaced seaward after a tropical cyclone (in February 1984) which caused a large rainfall, erosion in the river basin, and a large sediment supply through the river. During a very large river flood in July 1986 with peak river discharges in the order of 300 times the average flow of 8.5 m^3 s^{-1}, the barrier was completely eroded and washed out. Within about a year it was completely rebuilt.

The coastal plain that begins north of Mtunzini and broadens northward to a width of more than 80 km has been built up since the late Cretaceous with a series of beach ridges in an approximately north-south direction. The youngest beach ridge that forms the present coast dates from the Holocene; an older, partly Pleistocene ridge lies directly landward. Four large lake and lagoon systems were formed during the Holocene: the string of lakes and lagoons from Kosi Bay to Lake Kosi in the north, and Lake Sibayi, Lake St. Lucia, and Richards Bay farther south (Figures 4.23 and 4.24; Orme 1972, Heydorn and Flemming 1985). They are remnants of former extensive lagoons that were formed during the Holocene sea level rise in Pleistocene depressions connected with drowned valleys that were cut during the late Pleistocene. After present sea level was reached, the former lagoons were reduced in size by at least 60% through infill and segmentation. The latter was related to the formation of Holocene spits and beach ridges oriented approximately perpendicular to the coast, which was probably related to the development in the shallow lagoons of large circulation cells formed by the wind. The drainage outlets were increasingly obstructed by longshore sand transport along the coast and the formation of the present beach ridge. Now only Kosi Bay, Lake St. Lucia, and Richards Bay have an open and direct connection with the sea; the channel of Lake St. Lucia, until recently only seasonally open when river discharge was high, is now kept open continuously by artificial stabilization of the inlet. Richards Bay is considerably altered and stabilized by recent harbor development. The size of the lakes and lagoons may also have been reduced by a late Holocene minor lowering of relative sea level but this is not certain. Intertidal deposits are only present directly landward of the inlet of Lake St. Lucia, where they cover an area approximately 1.5 km wide.

4.10. THE PERSIAN GULF: QATAR TO OMAN

Intertidal and supratidal (sebkha) deposits occur along the south coast of the Persian Gulf between Qatar and Oman in separate embayments as well as along a 340-km-long zone between the Qatar Peninsula and Ras Granada, east of Abu Dhabi (Figure 4.25; Purser and Evans 1973, Drew 1985). The spring tide range is 1.5 m at Qatar, 0.9 m at Dubai, and 1.8 m at Oman (Musandam Peninsula). Waves are low about 75% of the time (height less than 0.9 m), and only 5 to 6% of the time are they larger than 1.5 m. There is almost no rainfall and salinities range from about 39‰ S in the open Gulf waters to over 60‰ S in coastal lagoons. There is no supply of terrigenous material except by wind, and up to 30% of the sediments may have been supplied as wind-blown sand and dust (Sugden 1963). Most of the sediment consists of skeletal carbonate detritus. The coast between Qatar and Ras Granada is in large part protected by reef barriers; the most frequent winds blow from the northwest.

Most of the coast is dominated by sebkhas which are situated just above normal high tide level and become submerged only during the higher tides and during storm surges. Low bluffs form the landward limit. Intertidal flats are present seaward of the sebkhas and at the eastern side of the "lagoon terraces" which are seldom flooded by more than 2 m of water and often fall dry at low tide. The sediment is mainly

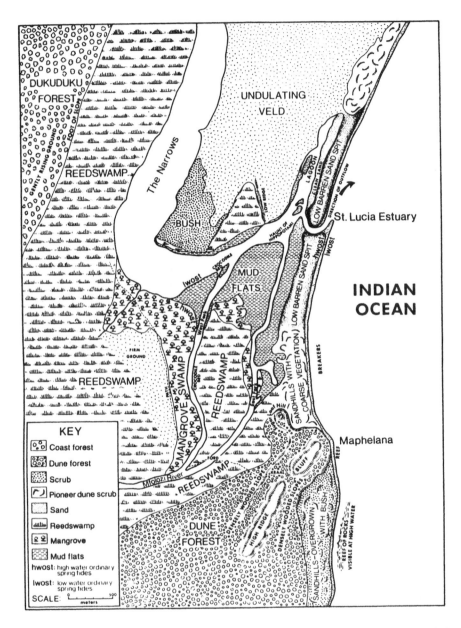

FIGURE 4.23   The St. Lucia estuary at the entrance to Lake St. Lucia. (From Orme 1972. With permission.)

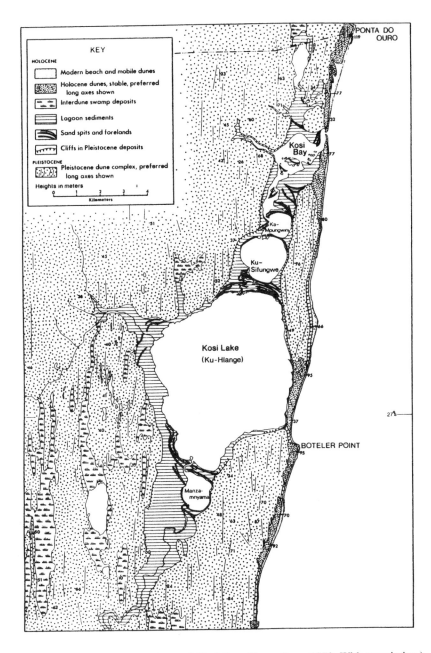

FIGURE 4.24   The Kosi Lakes and Kosi Bay. (From Orme 1972. With permission.)

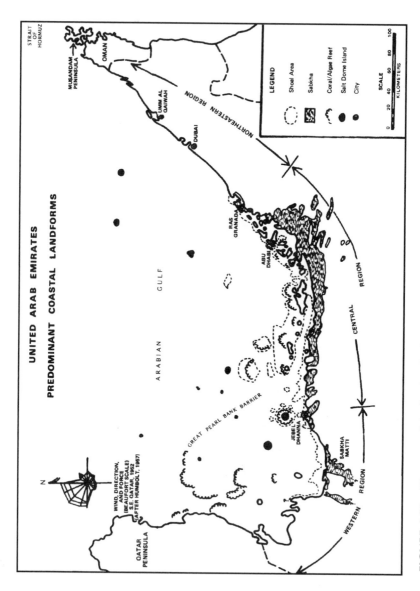

FIGURE 4.25   The coast of the Persian Gulf between Qatar and Oman. (From Drew 1985. With permission.)

unconsolidated detrital carbonate with minor amounts of quartz and other detrital minerals, and of evaporite minerals. The seaward limit of the sebkhas is usually formed by a beach ridge with older beach ridges more inland parallel to the shore. Beach ridges consist predominantly of gastropod shells. In the central region there are old strand lines and drainage channels. In the supratidal and intertidal flats, isolated low hills of Eocene dolomite crop out. These are easily eroded and fragments are dispersed over the adjacent flats.

Sebkha formation started about 7000 BP when detrital carbonate began to be deposited on top of aeolian sands. Around 4000 BP the sebkhas had their maximum extension when the (relative) level of the Gulf waters was about 0.5 m above present level. Between 4000 and 1000 BP relative sea level fell, probably because of the growth of offshore islands and reefs, which gradually reduced the tidal range and wave heights. After 1000 BP there was a regression of the Gulf waters and seaward outbuilding of the coastal deposits started. Sebkhas are flat with very low gradients (less than 0.5 m per km). Where water can rise through the sediment, pore spaces and sand grains become coated (reducing their friction), and soft quicksand can develop. Wet, soft, sticky calcareous mud is formed in the shallow coastal bays by algae and other organisms. Aeolian sand up to 1 m thick can be deposited on top of the mud. In the hot summers the mud dries and hardens. Wide intertidal shoals are present as well as narrow beaches, reefs, and islands. Those in the central area are separated from the coast by a lagoon up to 40 m deep and on its western side 140 km wide. Toward the east its width decreases to about 40 km. Mangrove swamps are small in size and consist mainly of *Avicennia* with muddy sediment below the trees, drainage creeks, and a dense population of burrowing crabs that build low mounds of sediment.

The Trucial Coast between Ras Al Khaf and Ras Granada consists of a series of lagoons with beaches on the seaward side separated by tidal watersheds with sebkhas, algal mats and aeolian dunes (Figure 4.26; Purser and Evans 1973). Eight channel systems are present. The lagoon channels are wide and almost straight with numerous, partly intertidal shoals and some wide distributary channels. The channel

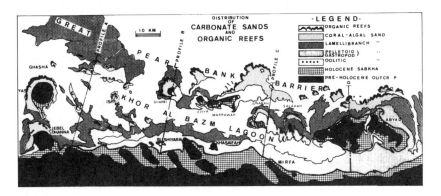

FIGURE 4.26   Reefs, sediments, and sebkhas along the central region Trucial Coast. (From Purser and Evans 1973. With permission.)

floors are sandy (skeletal debris) with ebb-tidal deltas extending into the Gulf. Between these deltas are barrier reefs. The channels are bordered by "lagoon terraces" as mentioned above, and by intertidal flats. The terraces and flats are bare and colonized by burrowing crabs as well as by surface-crawling mollusks. The sediment is generally sandy (skeletal debris of foraminifera, ostracods, and mollusks) but more landward it often has a high percentage of carbonate mud. The surface is extensively rippled with low bars with a steep face shoreward on the intertidal flats. The sediment surface becomes too hot for vegetation, so seagrass (and seaweeds) occur only from low tide level to a water depth of about 8 m. On the flats are burrowing worms, crabs, scattered bivalves, and abundant gastropods.

Along the inner margins of the lagoons the sediment surface is covered by blue-green algae that form algal mats, and locally some mangrove. Here also desiccation cracks are common. The tidal flats landward pass into the sebkhas, but usually there is a low step that marks the seaward edge of a former beach ridge along the seaward limit of the sebkha (Evans et al. 1969). On the intertidal and supratidal flats synsedimentary cementation of carbonate sediment takes place. Algal mats are often cemented with a thin crust, but in places they are also disrupted by the cementation. The surface of intertidal deltas often has become rock and forms a hard substrate for epifauna (bivalves).

Along the Qatar Peninsula coast small areas of intertidal flats up to 7 km wide are present in bays, with chenier beaches and spits in some bays (Figure 4.27; Shinn 1973). The flats are accreting. Cementation of tidal flat sediment occurs in layers of a few centimeters' thickness beneath the surface of the flats at the level of low tide in the sediment. The cement is aragonite (the sea water at salinities above 39‰ S is almost saturated with regard to aragonite), which can be replaced by microcrystalline magnesium calcite.

4.11. KERALA LAGOONS, ESTUARIES, AND MUD BANKS

Along the Kerala coast in southwest India between about 13° N and 9° N, a series of lagoons and estuaries have been formed behind low barriers. The mean tidal range is about 1 m, somewhat higher in the north and decreasing toward the south. Outflow from the lagoons and estuaries has been related to the seasonal formation of mud banks along the coast (Figure 4.28). Degeneration of laterites produce large quantities of soft mud that are released into the coastal sea by river runoff. Mud banks and laterite weathering products have a similar composition so that the banks were considered to be accumulations of laterite mud. Recent work has shown that there is probably no such relation. The location of the mud banks is not related to the outflow from rivers or lagoons, but to areas of wave convergence. The (elliptical and largely subtidal) mud banks are a seasonal phenomenon related to the monsoon, when a persistent swell develops along the coast. The rather sudden increase of persistent pressure variations on the seabed when the monsoon begins results in resuspension of bottom muds into a suspension of high concentration, which is moved shoreward over the bottom and becomes concentrated in areas of wave convergence.

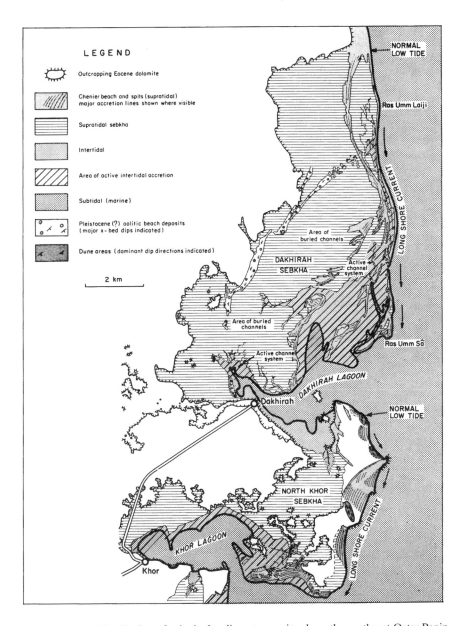

FIGURE 4.27 Distribution of principal sedimentary units along the northeast Qatar Peninsula. (From Shinn 1973. With permission.)

The mud suspension is highly viscous, wave energy becomes increasingly absorbed in the suspension, and waves are damped. Hindered settling, the result of the high suspended-matter concentrations, prevents rapid settling of the mud. As long as the swell persists, new mud is resuspended farther offshore, while near shore some mud settles. Once formed, the mud banks become highly stable, although it is a high-

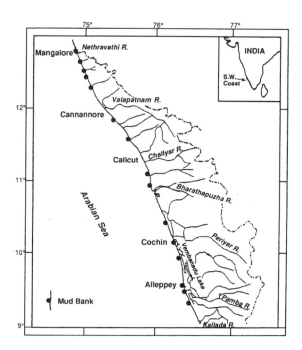

FIGURE 4.28   Mud banks along the Kerala coast. (From Mathew et al. 1995. With permission.)

energy environment. When the monsoon ends, resuspension by waves does not supply enough sediment anymore to maintain the mud banks. Concentrations decrease, more turbulence becomes possible and low-density suspensions are formed that flow downward over the near-shore seafloor. The mud settles in deeper water and is accumulated in semiconsolidated deposits for about 9 months until the next monsoon starts. The mud banks result in accretion of the coast, with erosion where the coast is not protected by the mud banks (Moni 1970; MacPherson and Kurup 1981; Mallik et al. 1988; Matthews et al. 1995a, 1995b; Faas 1995).

4.12. DELTAS, EAST COAST, INDIA

Between the Ganges delta in the north and Sri Lanka in the south four major river deltas are present along the east coast of India: those of the Mahanadi, the Godavari, the Krishna, and the Cauveri rivers (Figure 4.29). The first three have an arcuate shape with most of the coast formed by sand ridges and spits. The Cauveri delta has a straight north-south front of sandy ridges formed by the northeast monsoon waves, which are large because of the strong winds and the long fetch. The tidal range decreases from north to south. At the Mahanadi delta, there is a spring tide range of about 2 m and a neap tide range of 1.2 m; and at Madras the spring tide range reaches 1.2 m and the neap tide range, 0.6 m (Nagaraya 1965; Ahmad 1972). Only at the Godavari delta are intertidal flats of more than local extension present

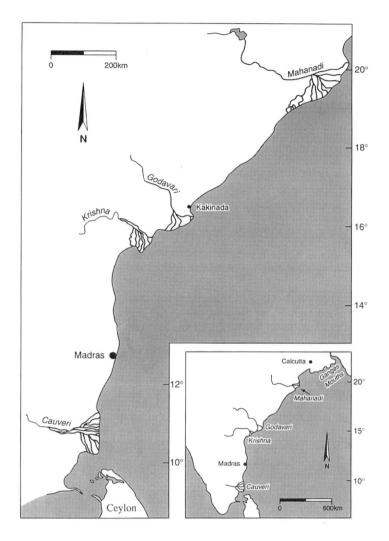

FIGURE 4.29   Deltas along the east coast of India.

behind a spit at the south side of Kakinada Bay and in smaller areas at the main river mouth and farther south along the coast at Pandi lagoon. The flats are covered with mangrove. At the other deltas, only very small areas with mangrove are present (Rao and Vaidyanadhan 1979; Figure 4.30).

4.13. SRI LANKA

The island of Sri Lanka lies in a quiet sea area with monsoon winds that rarely reach gale force and with a typhoon only once in 10 to 15 years affecting only the northern part. The spring tide range is within 1 m (0.6 to 0.8 m); the highest tide ever recorded was 1.32 m (Abeywickrama 1965; Swan 1985). Intertidal deposits are mostly of

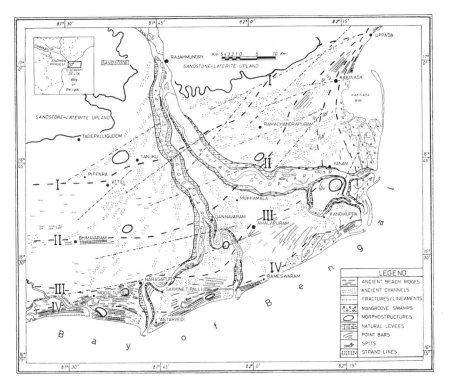

FIGURE 4.30 The Godavari river delta. (From Sambasiva Rao and Valdyanadhan 1979. With permission.)

limited extension and occur in lagoons and river mouths. These are usually covered with mangrove. Wide tidal flats open to the sea are present in the northwest in the Gulf of Mannar and in Palk Bay, where wave energy is very low; flats of smaller extension occur in the lagoons farther south down to Colombo and along a series of lagoons on the east coast (Figure 4.31). Salt flats, only flooded during high spring tides, are hypersaline. They are widespread in the dry zones in the north and northwest and along lagoons in the southeast. The vegetation consists of halophytic herbs (mostly grasses, low shrubs, and algae). In dry areas there are also some intertidal salt marshes, which were probably initiated when a storm damaged the mangrove, but also may have been caused by man's clearing the mangrove. When the flats became exposed to the sun and dry air, the salinity increases resulting in hypersaline conditions that are not suitable for mangrove. During prolonged dry periods, salt crystallization occurs and the sediment surface becomes encrusted.

4.14. GULF OF THAILAND

Along the western shores of the Gulf of Thailand the spring tide range varies from 2 m at Pratchuap Khiri Khan to 0.7 m at Songkhla and up again to 2 m in the south of Malaysia. The neap tide range varies between 0.6 and 1.1 m (Pitman 1985; Teh

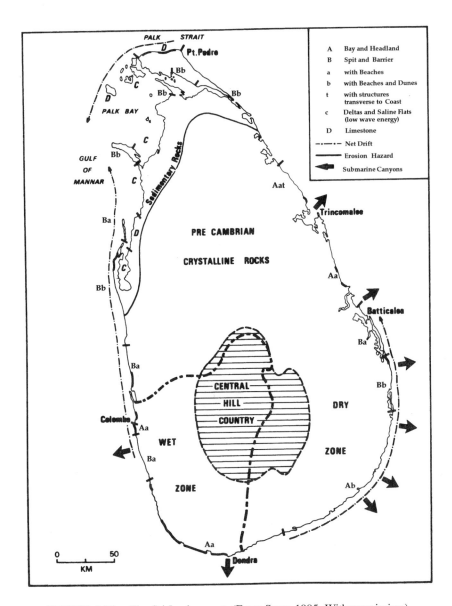

FIGURE 4.31   The Sri Lanka coast. (From Swan 1985. With permission.)

Tiong Sa 1985). Winds from December to April come from the north and east, between May and October from west and south. Longshore transport is not large and most sediment is stored in bays and lagoons. Mud flats, covered with mangrove where not cleared for fishponds and agriculture, occur in bays and lagoons and can locally be rather extensive, in particular where they are associated with river deltas (Figures 4.32 and 4.33). Toward the south in Thailand and in Malaysia the coast consists of sandy ridges, and intertidal deposits are rare.

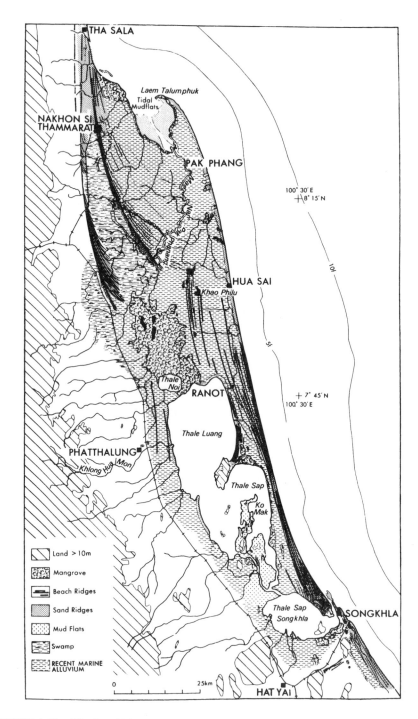

FIGURE 4.32   The Songkhla–Tha Sala coast along the western side of the Gulf of Thailand. (From Pitman 1985. With permission.)

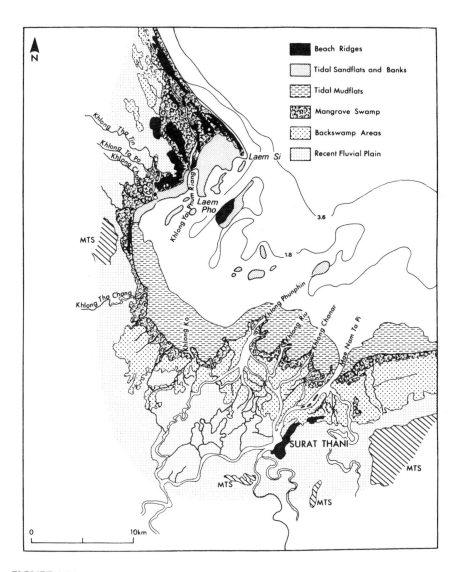

FIGURE 4.33 The Ao Ban Don embayment along the western side of the Gulf of Thailand. (From Pitman 1985. With permission.)

4.15. BOHAI SEA AND HUANG HE (YELLOW RIVER) DELTA

Deposition along the Bohai Sea is strongly related to the sediment supplied from the larger rivers: the Huang He or Yellow River, and the Liao He, Luan He, and Hai He rivers. The supply from the last three is small compared to the supply from the Yellow River but can be distinguished from it by a different mineral composition. Tidal flats are present along Bohai Bay (the western embayment of the Bohai Sea) in an approximately 40-km-long stretch of coast south of the Hai He river mouth, along the Yellow River delta, and in Laizhou Bay south of the Yellow River delta, which borders the Shandong Peninsula. Mud flats are also present north of the Hai He river mouth and around the Liao He river mouth, but these have hardly been studied. The flats around the Liaodong Peninsula consist of bare fine-sandy low flats with wave and current ripples, a central zone with benthos and a *Zostera* vegetation, and muddy high flats with *Suaeda* (Li et al. 1986).

The tidal range in Bohai Bay and around the Yellow River delta is about 1 m but increases northward toward the Liao He river mouth to more than 4 m. The Yellow River is characterized by a relatively low annual discharge (about 4.2×10^{10} m $3 \cdot y^{-1}$) which is less than 1% of the annual discharge of the Amazon, but with an annual sediment discharge about equal to, or 10% more than the Amazon. The sediment comes mainly from the loess area in north China in concentrations that at present reach up to a kg $\cdot$ l^{-1}. These high suspended-matter concentrations date from the Tang dynasty (618 to 907 AD) when the wooded grasslands in the loess area were increasingly converted into farmland, a process that had begun around 5000 BC when a hunting and pastoral life was changed into a more sedentary life based on agriculture (Ren and Zhu 1994). From about 3400 BC to 1128 AD the river mouth was located north of the present delta, and another delta (the ancient delta) was formed up to Tianjin. In 1128 AD the river dikes were breached for defense reasons, and the river started on a more southerly course: between 1158 and 1855 the river built up a delta south of the Shandong Peninsula on the Jiangsu coast, and the sediment was discharged into the Yellow Sea (see Section 2.11; Ren 1992, Ren and Zhu 1994). This delta has been eroded since 1855 when the river changed its course again to the north and the present delta was built up. Within the present delta the course of the river has wandered over about 180°, to which the many old river beds testify (Figure 4.34). Between 1953 and 1976 the river mouth shifted five times, or about every four to five years on the average. Since 1855 twelve major shifts have taken place, or one every 10 years. The present outflow is directed eastward. Around the mouth, sediment, mainly silt, is rapidly deposited in the form of spits, intertidal flats, and supratidal flats at a rate of 7 to 8 km $\cdot$ y^{-1}. When the river changes its course and the mouth shifts to another location, the spits and flats are eroded and the sediment is dispersed sideways so that the delta maintains a roughly arcuate shape with an irregular indented coastline which is the result of wave erosion on the northeastern side. The changes in the river outflow strongly influence the inter-tidal mud flats along Bohai Bay. The change in 1976 from a more northerly to a more easterly outflow, caused the sediment to go mainly to the south instead of to the northwest, which resulted in erosion along the Hai He river delta (Ren ed. 1986).

Along the Yellow River delta wide mud flats are formed along the interdistributary bays, called "mudbays," which are only 2 to 3 m deep and have a quiet environment. The bottom is covered with a layer of fluid mud 1 to 5 m thick. When a river mouth is abandoned, however, erosion begins and is particularly severe on the northeast side of the delta which is exposed to the strongest winds. At the former Diakou river mouth, which was abandoned in 1976, nearly 6 km was eroded between 1976 and 1981, while at the new river mouth the coast prograded 12 km in only one year (May 1976 to May 1977). Behind the two spits that border the present

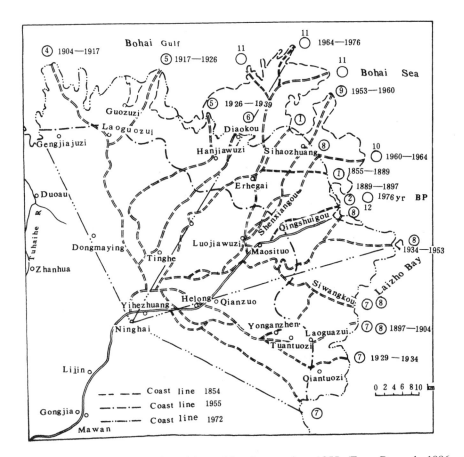

FIGURE 4.34 The Yellow river delta and its changes since 1855. (From Ren, ed., 1986. With permission.)

mouth, the mud flats on the north are 7 to 9 km wide and on the south side 4 to 5 km with a gradient of 0.45%. Supratidal swamps, only flooded during an exceptionally high tide and consisting of very soft, poorly consolidated mud, extend for another few km landward, covered with a vegetation of *Phragmites* and *Suaeda*. The mud flats farther south in Laizhou Bay are smaller and less prograding, or stable, because they receive less sediment than the delta: the flats in Shouguang County consist of 1.8 km of supratidal flat seaward of a dike that consists of silty mud with a smooth surface covered with vegetation (*Suaeda*), of 2.1 km of upper intertidal flats of muddy silt with small depressions on the surface, and 3.1 km of lower intertidal flats consisting of parallel laminated silts and fine sand. This surface is flat, covered with ripples and abundant crab burrows.

The flats to the north of the delta up to the Hai He river mouth have an average width of about 3 km with sediment that becomes somewhat finer in a landward direction. The flats are prograding; the flood velocity is larger than the ebb velocity. Seasonal variations are related to the frequency of storm winds. Mud bastions are formed in summer when winds and waves are stronger, and built up during the winter. They are several decimeters high and 3 to 18 meters long; after being eroded during a storm, they are buried below mud that is deposited when the waves subside. The supratidal flats are much smaller than along the delta and at Laizhou Bay, and have a width of only a few hundred meters. On the landward side they are bordered by a dike. The upper intertidal zone is about 700 m wide and consists chiefly of fine mud (5 to 10 micron). A middle intertidal zone is about 500 m wide and consists of coarse silt with shell fragments (Md more than 50 micron) interlayered with fine mud. The lower intertidal zone, 1.2 km wide, consists mainly of coarse silt and has many ripple marks on the surface. The zonation is most distinct in spring; in summer the mud flat surface is much reworked by waves. Around the Hai He river mouth the flats are retreating. The Hai He delta had been prograding until 1958 when the sediment supply from the river was reduced to less than 15% by the construction of a tide gate. Erosional scarps were formed with mud boulders along the front of the scarps. Following a further reduction of sediment supply because of the southward shift of the Yellow River mouth in 1976, the Hai He river delta was further eroded, aided by crustal downward movements and, at present, by subsidence because of ground water extraction (Ren 1993).

When a mud flat receives a reduced amount of sediment, such as occurs as the Yellow River mouth changes its position, a chenier can be formed, such as the one along the old Yellow River mouth along the Jiangsu coast (see Section 2.11). In the absence of sand along most of the Chinese coast, a chenier is usually formed out of mollusk shells of species that live near shore. Where the gradient of the flats is less than 1% (as is normal around the Yellow River delta), wave action is very weak and no chenier is formed, but older cheniers are present north of the delta, as well as along the delta itself, buried below 3 m of mud deposited on top of several decimeters of shells and silt. The mud dates from after 1855, and the most recent chenier dates from the period between 1194 AD and 1855 when the Yellow River mouth was south of the Shandong Peninsula. Older cheniers date from before 3400 BP (when for the first time the Yellow River shifted its mouth north to the area around Tianjin) and from 602 to 132 BC when the river mouth was again located south of the peninsula, as well as in 1194 to 1855 AD.

4.16. WEST AUSTRALIA

Along the coast of west Australia, which reaches from Eucla in the south to Bonaparte Gulf in the north, only the part up to Carnarvon north of Shark Bay on the west coast is microtidal. Here the spring tide range decreases from 2 m at Carnarvon to about 1 m around Perth and to less than 1 m along the south coast up to Eucla. Intertidal deposits and lagoons occur only along Shark Bay on the west coast and between Point Malcolm and Israelite Bay on the south coast (Figures 4.35 and 4.36; Woods et al. 1985). Shark Bay, between approximately 25° S and 27° S, with a mean tidal range of 0.5 to 1.0 m and a spring tide range of 1.0 to 1.8 m, consists of two bays. Where the coast is not rocky or the salinity of the sediment is not too high, the bays are bordered by tidal flats with mangrove. Older (Pleistocene) flats occur along the east side. Swell comes from the south-southwest from the

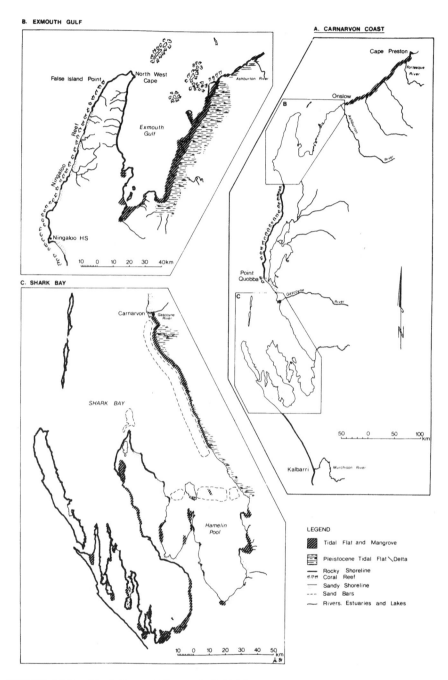

FIGURE 4.35   Shark Bay and Exmouth Gulf (West Australia). (From Woods et al. 1985. With permission.)

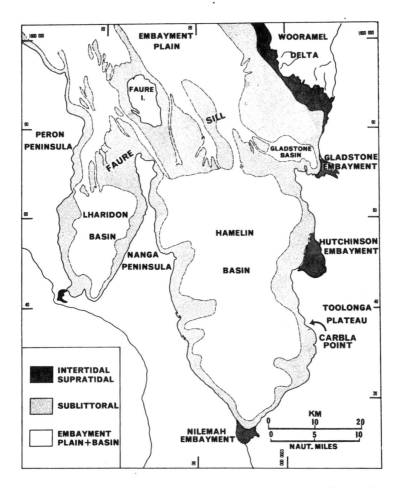

FIGURE 4.36   Basins and embayments at the southeast of Shark Bay. (*Source*: Brown and Woods 1974.)

Southern Ocean, while local waves are generated by strong southerly winds, but the inner parts of Shark Bay are sheltered against these waves. Evaporation is high because of the dry climate, and considerable carbonate deposition occurs where the water becomes hypersaline, as in Hamlin Basin. The supratidal flats along Shark Bay are bordered by shelly beach ridges (cheniers) that consist of mollusk shells formed in the subtidal zone and transported landward by storm waves. There are shallow channels, with a channel floor of carbonate mud, bivalve shells, and debris from older (Pleistocene) deposits. The flats have indurated crusts that may be broken up by internal pressure. The intertidal flats consist of flat pebble brecchias, fine sand, and fragments of aragonitic crusts, that consist of finely cemented skeletal pellets. The upper parts of the flats have desiccation cracks and gypsum crystals, aragonitic pellets, and algal mats with laminated sediment. The middle and lower tidal flats have well-laminated algal-bound sediment (stromatoliths, algal mats).

Intertidal algal mats develop where the salinity of the water is too high for mangrove to grow. Six structural types of algal mats have been distinguished (Logan et al. 1974), based on their size (from a few centimeters to several meters), morphology, and internal structure. Where the intertidal zone has a gentle gradient, there is a broad zonation of algal mat types; on headlands and in areas with an irregular topography, the algal mats are highly differentiated with a patchy development. The distribution is mainly determined by the elevation of the substrate, drainage, depth and nature of the interstitial ground water, and sediment influx, Primary depositional features, as described from the Gladstone and Nilemah embayments (Figure 4.37; Davies 1970), include algal filament molds, graded bedding, cyclic bedding, and storm features. The algal molds form the framework around the sediment particles; graded bedding reflects one tide (diurnal or semidiurnal). The cyclic bedding probably reflects seasonal flooding followed by regression: thick basal laminae are overlain by

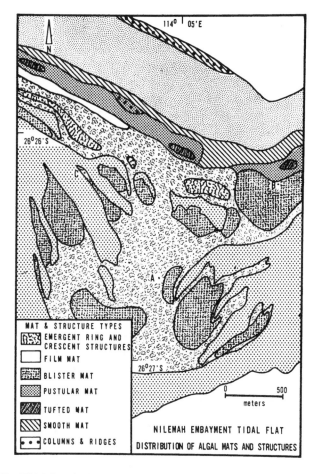

FIGURE 4.37   Tidal flats in the Nilemah embayment. (*Source*: Hagan and Logan 1974.)

a series of laminated algal sediment with the laminae decreasing in thickness toward the top. The top layer consists of several closely packed algal laminae (algal "peat") with little sediment. The basal layer is about 10 cm thick with 1.5 cm of laminae on top of it and up to 5 cm of thinly layered algal peat. The flooding that starts a cycle probably occurs during exceptionally high tides or during strong northwest winds.

Secondary (postdepositional) features of the algal mats are related to a variety of factors: desiccation, burrowing or grazing organisms, gas formation, enrollment by currents and waves, evaporation, and formation of crusts. Gas formation results in gas domes and a fenestral fabric. Crusts, pavements, deposits of flat pebbles consisting of conglomerate and wall-rock (spur and groove structures) occur on the inner areas of the intertidal zone and on the supratidal flats. The latter consist of an upper part, between about 1.6 and 2.6 m above high spring tide level, which is never flooded, and a lower part between high spring tide level and 1.6 m above it, which is occasionally flooded (in the order of once in 2 to 3 years). Storm deposits consist mostly of mollusk shells and fragments.

Surface crusts are cemented by aragonite, but also gypsum, halite, and perhaps dolomite can be formed. The crusts can become overgrown with a film of filamentous algae and some sparse vegetation (*Salicornia, Arthrocenum*; Logan and Cebulski 1970; Davies 1970; Logan et al. 1974).

The present flats developed after about 5100 BP when sea level reached its present position, but their formation may have started earlier around 7415 BP, when sea level was only about 1 m lower, but no data are available for Shark Bay. Deposition in Hutchinson Bay started when sea level was probably +2 m and continued while it was lowered to its present level. In Hutchinson Bay the modern intertidal and supratidal flats form a narrow wedge, up to 1.5 m thick, over earlier Holocene and Pleistocene deposits. During deposition of these formations sea level was between +1 and +2.5 m, and the sediments indicate transgressive conditions with only one thin layer of probably regressive intertidal deposits in the Pleistocene. During the latest Pleistocene depositional period, marine sediments were formed at normal marine conditions. Since then, during the Holocene, salinity increased from 35 to 45‰ S to the present 72 to 210‰ S. Present intertidal deposits in Hutchinson Bay consist of regularly laminated, unconsolidated to partly consolidated, fine-grained grey-to-white sediments with ovoid pellets. They lie on top of sublittoral deposits and similar deposits without internal layering that contain more than 75% aragonitic pellets. These are partly covered with stromatolithic-type deposits and, in a more than 1-km-wide zone, with evaporite deposits (mainly sediments containing gypsum). A similar sequence is present in the Nilemah Embayment at the southern tip of the Hamelin Basin. Here the Holocene-Recent sediment wedge is less than 50 cm thick, and the Pleistocene marine sequences are separated by calcareous soils (Hagan and Logan 1974; Brown and Woods 1974). Holocene tectonics may have regionally influenced relative sea level to a large extent (Semeniuk and Searle 1986).

Along the south coast, between Point Malcolm and Israelite Bay, at approximately 33° S, a 12-km-long lagoon/salt marsh lies parallel to the coast behind a sand barrier and is occasionally connected with the coastal sea. Local waves come from the southwest to southeast, and swell from the southwest. Because of the dry climate, salinities in the lagoon/salt marsh tend to be high.

4.17. SOUTH AND EAST AUSTRALIA: SOUTH AUSTRALIA, VICTORIA, AND NEW SOUTH WALES

Along the South Australia coast at about 35° S, there are two large bays, Spencer Gulf and Gulf St. Vincent, that extend up to 270 km inland with tidal flats and mangrove (Figure 4.38). Along the Victoria coast farther east at 38° S are Port Philipp Bay and Westernport Bay, which are much smaller, while northeast of Wilsons Promontory, Corner Inlet and the Gippsland Lakes are located, and farther north, along the coast of New South Wales, a series of small bays can be found. The tidal range along the entire coast is less than 2 m, except for a small area around Wilsons Promontory in Bass Strait, where the tidal range reaches 2.7 m. Within some of the bays the tidal range increases inward (in Spencer Gulf up to 3 m, in Westernport Bay to 3.3 m), in others it remains about the same (as in Corner Inlet where the tidal range is 2.4 to 2.7 m) or decreases (in Port Philipp Bay from 1.8 to 0.9 m, in the Gippsland Lakes from 0.9 m to almost 0, and in the New South Wales bays; Twidale 1985; Bird 1985; Thom 1985; Harbison 1984). Swell generally comes from the south and southeast and occasionally from the east, generated by a tropical cyclone. The bays are generally sheltered against swell. Local waves come from the southeast and north.

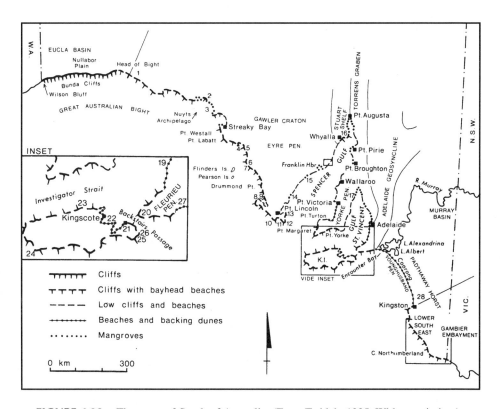

FIGURE 4.38   The coast of South of Australia. (From Twidale 1985. With permission.)

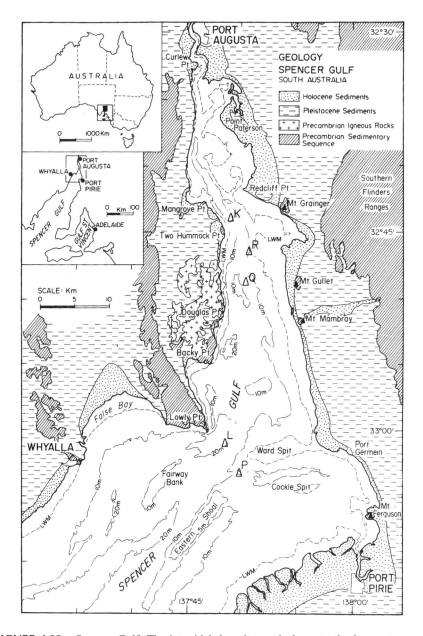

FIGURE 4.39   Spencer Gulf. The intertidal deposits reach down to the low water mark (LWM). (From Gostin et al. 1984. With permission.)

Spencer Gulf is bordered by scattered tidal flat areas, which are concentrated (and with a larger area) inside the bay north of Whyalia, where the tidal range is largest (Figure 4.39). They consist mostly of coarse to fine sand with mats of blue-green algae and mangrove on the higher parts. The climate is warm temperate with low rainfall and high evaporation. There is virtually no terrestrial runoff; channels are tidal and sediment in large part consists of biogenic skeletal carbonate (from bivalves, gastropods, foraminifera, echinoderms, coralline algae, and bryozoa) with a minor terrigenous component from erosion. The intertidal flats below mean sea level are bare or covered with seagrass; from mean sea level up to the level of spring high tide a dense mangrove with muddy sediment, which is only moderately calcareous, covers the flats. It grades inland into an open vegetation of halophytes and saltbush (*Atriplex*) with abundant algal mats which extend into the mangrove. The supratidal flats are bare carbonate flats with beach ridges and coastal dunes (Harbison 1984; Gostin et al. 1984).

Port Philipp Bay has a small entrance; the entrance to Westernport Bay is partly closed off by an island. Along Port Philipp Bay salt marshes and mangrove occur in small areas, but most of inner Westernport Bay the coast consists of extensive salt marshes fringed by mangrove. The salt marshes cover the upper intertidal flats with the mangrove fringe on the seaward side. The lower flats are mostly bare. While on the lower flats accretion and erosion alternate, accretion is sustained under the mangrove where sediment deposition is enhanced by the pneumatophore system of *Avicennia*. A depositional platform is formed up to high spring tide level, where the mangrove is succeeded by salt marsh. Where the mangrove dies or is cleared away, the terrace degrades into an intertidal wave platform. When the mangrove returns, the terrace is built up again (Bird 1985, 1986). Also Corner Inlet is mostly bordered by salt marshes and mangrove. The Gippsland Lakes are a lagoon that has been connected to the coastal sea by an artificial entrance since 1889. The former reed swamps since then have partly changed into salt marshes (Figure 4.40).

Along the coast of New South Wales are numerous bays and embayments, drowned during the Holocene sea level rise and partly filled in since then with sandy barriers, tidal flats, deltaic plains, and lagoons. Waves are exceptionally high along this coast: ocean waves are not much reduced in size before they reach the coast because of the deep and narrow shelf (Wright 1976). Waves frequently are higher than 4 m and storm waves may reach 16 m. There is a general northward longshore

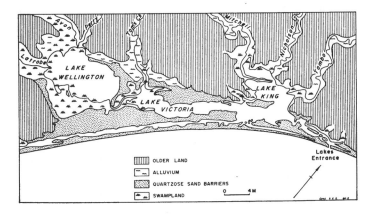

FIGURE 4.40   The Gippsland Lakes. (From Jennings and Bird 1967. With permission.)

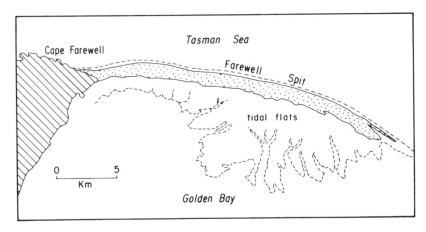

FIGURE 4.41   Farewell Spit, South Island, New Zealand. (From McLean 1978. With permission.)

sediment transport, which in the south is interrupted by headlands and deep estuary mouths, but farther north can continue, bypassing many embayments. The coast has remained stable after present sea level was reached about 6000 BP, with minor fluctuations up to about 1 m. Many bays are open, strongly influenced by the coastal waves, and mostly filled in with sand. Other bays are almost closed by sand barriers with a lagoon or floodplain behind the barrier. A third type of bay is a drowned river valley with fluvial processes dominating upstream. Local wind waves are insignificant because of the large water depth and the limited fetch. Here relatively small intertidal areas are present, fringed with salt marshes (Roy et al. 1980; Roy and Thom 1981; Thom 1985).

4.18. NEW ZEALAND

The tidal range along the New Zealand coast is 2 to 4 m on the west and north sides, and below 2 m along the southeast side (Davies 1972). Swell comes mainly from the south and west. Tidal flats occur in estuaries and these usually consist of muddy sand, and are behind spits, such as Farewell Spit on South Island (Figure 4.41). The estuarine flats in the extreme north of North Island are mostly covered with mangrove, their most southern occurrence. Introduction by man has brought the limit farther south than the natural one. Salt marshes and *Spartina* are common in all estuaries; shell banks are present on many tidal flats (McLean 1985).

4.19. PACIFIC ISLANDS

Most islands in the Pacific are volcanic (with fringing reefs) or atolls with a central lagoon. Intertidal deposits are only described from two large islands: New Caledonia at 20° S to 23° S, and Fiji at 16° S to 19° S (Bird and Iltis 1985; Bird 1985). The tidal range around these islands is less than 2 m; around New Caledonia it is mostly less than 1 m. Cyclones may pass over the islands; local waves are generated by the

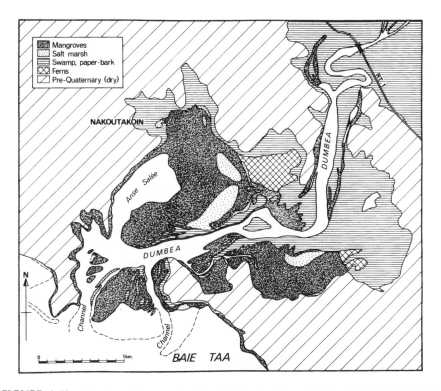

FIGURE 4.42 Dumbea river mouth, New Caledonia. (From Bird and Iltis 1985. With permission.)

trade winds and come from the southeast. Both islands are surrounded by fringing reefs; behind the reefs, shoals, algal deposits, and sandy flats with mangrove have been formed. Intertidal deposits occur in river estuaries and in bays located near a river mouth, from which sediment is supplied. At Lauthala Bay in Fiji on Viti Levu extensive tidal flats have been formed; on New Caledonia at the Dumbea river mouth extensive mangrove flats are present with salt marshes on the supratidal flats (Figure 4.42). Similar mangroves with salt marshes occur in river deltas on Fiji (Figure 4.43). In sheltered areas tidal flats with mangrove can develop rapidly, as at the Ne'poui delta in New Caledonia, where mud is accumulating as a result of mining operations upstream; between 1954 and 1976 tidal flats and mangrove advanced more than 400 m. At Mara in New Caledonia salt flats have been formed over dead coral reef.

4.20. ALASKA

Along Alaska (Figure 4.44) the tidal range is less than 2 m, and usually less than 1 m. It increases along the west coast to more than 2 m in Kuskokwim Bay and Bristol Bay. Along the north coast east of Point Barrow, a series of shallow lagoons extends to the Canadian border, interrupted by stretches of low cliffs and separated

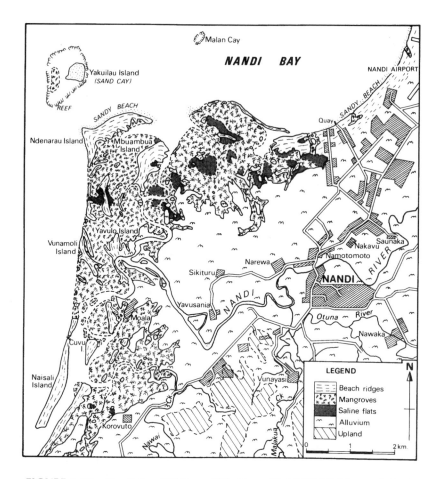

FIGURE 4.43   Nandi Bay, Viti Levu, Fiji. (From Bird 1985d. With permission.)

from the open sea by low barrier islands with wide inlets. To the east, in Canada, lies the large Mackenzie river delta. Around Point Barrow are flats and marshes and, farther to the west, a number of large bays and the Colville river delta as well as a number of smaller deltas. Lagoons continue westward up to west of Point Lay, at Point Hope and along Kotzebue Sound up to the most westerly point at Wales. Along the more sheltered parts of the Mackenzie river delta and along Kotzebue Sound, large salt marshes occur, but generally salt marshes are badly developed along the north coast of Alaska, where sediment and vegetation are regularly disturbed by ice. They contain numerous ice wedges, polygons, and thaw lakes. South of Seward Peninsula, embayments with flats and a few small lagoons occur along North Sound. Here on the south side, the large Yukon delta is located, which merges with the Kuskokwim river plain farther south. Both have extensive marshes and lakes, and at the Yukon delta interdistributary mud flats have been formed along Norton Sound (Walker 1985).

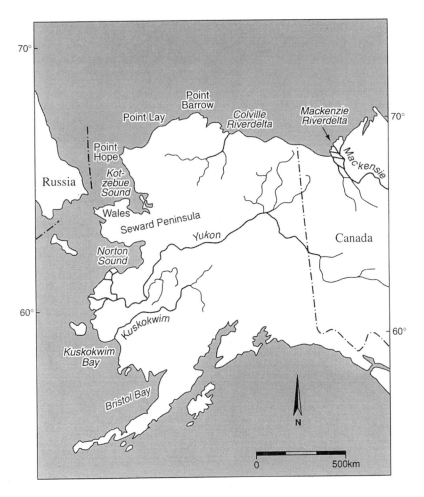

FIGURE 4.44 Alaska.

4.21. MAR CHIQUITA

Mar Chiquita is a lagoon on the Argentina coast at about 37° 40′ N (Figures 4.45 and 4.46). Several small rivers discharge freshwater into it, while sea water enters through a narrow inlet; a sandy barrier almost closes off the lagoon. The tidal range at the inlet is about 1 m and inside the lagoon is reduced to only several decimeters. The zone of intertidal sediments is therefore very small and consists of bare sandy mud with burrowing crabs. Supratidal flats with *Spartina* and *Salicornia* are separated by a low (40 to 50 cm) step from the intertidal flats. At the entrance there is an "inferior" and a "superior" upper marsh; the latter has a thick deposit of aeolian sand supplied from the coastal dunes. The lagoon is assumed to have been formed after a Holocene high-sea level period around 5000 BP, when relative sea level was about 2 to 2.5 m higher than at present. A rapid and minor fall of sea level led to

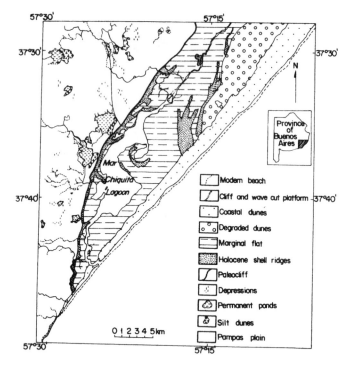

FIGURE 4.45   Mar Chiquita Lagoon. (From Schnack 1985. With permission.)

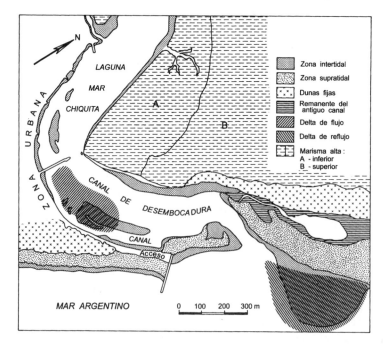

FIGURE 4.46   Inlet of Mar Chiquita Lagoon. (From Fasano et al. 1982. With permission.)

the buildup of a sandy barrier, which restricted the in- and outflow of the former estuary and a lagoon was formed. Because of northeastward longshore sediment transport, the inlet was displaced to the northeast and the inlet regularly became obstructed. Shortcuts through the barrier were made in 1904 and 1912 (Fasano et al. 1982).

4.22. VENEZUELA–COLOMBIA

In northern Venezuela the coast is a desert or semidesert up to west of Maracaibo at more than 10° N. Farther to the west the Magdalena river forms a broad delta with intertidal deposits that are largely covered with mangrove. The tides are mixed with a range less than 2 m; waves are usually low except during the passage of hurricanes, which rarely happen at these latitudes. Mangrove dominates, besides along the Magdalena River delta, along the adjacent bays and in the shallow coastal areas near Maracaibo and Coro on the north coast of Venezuela. Along the dry parts of this coast, salt flats occur that are occasionally flooded and are often covered with freshwater during the rainy season. In addition to the Orinoco delta, there are small river deltas such as the delta of the Rio Mitare.

4.23. THE LOUISIANA COAST–MISSISSIPPI DELTA

Along the western part, the Louisiana coast consists of a chenier plain that extends up to Vermillion Bay, and along its eastern part (up to the Chandeleur Islands) of the Mississippi deltaic plain with active and abandoned deltas (Figure 4.47). The chenier plain, up to 15 km wide, consists of a series of cheniers that began to form around 2800 BP with flat mud deposits between (and below) the cheniers (Figure

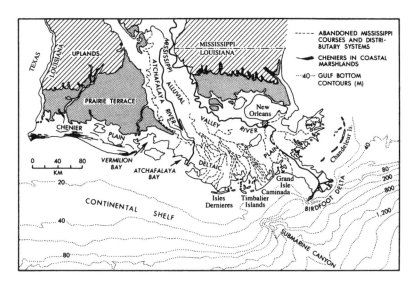

FIGURE 4.47   Louisiana–Mississippi coast. (From Nummedal et al. 1985. With permission.)

4.48). Former shorelines are indicated by the cheniers and by surface lineations. The mud flat deposits consist of clayey silt with an abundance of root fibers and often authigenic pyrite. Marsh deposits, on top of the mud flat deposits, cover most of the chenier plain and continue to accumulate between the cheniers. They are formed in brackish water (Byrne et al. 1959). The present mud flat progradation, which started recently, is related to mud supply from the Atchafalaya river. This river, partly a distributary of Mississippi river water, started to build up a subaquaceous delta before 1960. This then became subaereal during the high river discharges of 1973 to 1980 (Van Heerden et al. 1983). Normally the higher levees and the lower areas between these levees (back-bar flats) are intertidal and consist of parallel laminated silts and clays, and algal mats. On the back-bar flats, sand is deposited in cross-laminated lobes and as sheets only a few centimeters thick by levee-overtopping and weak tidal currents. Within the silt and sand beds the algal mats are interbedded as reduced organic layers (van Heerden et al. 1983). Mud, supplied from the river, is moving westward along the Louisiana coast in a zone up to 20 km wide, as has been observed by airplane and satellite. Part of this mud is accumulating on the western Louisiana coast as a gel-like fluid. Incoming waves are damped because of the high suspended-matter concentrations, which favors further sedimentation. The volume of sediment that moves westward is about half the amount discharged by the Atchafalaya river and is about one order of magnitude larger than can be accounted for in yearly mud flat accretion (Van Heerden et al. 1981; Wells and Kemp 1981; Wells and Roberts 1981). The cheniers consist of sand and mollusk shells that have been winnowed from marsh and tidal flat deposits. They are assumed to be correlated with a more easterly outflow of the Mississippi river. When the outflow is more westerly, mud is transported westward along the Louisiana coast and deposited in front of the youngest chenier: mud flats and marshes were formed in periods of coastal progradation, cheniers in periods of coastal stability or retreat (Byrne et al. 1959; Gould and McFarlan 1959). However, the corre-lation of the cheniers with the Mississippi outflow as indicated by the different delta lobes described by Kolb and Van Lopik (1958) is not clear.

In the Mississippi delta the different lobes formed since about 5000 BP reflect the shifting position of the river mouth (Figure 4.49). The shifts started upstream and occurred when the increasing meandering of the river in its lower course led to an increasing length of the river and a smaller gradient. This caused deposition of sediment on the river bed which raised the bed above the level of the adjacent plain. Normally the river remained within its levees, but when the river broke through a levee during a high flood, it could shorten its course to the sea considerably by a new route. The river plain is subsiding (and has done so in the past) because of crustal downwarping under the deltaic sediment load, consolidation of the sediment, which resulted in a smaller sediment volume, and local tectonics. The subsidence decreases inland and with the increasing age of the deltaic lobes. The tidal range of the Mississippi delta is low: 15 to 61 cm. Easterly winds prevail but waves remain low because of the shallow depth and the flat offshore slope. Waves become locally higher where the water is relatively deep and the slope steeper. Hurricanes, occurring on the average once every four years, produce "wind tides" flooding the marshes along the coast with up to 1.5 m of water (up to 3 m has been recorded). The more regular winds from the north and west push the water out of the marshes.

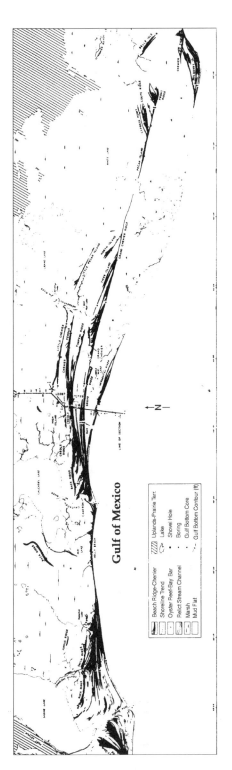

FIGURE 4.48   Cheniers along the Louisiana coast. (*Source:* Kaczorowski 1980.)

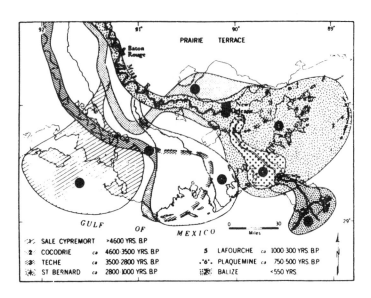

FIGURE 4.49   Mississippi delta lobes. (*Source*: Kaczorowski 1980.)

Intertidal deposits along the Mississippi delta plain include mud flats, marshes, and mangrove swamps. Mud flats are present where sediment supply along the coast dominates over beach-forming processes: where mud supply is strongly reduced or stopped, beaches and cheniers are formed. Marshes and vegetation develop where the top of the sediment comes within 30 cm of the average water level. The vegetation traps sediment, and mud flats can develop which can grade into grass-covered supratidal flats. The mud flats consist of watery clays and silts which can be homogenized or, through winnowing, can be turned into clays with silt laminae. They usually contain reworked plant material, and they have a relatively high organic content compared to other deltaic sediments. Mud flats cover wide areas and mud flat deposits can be distinguished from marsh deposits because they do not contain peat or *in situ* peat pockets. On the flats are many small lakes connected by tidal channels which lose their significance when the lakes gradually grow larger. The channels meander but only in a limited number of channels are there indications for migration of meanders: comparison of maps has shown that most meanders have changed little in 50 years. Where they have migrated, bends have been enlarged and oxbow lakes were left, filled in with fine sediment (Kolb and Van Lopik 1975). On the delta plain are thousands of abandoned tidal channels up to 10 m deep and 110 m wide, but many are minor and insignificant.

Marshes (Figure 4.50) are nearly flat and covered with grasses and sedges. There are four types of marshes characterized by vegetational units that reflect the salinity of the marshes: freshwater marshes, floating marshes, freshwater to brackish marshes, and saline-brackish marshes. The last three types are intertidal. The floating marshes consist of a vegetation mat, 10 to 40 cm thick, on top of 1 to 5 m of organic ooze which grades downward into clay. Consolidation results in a black organic

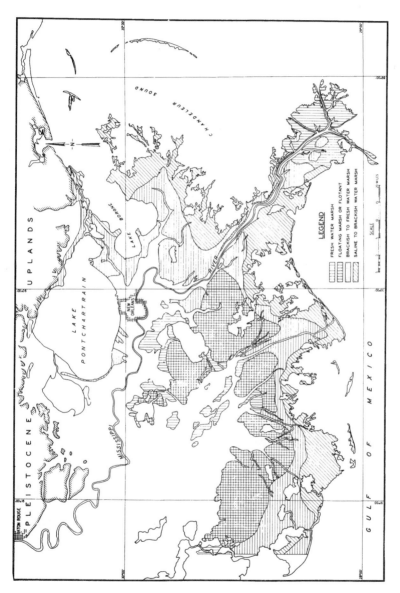

FIGURE 4.50 Distribution of marsh deposits in the Mississippi delta. (*Source:* Kolb and Van Lopik 1966.)

clay/peat layer with more than 50% organic matter. In the low-salinity marshes a vegetation mat of roots, 12 to 25 cm thick, with other vegetational fragments and fine mucky material, lies on top of 0.3 to 3 m fibrous peat. Below this, a layer of blue-grey clay and silty clay with thick lenses of dark-grey clays and silty clays are present that have a high organic content and only 10 to 20% inorganic material. The high-salinity marshes have a 5- to 25-cm-thick mat of roots, stems, and leaves on top of fairly firm blue-grey clay with tiny organic flakes and particles, and with few roots and plant remains. With depth, the clays become less organic and firmer. These marshes form a belt 800 m to 12 km wide along the present shoreline and have a rather high percentage of inorganic material. Therefore, they are more stable than other marsh deposits: 30% is silt and fine sand; organic clays average 50%; and peat layers 15 to 30%. Deposition on the marshes takes place mainly during southerly winds, which raise sea level along the coast, enhance resuspension of bottom sediment, and increase flooding. The occasional hurricanes increase sediment deposition but the heavy rainfall afterward tends to remove much of it. Usually, meteorological effects on water level and the formation of waves mask the tidal variations in water level (Reed 1989).

Swamps (Figure 4.51) are distinguished from marshes because they have a dense growth of trees, which are absent on marshes; the intertidal swamps are mangrove swamps which form a narrow belt along the coast and islands. On the Chandeleur Islands, the mangrove is dense and a mile wide, 60 cm to 1 m high and extending along the length of the islands. They are at the back of a landward retreating beach: the barrier passes over the mangrove and stumps appear on the foreshore. The mangrove is seldom higher than 1.3 m here. Below the trees is a thin layer of dark-grey to black, very soft organic clay on top of (and forming the matrix of) a tangled interlocking root zone, 15 to 30 cm thick (the mangrove is *Avicennia*). Pneumatophores project 5 to 10 cm above the sediment surface. Below the roots are at least a few meters of organic rich clays, silts, sands, and mangrove fragments, but there are few data on this. Along the barrier beaches mangrove covers overwash fans: the mangrove traps fine sediment so that organic clayey sands and clays are formed, mixed with shell material. Organic clayey sands lie on top of organic sands and silt.

The subsidence of the delta coast is in the order of 9 mm · y^{-1}; the vertical accretion (measured by ^{210}Pb and ^{137}Cs) varies from 4.4. to 15.2 mm · y^{-1} and is higher on the levees along the stream channels. Accretion decreases at greater distance from the channel. On the levees the sediment deposition keeps pace with subsidence, but at greater distance from the levees, subsidence dominates so that wide areas gradually become flooded. Within the past century the coast has retreated at an accelerating rate: in 1910 there was about 0.12% loss of wetlands per year, in 1975, 1.5% (Hatton et al. 1983; Salinas et al. 1986; DeLaune et al. 1987; Rejmanek et al. 1988).

4.24. NORTHWEST FLORIDA

Along the west coast of Florida between 29° N and 26° N tidal marshes are exposed to the open sea for a distance of about 300 km (Figure 4.52). There is no recent terrestrial sediment supply except reworked Pleistocene sand, which consists of

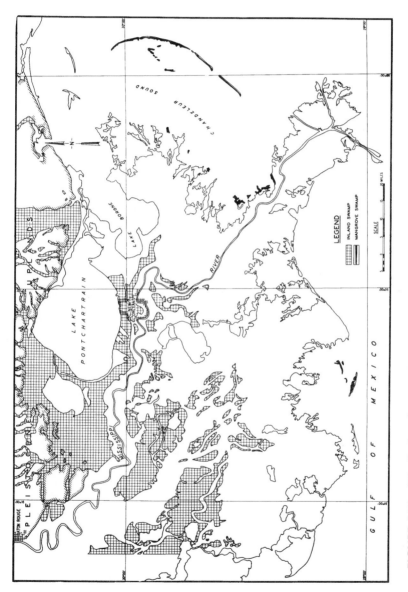

FIGURE 4.51 Distribution of swamp deposits in the Mississippi delta. (*Source:* Kolb and Van Lopik 1966.)

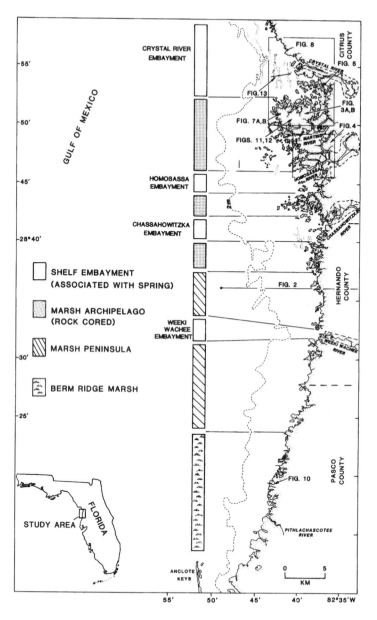

FIGURE 4.52   The northwest Florida coast. (From Hine et al. 1988. With permission.)

reworked early-Tertiary material. The intertidal sediments are thin, up to about 2 m only, and lie on top of a carbonate platform that consists of Eocene-Oligocene shallow-water carbonate. The coast grades with numerous islands and embayments into a broad flat continental shelf. Because of the shallow depth over 100 to 150 km from the coast and a relatively short fetch (compared to the ocean along the east Florida coast), waves are generally low, mostly lower than 45 cm; only during winter storms may waves reach a height of 2 m or more. Storms are relatively weak, swell generally low, and hurricanes (which blow in summer) infrequent. The tides are semidiurnal with a strong diurnal inequality. The neap tide range is about 60 cm, the spring tide range 1.2 m at most, but during storm surges sea level may be temporarily raised by more than 2 m. The intertidal deposits consist of marsh deposits with small areas of bare tidal flats and, on the seaward side, often a sandy berm or overwash deposit. The coast is slowly eroding and has been slowly submerging during the past 5000 years. Recently the rate of relative sea level rise has increased from 0.4 mm · y^{-1} (based on radiocarbon measurements) to 1.6 mm · y^{-1} (based on recent tide gauge data). No accretion has been observed (Hine et al. 1988; Leonard et al. 1995).

The marsh deposits consist of black, fine, sandy, organic-rich mud with plant roots. The bare flats are more sandy; the berms and overwash deposits consist of reworked quartz sand with mollusk shell hash. They have been transported landward and onto a beach by (low) waves and have been pushed over the near-shore marshes during storms. The marshes are dissected by tidal creeks which follow the dissolution zones in the carbonate platform and connect sinkholes. Carbonate dissolution gives a variety of topographic features, ranging from an irregular array of small pits, depressions, and etchings by marsh water and plant roots to 10- to 100-m-long linear depressions following rectilinear features in the carbonate, and to larger-scale elevated bedrock highs where the bedrock is not fractured and therefore less easily dissolved, as well as to coastal embayments on a km-scale near points of freshwater discharge. On the topographic highs (marsh hummocks) usually more sediment is present; it retains freshwater (rain water) and is covered with trees and shrubs. Rocky outcrops are often bare, and where they are located on coastal headlands, covered with a marine epifauna. Islands are formed where the topography has become very irregular because of rock dissolution. In the deeper karst holes the marsh deposits lie on top of peaty mud (often with macroscopic marsh plant detritus) and muddy sand. Below this Holocene deposit often lies Pleistocene quartz sand on top of a basal carbonate weathering surface.

Sediment is only transported onshore in significant quantities during periods of strong wind in the winter, when suspended-matter concentrations may increase by two orders of magnitude; the summer hurricanes are too few in number to have a significant effect. The net sediment transport depends on a coupling of a storm-raised water level and spring tide, which results in a large tidal volume. Measurements at Cedar Creek in the northern part of the marshes, indicate peak flood currents to be 10 to 20% higher than peak ebb currents, while the ebb currents are of longer duration than the flood currents. Maximum flood velocities occur near maximum high tide, maximum ebb velocities around mid-tide. This results in a net landward transport of suspended matter, which is strongest during spring tides (Leonard et al. 1995).

4.25. SOUTHWEST FLORIDA–BELIZE

In southwest Florida intertidal deposits are present in the Ten Thousand Islands area at 25° 40′ N to 26° N, around Whitewater Bay at 25° 10′ N to 25° 25′ N, and in

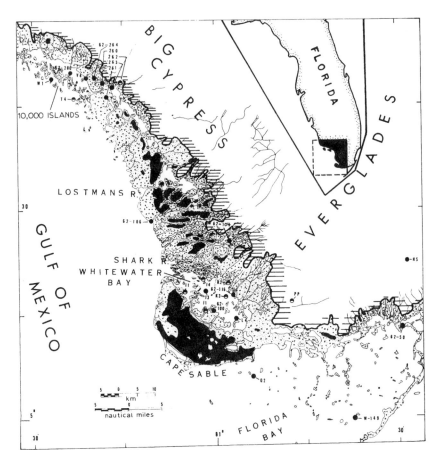

FIGURE 4.53   The southwest Florida coast and Florida Bay. (From Scholl 1964. With permission.)

Florida Bay around 25° N (Figure 4.53). The Ten Thousand Islands area is about 35 km long and 4.5 km wide and consists of numerous mangrove-covered islands about 1.5 km offshore from a shallowly submerged coastal mangrove forest. The islands consist of quartz sand with shells and shell fragments, and of calcareous silts and organic-rich (peaty) calcareous quartz sands. Quartz dominates and is supplied by shoreward drift from the Cape Romano shoals. The coastal swamps consist of shallow bays, anastomosing waterways, and mangrove-crested and overgrown sand bars, oyster banks (*Crassostrea*), and reefs of *Vermetus*. The calcium carbonate is 70% to 80% aragonite and consists mostly of mollusk shells and shell fragments. The organic matter content varies from 2 to 25% and consists largely of mangrove detritus.

Whitewater Bay is a 20-km-long and 9-km-wide embayment with its length axis in a southeast to northwest direction and bordered by broad mangrove forests on its north and northeast side. The sediment is mostly organic-rich shell debris with very little quartz (5 to, at most, 15%), as there is no large supply from a nearby source.

The calcium carbonate in the coarser size fraction is 70% aragonite, but in the fine fractions it is up to 90% calcite. The aragonite is formed in the bay; the calcite comes as lime mud from the Everglades and from the limestones below the coastal forest. Runoff from the Everglades discharges through the coastal mangrove in summer and early fall (the rainy season) but is absent in winter and early spring so that during that time the water is fully marine and can be hypersaline (Scholl 1963). The mangrove forests started to grow about 4000 years BP during a continuous, gradually decreasing relative sea level rise that began at least around 7000 BP (Scholl 1964a,b; Scholl and Stuiver 1967; Scholl et al. 1969). Deposition started with freshwater muds and peat, followed by a basal layer of autochthonous *in situ* peat of mangrove and other rooted halophytes with peaty and calcareous shell debris on top of this in the Whitewater Bay area, and shelly quartz-rich sand and silt in the Ten Thousand Islands area. The mangrove peat deposits are generally transgressive.

In Florida Bay, at a tidal range of 45 to 50 cm, intertidal deposits are present on mud banks that are exposed at low tide and on islands that vary in size from a few tens of square meters to a few square kilometers. The mud banks are mostly covered with turtle grass (*Thalassina*) and have patches that are bare or covered with green algae. The sediment consists of sheets of pellets and mud with nearly horizontal laminae that are partly destroyed by crab burrows. The smaller tidal channels through the banks can be choked with sediment; the large channels are incised and meander with a lag-deposit of shells at the bottom and well-laminated muddy banks. The islands are partly or entirely flooded at spring tide and consist mostly of a central supratidal flat with a fringe of (intertidal) mangrove. The sediment on the supratidal flats is laminated fine silt or mud with pellets and skeletal grains (foraminifera, gastropods) with stromatoliths of blue-green algae, mud cracks, animal burrows, and other desiccation marks such as upcurling algal mats. Lenses of *Halimeda* debris (a calcareous algae that lives in the shallow waters around the islands) are probably storm deposits. During a storm up to 5-cm-thick sheets of pelleted lime mud and skeletal particles can be deposited. With time, a deposit of dense mud layers is formed, separated by organic-rich concentrations of algal mats with scattered gastropod shells (*Cerithium*, which grazes on the mats).

The mangrove consists of *Rhizophora* on the outside and *Avicennia* on the inside. The tangle of mangrove roots and trunks reduces the water flow, binds sediment, and stabilizes the substrate. The sediment consists of lime mud and peat of black to dark-grey plant fibers, burrowed by crabs and bivalves. The coastal mangrove swamps are dense forests extending for several kilometers inland. They are best developed on thick sediment or peat that is not too compact for root penetration. The sediment is lime mud with mollusk shells and dark fibrous to plastic peat with burrows that are filled with lime mud. The intertidal deposits have been formed during a continuous rise of relative sea level and are transgressive: freshwater deposits were first formed on the Pleistocene surface followed by coastal mangrove swamps with shallow bays. In these bays, mud banks were formed. Some of them developed into islands with supratidal flats; others, where mangrove could not settle, remained mud banks. The islands show a wide variety in depositional history: some supratidal sedimentation persisted from an early stage; others became supratidal very late. The accumulation of mangrove peat began in depressions in the Pleistocene surface: here the presence of water enhanced the growth of mangrove, while dissolution of the Pleistocene limestone by the acid mangrove water may have deepened the depression so that peat was accumulated. On the sedimentary sequence formed during transgression a — usually much thicker — regressive sequence has been deposited. The present mangrove is expanding seaward and the trend is toward more and larger islands (Basan 1973; Enos and Perkins 1979).

Along the south coast of Florida, in the Reef Tract and protected by shoals against winds and waves from the southeast (in the summer) and the northeast (in the winter), lies a Holocene mud bank, Rodriguez Bank (Turmel and Swanson 1976). The mean tidal range is 66 cm, the spring tidal range 80 cm. The bank consists of unconsolidated calcareous sediment, deposited during a relative sea level rise without strong waves and with its surface approximately 10 to 30 cm above mean low tide level. The sediment consists of skeletal sand and gravels, and lime muds with worm burrows. Much of the skeletal material consists of fragments of *Halimeda*, which grows around the bank. The higher parts of the bank are covered with mangrove, which produces organic sediment, and the lower parts with marine grasses (*Thalassina*), green and red algae, and coral. The bank was formed in an embayment in the Pleistocene surface in an initially quiet area with restricted circulation, and remained in existence because of its sheltered position. The oldest peat dates from 5500 BP.

Along the coast of Belize at 16° to 18° 30′ N, at a spring tidal range of about 50 cm, Holocene, mostly intertidal sediments are present in narrow discontinuous bands up to 100 m wide. Landward they are bordered by Pleistocene limestones and clays, and seaward by marine and lagoonal deposits. North of Belize City are numerous cays and a few streams at a coast consisting of barrier lagoons and marshes. South of Belize City the open coast is exposed to heavy surf, and there is an abundance of small streams with numerous small deltas. Farther south the area with marshes increases again. Intertidal deposits include deposits along the margins of the lagoons, and coastal marshes along the estuarine delta coast. The lagoonal mud flat deposits are relatively extensive and have a mangrove fringe. The mud deposits start as low flats that increase in height until they reach a level where they are no longer flooded and mangrove growth is inhibited, which occurs at a height of +30 cm. The higher flats are bare and supratidal with sediment deposition only during storms. Because of the consolidation of the sediment during the period of subaereal exposure, and because the mangrove fringe breaks the force of the wind and waves, the supratidal flats are not eroded during the storms and suspended matter can settle on the flat surface: the sediment consists of sandy carbonate mud with scattered plant debris and salt crystals at the surface. Locally on the mainland side of the lagoons, dark red-brown peat is formed. Where local runoff flows over the flats — as happens along the barriers with dunes that store freshwater — dense, tall sawgrass grows. Mangrove islands (*Rhizophora* and *Avicennia*) in the lagoons develop from former *Thalassina* banks and a muddy mangrove peat is formed.

The coastal marshes along the estuarine delta coasts are low and flat with a sparse growth of small *Rhizophora*. Sediment is deposited through entrapment and stabilization by the mangrove. In the lagoons the tidal deposits extend inward, whereas along the estuarine delta coast they extend outward toward the sea. The lagoons are connected with the sea by deep tidal channels which are stabilized by mangrove along the banks. Little sediment and mostly coarse lag deposits occur in the channels because of the strong currents.

The Holocene sediments have been deposited transgressively over poorly indurated Pleistocene limestone. At the basis is a stiff, terrigenous, pale blue-grey, peaty and sandy clay, deposited before 3000 to 4000 BP on grass meadows similar to the ones that now exist inland from the present lagoons. Above this lies a fibrous, dark red-brown, massive marine mangrove peat, generally undisturbed, but locally reworked and transported. Its average thickness is 30 cm, its maximum thickness is 1.2 m, but in many places it is not well developed. It is covered by an upper layer of calcareous mud and fine-grained skeletal sands with a maximum thickness of 1.8 m. This unit was formed at supratidal conditions. The present intertidal deposits are formed as different units of limited extension that were formed at the expense of former extensive mud flats. Locally the supratidal flats are expanding, which produces a regressive sequence on top of the Holocene transgressive deposits (High 1975).

4.26. ANDROS, CAICOS, AND GRAND CAYMAN ISLAND (BAHAMAS–BRITISH WEST INDIES)

On the islands of Andros and Caicos, at about 24° 30′ and 21° 30′, respectively, and with a tidal range of less than 1 m, wide tidal flats are present at the leeside of the islands and adjacent to a broad, shallow carbonate platform (Figure 4.54). Although both islands lie in the same area of the subtropics and in many respects are very

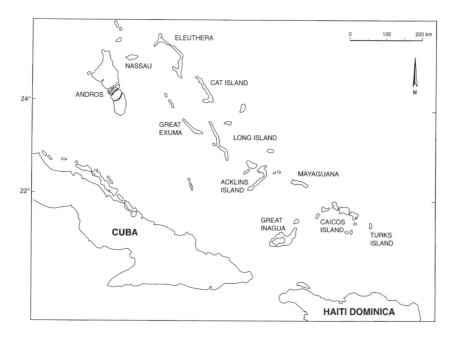

FIGURE 4.54   The Bahamas and Caicos Island.

similar, they are situated within very different wind regimes, with different wave regimes around the islands and considerable consequences for sediment transport and deposition.

The flats on Andros Island are partly exposed to large waves from the northwest that are generated during winter storms. Resuspension of bottom sediment by these waves and subsequent deposition results in the formation of shoreline levees and in-channel levees along the tidal channels that dissect the lower flats. Both kinds of levees are broad with strongly laminated sediment with millimeter-thick laminae consisting of pellets and skeletal fragments formed on the carbonate platform.

Caicos Island lies within the trade winds and is affected not by storms from the northwest but by infrequent hurricanes. Both the persistent trade winds and the hurricanes come from easterly to northeasterly directions. The waves generated by the trade winds produce a regular turbidity in the near-shore waters through resuspension of bottom sediment in relatively low concentrations, so that on Caicos, sediment supply is low but persistent (except during the passage of a hurricane), while on Andros sediment supply is infrequent and seasonal but in large quantities when it does take place. Therefore, on Andros the levees are conspicuous, but on Caicos they are low and often poorly developed, although the flats are regularly flooded. On Andros, however, there is a middle tidal flat zone dominated by ponds that are hardly being filled in, whereas on Caicos they are gradually being filled with sediment supplied during the regular flooding. When filled up they become covered with algae, as are the higher flats both on Andros and Caicos, as well as the levees. On the levees algal mats are formed at or just below the sediment surface, and the filaments stabilize the surface sediment after deposition. During heavy storms a lamina couplet is being formed, consisting of a basal mud lamina topped by a thin layer (less than 1 mm thick) of sand. On the higher flats the algal mats have a more permanent character. Hurricane deposits on Caicos consist of a 0.5- to 2-cm-thick layer of fine- to very fine-grained pellets and skeletal fragments derived from the carbonate platform. The storm layers are interbedded with organic-rich layers, formed during the quiet periods with algal growth between hurricanes. On the higher flats, where no storm deposits are formed, algal mats dominate and can grow to several centimeters' thickness. Where rainfall is very low, gypsum is formed in cemented crusts around high tide level. This is absent on Andros, where rainfall is about twice as high as on Caicos (Wanless and Dravis 1984; Wanless et al. 1988).

On Andros the tidal flats are located on the west side in a general north-south direction, while the wind comes primarily from the east. The maximum spring tide range is 48 cm, the neap tide range 17 cm. The water on the carbonate platform in front of the flats is highly saline to hypersaline. The tidal flat sediments consist of pelleted carbonate mud: aragonite is aggregated into silt- or sand-sized ellipsoids or rounded irregular peloids in varying degrees of induration. The peloids on the flats are probably resuspended offshore and washed onshore during storms. The sediment is finely laminated or consists of thin beds, where it is not bioturbated. Locally, coarse lag deposits of mollusk shells and foraminifera occur and blue-green algae can form a Mg-calcite. In this way an *in situ* algal dust sediment can be deposited. The maximum thickness of the sediments is 3 m; they are formed on top of Pleis-

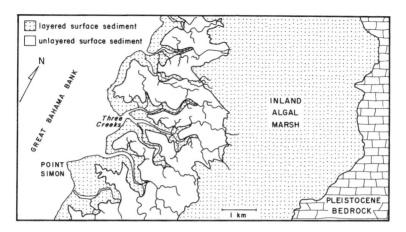

FIGURE 4.55 Channels and algal marsh along western Andros. (From Hardie and Ginsburg 1977. With permission.)

tocene limestone. The flats have a variable morphology and can be divided into three types: 1) a broad, flat, exposed coastal plain, 4 to 8 km wide with or without a few active channels but with scars of abandoned channels, and inland with shallow ponds and lakes (south of Williams Island); 2) flats that are strongly influenced by the Pleistocene surface below, so that "island ridges" are formed, with only a few branching channel systems that are not well developed and are connected with wide tapering waterways that end in wide shallow ponds (east of Williams Island); and 3) a transgressive flat with a 3- to 4-km-wide belt between the shoreline and the inland algal marsh, which is dissected by very active branching, sinuous, or meandering channels, 30 to 100 m wide and up to 4 m deep (north of Williams Island). Landward they end in narrow curved or meandering gullies that are less than 1.5 m deep (Figures 4.55 and 4.56).

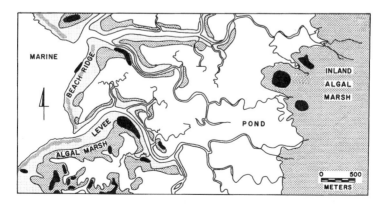

FIGURE 4.56 Details of channels in the Three Creeks area. (From Hardie and Ginsburg 1977. With permission.)

All active main channels are cut into the underlying Pleistocene limestone, but the gullies are cut only into recent sediment. The belt of flats with channels is 1.5 to 3 km wide and lies parallel to the general direction of the flats. The sediments in the channels vary from mud to shell debris and are generally coarser where currents are stronger (in meanders and curves) and finer where currents are slower (in the straighter parts of the channels). Bars in the channels have a relief of up to one meter and are covered with ripples, with fragments of hardened (dolomite) crusts eroded from the channel walls, and soft pellets of mud, silt, and shell debris (foraminifera, gastropods, and silicious sponge spicules that are only found in the channels). Channel deposits show burrows, crossbedding, and inclined bedding.

Along and between the channels a zonation of distinct environments with typical characteristics is present: the channel banks, levee crests, levee back slopes, high algal marsh, low algal marsh, and ponds (Shinn et al. 1969; Hardie and Garrett 1977). Between mean tide level and mean high tide level, the channel banks are covered with algal mats (*Schizotrix*), which form large knobs where they are draped over *Uca* burrows. In meander bends, active undercutting gives steep sides without mats. The levees along the channels are higher on the seaward side, where the tidal range may reach 46 cm, and lower inward where the tidal range may be only 29 cm. The levee crest is flat and smooth, with a sparse vegetation of halophytes. There are no mud cracks because the sediment is held together by thin algal threads (*Schizotrix*). The upper 5 to 15 cm of the sediment are well laminated. Entrapment of air can lead to the formation of air domes when ground water comes up at rising tide. Current lineations are common on the sediment surface. The levee back slope has shallow mud cracks that are closely spaced, and crab burrows. The formation of mud cracks can be arrested by the growth of a *Schizotrix* mat. A fine (mm-size) lamination is mostly destroyed by burrowing. The high algal marsh that lies outward from the levee back slopes is covered with *Scytonema* mats, stunted mangrove, and halophytes. Near high tide level are crab burrows. The sediment surface often has current lineations. Patchy crusts are formed by cementation with Mg-calcite and aragonite. The low algal marsh has isolated tufts of *Scytonema* and is continuously wetted in summer but mostly dry during the winter. There is a sparse occurrence of dwarf mangrove; crab burrows (*Uca, Sesarma*) are common. The ponds between the channels have an irregular shape, which is determined by the channel configuration, and in the summer contain several decimeters of water. In the winter, the water level is 10 to 15 cm lower, and the ponds may remain for weeks without flooding, so that most of the remaining water evaporates. The sediment consists of fecal pellets with a loose algal mat. There are worm burrows, and the sediment is not layered. Ephemeral cracks are formed which soon disappear when the water returns. The mats do not give protection against erosion or resuspension: turbidity can increase considerably when the wind moves the water, and lumps of mat may be torn loose and transported. The fecal pellets are produced by worms and gastropods.

The ponds are closed off from the sea by a low beach ridge 10 to 30 m wide that merges landward into washover lobes. The ridge (a "shoreline levee") is covered with a *Schizotrix* algal mat and some tufts of *Scytonema* and *Rivularia* (another filamentous algae) on the seaward side. On the flanks and washover lobes, crusts are formed in patches. The higher parts of the ridges are supratidal and called "hummocks": discontinuous, narrow, low, sandy mounds, up to 1.3 m above mean tide level and only flooded during a hurricane. It is a terrestrial, freshwater environment with pine trees (*Casuarina*), and burrowing ants, earthworms, and land crabs (*Cardisoma*). The sand is coarse, crossbedded, with gastropods and partly cemented: it is deposited during storms and derived from the shallow sea directly offshore. Lower ridges, rising less than 40 cm above low tide level, occur on headlands and consist of sand-sized pellets and skeletal fragments, deposited in laminae that are graded (finer at the top) and laterally discontinuous. The overwash lobes behind the ridges are also covered with a *Schizotrix* algal mat together with some halophytes, grasses, and mangrove (*Avicennia*). The washover deposits are characterized by thin sheets of rippled sand and patches of flat pebbles (plates, slabs) of eroded crust. The ripples are cuspate and very flat, only a few millimeters high (up to 7 mm) and up to 30 cm apart.

On the tidal flats numerous abandoned channels and gullies are visible as faint scars on the surface. Mangrove grows on the abandoned beds, while old levees are overgrown by palms: because of sediment trapping the elevation has become higher (palm "hammocks"). The sediment in the abandoned channels is very similar to that in the active channels and consists of pellets ranging in size from 40 to 80 micron to more than a millimeter. The largest pellets — of 1 to 3 mm — are formed by a polychaete worm. The sediments are burrowed by worms and crabs as well as by mangrove roots (which make holes of about 1 cm diameter) and root hairs (holes of about 1 mm). Higher in the intertidal area, burrowing is less abundant because of a longer exposure during ebb. Algal mats (*Scytonema*) occur, again, at the transition to the supratidal flats above normal high tide level.

These flats, landward of the tidal flats, cover the largest area, are 4 to 8 km wide, and are covered with algal mats (*Scytonema*). The sediment is similar to the sediment on the tidal flats but locally has a large admixture of organic matter in distinct laminae, which are abundant and accentuated by color differences ranging from white to light tan, grey, and dark brown. Laminae range from 1 mm to 1 cm in thickness; within the thicker laminae there are sometimes finer laminae, some inclined or crossbedded. There is a considerable lateral variation. The algal mats have algal heads locally, some of them cemented with calcite. Surface crusts are formed a little above high tide level and may contain up to 80% dolomite (nonlithified sediment contains 5 to 20%).

On the east coast of Andros Island, on the windward side, the near-shore water has a normal sea water salinity. There is little precipitation or formation of crusts. Sediment accumulates mainly through entrapment of particles by algae. The algal mats (*Rivularia, Dichotrix, Schizotrix*) are not laminated. Where *Schizotrix* occurs in pure stands, finely laminated domes and mats are formed in which the laminae reflect a daily cycle (Monty 1972).

The hurricane shoreline levees on Caicos are only 30 to 40 cm high and have an erosional scarp on the seaward side with eroded mangrove bushes and roots and no algal mats (Figure 4.57). On the crest, sediment is deposited on top of mats that consist of sand and eroded crust pebbles. On the landward flank there are hurricane deposits of two coarse layers with a fine layer between. The coarse layers are assumed to have been formed by wave-overwash during storms, and the fine layer

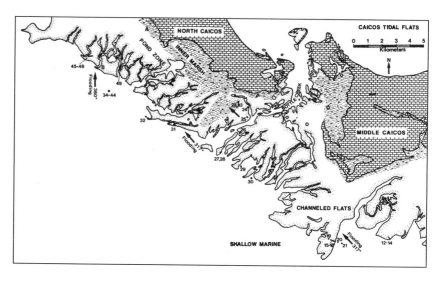

FIGURE 4.57   Caicos platform with tidal flats (marsh pattern) and channels. (From Wanless et al. 1988. With permission.)

during maximum flooding when the waves were passing over the crest. Erosional gullies cut through these deposits, draining landward, with surface current lineations and ripples in the same direction on the levee. The flat areas behind the levees are overgrown with scattered mangrove bushes, and the sediment has usually hardened. The overwash shoreline levees and channel levees, formed during less extreme conditions, are generally few and narrow with laminae of millimeter thickness. The storm deposits may reach a thickness of 2 cm at 3 to 6 m from the channel margin and in addition to being finely laminated may be faintly rippled. The channels that are present in the outer part of the flats are straight or sinuous with a variable width, sometimes with a pool at the landward end. Some are interconnected. The tidal flat deposits are all formed out of suspension and mixed with pebbles eroded from former tidal flat deposits. The inner zone with algal mats consists of *Scytonema* cushions formed on top of hardened crusts. During storms the cushions are lifted and redeposited in chaotic or imbricated conglomerates. In depressions these storm deposits can reach a thickness of 4 cm. After deposition, sediments can be modified by desiccation and hardening as well as by algal recolonization. Desiccation occurs where new sediment is deposited over algal cushions. Concave upward polygons are formed within two weeks after deposition, usually with raindrop marks. Hardening takes place where the algal mats are less pronounced. Algal recolonization can take place within 15 days where a thin sediment layer covers a *Schizotrix* mat: a similar mat is formed over the new sediment. Where the new sediment is thicker, colonization starts from the parts of the mats that have not been covered and takes more time. For *Schizotrix* this may take a few months, but for *Scytonema*, recolonization after five months is still restricted to areas near the surviving mats.

On **Grand Cayman Island**, some 300 km south of Cuba at about 19° 30′ N, intertidal deposits consist of mangrove peats with a thickness of up to 4 m (Woodroffe 1981). The tides are semidiurnal with a distinct diurnal component, have a mean range of 35 cm, and a maximum range of 60 cm. The peats lie on top of plastic muds, deposited in a seasonally flooded environment, and on subaereal crusts. The peats reflect a period of relative sea level rise: some peat was later drowned and covered with shelly near-shore marine mud; in other areas a similar shelly mud was covered with mangrove peat and locally seagrass peat. There is no intertidal or supratidal carbonate, and there is hardly any supply of inorganic material to the mangrove, as there is only pre-Holocene carbonate rock and no terrestrial drainage system. Most of the island is covered with brackish water during high tide, and mangrove covers most of this area. The mangrove peat is almost entirely autochthonous and consists of the organic remains of the mangrove themselves: the deposit is fibrous with abundant roots and rootlets. Mollusks are rare, although they are abundant in the adjacent shallow waters. ^{14}C dating has indicated that the oldest peat was formed in at least 2100 BP.

4.27. EAST COAST UNITED STATES: SOUTH CAROLINA TO MASSACHUSETTS

From northern South Carolina to Massachusetts, between about 33° N and 42° N, the coast consists of sandy barriers or barrier beaches with inlets, estuaries, and lagoons (some very large, such as Pamlico Sound and Albemarle Sound), and of numerous small and several very large embayments (Chesapeake Bay, Delaware Bay, Long Island Sound). South of Cape Hatteras the coastal barriers form large concave bays — Long Bay, Onslow Bay, Raleigh Bay; north of the Cape are large convex barrier coasts separated by the large embayments. The tidal range along this part of the U.S. east coast is less than 2 m, and often less than 1 m. Inward from an inlet or a bay entrance the tidal range is reduced and in wide areas becomes about 10 cm. Intertidal deposits are formed behind the barriers and inside the bays and embayments, with bare tidal flats below mid-tide level and salt marshes starting a little below that level. The salt marshes at that level consist mainly of *Spartina alterniflora* with other species on the higher marshes). Where the tidal range is very low, the bare flats become small and insignificant, and the intertidal zone is dominated by marshes.

At North Inlet in South Carolina (Figure 4.58), at about 33° 20′ N, the mean tidal range is 1.37 m, the spring tide range 1.62 m, with a diurnal inequality of, on the average, 37 cm. There is no permanent freshwater influx, and the only sediment source is the nearshore sea. On the seaward side is a beach barrier with an inlet and a large ebb tidal delta; the flood tidal delta is much smaller and is strongly influenced by swash at high tide. The ebb tidal delta has a variety of large ripple systems, but the flood tidal delta only a few, poorly developed sand waves and generally few flood-oriented bedforms. The marsh behind the barrier is dissected by interconnected meandering channels and creeks. Aerial photographs show hardly any abandoned

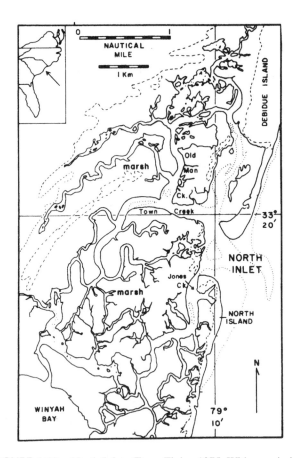

FIGURE 4.58   North Inlet. (From Finley 1975. With permission.)

meanders, and about 90% of the tributary creeks join the larger channels at an angle of about 90°. So it is probable that there has not been any active meandering during the recent past (Finley 1975; Gardner and Bohn 1980). The coast is slowly subsiding, with the marshes gradually growing landward as well as vertically.

A little farther north lies Murrell's Inlet, which is similar to North Inlet with a Holocene barrier and a salt marsh on the landward side. The spring tide range is 1.8 m. The inlet is the sole source of sediment: inward grain size decreases, while mud content increases. Point bars near the inlet occasionally have mud drapes in ripple or megaripple troughs, and fecal pellet debris is present on the slipface of bars. Farther inward there is little sand, and thixotropic mud bars develop in the channels. Carbonate content is highest near the inlet because of the admixture of coastal marine bivalve debris (Barwis 1978). The meanders in the channels and creeks vary from very tight bends to gentle loops. The tight bends have relatively small point bars and no chutes, while the gentle meanders have long, narrow point bars with relatively deep flood-dominant chutes. Intermediate meanders have multilobed point bars and also flood-dominant chutes. Erosion by the meandering channels and creeks results in bank slumps. The marsh surface sediment is protected by dense, 30- to 80-cm-thick mats of *Spartina* roots. The mud below this is less stable: it has an angle of repose of 15° to 45°, whereas the banks formed by roots have an angle of repose of 45° to 90°. Slumping is enhanced by the burrowing of mud crabs (*Panopeus*). Slumping of undeformed root/sediment/grass blocks is common and results in chaotic bedding. Smaller-scale erosion results in mud balls or mud pebbles of up to 20 cm in diameter. They are usually partly armored with shell fragments and fine sand. Along channel margins and on point bar crests, bioturbation by mud crabs, as well as by hermit crabs (who can make 150 burrows per m²) is significant. Birds, rays, alligators, and surface snails leave tracks on the sediment surface.

As in other tidal areas along the U.S. east coast, the channels in Murrell's Inlet are ebb-dominated. Somewhat to the south, at a Charleston harbor creek (Figure 4.59) at approximately 33° N, with a tidal range of 1.5 to 2.0 m, an ebb-dominated channel exports organic suspended matter from the salt marsh, but imports inorganic suspended matter (Settlemyre and Gardner 1977). Import takes place during the winter when the winds maintain high suspended-matter concentrations during and, in particular, at the end of the flood; export takes place mainly in summer. On the whole, there is a balance between import and export during the winter, so there is a net export over the year.

The large Pamlico and Albemarle lagoons at about 34° 40' to about 36° 30', are separated from the ocean by long, narrow, extended barrier islands separated by inlets (Figure 4.60). The central basins are shallow (little more than 6 m deep), with fringing embayments and four large, wide river estuaries (of the Roanoke, Alligator, Pamlico, and Neuse rivers). The tidal range is in the order of 1 m near the inlets (spring tide range at Cape Lookout 1.34 m, neap tide range 1.13 m; Wells 1988), but in the lagoons is rapidly damped to about 10 cm. At the inlets the bottom sediment is medium sand; in the lagoons fine silt and organic-rich clays. Intertidal sediments along the lagoons and embayments consist mainly of marsh deposits, whose characteristics are primarily related to the degree of flooding. At Oregon Inlet (about 35° 50' N), the tide is semidiurnal with a range of 30 cm, and the marshes are flooded 40 to 50% of the time. The vegetation consists of *Spartina alterniflora*, and sediment deposition rates, determined with ^{137}Cs, are 2.7 ± 0.3 mm · y^{-1} along the stream sides, and 0.9 ± 0.2 mm · y^{-1} on the back marshes. At Jacob's Creek (halfway along the south bank of the Pamlico River estuary at about 35° 20' N) the tidal range is less than 10 cm, and the marshes are irregularly flooded during spring tide and winter storms. They are flooded less than 10% of the time; the vegetation is a mosaic of *Juncus*, *Distichlis*, and two *Spartina* species. The sediment deposition rates are 3.6 ± 0.5 mm · y^{-1} along the stream sides and 2.4 ± 0.2 mm · y^{-1} on the back marshes. At Oregon Inlet, mainly mineral particles are deposited and relatively little organic matter. At Jacob's Creek the vertical accretion is by *in situ* production of organic material as well as accumulation of deposited organic matter. The rate of sea level rise along the North Carolina coast was about 1.9 mm · y^{-1} in 1940 to 1980 so that the marshes near Oregon Inlet are slowly drowning (Craft et al. 1993). The barriers that form the coast are transgressive overwash storm deposits: they are eroded on the seaward side and migrate landward. During the slower rates of relative sea level

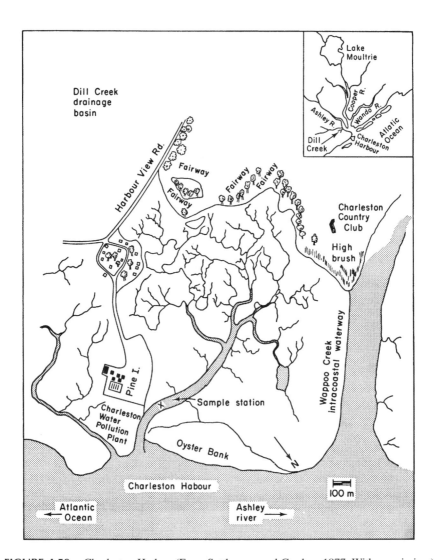

FIGURE 4.59   Charleston Harbor. (From Settlemyre and Gardner 1977. With permission.)

rise during the past 4000 years (as compared to the postglacial sea level rise), the rate of landward migration of the barriers slowed down. Longshore transport became more important, and subaquaceous shoals, as well as Cape Lookout, were formed.

Chesapeake Bay (Figure 4.61), 315 km long and 5 to 56 km wide, has at its entrance at 37° N a tidal range of about 1.20 m which decreases inward to about 50 cm or less. Shoreline erosion in the bay is in the order of 20 cm · y^{-1}, which may reach locally (Tangier island) 3 m per year. Deposition is subtidal; during a rapid sea level rise up to approximately 3000 BP, older, late-Pleistocene channels were filled in (Hobbs et al. 1992). The present mean water depth is about 9 m. The shallow margins of the bay consist of beaches or low eroding cliffs, barrier beaches consisting

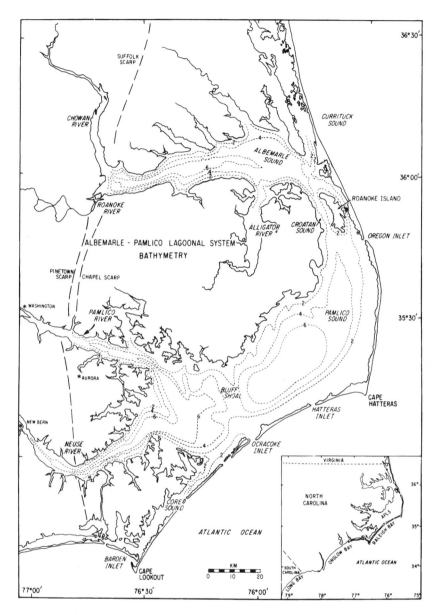

FIGURE 4.60 Cape Hatteras and Albemarle-Pamlico lagoons. (From Wells and Kim 1989. With permission.)

of a veneer of sand over marsh peat (17% of the coastline), and marshes (20% of the coastline). The marshes occur along the northeast and southwest margins of the bay where subsidence is largest (Rosen 1980). The marshes with their extensive root system retard shoreline erosion. Along the intertidal zone at the back of the beaches fringing marshes of *Spartina alterniflora* occur parallel to the beach. In the Black-

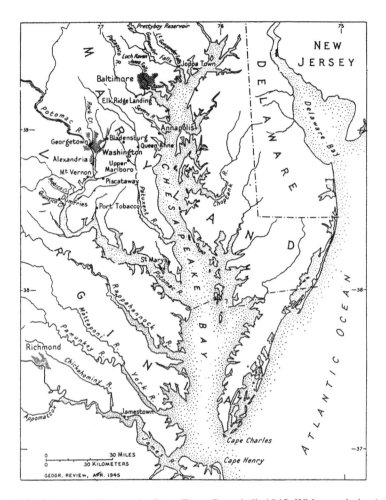

FIGURE 4.61   Chesapeake Bay. (From Gottschalk 1945. With permission.)

water area (Figure 4.62), marsh deposits are present in the form of unconsolidated organic ooze, usually less than 1 m, but up to 3 m thick, and covered with a rhizomatous mat. The area is drained by tidal channels and creeks, sinuous to strongly meandering, and up to 4 m deep. The tidal range is about 50 cm, but storm winds may raise the water level to 2 m for 2 to 5 days. The ebb discharge, because of an additional freshwater supply, is larger than the flood volume, and ebb current velocities are higher (up to 50%). Vertical accretion is through deposition of peat, the supply of inorganic material is of little influence, although there is more supply now than before because of agriculture on the adjacent watersheds. The sediment deposition rate is 1.7 to 3.6 mm · y^{-1}; because the sea level rise (since 1940) has been 3.9 mm · y^{-1}, the marsh is slowly subsiding. Erosion of the marsh starts with deterioration of the 10- to 50-cm-thick marsh mat, which is probably enhanced by the grazing of muskrats. The interior ponds gradually grow larger by rotational slumping at the sides where the mat breaks down. When the ponds have a diameter

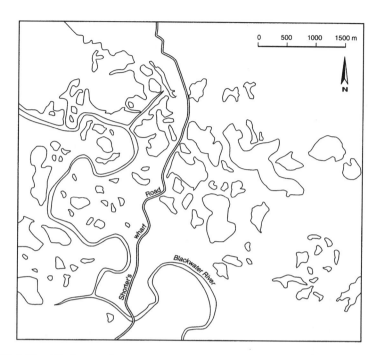

FIGURE 4.62 Blackwater river area. (From Court Stevenson et al. 1985. With permission.)

of about 1 km, wind waves are able to undercut the mat, and during storms from the northwest the ponds are elongated in a northwest-southeast direction. Long-term degradation of deeper peat layers is likely to be related to the coastal subsidence and the ebb dominance. Near its inner margin the marsh peat is intact over its entire thickness, but away from the margin, it grades into a structureless ooze. The process of breakdown is not known but is probably related to carbon loss by methane formation or to muskrats fragmentizing the peat (Stevenson et al. 1985).

Along the ocean coast of the Delmarva Peninsula that forms the eastern margin of the Chesapeake Bay in Virginia, a 3- to 12-km-wide belt of tidal flats, marshes, and channels lies behind a 100-km-long chain of barrier islands, ranging from the Chesapeake Bay entrance to beyond the Maryland border. Salt marshes and tidal flats occur along embayments interconnected by numerous tidal channels. As there is hardly any freshwater inflow, the circulation is almost entirely tidal. The tides are semidiurnal with a mean tidal range of 1.06 to 1.25 m. The spring tide range is 1.5 to 2.1 m along the ocean coast and decreases inward to 10 to 20 cm at the mainland. Storm winds can raise sea level temporarily by 3 to 4 m. The tidal flats are bare from low- to about mid-tide level, where *Spartina* starts to cover the flats. Extensive areas of flats occur marginally along channels as channel levees of km length. Intertidal deltas are formed behind temporary storm inlets and landward of washover areas on the coastal barriers; on the intertidal parts of the deltas and washover areas, laminated sands occur with algal mats. Away from the barriers, mud deposition increases as well as burrowing by crabs, worms, and bivalves. The salt marshes consist mainly of *Spartina* and some other halophytes and are completely flooded during most spring tides. The marsh deposits are structureless or show faint laminations or undulations along channel margins. *Uca* (a hermit crab) is common on the marsh edges, as well as *Littorina*, a snail, and *Modiolus* and *Crassostrea*, both bivalves, on the marsh itself. Through the marshes and flats go small meandering creeks that discharge into the larger channels (Harrison 1975; Boon III 1975).

Delaware Bay (Figure 4.63), with its entrance at about 38° 50′ N, about 100 km long, up to 50 km wide, and a tidal range of 1.2 to 1.3 m, is bordered along about

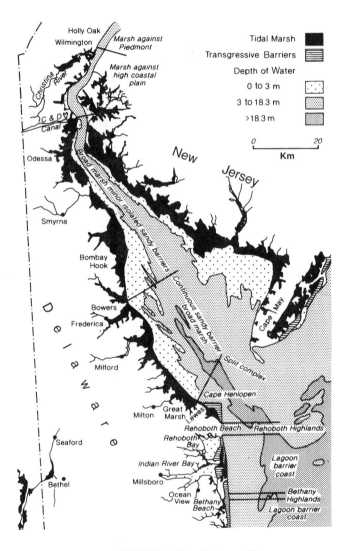

FIGURE 4.63   Delaware Bay.

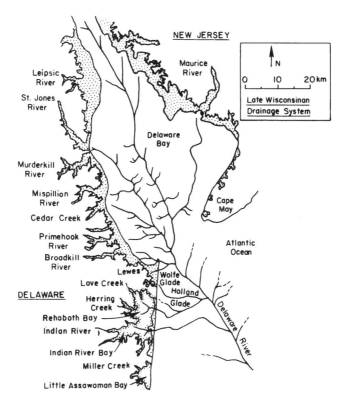

FIGURE 4.64   Delaware Bay: pre-Holocene drainage system. Shaded areas = tidal wetlands. (From Fletcher et al. 1992. With permission.)

65% of its coastline by tidal wetlands, mainly on the north side (Figure 4.64). The Delaware river and numerous smaller coastal plain rivers are tidal over most of their length with some small intertidal areas. The intertidal deposits are mostly fine-grained with a high content of organic matter. The supratidal deposits are often coarser grained because of storm overwash. Most of the intertidal deposits are covered with salt marshes which have higher deposition rates (9.7 to 15.5 mm $\cdot$ y^{-1}) in the inner bay than in the outer bay (1.7 to 5 mm $\cdot$ y^{-1}). The salt marshes grade inward into freshwater marshes (palustrine wetlands) that were not tidal 1000 to 2000 years ago (Fletscher III et al. 1992; Pizzutto and Rogers 1992).

Along the coast of New Jersey, from about 39° N to about 40° 20' N, short (12 km), narrow (0.4 km) sandy barrier islands separated by tidal inlets occur with back barrier areas up to 5 km wide at a mean tidal range of 1.3 m and a mean spring tide range of 1.4 to 1.5 m. The back barrier areas are filled with marshes dissected by channels and creeks, and by lagoons with intertidal delta shoals (Figure 4.65). Freshwater inflow is insignificant. Along the ocean coast there is a net longshore drift of sand to the south, winds usually blow from the west or southwest — in the summer from the south — while storm winds come from the north-northeast. The barriers and back barrier areas were formed during the Holocene sea level rise:

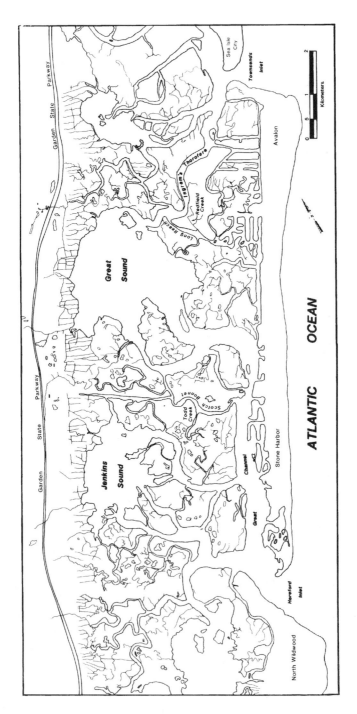

FIGURE 4.65   Great Sound area, New Jersey coast. (From Ashley ed. 1988. With permission.

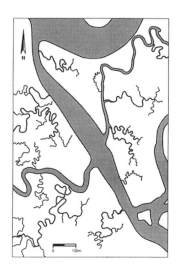

FIGURE 4.66   Channels and creeks in the Great Sound area. (Adapted from Zeff 1988.)

Pleistocene relief was filled in, a series of shore-parallel or shore-oblique sand ridges were formed, and finally the marshes were built up. The oldest salt marsh deposit dates from about 6000 BP (found at 12 m depth). The barriers were formed on the old interfluves; the inlets in the former valleys (Ashley 1988). The channels can be divided into two main groups: through-flowing channels that 1) connect the lagoons (area 3 to 6 km²) with the ocean, or 2) connect two tidal channels, and dead-end channels that flood and drain the marshes and terminate in the salt marsh (Figure 4.66). The through-flowing channels that connect the lagoons with the ocean are largest, 150 to 200 m wide and 3 to 10 m deep. They meander with wave lengths of 1.5 to 2.0 km, have a fine-to-medium sand bottom and ebb-dominant flow. The through-flowing channels that connect other channels are 50 to 100 m wide with no systematic variations in width, are meandering with a wave length of 800 m, form point bars, have pools 2 to 5 m deep, and sandy bottom sediment with 4% organic matter. Of the dead-end channels there are about 10 for every channel-connecting through-flowing channel and about 100 for every lagoon-to-ocean channel. They are 1 to 20 m wide, terminate in the marsh, meander with wave lengths of a few to a few tens of meters, and have a fine muddy sand bottom with 90% organic matter and no, or little, bedload transport. The small channels are drained during ebb with, at low tide, only some standing water in the deepest parts. Both the channel-connecting through-flowing channels and the dead-end channels have a (very small) ebb dominance. The total amount of sediment carried simultaneously by all creeks in a system is of the same order of magnitude as the amount transported by a single ocean-to-lagoon channel.

The through-flowing channels have a hydraulic geometry that is very similar to the geometry of other tidal channels, have a low silt/clay (mud) content (average 9%), and a high width:depth ratio (34:129), while the dead-end channels have a hydraulic geometry similar to rivers, a high silt/clay (mud) content (average 78%), and a low width:depth ratio (5:21). There is a quasi-equilibrium between the erosive forces of the flow and the resistance of the channel banks to erosion. In the larger channels, an increase in discharge leads to an increase in flow velocity; in the smaller channels to an adjustment of channel width and depth, with the depth increasing at a greater rate than the width. This is enhanced by the cohesiveness of the bank sediment and the marsh vegetation, which also makes very tight meander bends possible compared to the bends in rivers. The through-flowing channels evolved out of flood tidal delta channels that became fixed in position when the delta shoals became stabilized by vegetation and cohesive channel banks. The dead-end channels originated as ebb drainage patterns on bare flats and also became fixed by vegetation and cohesive sediment. The high sinuosity developed while the surface was vegetated: after stabilization there was little or no lateral migration of the channels. Also rills along larger channels on the bare lower banks and oriented at approximately 90° to the channel axis could have developed into dead-end channels. The sediments generally have little primary layering. The bare or sparsely vegetated channel margin flats are burrowed by crabs, while in the channel margin marshes, 1 to 2 m above the channel margin flats, the marsh plant roots form a very dense network that destroys any primary layering. The through-flowing channels had a more dominant role during the early development of the back barrier area, when there was a change from predominantly open water to predominantly lagoons. The dead-end channels became increasingly more important with time when less open water remained (Ashley and Zeff 1988; Zeff 1988).

Along Long Island Sound up to Cape Cod, intertidal areas are small. In a tidal area on the north coast of Long Island, Flax Pond near Stony Brook (Figure 4.67),

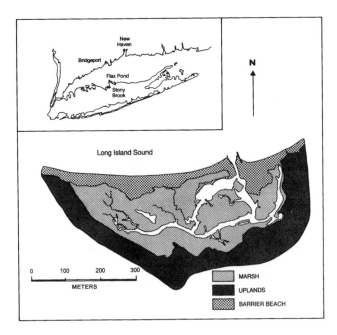

FIGURE 4.67 Flax Point Long Island. (From Richard 1978. With permission.)

the lowest flats are bare, with around mid-tide, areas that have recently been colonized by *Spartina*, and at high tide level a *Spartina* marsh with peat. Sediment accumulation rates, estimated with brick dust and aluminum powder, gave the highest rates on the bare lower flats (20.5 to 45.5 mm · y^{-1}), lower rates on the newly colonized areas (9.5 to 37.5 mm · y^{-1}), and the lowest rates on the marshes (2.0 to 4.25 mm · y^{-1}). Reduced tidal submergence and reduced water depth result in less deposition. Winter ice and storms cause erosion or reduced accretion, and there are probably substantial annual variations in deposition rate: the average rate over 174 years is 3.4 mm · y^{-1} (Richard 1978). At the Barn Island marshes, at the border between Connecticut and Rhode Island, the mean tidal range is 80 cm. Between 1947 and 1988, the marsh vegetation has changed from typical *Juncus* and *Spartina* belts to a mosaic of *Triglochin* and *Spartina* with a complex of stunted halophytes. This is primarily related to different rates of accretion and sea level rise, maybe also to changes in sediment supply and to anthropogenic modifications. The present marsh vegetation is only 4.6 cm above mean tide level, whereas the older vegetation was 13.9 cm above mean tide level. The rate of vertical accretion in the stable marshes is about equal to the sea level rise, which is about 2.0 to 2.5 mm · y^{-1}, but in the modified areas the rate of accretion is about half that rate, which results in more frequent and longer flooding and wetter marshes with a more open vegetation (Warren and Niering 1993). In Narragansett Bay the semidiurnal tides have a mean range of 1.1 m, which increases inward to 1.4 m at the head of the bay. The marshes are covered with *Spartina* and were formed during the past 3000 to 4000 years. Sedimentation rates, determined with ^{210}Pb and based on anthropogenic copper waste that began to be deposited in 1865 to 1885, are for the lower marshes 2.5 to 6.0 mm · y^{-1} (average 4.3 mm · y^{-1} based on ^{210}Pb) and 3.0 to 5.9 mm · y^{-1} based on the copper concentrations. For the higher marshes the averages are 2.4 mm · y^{-1} and 4.7 mm · y^{-1}, respectively. For the lower marshes the accretion rate is about 1.5 times the local relative sea level rise; for the high marshes it is about equal. The deposited peat has a high water content: about 9% dry solid matter contributes to the accretion (Bricker Urso et al. 1989). At Barnstable, Massachusetts, salt marshes lie above mid-tide level along open flat areas bordered by bare sand flats below mid-tide level, that are drained by shallow channels (Figure 4.68). At Chatham Harbor, Cape Cod (Figure 4.69), flood and ebb tidal deltas are partly intertidal. The semidiurnal tides have a range of 2 m in the adjacent ocean, 1.3 m at the inlet, and 0.90 m west of the inlet in Nantucket Sound. The flood tidal delta and adjacent spits are covered by fields of sand waves and megaripples, while on the ebb tidal delta, which is only in small part intertidal, swash bars are formed (Hine 1975).

4.28. ICELAND

Around Iceland the mean spring tide range is 1.5 to 4 m but is strongly influenced by the local configuration of the coast, with many narrow gulfs and fjord-like indentations. The coast is rocky or sandy with coarse sand and gravel of volcanic origin, with sandy barriers, and along the south coast, coarse sandy flats with a high rate of fluvial and littoral sedimentation without tidal flats or salt marshes. Tidal flats only fringe some shallow lagoons on the southeast coast (Figure 4.70; Bodéré 1985).

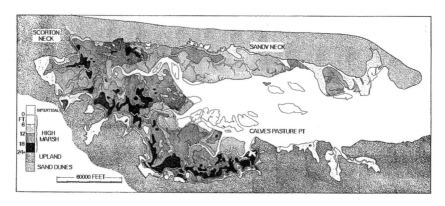

FIGURE 4.68   Barnstable area. (From Redfield 1967. With permission.)

4.29. MICROTIDAL DEPOSITS

The microtidal deposits along the northern coasts of Eurasia, in Alaska, and in Iceland have been relatively little studied. In Europe and South America, microtidal areas are rather rare or are part of large mesotidal areas like the Wadden Sea and the Amazon mud dispersal system and are discussed in Chapter 3. Where the tidal range is very low and considered to be absent, as in the Mediterranean, the Baltic, and along large parts of the Caribbean coasts, wind effects dominate and will be discussed in Chapter 5. The relatively well-studied microtidal areas are mainly in Africa, south and east Asia, Australia, and North America (except Alaska). The microtidal areas in the Persian Gulf, the Bahamas and the British West Indies, and in Shark Bay, West Australia, have attracted special attention because of the present

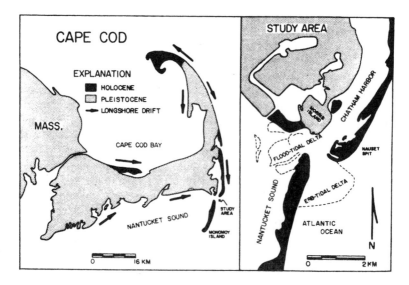

FIGURE 4.69   Cape Cod and Chatham Harbor estuary. (From Hine 1975. With permission.)

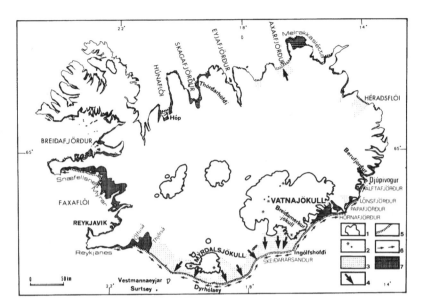

FIGURE 4.70 Iceland. 1 = glaciers; 2 = subglacial volcanoes and/or high temperature fields; 3 = outwash plains; 4 = tracks of main glacier bursts; 5 = beaches and barrier spits; 6 = predominant longshore drift; 7 = strand flats. (From Bodéré 1985. With permission.)

formation of carbonate and algal mats (stromatoliths). Based on the well-studied areas, the following observations can be made:

Sandy barriers, barrier islands, and/or spits are present along all microtidal coasts except where protected by reefs (Trucial Coast, southwest Florida, Pacific Islands), or where the sediment consists almost exclusively of silt and mud, and sand is rare (Yellow river delta, Bohai Sea).

Cheniers are present in Louisiana (sandy) and in Qatar-Oman, Bohai Bay, and Shark Bay (carbonate).

Channels and creeks occur in a wide variety, as on the tidal flats and marshes with a larger tidal range.

Straight channels: in the Niger delta, the Zaire river mouth, on the Farewell spit flats, and on Andros Island, but elsewhere they are rare. In northwest Florida they follow the carbonate dissolution zones.

Curved or sinuous channels: in the Dyfi estuary, in Moulay Bou Salham and Oualidia, in Sine Saloum and Casamance, Sierra Leone, the Niger delta, the Zaire river estuary, Langebaan Lagoon, Qatar-Oman, the Godavari delta, the Farewell spit flats, the Mississippi delta, northwest Florida, Andros and Caicos islands, in Charleston Harbor, and along the Virginia–New Jersey coast.

Meandering channels: in Oualidia, Sine Saloum, Casamance, the Volta river delta and Benin–Niger delta, the Zaire river estuary, Langebaan Lagoon, Mississippi delta, Andros and Caicos Islands, Charleston Harbor, North

Inlet, Murrell's Inlet, Chesapeake Bay, along the Virginia–New Jersey coast, and Barnstable harbor.

Dendritic channel systems: in the Dyfi estuary, Sierra Leone, the Niger delta, Andros and Caicos Islands, Charleston Harbor.

Elongated dendritic channel systems: the Zaire river estuary.

Interconnecting channels: Sine Saloum and Casamance, the Niger delta, the Zaire river estuary, Andros Island, North Inlet, the Virginia–New Jersey coast.

Distributary channels: Trucial Coast.

Parallel and **braided** channels have not been observed in microtidal areas. Interconnecting channels cover large areas; dendritic and distributary channel systems are rather small or not well developed. Within the interconnecting and meandering systems a difference can be made between stationary systems with few or no abandoned channels or meanders and with little or no erosion (e.g., North Inlet, New Jersey coast), and gradually changing systems with lateral erosion and abandoned channels and meanders.

Point bars and channel levees have been observed in the Niger delta, along the Trucial Coast, at the Mississippi delta, the Andros and Caicos Islands, and along the Delmarva Peninsula, where the tidal range is relatively high. On the Niger delta flats tidal watersheds occur, and at the Delmarva Peninsula small intertidal deltas. Slumps of channel walls occur where active meandering takes place (e.g., Murrell's Inlet).

Sediment structures are present, as in the macro- and mesotidal deposits, in a wide variety (and often not primarily related to the tidal range). Parallel lamination is present in the Yellow river delta, Shark Bay, the Mississippi delta mud flats, and Andros and Caicos islands, but in most microtidal areas it is rare or absent. This points to a diminished influence of tidal sheet-like flow and a predominance of unidirectional currents, waves (in particular storm waves), and bioturbation on sediment structures. Sand/silt/mud beds and interlayered sediments, including plant debris, are described from the Niger delta, the Mtamvuna river mouth, the Yellow river delta, the Mississippi river mouth marshes, Andros and Caicos islands, and salt marshes along the U.S. east coast. Algal mat deposits are laminated and usually interlayered with sediment (Shark Bay, Andros Island). Hurricane deposits on Caicos consist of two coarse-grained layers with a fine-grained layer between. Megaripples have been observed in Murrell's Inlet (along channels and the ebb-tidal delta) and in Sierra Leone (on the flats), where also, in the channels, sand dunes are formed. Small ripples and/or ripple structures have been observed in most microtidal areas and are probably present in all intertidal areas. Current lineations have been seen in Sierra Leone and Langebaan Lagoon; cross-stratification is known from Sierra Leone, the Niger delta (small and large), Langebaan Lagoon, and Andros and Caicos islands; and herringbone and runoff structures (rill marks, meandering runnels, miniature deltas) from Langebaan Lagoon. Other erosional features are rare: mud bastions left by erosion are described from the Hai He river delta (where the erosion is man-induced); a soft ooze, covered with a rhizomatous mat, formed by salt marsh vegetation, occurs in eroding areas in the Blackwater river area (Chesapeake Bay). Scour pits are described in Sierra Leone.

A number of phenomena on microtidal flats are related to a very dry climate or a very dry season: mud cracks (Sierra Leone, Langebaan Lagoon, the Trucial Coast, Shark Bay, Andros and Caicos), crusts (Trucial Coast, Shark Bay, Andros, Caicos) and crust flakes or pebbles (Shark Bay), sebkhas (Northwest Africa, Persian Gulf) and high bare (salt) flats (tanns: Senegal to Liberia), and algal mats (Sierra Leone, Langebaan Lagoon, Trucial Coast, Shark Bay, Spencer Gulf, Mississippi delta, Andros, Caicos). Fluid mud was seen only on the Louisiana coast. Microtidal river mouths and estuaries do not have a characteristic shape. Rivers tend to form deltas on microtidal coasts (Niger, Mahanadi, Godavari, Yellow river, Mississippi) but may not do so when the sediment is easily transported away toward deeper water (Zaire), or where longshore transport dominates (Cauveri), or where the sediment is dispersed (Chao Phraya). The rivers usually flow to the sea through a number of distributary channels (but not in the case of the Chao Phraya).

5 Wind Flat Deposits

Wind flat deposits are formed along coasts where the tidal range is small, in the order of several decimeters at most, and onshore winds push the water regularly toward the coast. The area that is flooded depends on the flatness of the coast and the temporal and local rise in sea level; flooding can cover an area up to 10 or 20 km inland. The flooded area is temporarily covered with salt or brackish water and a vegetation may develop (if at all) that is adapted to this. In the following sections an overview is given of the most important wind flat coasts in the world. The inaccessibility of most of the Russian and east European literature on this subject makes the sections on the Baltic, Black Sea, the Caspian, and east Siberia, including the majority of wind flat coasts, shorter than desirable. The main difference between tidal flats and wind flats is the more cyclic character of the tidal flats where flooding occurs in daily, fortnightly, and often also seasonal cycles, whereas the flooding of wind flats is more variable and episodic, but often also with a seasonal variation.

5.1. THE BALTIC

Along the Baltic, where the tidal range is less than 0.5 m, westerly winds push the water to the eastern shores, and northerly winds to the south shore. Large parts of these coasts consist of barriers, spits, and beaches with fore dunes and cliffs. Along the eastern shores longshore transport of sand is considerable between Kaliningrad and the Gulf of Riga, reaching a maximum of about $1 \times 10^6 \ \mathrm{m}^3 \cdot \mathrm{y}^{-1}$ at the entrance to the Gulf. Swell is negligible in the Baltic, but heavy storms, which occur every 6 to 10 years (or 20 to 25 years if only the most severe are counted), together with a simultaneous lowering of the atmospheric pressure, give a temporal rise in sea level up to 2.5 m. Along the flat east coast of the Gulf of Riga, this results in the formation of the so-called meadow beaches with fine-grained sediment, and along the Gulf of Finland, in low coastal marshes (Gudelis 1967, 1985; Alestalo 1985). Along the Polish coast storm surges cause only local flooding, besides the destruction of barriers, formation of overwash fans, intrusion into coastal lakes, and formation of inward inlet deltas. Flooding also enhances the longshore transport of sand and contributes to an intensive cliff erosion of 0.6 to 1 m $\cdot$ y^{-1}, locally up to 2 m $\cdot$ y^{-1} (Borowka 1985). Swampy shores that are flooded during episodic (seasonal) periods of high water level, occur along 28% of the Polish coastline (Figure 5.1). In the western Baltic, storm surges may raise sea level by up to 3.5 m. This also affects bays and lagoons not directly influenced by the wind, and in sheltered areas

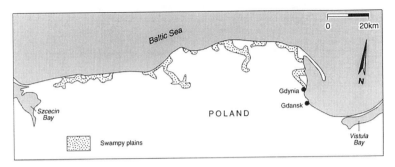

FIGURE 5.1 The Polish coast with low coastal plains. (Adapted from Borowka 1985.)

behind spits and barriers about 25 km² of salt marshes have been formed (Gierloff Emden 1985a,b).

5.2. THE MEDITERRANEAN

In the Mediterranean, temporary higher sea level occurs mainly through a combination of increased tidal range, induced by the shape of the basin, and strong onshore winds. In the Adriatic, which is practically a closed basin, the tides reach a range of 1 m, but during high tide in combination with strong winds from the southeast, low atmospheric pressure, and seiches, sea level may rise by more than 2 m. Along the northern Adriatic around Venice, this has resulted in the formation of extensive coastal swamps (Mosetti 1971; Zunica 1985). In the Gulf of Syrte along the coast of Libya, which also has the character of a more or less closed box, the mean tidal range reaches over 50 cm, while the spring tide range reaches a little below 1 m and even higher during onshore storm winds. Salt flats (sebkhas), that normally remain dry, are flooded during such periods (Davies 1973; Schwartz 1985).

5.3. THE BLACK SEA

In the Black Sea, the tidal range does not exceed 8 cm, and fluctuations of sea level are almost exclusively caused by variations in wind force and atmospheric pressure. Wind flats occur where the coast is low and flat along the northern shore between Dniepr's Liman and Bakal Spit. Winds blow more from northerly and easterly directions (about 45% of the time) than from southerly (southeasterly) directions (about 33% of the time). Also the stronger winds come from the north, in particular during the winter. Wind-induced lowering of sea level therefore prevails over wind-induced rising of sea level. Lowering of sea level is normally by less than 1 m, but can be up to 2 m; wind-induced rise in sea level is by 50 to 70 cm. This results in the flooding of an area along the coast 0.5 to 2 km wide. During onshore storms, suspended-matter concentrations in the coastal waters increase by two orders of magnitude because of erosion of fine-grained coastal deposits. The bulk of this resuspended or eroded material is transported seaward by currents reaching

1.2 m · s⁻¹. This is similar to what happens along tidal coasts during onshore storms. On the flats, a veneer of reworked sediment is left: the flats are predominantly erosional (Shuisky 1985).

5.4. THE CASPIAN SEA

Wind flats in the Caspian Sea are present along the northwestern and northern shores (Figure 5.2). The flat coastal plain allows flooding several km inland during southerly winds. During northerly winds, the water level is lowered and the shoreline retreats several km southward. The variations in sea level are accompanied by strong currents up to 1.5 m · s⁻¹ that erode the coastal plain and the shallow seafloor: channels and

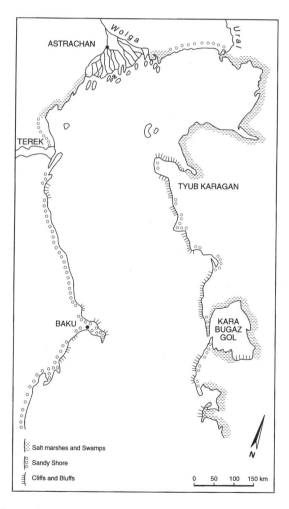

FIGURE 5.2   The northern shore of the Caspian Sea with low flats. (Adapted from Leontiev 1985.)

mud flats are formed, which are very similar to those formed in tidal areas. Debris cones are formed at the lower end of the channels (Leontiev 1956; Zenkovich 1962). The tidal range is only a few centimeters but the variations in sea level because of storm winds can reach 3 m. The southerly winds also give steep waves in addition to occasional storm surges. In the open sea, storm waves can reach 6 m (with 11 m as the recorded maximum). The entire north shore of the Caspian Sea on both sides of the Volga river delta from the Terek river delta to Tyub Karagan consists of salt marshes and swamps and has advanced several km because of a general fall in sea level. Between the Volga river delta and the Ural river and east and south of the Ural river are wide tidal flats. Because of the arid climate, the sediment dries out between floodings, and clay crusts are formed which break into fragments that can be loosened by the wind, and clay dunes are formed. The lowering of the sea level of the Caspian Sea began after the maximum level was reached after the Holocene sea level rise. This was in the 16th century. A sharp drop of about 2 m occurred between 1919 and 1945; present sea level is 28.5 m below the level of the Black Sea. Along the east side of the Caspian Sea is Kara Bugaz Gol, a hypersaline lagoon of about 18,000 km² with vast salt flats, in particular on the north side. It was formed after 1930 as a result of the lowering of the sea level and is now closed off with a dam (1981) (Leontiev 1985).

5.5. EAST SIBERIA

Along the East Siberian Sea lies a vast flat coastal plain bordering a very shallow coastal sea. The tidal range is only several centimeters. During strong northerly winds, the shore areas are flooded up to 30 km inland; during strong southerly winds, the shoreline retreats seaward over a similar distance. The sediment on the flooded plain is muddy (Sovershaev, 1977; Zenkovich 1985).

5.6. LAGUNA MADRE, MEXICO/TEXAS

The Laguna Madre, a large lagoon, extends along the coast of northern Mexico and southern Texas behind a narrow barrier island (Figure 5.3). The lagoon is bordered by bare mud flats, salt marshes, deltaic deposits, and natural levees along the channels. The deposits formed along the Rio Grande river mouth, which forms the Mexico–U.S. border, divides the lagoon into two parts. The tides have an average range of 60 cm and a maximum (spring tide) range of 90 cm, but sea level along the Gulf coast may be raised to +4.8 m during storms (hurricanes, which mostly come from east-southeasterly directions), and in the bay, may rise to +6 or +7 m. The southern part of the Texas section of the Laguna Madre, separated from the Gulf by the long and narrow Padre Island, is less than 1 m deep with a maximum depth of 2.4 m and relatively protected against hurricane flooding. The tidal range here is 15 cm (in the adjacent Gulf, 40 cm); wind and atmospheric pressure can give a variation in level up to 60 cm. Most of the lagoon is bordered by wind flats approximately 100 km long and up to 20 km wide that are flooded during strong northerly winds and during hurricanes (Figure 5.4). The flats are broadest between

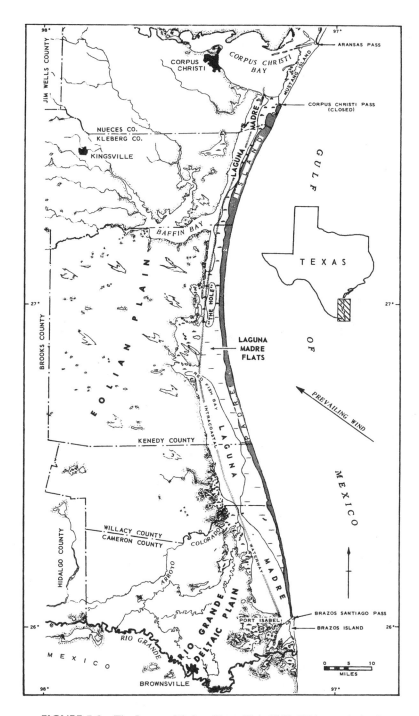

FIGURE 5.3   The Laguna Madre. (From Fisk 1959. With permission.)

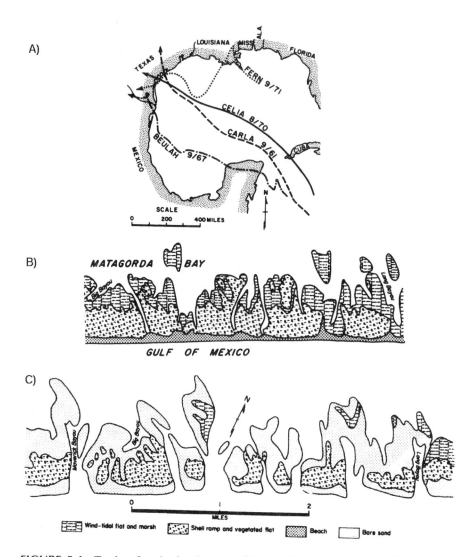

FIGURE 5.4 Tracks of major hurricanes striking the Texas coast (1961–1971) (A) and effects of hurricane Carla (1961) on Matagorda Peninsula, northeast of Laguna Madre (B, C). B = Matagorda Peninsula in 1957; C = Shortly after passage of Hurricane Carla. (From McGowen and Scott, 1975. With permission.)

26° 24′ N and 27° N and form a bridge about 10 km wide between the Texas coast and Padre Island. On the inner parts of the flats, clay dunes have been formed out of clay flakes that are the result of intermittent drying and wetting of the flats and the semi-arid climate. The irregular inner contours of the lagoon are caused by these dunes as well as by washover channels and deposits, and by aeolian sands blown inward from the Gulf beaches. The water in the lagoon is usually hypersaline with salinities that are regularly above 50‰; the turbidity is mainly related to the wind force (resuspension of bottom sediment) and direction (suspended matter dispersal). The elevation of the flats above the mean water level determines the extent and frequency of flooding, and together with variations in the supply of (mainly aeolian) sand and suspended matter, a variety of deposits is formed (Figure 5.5).

On the frequently flooded lower and middle flats blue-green algal mats and layers of olive-grey clay are common and well developed on top of lagoonal sand and poorly laminated clay. Desiccation and burrowing produce contortion of the clay bands and mats. Higher on the flats where flooding is less frequent, sand becomes a major component, and gypsum is being formed, filling in burrows and voids formed by trapped gas bubbles or by desiccation. Ponds of usually less than 25 cm depth are formed in shallow depressions and can develop into saltpans with ephemeral crusts of halite. After dissolution of the salts, a layer of organic black clay remains, containing decaying grass and filamentous green algae. Lens-like crystals or crystal aggregates of selenite occur in the sediment, in many cases oriented parallel to the bedding. In the clay-algal mat sediment they occur at approximately 1 m depth in the sediment where the content of interstitial water is high, and in sandy deposits well below the water table at more than 2 m depth. Minor amounts

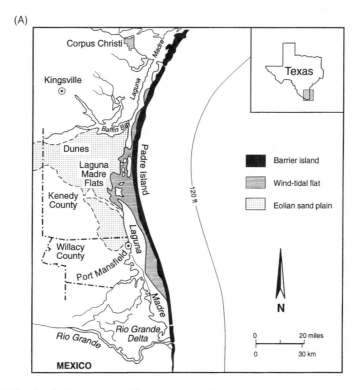

FIGURE 5.5   South Texas coast with major depositional environments (A), and the distribution of sedimentary facies, Padre Island and Laguna Madre flats (B). (Adapted from Miller 1975.)

(B)

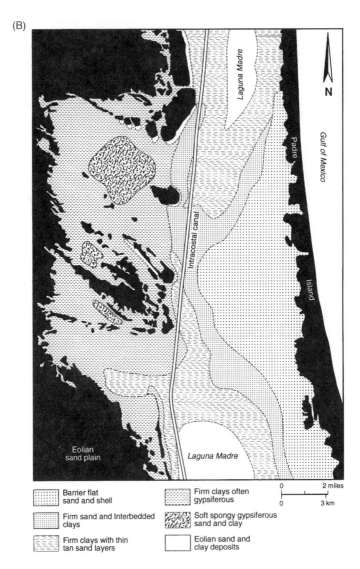

	Barrier flat sand and shell		Firm clays often gypsiferous
	Firm sand and Interbedded clays		Soft spongy gypsiferous sand and clay
	Firm clays with thin tan sand layers		Eolian sand and clay deposits

FIGURE 5.5 *Continued.*

of authigenic aragonite, calcite, and magnesium carbonate occur in low percentages (<5%) but locally may become 100%. Bacteria probably produce the soil conditions that are suitable for the formation of authigenic carbonate and gypsum (and iron sulfides) (Fisk 1959, Miller 1975, McGowen and Scott 1975, Aronow and Kaczorowski 1985).

6 Morphology of Intertidal Areas

Intertidal depositional areas are predominantly flat, which reflects the sheet-wise movement of the water propagated by tides and waves. This also applies to the smaller intertidal areas such as the intertidal parts of largely subtidal deposits like the tidal deltas at inlets and the sand- or mud banks in tidal estuaries (Figures 6.1, 6.2, and 6.3). The steepest intertidal areas are the narrow intertidal strips along tidal beaches. The flatness of intertidal areas is retained when they become covered with vegetation and in supratidal flats that are still occasionally flooded by sheet-wise moving water. The flatness is also reflected in the flat, parallel horizontal or nearly horizontal bedding that dominates in the surface sediments of macrotidal and mesotidal areas. In most microtidal deposits, horizontal layering is rare or absent. This points to a diminished influence of tidal sheet flow and an increasing predominance of unidirectional (tidal) currents and bioturbation.

The overall flatness is interrupted by 1) ridges that are transverse to the main direction of flow (barriers, cheniers, large ripples, dunes, erosional steps) and by (much rarer) longitudinal ridges, 2) channels, creeks, and gullies of different types and dimensions with associated features such as channel levees, 3) by vegetation (shrubs, trees, algal domes), 4) by outcrops of older deposits or bedrock, and 5) by man-made constructions. Because of these morphological interruptions and by a

FIGURE 6.1   Sandy tidal flat (Wadden Sea) at low tide. (Photograph courtesy of J. van de Kam.)

FIGURE 6.2   Fine sandy-muddy tidal flat (Wadden Sea) at low tide. With *Arenicola* and *Mytilus* (see Chapter 9, Part I). (Photograph courtesy of NIOZ.)

FIGURE 6.3   Sandy tidal flats around a tidal inlet (Wadden Sea). (Photograph courtesy of J. Abrahamse.)

variety of small-scale features, tidal areas, in spite of their general flatness, show a large morphological variability which is accentuated by differences in sediment composition (sand, silt, clay, carbonate), vegetation, and climate. Only few of these morphological interruptions, however, are related to the tides: barriers, cheniers, and dunes are primarily related to waves, longshore currents, winds, and variations in sediment supply. Older outcrops are related to the geological and morphological history of the area, while human activity may often be, but is not necessarily related to the intertidal character of the area.

All intertidal deposits, by definition, are located in tidal areas that are predominantly depositional. Usually, however, there is a mixture of local, regional, and temporal deposition alternating with erosion with an overall net sedimentation. Even in King Sound (Australia), where erosion seems to dominate, recent intertidal deposits are formed. The morphological features dominated by erosion include channels, creeks, gullies, erosional steps and pits, hollows, and low points in the sediment surface. Channels or creeks need not be erosional and can be formed simultaneously with adjacent tidal flat deposits. Depositional features include, in addition to a flat sediment surface, ripples of different dimensions, channel levees and undulations.

Generally tidal flats and intertidal deposits are related to tidal elevation: high or upper tidal flats are located above mean neap high tide level, middle flats between mean neap high tide and mean neap low tide level, and lower tidal flats below mean neap low tide level. Supratidal flats are above mean high spring tide level, and subtidal flats, which can be present as a seaward extension of the intertidal flats, below mean low spring tide level. These distinctions, however, are not everywhere made in this way and often are less precise when only high and low tide levels or mean sea level are indicated. Also sediment type is often related to tidal elevation: the lower and middle flats are usually sandy, except at the Colorado river mouth and along the Chinese coast where entire flats can be muddy with some sandy deposits on the middle tidal flats. The higher flats are usually muddy and, at higher latitudes, covered with salt marsh. The upper part of the middle flats is, in the tropics and subtropics, usually covered with mangrove and often muddy (see the summary for siliciclastic flats in Amos 1995).

In analogy to the distinctions made by Thom (1982) and Woodroffe (1992) in the occurrence of mangrove based on the classification proposed by Wright et al. (1974) for deltas and by Roy et al. (1980) for bedrock embayments (rias), intertidal areas can be:

- River-dominated (at the larger river deltas)
- Tide-dominated (along many meso- and macrotidal coasts, e.g., in China and Korea)
- Wave-dominated (temporarily during the passage of storms, typhoons, hurricanes; more permanent along microtidal and sheltered coasts (lagoons), e.g., along the U.S. east coast and the Gulf of Mexico)
- Composite river-wave-dominated (where a large river supply is distributed along the coast by waves, e.g., Niger delta)
- Ria-type (mainly bayhead deposits, e.g., Atlantic coast of France and Spain, northwest Australia)
- Carbonate type (with mainly locally produced carbonate sediment, but terrestrial supply can be considerable; Florida, Bahamas, Australia)
- Ice-dominated (where ice is present during a significant period, reduces the sediment supply, and induces supply of a wide range of grain-sizes)

Relatively few comparisons have been made between intertidal flats in different areas. deVries Klein and Sanders (1964) compared intertidal sediments from the Bay of Fundy and the Dutch Wadden Sea, while deVries Klein (1967) made a wider comparison including ancient tidal flat deposits, carbonate sediments, estuarine sediments (Recent and ancient), and some evaporite deposits. Marshes and estuaries at different latitudes were described by Guilcher (1979), who emphasized the strong influence of differences in climate on the intertidal areas: the distribution of salt marshes, saltpans, flats with boulders and gravel, mangrove, tropical salt marshes, and dry salt flats to a large extent is determined by climate. Alexander et al. (1991), from a comparison of the flats in Korea, the Gulf of California, Louisiana, the Chinese coast, and Suriname, concluded that muddy intertidal environments can be formed and persist with or without the presence of a seaward barrier and within a wide range of environmental conditions: from micro- to macrotidal, with weak to strong currents, and low to high wave energy. Wells (1983) compared mud flats and fluid muds in Louisiana (subtropical, microtidal), Suriname (tropical, low-mesotidal), and the west coast of Korea (temperate, macrotidal): all muddy coasts with high suspended-matter concentrations in the water, fluid muds, and mud shoals. The only major requirement for their formation (and for the muddy intertidal deposits) is a large source of fine-grained sediment. Similar sediment structures are observed in the (more sandy) Korean flats and the Wadden Sea flats, as well as along the U.S. east coast and in the Bay of Fundy, but ancient muddy tidal flats are difficult to recognize because they have few distinctive features, and probably for that reason have not been much reported.

A general division in types of intertidal areas can be made in terms of tidal range, climate and geographical/geological setting (flanking rocky coasts, or along a low-relief soft-sediment coast; deVries Klein 1967). This gives a kaleidoscope of about 20 different types, to which can be added carbonate flats and evaporite flats, insofar as these do not belong to a climatic subdivision. The comparisons by Wells (1983) and Alexander et al. (1991) indicate that even more subdivisions are needed than are indicated in such a scheme, to which more subdivisions may be added based on channel type (see following paragraph). This is not very practical, and it is unlikely that such a classification of intertidal areas and deposits would lead to more insight in tidal flats and intertidal deposits than a consideration of the processes involved in their formation. This is even more the case because generalizations lead to very uncertain results when only part of the intertidal areas has been studied to some extent and only relatively few have been studied well.

6.1. THE INTERTIDAL SURFACE

Wright et al. (1973) and Pethick (1996) have indicated that the tidal flat and channel morphology and the energy of tides and waves are interrelated: the morphology depends on tides and wind, and the tidal properties and wave activity depend on the size and shape of channels and flats; any consideration of relationships has to take this interaction into account. Thus tidal resonance results in funnel-shaped estuaries with a wide shallow cross-section. The form of the tidal wave in a channel depends on the water depth: at shallow depths, the water depth is significantly larger during flood than during ebb with a faster propagation of the tidal wave during flood, which results in a tidal asymmetry with a stronger flood of shorter duration and a weaker ebb of longer duration. The initial deposition results in extensive mud flats which gradually become higher and accommodate less water during high tide. This reduces mean channel depth and the tidal wave length. Continuing deposition on the flats finally results in a tidal wave length that is approximately a multiple of four times the length of the tide-influenced channel or estuary. Resonance then produces a standing wave with a nodal point (with maximum flow velocities) at the channel mouth and an antinode with zero velocities at the inner end (Pethick 1996). Mud flats then extend to the inner end of the channel, as is found in the estuaries along eastern Britain, which also have a similar ratio between tidal channel length and tidal wave length. Pethick (1996) indicates that there is also a relation with the tidal range: macrotidal estuaries are long and more funnel-shaped, while mesotidal estuaries tend to be short and broad. The data given in the previous chapters also indicate that funnel-shaped channels are more common in macrotidal estuaries. At the larger tidal amplitudes in macrotidal channels, the effect of water depth is already present at greater water depths so that the development of a funnel-shape is favored and will start earlier.

Another important factor is the total volume of water that is exchanged during a single tide through a channel cross-section, which determines the discharge and therefore the flow through the cross-section. This applies to any cross-section in a tidal channel and has been used to model an equilibrium morphological framework of a tidal basin (van Dongeren and de Vriend, 1994). Development of channels and flats in plan-form are paralleled by changes in cross-sectional area and depth of the channels. Shallowing of the channels results in increasing tidal asymmetry (flood dominance), with increasing deposition on the intertidal flats. Most of the deposition takes place at higher levels on the flats, where current velocities are low, which gives the flats a flat, nearly horizontal upper surface with a break in the slope around or below mid-tide level (Figure 6.4; Evans 1965). The deposition reduces the water volume over the flats during high water, which reduces the progression of the tidal wave crest during the flood, so that eventually an ebb-dominated channel may develop with export of sediment. Interfering with this is an increasing erosion of the flats by waves, when they get higher, but also the degree of exposure to larger waves plays a large role. Pethick (1996) describes the Colne estuary in eastern Britain, which was flood-dominated in 1918 (with a shallower channel) and ebb-dominated with a deeper channel because of increased erosion in 1949. The deepening of the channel will eventually reverse the estuary into a flood-dominated system, which indicates the possible existence of long-term oscillations in tidal systems. Waves can easily resuspend sediment on the tidal flats, but the horizontal displacement is small, so that erosion and lowering of the flats only takes place when the tidal currents remove the resuspended sediment during the ebb. This is enhanced when the water level is also raised by the wind and a backflow develops. Lower tidal, or backflow, current velocities are required for removal than for erosion as the sediment has already been resuspended by the waves. Wave resuspension is intermittent, depending on the variations in velocity and direction of the wind. Pethick (1996) distinguishes between (estuarine) intertidal areas that are exposed to open sea waves that can be propagated in the channels, and sheltered areas that are only exposed to local waves. Generally, larger waves are less frequent than smaller waves, but it is mainly the larger ones that cause erosion and produce the flat, wide surface profiles of the flats. Eroded flats can be built up again, when the larger waves have subsided, provided sufficient sediment is available. On inner flats the wave heights are largely determined by the area over which the waves are generated. The flats will tend to develop a shape and width capable of fully attenuating the waves with maximum wave height and will respond to variations in waves by erosion or deposition. The inner flats tend to become higher (and narrower) and with a steeper

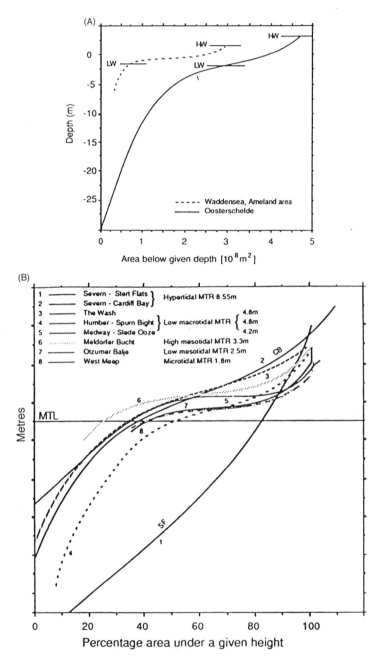

FIGURE 6.4   Hypsometric curves for an eroding estuary (Ooster Schelde) and an accreting estuary (Wadden Sea) (A); Hypsometric curves for some British and German estuaries (B). (From Pethick 1996. With permission; and adapted from Dronkers 1986 and Kirby 1992.)

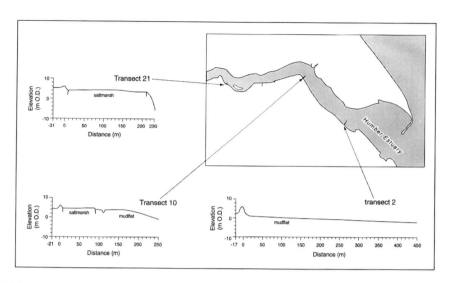

FIGURE 6.5 Intertidal cross-sections from the Humber estuary. (From Pethick 1996. With permission.)

outer slope than the more exposed flats, as is demonstrated by the flats in the Humber estuary (Figure 6.5).

Because of the seasonal variations in wave intensity, intertidal flats usually show seasonal variations in elevation and gradient, with low elevations and low gradients during stormy periods. Mud flats, which are more cohesive and consolidated than sand flats, respond more slowly to seasonal wave variations, but sand flats are formed more in the exposed areas, where suspended matter does not easily settle and is easily resuspended. Even small (ebb tidal) gullies and channels can survive the winter storms in the more sheltered Dollard area of the Dutch Wadden Sea, whereas such gullies are obliterated on the more exposed sand flats. Salt marshes as well as mangrove are closely linked with the bare flats within the same intertidal area as two parts of the same morphological unit. The seaward salt marsh or mangrove edge marks the line where wave energy and water depth become so low that vegetation can settle, and shift with changing wave conditions. Erosion of the salt marsh releases large amounts of sediment that are deposited on the bare flats in front, and conversely, sediment is transported from the upper bare flats into the salt marshes, which then expand farther seaward. In mangroves this shifting is less obvious because the sediment covered with mangrove is less likely to be eroded (see Section 9.2 and 9.3).

6.2. CHANNELS, CREEKS, AND GULLIES

The ten types of channels and channel systems that can be distinguished (where channels are mentioned, creeks and gullies are included) can be divided into single channel types (straight, sinuous, or meandering), types of channel systems (parallel channels, dendritic and elongated dendritic, distributary, braided, interconnecting), and few or no channels (Figure 6.6). The factors involved in the formation of tidal

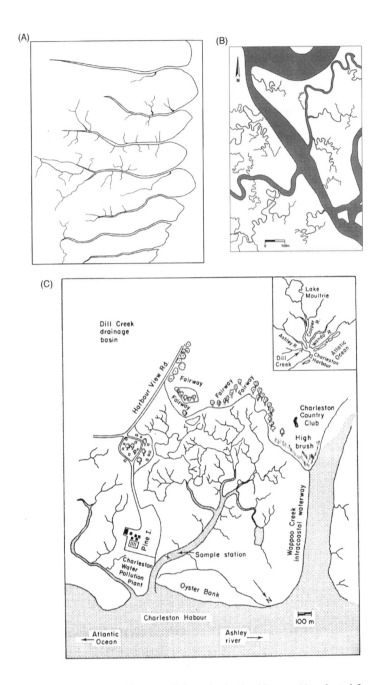

FIGURE 6.6 Channel types. Single: straight and wiggling/sinuous (A; adapted from Pethick 1992) and meandering (B; adapted from Zeff 1988). Channel systems: parallel (A), dendritic (C; from Settlemyre and Gardner 1977. With permission), elongated dendritic (D), meandering and interconnecting (E; *Source*: Allen 1965b.) (D and E on following page.)

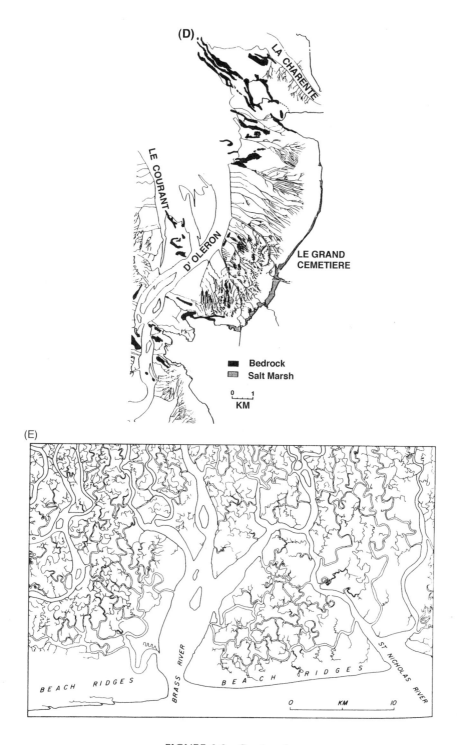

FIGURE 6.6 *Continued.*

channels include the tidal characteristics and the characteristics of waves and flow, the volume of water involved, the slope, type of sediment, sediment consolidation, the amount of sediment transported in relation to the water volume transport, vegetation, and faunal activity (burrowing, pellet formation). This list is by no means exhaustive: among other aspects, instabilities and lag-effects may play a large role locally, if not in a more general way. The different types of tidal channels, creeks, and gullies have been relatively little studied compared to the work that has been done on sediment transport and sediments. Here, what is known about the different channel types, as given in the previous chapters, is discussed. In rivers and streams, a general distinction has been made between contributive and distributive channel systems (Allen 1965), and in contributive systems (which form the majority), between dendritic, rectangular, and radial systems (Zernitz 1932). These distinctions are usually related to geological structure and history, the original relief (slope), and climate. Channels within these systems can be braided (anastomosing), meandering, or straight. Schumm (1963) distinguished five groups of channels based on their sinuosity: straight, transitional (both sinuous and regularly meandering), irregularly meandering, and tortuously meandering. Straight channels can wiggle and have irregular undulations or bends which become more pronounced in the transitional type. Leopold et al. (1964) termed channels meandering when the ratio between channel length and downstream distance (the "sinuosity ratio") is 1.5 or greater, sinuous or straight when this ratio is below 1.5. The development of a braided channel is considered to result from either extremely coarse bed material (competence limited), or sediment overloading (capacity limited; Brierly 1989). The interconnecting channels, observed in intertidal areas, can be regarded as a special type of braided channel system, being originally a capacity-limited channel system with channels and banks, developed by tidal currents in a tidal depositional area. The overloading has been caused, however, by the (slow) supply and deposition from the same flow that becomes overloaded. Brierly (1989) also distinguishes wandering channels, which are unstable and are a transition from braiding to meandering channels. The different types of channels can occur within the same channel system.

As seen in the preceding chapters, straight single channels occur in macrotidal and mesotidal areas, but somewhat less in microtidal areas. Sinuous (curved) and meandering channels occur in all three types of tidal areas. Parallel channels, distributary and braided channel systems, and flats with few or no channels occur, with a few exceptions, only in macrotidal areas. Dendritic channel systems and interconnecting channels occur in all three types of tidal area, but elongated dendritic systems only (with one exception) in macrotidal areas, while interconnecting channels cover large microtidal areas. Tidal chutes, tidal watersheds, and channel levees occur mainly in meso- and microtidal areas. Lateral shifting of channels is mainly described from macro- and mesotidal areas. In some interconnecting channel systems (in North Inlet and along the New Jersey coast), it is observed that the channels do not shift laterally.

Brierly (1989) noted that there is no relation between the one-dimensional river plan-form and the differentiation of fluvial deposits. The data given in the previous chapters indicate that there is also no relation between intertidal sediment structures and channel or channel system plan-forms, but, as will be discussed in more detail in Chapter 7, there is a relation between internal sediment structures and the tides.

6.3. SINGLE CHANNELS: STRAIGHT SINGLE CHANNELS

Straight single channels are not mathematically straight but have irregularities and undulations that make the channel deviate from a straight line: they wiggle but the general channel direction is straight and different from a sinuous (or meandering) channel. Straight channels and long straight channel sections occur predominantly in comparatively large tidal channels, often the largest ones in their area (the Wash, Wadden Sea, Mahakam delta, Firth of Thames, Niger delta, Zaire river estuary, Farewell spit flats, Andros Island). Usually small gullies or creeks, coming from the flats, flow into them at right angles. On the Taizhou Bay and Wenzhou Bay flats, straight parallel gullies, separated by low ridges, go downward on the middle part of the flats following the tidal flat slope (Ren ed. 1986; Wang and Eisma 1988). They are initiated where the ebb flow is strong enough to start eroding the sediment surface. The flow is concentrated in parallel zones so that gullies are formed. They end near the line of low tide where the flow is reduced. During the flood and high slack tide the gullies are partly filled in, followed by erosion (resuspension) during the next ebb, but not all the deposited sediment is resuspended during the ebb so that there is a net sediment accumulation (accretion). In San Sebastian Bay (Argentina) the straight channels occur only in sandy sediment; in Broad Sound (Australia), in the mangrove, the smaller channels have straight sections connected with sharp bends (the large channels are arcuate); in the Anse de l'Aiguillon (France) and on the Jiangsu coast, straight channels are continuations of man-made (or man-modified) channels that drain reclaimed areas.

Tidal channels are generally formed initially by the ebb on bare tidal flats with little influence of vegetation. The initial channels are small and start to form where the runoff is sufficient to erode the sediment. It takes a certain length of flow to reach sufficient water volume and velocity: the minimum length that is required is called the critical length (Horton 1945) and depends on the surface slope, the intensity of runoff, the infiltration capacity of the sediment, and the resistance of the sediment to erosion. Initial rills and gullies follow the original slope, which on the flat tidal flat slopes can be maintained for a considerable distance. Above the initial rills the ebb water moves as sheet flow.

Once formed, the smaller creeks and gullies are predominantly maintained where the sediment is cohesive or protected (and reinforced) by vegetation. Where they are formed in bare, sandy sediment, they can easily be eroded during the flood, as has been observed in the Wadden Sea (Van Straaten 1954), and the gullies in sandy sediment are smaller and shallower than those in (consolidated) muddy sediment; on sandy flats they are often absent. Vegetation leads to accretion, in addition to stabilizing the channels and creeks the vegetation itself induces sediment deposition by reduction of the flow between the plants.

In large channels, the channel shape takes longer to change than in small channels because more sediment has to be displaced to change the shape. On Andros Island the flow in the straight parts of the channels is slower than in the sinuous or meandering parts, which also would make the straight parts more stable. Large channels may take centuries instead of decades to change their shape and position significantly, as is well demonstrated in areas where historical records (maps, soundings) covering several centuries are available (as in the Wadden Sea). Most large changes are probably related to episodic catastrophic events. The rather straight gullies in Taizhou Bay and Wenzhou Bay do not change into sinuous or meandering gullies or channels (as in many tidal areas, e.g., along the Jiangsu coast and in the Wadden Sea; Zhang and Wang 1991, van Straaten 1954) because of the large amount of sediment that is deposited between the ebb tides: the gullies are partly excavated anew during each ebb period, and no stable gullies can be formed that can start meandering. In both large channels and small gullies, the kinetics of the processes involved prevent the channels or creeks from developing deviations from the nearly straight channel form, but the straight form itself probably reflects an equilibrium between erosion and deposition, over decades in the case of channels, over a tidal cycle in the case of tidal gullies.

As tidal flats can be regarded as an intermediate stage in the filling-up of a coastal depression (Frey and Basan 1982), channels and creeks — and channel and creek systems — can be inherited, in particular where a bare tidal flat with channels and creeks is transformed into a salt marsh or mangrove and current velocities as well as sediment deposition are reduced (Pye 1992, Perillo et al. 1996). Gardner and Bohn (1980) suggest that a tidal creek in a small marsh basin originally was a freshwater stream. A distinction therefore can be made between fixed, inherited channels and creeks, and those that have been formed under present conditions. Such a difference is not everywhere obvious: inherited or incised channels and creeks can also be subsequently laterally displaced and changed by headward erosion. Most channels in salt marshes and mangroves are probably initially inherited; a subsequent change by lateral erosion is opposed by the root systems of the vegetation as well as by sediment consolidation. New formation of channels on an existing salt marsh is described by Perillo et al. (1996) in Loyola Bay, Rio Gallegos, in southern Argentina. This begins in saltpans that collect water and become hypersaline because of evaporation, which retards and inhibits plant colonization. The winds, which are permanent in this area, enlarge the ponds by producing small steep waves, that erode the downwind side of the ponds and form little cliffs there (microcliffs). This is little opposed by the low vegetation around the ponds. The ponds grow larger by collapse of the microcliffs, and interconnecting sections develop between the ponds, which form proto-channels that can develop into channels by further wind erosion and by headward erosion from the already existing channels. Once a channel has developed, new vegetation may colonize its borders because water of lower salinity comes in.

6.4.   SINUOUS SINGLE CHANNELS

Sinuous (curved) channels cannot be sharply distinguished from (almost) straight channels, but as a group can easily be recognized as being different from either straight or meandering channels. Of the sinuous channels that are described, it cannot be assessed to what extent they are stable, except for those in the Wadden Sea where older charts show that sinuous channels can be relatively stable over periods of decades: also where they change their position, they remain sinuous and do not change into meandering channels. This also applies to sinuous creeks in the Dollard where only the smallest gullies (that usually meander) may change within a few years.

The presence of sinuous channels in addition to straight or gently undulating or wiggling channels and meandering channels can be understood by starting from the results of Bejan (1982), who has shown that the equilibrium shape of a stream bed is sinusoid with a wave length that is proportional to the width of the stream. Sinuosity can be induced by irregularities in the bed, caused by differences in erodibility. This leads to an increase of turbulence in the water, a sideways movement of the flow, and lateral excavation of bends, which eventually can result in a sinuous channel. If the bends are formed at higher discharge with a higher water level and stronger current velocities in the channel (as long as no adjacent flats are flooded), their development can be reduced or completely stopped when the water level and current velocities are lowered. Local erosion of the banks can result in the formation of bends which is reduced or stops when the discharge decreases. In this way a possible development of bends into meanders is retarded. Regular and episodic variations in discharge are normal during the tidal cycle, as are current reversals which may reduce the tendency to form bends, or meander. Many tidal channels meander, however, so that sinuosity may also be induced only by variations in the resistance of the banks against erosion (e.g., because of variations in sediment consolidation or vegetation). This would result in the formation of irregular bends in the channel that remain in existence as long as the current velocity does not increase (or the banks do not collapse). When the bends have been formed (or enlarged) only during episodic high discharge, enlargement and displacement of the bends, as well as a transition to meandering, may proceed very slowly. When channels and creeks become entirely submerged during high tide, bends or even meanders may be flattened by the flow over the submerged flats and (partly) re-eroded by ebb currents at falling tide. Also waves may flatten the channels or gullies during submergence, and strong waves may flatten them out completely on bare, sandy, unconsolidated sediments (as in the Wadden Sea).

In channels that continue from a salt marsh into bare flats with loose sediment, the sinuosity decreases (Pestrong 1965): vegetation increases the resistance to erosion so that bends or meanders are less easily flattened on the salt marsh than on the bare flats, while meandering may continue or even increase by undercutting and slumping. Differences in vegetation can have a similar effect: lateral migration of channels is slower (by about 30%) in salt marshes than in neighboring freshwater marshes and have a higher sinuosity (Garofalo 1980), while the channels tend to become entrenched.

6.5. MEANDERING CHANNELS

There is as yet no generally satisfying explanation for the formation of meanders (Figures 6.7, 6.8, and 6.9). Scheidegger (1991) has summarized the existing ideas on meander formation, which can be divided into hydraulic, mechanistic, and stochastic theories. The first are based on the idea that the bed of streams that are not at grade is lengthened by meanders until equilibrium is reached, which may be related to a number of hydraulic variables. In general, meanders will become wider until the flow is so far reduced that lateral erosion stops. This occurs when the meander radius is about two to three times the channel width (Bagnold 1960). The ratio of meander wave length to channel width is 2 to 3 for straight (young) channels

FIGURE 6.7   Salt marsh creek (Terschelling, Wadden Sea). (Photograph courtesy of J. Abrahamse.)

FIGURE 6.8   Meandering gully and channel (Wadden Sea). (Photograph courtesy of J. van de Kam.)

and 6.5 to 11 for natural rivers with a long history (Leopold and Wolman 1960). This points to the possibility that many sinuous or straight channels that are often very young have had no time to develop meanders. Because of the relation with hydraulic parameters, meander patterns change downstream as well as in response to human activities (woodcutting, grazing, industrial or urban development).

FIGURE 6.9   Gully in mud flats (San Sebastian Bay, Argentina). (Photograph courtesy of Th. Piersma.)

Mechanical theories are based on the relation of meanders to secondary currents and the helicoidal flow that develops in the stream when a disturbance is present. High velocity water is pushed sideways and erodes the outer bank, while deposition takes place on the inner bank (formation of point bars). Stochastic theories describe the development of the most probable or expected meanders under given conditions of flow and environment. The results have not been satisfactory up to now because either the meander systems produced were too regular, or the calculations were too complicated. Like the formation of bends, the formation of meanders is probably related primarily to the periods of strongest flow, as was shown by Ashley (1980) for the Pitt river in Canada. When during the highest water level the flow remains within the channel, a bench may be formed along the channel banks, or two benches when the tides are semidiurnal and there is a significant difference between the levels at high tide. Benches may also form because of slumping along the banks when they are undercut. In mesotidal marshes in South Carolina, where the channels are ebb dominated, and where also a freshwater influx is absent, Barwis (1978) found that all meander point bars are skewed toward the ebb direction. They are relatively small with steep flanks and no significant differences between the ebb- and flood-side when the ratio of meander radius to channel width is less than 2.5 (tight meanders). At intermediate meanders with a ratio between 2.35 and 3, the point bars are multilobed and complex, with flood-oriented chutes. In gentle meanders with a ratio above 3, the point bars are long and narrow with deep flood-oriented chutes and fields of sand waves and megaripples at both ends.

6.6.   CHANNEL WIDTH

Tidal channels have a roughly constant width, or the banks may converge inward so that the channel becomes funnel-shaped, or the channel may show a widening and narrowing with the narrowest points at the bends (Figure 6.10). Creeks and channels are usually wider and shallower on bare tidal flats and become narrower and relatively deeper in salt marshes and mangrove where the channel banks are enforced by plant roots (Figure 6.11). Lateral erosion can result in the formation of

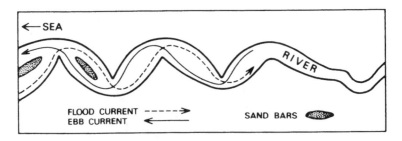

FIGURE 6.10 Estuarine meanders. (From Bird 1984. With permission.)

FIGURE 6.11 Gullies in salt marsh and on bare flats flowing into a channel (Wadden Sea). Many small, meandering creeks in the salt marsh, virtually none on the bare flats. (Photograph courtesy of S. Tougaard.)

small terraces where consolidated sediment along the channel banks after slumping accumulates on the bed along the bank. For channels with inward converging banks, Wright et al. (1973) have shown that the channel banks converge exponentially inward from the mouth, when the tidal prism is considerably larger than the fresh-water discharge, a standing wave is formed, and the tidal amplitude decreases inward at a cosinus rate. This develops when the length from the mouth to the inward limit of tidal influence is about 25% of the tidal wave length. The presence of converging channel banks, however, does not imply that the tidal amplitude decreases inward. A notable exception is the Gironde estuary (at Bordeaux, France), where the tidal amplitude first increases inward and then rapidly drops to zero (Figure 8.6). This is attributed by Allen et al. (1980) to bottom friction and friction along the channel's banks which makes the crest of the tidal wave travel inward faster than the trough. Channel banks remain about parallel when the tidal wave behaves more like a progressive wave, the tidal amplitude remains constant up to near the limit of tidal influence, and the tidal prism decreases inward at a constant rate. This develops where the channel length is small relative to the tidal wave length and the channel has a low width:depth ratio. Funnel-shaped channels are observed at river mouths

(Ord river, Australia; the Shatt-el-Arab, Irrawaddy and Hooghly river mouths; Wright et al. 1973) as well as in tidal creeks (North Norfolk marshes; Pethick 1992). Parallel banks, for instance, are present at the King river mouth (Australia; Wright et al.1973). In the North Norfolk marshes the creek width decreases inland exponentially at each subdivision between bifurcations: the total length of the salt marsh creeks depends on the width of the creek (or the tidal prism) at the mouth. The channels end abruptly near the landward edges of the marshes.

6.7. LEVEES

Where the flow during rising tide goes over the channel banks and over the flats, sediment is dispersed laterally and deposited on the flats. Dronkers and Zimmerman (1982) have shown that this lateral dispersal also plays an important role in the longitudinal dispersal in a tidal system. Lateral dispersal is related to the turbulence generated at the bottom as well as to the channel width and the development of horizontal eddies, which may develop behind irregularities along the channel. Wind effects may be important, especially when tidal currents are weak, both by raising water level and by generating an additional circulation. When, during the overflow over the flats, the currents are reduced, sediment is deposited, which can occur along the inner margin of the flats as well as at the channel edge. The latter results in the formation of levees, which is enhanced by vegetation (salt marsh, mangrove) that reduces the flow near the bottom. The coarser (more sandy) material is deposited predominantly along the marsh edge and forms a marsh levee, while the fine material is dispersed over the flats. Along channels and creeks, levees are formed where there is an inward flow that rises over the banks during the flood; during ebb the flow concentrates from the flats toward the creeks or channels so that ebb gullies are often formed. Smaller creeks with a weak flow during the flood do not have levees. Shoreline levees (as present along salt marshes and, e.g., on Andros Island in the Bahamas) are related to a temporary sea level rise during onshore storms, or during exceptionally high tides, when the seaward part of the flats is flooded and sediment is deposited along the edge.

6.8. CHANNEL SYSTEMS: SYSTEMS OF PARALLEL CHANNELS, OR NO CHANNELS AT ALL

Flats with systems of parallel (single) channels only are described from macrotidal flats: Korea, China (Jiangsu, Taizhou Bay, Wenzhou Bay), Koojesse Inlet (Baffin Island); flats with very few or no channels from macrotidal areas: Baie de Bourgneuf (France), Inchon Bay (Korea), the north shore of Hangzhou Bay (China), and from mesotidal areas: salt flats at the Indus river delta area and along the Gulf of Carpentaria (north Australia), and along James Bay (Hudson Bay). The macrotidal flats are regularly reworked by storms as well as by the flood tide. The parallel single channels along the Jiangsu coast, China, are cut into fine silt and sand. They are sinuous but not meandering and show lateral displacement. Channel deposits are regularly observed within the tidal flat sediments (Ren ed. 1986). The channels in this area are not, or not entirely, destroyed by the tides and the summer storms, probably because this part of the coast is protected from large storm waves by offshore sand banks. The rills and gullies formed on the Taizhou Bay and Wenzhou Bay flats are partly eroded and filled in during the flood, but eroded again during the ebb; when they are destroyed by storm waves, they are quickly formed again. In the Bay of Inchon (Korea) the flats have only temporary channels during the summer, which disappear during the fall; only one channel has a more permanent character. In Keonggi Bay (Korea) the flats that are exposed to the winter storms have parallel channels, whereas on the more sheltered flats, the channel systems are dendritic. On the exposed flats, 10 to 30 cm of sediment is eroded during the winter and (re)deposited during the summer. The salt flats at the Indus river mouth and along the Gulf of Carpentaria are rarely flooded and are located in a dry climate so that water flows over the flats only irregularly, probably mostly as sheet flow. In James Bay (southern Hudson Bay) ice covers the flats for up to six months, and the sediment is covered with boulders and stones: both conditions are unfavorable for channel formation. The parallel channels, in Keonggi Bay and along the Jiangsu coast, although they have a more permanent character, do not reach a meandering stage, probably because they are cut in fine sand and silt which can easily be eroded. They are partly destroyed during storms, but on the Jiangsu coast some of the channels are an extension of channels draining reclaimed land so that water regularly flows through them. However, the other channels on these flats, which originate mostly near a dike, are not different from the channels with regular flow, so it is most likely that these channels are also regularly (partly) eroded and do not reach a meandering stage. Also they do not develop into a dendritic channel system or any other system, although at their upper end they have a few tributary rills or gullies.

6.9. DENDRITIC CHANNEL SYSTEMS

Dendritic channels occur at all tidal ranges (Figures 6.12 and 6.13). Pestrong (1965) found that the channels on the tidal marshes of San Francisco Bay very much resemble the terrestrial dendritic drainage systems. According to Horton (1945) and Strahler (1957), channels are divided into segments following a hierarchical order of magnitude (Figure 6.14). Small rills and tributaries at the upper margin of the system are designated as first-order channels. Two first-order channels meet to form a second-order channel, etc. The number of orders reached depends not only on the characteristics of the stream, but also on the scale of the maps that are used (Leopold et al. 1964). When very small rills can be designated as first-order channels, a large river channel may reach a tenth-order rank. The stream order is related to the number of segments, to the length of the segments, and to the drainage basin area. With increasing segment order, the relations found for tidal channels in San Francisco Bay deviate increasingly from the relations found for terrestrial drainage basins: in the tidal channels the lower order segments have only significant flow velocities during the ebb, the higher-order segments both during ebb and flood (Pestrong 1965). In tidal creeks along the Van Diemen Gulf (north Australia) a much closer agreement between tidal channel networks and terrestrial networks was found (Knighton et al. 1992).

FIGURE 6.12 Channel/gully system in fine sandy flats (Dollard, Wadden Sea). (Photograph courtesy of K. L. M. Aerocarto.)

Dendritic drainage systems have a wide distribution on land and different types develop in response to structural, topographic, and lithologic conditions, which can be distinguished by the junction angles of the segments, which can range from 90° to very low angles. Horton (1945) found that the junction angle between two segments is a function of the gradients of the segments: when the gradient of a tributary is much larger than that of the main stream in which it flows, the junction angle is approximately 90°. This is found along many tidal flats where creeks and gullies flow into a much larger main channel with a much lower gradient. On the San Francisco Bay marshes the channel gradients decrease with increasing segment order. The junction angles between first- and second-order channels can be up to 90°, but for higher-order channels, this was not found (Figure 6.15). Pestrong (1965) points out that this may be influenced by the lower number of data so that anomalous data can have a large effect on a mean value, but it is more likely that it is related to the alternation of ebb and flood in the larger channel segments, which will influence the junction angles, while tributaries even may be extended in the flood direction.

Dendritic channel networks can also be characterized by the distribution of branches in the network (Horton 1945; Giusti and Schneider 1965). This is expressed in the bifurcation ratio, which is the ratio between the number of streams of a given order to the number of streams of next higher order. Bifurcation ratios around 2

FIGURE 6.13   Channel and creeks in a salt marsh (Ooster Schelde). (Photograph courtesy of J. van de Kam.)

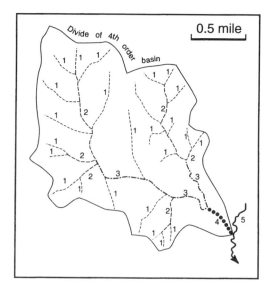

FIGURE 6.14   Method of designating stream channel orders. Order nr. 1 = 25 streams; order nr. 2 = 6 streams; order nr. 3 = 2 streams; order nr. 4 = 1 stream. (From Strahler 1957. With permission.)

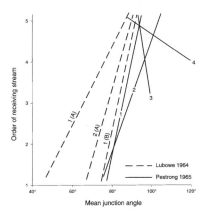

FIGURE 6.15 Mean junction angle versus channel order for terrestrial stream channels (Lubowe 1964) and tidal marsh channels in San Francisco Bay. (From Pestrong 1965. With permission.)

occur in flat or rolling areas. Ratios of 3 to 4 occur in highly dissected basins. Shreve (1966, 1967) found that bifurcation ratios around 4 are the most probable for a dendritic network. Knighton et al. (1992) found bifurcation ratios around 4 in the tidal creek systems at the Van Diemen Gulf but with large variations and much higher ratios during intermediate stages of development. Random topological channel network models, as developed by James and Krumbein (1969), Dacey and Krumbein (1976), and Van Pelt et al. (1989), assume that natural channel networks are topologically random in the absence of strong environmental controls. The creek networks along the Van Diemen Gulf do not differ significantly from randomness.

The development of dendritic channels starts, like for all channels, with the initial formation of parallel rills following a predetermined sediment slope. Erosion and rill formation start where the critical flow length is exceeded. The development of a drainage system goes through cross-grading and micropiracy, whereby slightly larger rills grow by overtopping and breaking down the low ridges between the rills (Horton 1945; Leopold et al. 1964). Because of small (natural) variations in depth and elevation between the rills, a sideways component toward the larger rills develops in this way, which is enhanced during periods of higher water level. The sideways gradient across the main gradient (cross grading) eventually obliterates the original parallel rill pattern and a dendritic pattern develops. The formation of rills, micropiracy, and cross-grading, and the development of a small gully drainage system may take several tides, even in loose sediment. Destruction, entirely or partly, may occur during the flood or by waves during storms, but the gullies will develop again during the next ebb tide. Deposition of sediment during the flood or at slack high tide may partly or entirely fill the system with sediment, followed by erosion during the next ebb, which results in net-deposition when the erosion during the ebb is less than the deposition earlier. In this way a tidal flat may be accreting upward as well as laterally while the drainage system remains in existence. The tendency of larger (longer) gullies or channels to grow more than the shorter ones is related to their elevation, which for the larger channel is lower than for the shorter channel. So when the divide between the two is breached or broken up, the water will flow from the drainage basin of the shorter channel into the basin of the larger channel (Hack 1957, Leopold et al. 1964).

Drainage patterns tend to develop with time into a regular pattern: it has been established that a pattern with a logarithmic relation between stream length and number of streams with stream order numbers is the most probable one. With the restriction that in higher-order tidal channels both flood and ebb flow takes place, it can be assumed that the evolution of a simple drainage network into a more complex dendritic system, even on loose sediment, may take a long time, as can be seen by comparing the development of a drainage network on glacial deposits of different age (Ruhe 1952; Leopold et al. 1964). Increasing with the age of the deposits from 13,000 BP to 17,000 BP the number of first-order tributaries increases from 2 to 22 and the number of third-order tributaries from 0 to 2. This is based on maps with a scale of about 1:60,000. Smaller-scaled tidal drainage networks will develop more quickly, but comparison of charts and soundings indicates that the formation of tidal channels may take decades to centuries.

Rapid expansion of tidal creek networks was found along the Van Diemen Gulf (Mary river coastal plains, north Australia), where in 50 years creek networks expanded inland over 30 km (Knighton et al. 1991, 1992). Growth was exponential by extension of the creeks through headward erosion, by the development of tributaries and by widening of the creeks. The rate of development was enhanced by the large tidal range (5 to 6 m in the Gulf), the flatness and low elevation of the coastal plain, and the existence of earlier channels: palaeochannels, freshwater channels, and buffalo swim channels. Headward extension starts with saltwater invasion in the slightly lower areas and the formation of seepage zones. Then the central part of such a zone is scoured by tidal (ebb) currents and creeks are incised. The network develops as the creeks extend headward and expands by addition of exterior links, which occurs through bifurcation at the head of an exterior link (branching), and by the development of tributaries at an exterior or interior link. Where channels expand rapidly along a paleochannel or go beyond paleochannel boundaries, tributaries develop later. Additions to the network by branching usually begin near the mouth before spreading upstream. Further expansion of the network in the lower (older) 4 km is arrested by divides and competition for space as well as by the growth of mangrove (*Avicennia*), which enhances sediment deposition so that smaller tributaries tend to be filled in. This results in an increased role for the main creeks. The network is stable here, while more inland it is expanding. Generally, a slightly higher ground inhibits further development of the network. Headward erosion does not necessarily start with the formation of a seepage zone. Where the sediment is consolidated, the water, following a small depression on the marsh surface, erodes an incision when flowing out into a creek, sometimes forming a small cascade.

A special form of dendritic drainage pattern is the elongated dendritic pattern that is present on several French macrotidal flats (at the Perthuis d'Oléron and the Anse d'Aiguillon; not found in other macrotial, meso- or microtidal areas). This type has small junction angles: the channels are directed more in a downslope direction than is normal in dendritic systems, which probably is related to the relatively large tidal range, as well to some unknown local conditions.

FIGURE 6.16   Braided pattern at the widening downstream end of a meandering gully (Wadden Sea). (Photograph courtesy of NIOZ.)

6.10. BRAIDED AND DISTRIBUTARY CHANNELS

Braided channels (Figure 6.16) have only been observed in three macrotidal areas: the Baie de Somme (France), the Meghna river estuary (Bangladesh), and the Bay of Fundy (Canada), and in the salt marshes of the mesotidal Wadden Sea and Ooster Schelde (the first two are river mouth channel systems, the third tidal channels and gullies, the fourth salt marsh creeks), but small braiding channel or gully sections may occur locally and on a limited scale on many bare flats. Braided channels are related to the ratio between the slope of the stream bed and the bankful discharge: the channels become braided when the slope is relatively steep (Leopold et al. 1964). Van Straaten (1951) associated braided channels with strong tidal currents and firm creek banks enforced by vegetation (roots), but they occur also on completely bare flats. Steeper bed and bank slopes are often associated with coarser, heterogeneous bed material and increased bank erosion.

Braiding is related to the initiation of bars on the bed. During low flow the channels wind between the low banks; at high flows the banks may be submerged. Bank erosion, resulting in a large amount of loose sediment and transport of a relatively large bedload (but a low transport of sediment per unit width of the channel) are essential for braided channels. Braiding seems to be a response to a debris load that is too large to be transported through a single channel. In both the Meghna and the Somme river mouths this is probably the case because of a relatively large sediment deposition in the mouth. In the Bay of Fundy, the few rivers that cross the tidal flats form very shallow braided channels about 5 cm deep. Large amounts of loose sediment eroded from the nearby cliffs are accumulated on the tidal flats by tides and waves, and overload the streams. Although braided streams are often considered overloaded and associated with aggradation (Leopold et al. 1964), the aggradation may have been done by other agents and not by the rivers, as the Bay of Fundy shows. In the Wadden Sea braided channels are often a transition between two nonbraiding systems.

Distributary channels are found along King Sound (Australia), at the Baie du Mont-St. Michel (France), at Koojesse Inlet (Baffin Island), and at the Trucial Coast in the Persian Gulf: the first three areas are macrotidal, the last one microtidal. Distributary tidal channels are associated with small internal deltas or fan deposits and start fanning out from one central channel, although this is not always clear in the areas where they occur. All distributary channels are in the ebb direction. It is likely that the ebb flow, following the central channel, fans out over the flats where the slope becomes less than in the channel.

6.11. INTERCONNECTING CHANNELS

Interconnecting channels are channels that both begin and end at another channel. They are found in macrotidal areas (the Bay d'Arcachon, France, the Sundarbans in Bangladesh, the Irrawaddy delta area, the Chongming Island flats in the Chang Jiang river mouth), in mesotidal areas (the Ooster and Wester Schelde in The Netherlands, the Ria de Formosa in northern Spain, and the Bahia Blanca in Argentina), and in microtidal areas (the Sine Saloum and Casamance tidal areas, the Niger delta, Zaire estuary and along the Virginia–New Jersey coast, on Andros Island, and in North Inlet). In the mouth areas of the Niger, Irrawaddy, Sine Saloum, Casamance and along the Virginia–New Jersey coast, interconnecting channels occur in combination with dendritic channels in extensive networks that cover large areas. In spite of this, only at the Niger river delta (Allen 1965) and along the New Jersey coast (Ashley and Zeff, 1988; Zeff 1988) have these channel networks been studied in detail. Both areas are microtidal with tidal ranges below 2 m.

Along the New Jersey coast, Ashley and Zeff (1988) distinguish dead-end branching channels that have a more or less dendritic pattern and end (or begin) in salt marshes, and through-flowing channels that either connect a lagoon with the coastal sea or connect two tidal channels. All types of channels can meander. In the Niger delta, Allen (1965) distinguishes feeder channels which (mainly in the eastern part of the delta) can be subdivided into predominantly saltwater tidal channels with a "blind" inlet and little or no freshwater influx, and feeder channels that receive a freshwater influx from the Niger or from another, smaller river. The feeder channels are connected by subsidiary open-end channels which follow a broad reticulate pattern. All channels, and in particular the subsidiary channels, meander: both along the New Jersey coast and at the Niger delta, meandering dominates. Relatively few channels are sinuous, and even fewer are straight. Because of the rivers discharging into the tidal area, freshwater flow is an important component in the Niger delta, but is negligible along the New Jersey coast where freshwater comes only from rain. The through-flow ocean-lagoon channels are 150 to 300 m wide, mostly 7 to 10 m deep; those that connect channels are 50 to 100 m wide with no systematic variation in width, and 2 to 5 m deep. In both types of channels the flood velocities are higher than the ebb velocities, but in those that interconnect channels, the ebb velocities are higher during a storm. The dead-end channels form discrete branching networks that become abruptly smaller inland and end in a salt marsh. The shape of the channels suggests that ebb and flood follow different paths through the channels.

The present pattern of channels is explained by Ashley and Zeff (1988) and Zeff (1988) by its (hypothetical) historical development. With the formation of the coastal barriers, a back barrier bay developed between the barriers and the coast. Narrow fringe marshes with small dead-end channels were formed along the inner margins with flood-tidal deltas, with subtidal bars and channels at the inlets between the barriers. When sedimentation continued, the banks near the inlets became subaereal and partly covered with salt marsh, while along the inner margins of the bay salt marshes prograded. With continuing salt marsh growth, the open waters of the bay were gradually dissected into discrete lagoons with through-going channels connecting the lagoons with the coastal sea. With further expansion of the salt marshes the lagoons decreased in size, and dead-end channels became increasingly important, while the importance of the through-going channels was reduced. For the future, it is foreseen that the through-going channels will be further reduced and become subordinate to nonexistent, while some rudimentary lagoons will remain.

In the Niger delta the channel system changes by lateral migration of channels and capture of adjacent channels, as well as by the formation of new tidal flats. Capture of a meandering open-end channel by another meandering open-end channel results in an enlargement of a tidal flat. Capture of one blind channel by another results in a new open-end channel that subsequently can be enlarged by the tides, and to a division of a tidal flat into two parts. Capture of a blind channel by an open-end channel results in a new open-end channel and division of a tidal flat. Lateral deposition on the inner side of a meander is balanced by erosion on the outer side; net deposition occurs only by vertical deposition on the flats, except directly landward of the coastal barrier islands and along the inner margins, i.e., near the sources of sediment supply (runoff and riverflow from inland and inflow from the coastal sea). The entire system moves slowly seaward with the growth of the delta. Over a long period, all mud flats will be approximately the same size.

Both along the New Jersey coast and in the Niger delta, new flats are formed from banks that grow above low tide level: in the Niger delta this is happening now and being observed; along the New Jersey coast, it has been postulated to explain the filling in of the original back barrier lagoon and the present channel pattern (Ashley and Zeff 1988; Zeff 1988). Infill of the lagoons takes place through the formation of small internal deltas. In the Niger delta the new banks that are formed in the broad channels become covered with mangrove when they rise above mean tide level. This enhances sediment deposition, and the bank becomes a tidal flat that continues growing laterally until deposition is balanced by erosion. The Niger delta receives a regular supply of sediment from land as well as from the sea and the whole system is slowly moving seaward. Along the New Jersey coast there is only sediment supply from the sea, and the whole system is stationary (or, perhaps, slowly reduced because of the slow landward displacement of the barriers).

6.12. EROSIONAL STEPS AND MARSH PONDS

An erosional step or slope is often found at the seaward edge of a salt marsh or mangrove. It is related to erosion during normal high tide, mostly by waves, A step or low cliff, up to several meters high, is formed where the sediment is consolidated or enforced by roots, a slope is formed where the sediment is weaker. A cliff can be undercut and can retreat by slumping. The material eroded from the cliff is partly deposited by the waves on the marsh on top of the cliff, and there contributes to the formation of a marsh levee. Material slumped from the cliff and not removed, can enclose lower areas at the base of the cliff that become shallow ponds, which remain when the lower flats become vegetated. Ponds on the marshes can be formed when, during development of the marsh, areas remain without vegetation and where water is enclosed that cannot flow off. Also abandoned or closed off creeks or parts of creeks can become ponds.

6.13. BARRIER BEACHES AND CHENIERS

Although barrier beaches and cheniers are not intertidal deposits, they are often present along or within intertidal areas. The formation of a beach barrier can induce the formation of intertidal deposits where a sheltered bay or lagoon is formed. Barrier islands and beach barriers are generally considered to be formed by waves where sand is supplied, or available for reworking, and the shelf gradient is low (Oertel and Leatherman eds. 1985). These conditions are predominantly present along trailing continental margins (along the Atlantic and large parts of the Indian Ocean), but barrier islands are also found along marginal seas and collision margins, where there is less sediment and the barrier islands are related to headland erosion and the formation of spits (Glaeser 1978). Along the U.S. east coast, where the tidal range is relatively high (2 to 4 m), wide complex lagoons occur that are strongly correlated with shallow, convex offshore submarine slopes (that reflect low-energy conditions), while narrow and simple lagoons occur where the offshore profiles are steep and concave and reflect high-energy conditions. Along the coastal section where the tidal range is only 1 to 2 m, the reverse is present: the lagoons are wide and complex where the offshore profile is steep and concave, while simple lagoons are associated with shallow convex offshore profiles (Hayden and Dolan 1979). This distribution reflects the distribution of wave energy along the coast and the interaction with the tides. Usually the building of a barrier island has been quicker than the infill of the lagoon behind it, either through the formation of sand banks at the inward side of a tidal inlet or through the formation of intertidal deposits along the landward margin of the lagoon. The tidal flats and marshes behind a barrier, however, are not necessarily the result of lagoon infill only, but may be the result of coastal subsidence with the marshes expanding not only into the lagoon and growing upward, but also expanding landward over formerly terrestrial areas. The latter process is considered to take place at present along the U.S. east coast, although the earlier development of the marshes was probably due more to lagoon infill only (Gardner and Boon 1980).

A very different situation is present where cheniers are formed, which occurs when mud accretion along the coast is interrupted and a chenier plain is formed consisting of a mud plain with chenier ridges. For the formation of such a plain the presence of a mud-discharging stream is required (Otvos and Armstrong Price 1979); a chenier plain can be distinguished from a beach ridge — strand plain that is partially buried by salt marsh — and deltaic deposits. A chenier plain can be formed along a large-scale bight, as well as in a bay head. Two types of cheniers can be formed (Augustinus 1989a,b): those formed by winnowing of sand, mollusk shells, etc., out of the muddy sediments and subsequent concentration of the coarser material in a longshore bar or ridge, and those formed with material supplied by longshore drift. When cheniers have been formed, they migrate farther landward by washover processes during spring tides or storm surges. Cheniers indicate periods of coastal standstill or retreat and occur in many coastal regions of the world from the tropics to the subarctic. Their formation can be strongly influenced by a shifting river mouth, changes in sea level and/or climate or storm frequency, by the mortality of shellfish, and by the coastal configuration. They usually can be dated with some accuracy, although reworked carbonate may create problems in interpretation. They can provide a very accurate record of the sea level during their formation. When mud deposition resumes, the chenier is left to rest, with muddy deposits in front and behind as well as below. This is different from (former) barrier beaches, which have a (sandy) seaward extension down to the level of the wave base during their formation.

6.14. QUALITATIVE COMPARISON OF MACRO-, MESO-, AND MICROTIDAL AREAS

On the basis of the data presented in Chapters 2, 3, and 4 qualitative comparison of morphological characteristics in macro-, meso-, and microtidal areas can be made. This is summarized in Table 6.1; the value of the relative abundance of the morphological features that are listed can be checked by using the subject index. Keeping in mind the limited number of intertidal areas that have been studied, as indicated in the Introduction, the following observations can be made.

Bare flats (both sand flats and mud flats) are most common in macrotidal areas, where also features of accretion/sedimentation or erosion, megaripples, cheniers, low steps or scarps (along the edges of salt marshes or, less commonly, mangrove), and slumping are present (relatively) more frequently than in meso- and microtidal areas. Beach ridges, barriers, barrier islands, peat, and features of wave erosion and deposition in general are relatively rare, while intertidal mud banks and lagoons have not been described from macrotidal areas, although fluid muds and mud flats can be formed at all tidal ranges, as observed by Wells (1983).

Also in mesotidal areas megaripples, cheniers, low steps, scarps, and slumps are relatively common, while beach ridges, barriers, barrier islands, and peat deposits are more commonly found in mesotidal areas than in macrotidal areas. Less common are bare flats and features of erosion and deposition/accretion as well as wave effects, mud banks, and lagoons (mud banks are rather rare anyway). This distinction, however, is not very sharp. A large mesotidal area like the Wadden Sea is characterized by extensive bare flats (sand flats as well as mud flats) with marked features of accretion/deposition or erosion and wave effects, while the Wadden Sea may also be considered a lagoon.

TABLE 6.1
Qualitative Comparison of Morphological Characteristics in Macro-, Meso-, and Microtidal Areas

Feature	Macrotidal areas	Mesotidal areas	Microtidal areas
Bare flats	+++	+	+
Sand flats	++	+	+
Mud flats	++	+	+
Accretion/deposition	++	+	+
Erosion	++	+	+
Megaripples	++	++	+
Cheniers	++	++	+
Low steps, scarps	++	++	+
Slumps	++	++	+
Beach ridges	+	++	++
Barriers (islands)	+	++	++
Peat	+	++	++
Wave action	+	++	++
Mud banks	o	+	+
Lagoons	o	++	+++
Deltas	+	++	++
Funnel-shaped river mouths	++	+	o

+++ relatively very common
 ++ relatively common
  + relatively rare
  o not observed

Almost all lagoons are found in microtidal areas where beach ridges, barriers, barrier islands, and other wave-built features are also common as well as peat deposits, while slumps, scarps, cheniers, megaripples, bar flats, and erosion/accretion features are relatively rare. While in macrotidal areas the tides, with strong currents, dominate, which results in rather vigorous erosion and deposition, in microtidal areas waves are more important and lagoons tend to be formed. Mesotidal areas have an intermediate character and show features that are also relatively common in macrotidal areas (megaripples, cheniers, low steps, scarps, and slumping), as well as features that are common in microtidal areas (beach ridges, barriers, barrier islands, peat deposits, features of wave action). Deltas and funnel-shaped river mouths, which are partly intertidal like beach ridges, barriers, and cheniers, also show a clear relation to tidal range: deltas are relatively rare in macrotidal areas but common in meso- and microtidal areas. The presence or absence of some conspicuous intertidal features such as salt marshes, mangrove, channels, creeks, and (fluid) mud deposits do not show a relation with tidal range, or only in a very limited way. As was discussed above, parallel single channels are only described from macrotidal areas, and flats with no or very few channels from macrotidal and mesotidal flats only, while other types of channels occur independently of the tidal range. Tidal flats in general tend to be more common in macrotidal areas.

7 Intertidal Sediments: Composition and Structure

P. L. de Boer

The commonly twice-a-day flooding and drainage largely defines the character and composition of sediments of the intertidal zone. Sediments are supplied through nearby delta distributaries, by the erosion of headlands, from the adjacent sea floor, and, from greater distances, in suspension. *In situ* production of sediment by biological activity (carbonate, organic matter) may be an additional (temperate climates) to dominant (dry tropical and subtropical climates) source of sediment.

Depending on the available sediment and the (varying) energy levels, intertidal areas consist of sandy, silty, and muddy sediments, and mixtures of these. Grain size commonly decreases from high-energy toward low-energy, more sheltered parts. Vegetation, which can be abundant in the more quiet parts of the intertidal zone, may not be directly apparent in lower-lying, higher-energy zones, but it is there, normally represented by algae, as is testified by the common abundance of burrowing and grazing organisms which feed on them or their remains. Relatively quiet parts of intertidal systems may be occupied by organic buildups, such as oyster and mussel beds. Sedimentary structures (Table 7.1) reflect the various energy levels.

All tidal systems show a neap-spring cycle with a regular fluctuation of high-water and low-water levels in about 14 days. As seen in Chapter 1, the most frequently occurring tidal system is the semidiurnal one, with two tidal cycles in slightly more than a day (24 h 40 min). A limited part of the coasts on Earth (about 10%, Lisitzin 1974) is characterized by diurnal tides, with only one tidal cycle a day. Mixed tidal systems experience an alternation of semidiurnal and diurnal characteristics during the neap-spring cycle. Semidiurnal systems generally have a larger difference between low water and high water (more than 10 m in extreme cases, e.g., the Bay of Fundy) than diurnal and mixed systems. Such a difference in tidal amplitude defines differences in current velocities in diurnal and semidiurnal systems in two ways. First, the generally greater tidal amplitude in semidiurnal systems makes their tidal-current strength greater than in diurnal ones. Second, in semidiurnal tidal systems six hours separate high water from low water and vice versa, while this period is doubled in diurnal systems. Thus, in cases of the same tidal amplitude, maximum current velocity in the semidiurnal system will be twice that of diurnal

TABLE 7.1
Selection of Features Typical for Intertidal Deposits

Unidirectional

Lower phase plane bed

Small-scale ripples — straight and sinuous: immature; linguoid: mature; Baas et al., 1993

Erosive ripple forms — Reineck and Singh, 1973, their Figure 7.8

Back-flow ripples — small-scale ripples formed in troughs of large-scale ripples through backflow; Boersma et al., 1968

Megaripples, dunes — straight-crested, undulatory/sinuous, lunate

Tidal bundle sequences — series of, ideally, about 28 megaripple foresets of increasing and decreasing thickness, formed due to the increasing and decreasing transport capacity of tidal currents during a neap-spring-neap-neap period; in intertidal environments only one (high-slack water) mud drape can be deposited

Upper phase plane bed — current lineation

Antidunes — extremely low preservation potential

Graded beds — deposited during storms

Bidirectional

Ebb caps on megaripples; Boersma and Terwindt, 1981

Flaser-linzen bedding; Reineck and Singh, 1973

Wave ripples

Combined ripples produced by the combined action of waves and currents with different directions

Interference ripples; combination of wave and/or current ripples with different direction; Reineck and Singh, 1973

Adhesion ripples; accretion of generally pre-existing ripple forms due to aeolian transport during low water; van Straaten, 1953

Runoff Structures

Microdeltas in channels along shoals with runoff through small gullies

Rill marks; Reineck and Singh, 1973

Erosional Structures

Shell lags; scour marks, Figure 7.10

Tool Marks

Figure 7.11; see also Dionne, 1988

Deformation Structures

Load cast

Cavernous sand, Emery, 1945

Convolute lamination; de Boer, 1979

Mud volcanoes

Mud cracks produced by desiccation

Slumps and Faults

Formed due to slope instability, e.g., along the margins of channels

Bioturbation

U burrows (*Arenicola*, certain shrimps); Figure 7.7

Pressure structures; Figure 7.12

TABLE 7.1
Selection of Features Typical for Intertidal Deposits *(continued)*

Biological Life

Manyfold; shell bioherms; see text and other chapters

Typical Sediment Types

Ooids; sand-sized concentric carbonate grains, formed in agitated waters in the subtidal and intertidal
  zone; Bathurst, 1971

Broken and complete skeletal carbonate of organisms typical for the intertidal (and often also subtidal)
  zone

Clay flakes, formed due to shrinkage of thin mud layers and/or fine-grained layers stabilized by algae
  and/or early cementation

Organic remains of inter-supratidal plants

Additional

Raindrop imprints
Foam impressions
Water-level marks
Crystal imprints in evaporites environments, etc.

---

systems where high and low tide are 12 hours apart. The effect of this difference
on sediment transport is, of course, much greater, as the transport capacity of water
currents is related to the current velocity with a power between 3 to 5 (Allen, 1984).

7.1. SILICICLASTIC SYSTEMS

Intertidal sediments occur along all coasts, but thickness of the deposits and the
areal extent vary (Figure 7.1). Grain size is a reflection of the distance to source
areas and of the transport capacity of the hydrodynamic system. Thus, eroding cliff
coasts along uplifted bedrock may be characterized by winnowed (narrow) pebble
beaches, while in subsiding areas rivers and marine processes supply large amounts
of sandy, silty, and clayey sediments.

In siliciclastic systems, lithofacies zonations strongly depend on hydrodynamic
energy. Coastlines with strong wave activity and tides are characterized by a concentration of sands, while in the more sheltered parts silt and clay dominate. In
higher intertidal salt marshes and mangroves the sheltering effect of higher plants
plays a large role. Microbiota, although commonly not visible with the naked eye,
may strongly increase the erosion resistance of clayey as well as sandy sediments
(see Section 9.1, and references there).

7.2. CARBONATE SYSTEMS

Carbonate intertidal systems (Figure 7.2) differ fundamentally from siliciclastic ones,
in that sediment is not supplied from external sources but is produced within or

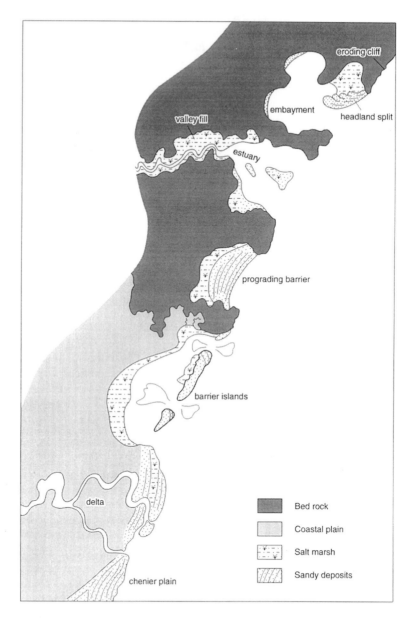

FIGURE 7.1 Different coastal environments dominated by terrigenous sediment supply. Erosion due to e.g., tectonic uplift in the north, and toward the south extensive deposition of fluvial and longshore-transported sediment. Fine-grained suspended matter (clay, organic matter) is carried ashore by marine currents. The abundance and preservation potential of intertidal sediments increases from north (eroding bedrock) to south (subsiding coastal plain). (Redrawn from Intasen and Roy 1996.)

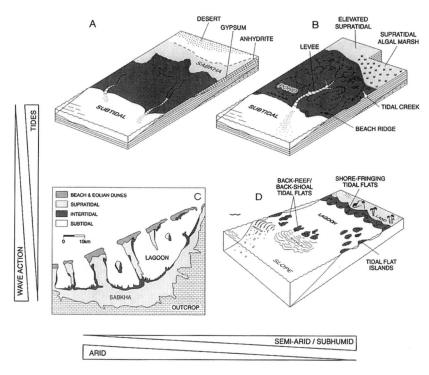

FIGURE 7.2 Carbonate coastal environments. A (semi-)arid climate and a (near-)absence of terrigenous sediment input are the main prerequisites for carbonate coast development. The wave and tidal climate and the aridity form the main controls for the different types of carbonate coast. (From Pratt et al. 1992. With permission.)

close to the site of deposition (Purser 1973). Biogenic buildups, such as reefs, form hard, erosion-resistant elements. The *in situ* produced carbonate particles are subject to hydrodynamic forces similar to siliciclastic sediments in noncarbonate settings.

In addition to the features controlling siliciclastic systems (see above), climate and aridity are of major importance for the character of clastic carbonate intertidal systems. Also the higher salinity exerts a great influence on the chemistry of the sea water, on the character of the carbonate-producing organisms and on the resulting clastic products (Bathurst 1971, Lees 1975). Prerequisites for pure carbonate intertidal sediments are a warm and dry climate permitting the production of biogenic carbonate and an absence of siliciclastic input, which hampers the production of, and also dilutes, the biogenic carbonate.

Microbial activity defines the character of the intertidal zone in carbonate systems to a much greater extent than in siliciclastic systems. It may form mm- to cm-thick carpets, consisting of carbonate mud and organic matter, which may lead to the formation of desiccation cracks upon drying. Under evaporating, saline conditions early cementation and transformations of carbonate to dolomite may occur.

7.3. COLD REGIONS: INFLUENCE OF ICE

A special case are intertidal sediments in cold regions. Dionne (1988) provides an extensive review: during the cold season, ice cover protects the intertidal flats against wave action, while below the ice abundant mud may settle from suspension, up to centimeters to decimeters per season. Shore ice can strongly erode tidal flats by disturbing the sediment surface and subsequent resuspension, as well as by ice-rafting. In the latter process, all sediment sizes, up to boulders, can be transported and redeposited on the tidal flats or offshore. Scouring by ice and deposition of ice-drifted sediments can produce many different, often spectacular structures (Dionne 1988). The preservation potential of recent high-latitude intertidal flats is limited as in such areas uplift due to isostatic rebound after the melting of polar icecaps still continues.

7.4. SEDIMENTS

Due to the continuous activity of tidal currents and waves, sediments in the intertidal zone are commonly well sorted. Apart from selection in size, also the form (rollability) of sand grains is a selection criterion (Winkelmolen 1971). Irregular particles with sharp edges are picked up by currents much more easily than round particles of similar size. Thus currents will carry away the more irregular grains and leave behind the rounder ones. In this way Winkelmolen and Veenstra (1974) recognized sediment transport patterns in tidal systems.

In the supply of **terrigenous sediments** an important selection occurs related to the method of transport — as bedload or as suspended load. Sediment transported as bedload can generally be traced back to relatively nearby sources, whereas fine-grained sediment in suspension may be transported over long distances.

Sandy and coarser sediments and their sedimentary structures in the intertidal zone commonly are a direct reflection of the dominant hydrodynamic regime. Fine-grained sediments sampled from recent or fossil deposits, and analyzed in the laboratory, may, however, show a textural composition quite different from the *in situ* material. The reason is that individual clay particles may aggregate through flocculation into larger particles which settle much faster than the original ones (Eisma 1995), and that animal life produces fecal pellets and pseudofeces which may behave as sand-sized sediment (Oost 1995; Figures 7.3 and 7.4).

This explains why mm- to cm-thick mud drapes are found in subtidal deposits (examples of Terwindt and Breusers 1972; Visser 1980) deposited during one slack-water period of the order of 20 to 30 min. It is obvious that settling from suspension cannot have been the (only) source in such cases. Apart from the fact that the 20 to 30 min slack-water period is too short for the settling of several centimeters of mud, also the water column would have had an unrealistically high concentration of mud.

Another mechanism to produce clasts consisting of clay is the formation of clay flakes due to desiccation and of clay pebbles upon erosion of older sediment by, for example, laterally migrating tidal channels.

In **carbonate systems,** the grain-size distribution is defined not only by current activity, but biogenic, *in situ* production of carbonate (shells, reef debris) may

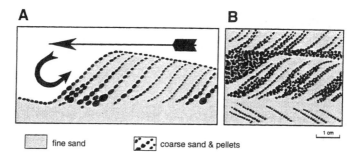

FIGURE 7.3 Biodeposits (fecal pellets and pseudofeces) may be several hundred microns in diameter, and behave hydrodynamically like mineral grains of sand size. They consist of clay and silt aggregated by filter feeders and deposit feeders. (From van Straaten 1954. With permission.)

FIGURE 7.4   Small-scale ripples largely consisting of fecal pellets. Dollard estuary, northern Netherlands.

introduce coarse-grained elements which are not representative of the local hydro-dynamic regime. In most shallow marine and intertidal depositional environments there is an abundance of organisms, part of which produce carbonate skeletons which may contribute up to about 10% of the sediments in temperate climates and up to 100% in (semi)arid (sub)tropical areas which lack terrigenous input. Carbonate sediment may consist of complete or broken tests of all kinds of carbonate-producing animals, and of carbonate mud produced by algae and by destruction of recent and fossil carbonates by grazing and boring organisms (Bathurst 1971). Carbonate build-ups, such as reefs, often form hard, erosion-resistant features which dominate and

FIGURE 7.5   In the background algal stromatolites; in front carbonate sand, partly protected by algae and slight cementation, and partly unconsolidated and rippled. Shark Bay, Australia.

define the character of the adjacent environments (Figure 7.2). Algal activity can form stromatolites (see Section 9.4), and stimulate early lithification of the sediment surface (Figure 7.5). In cases where algal structures are not fixed to a substrate and are moved and rolled by waves and currents from time to time, spherical algal structures of up to about a decimeter in diameter (oncolites) may form. Early cementation by purely chemical processes may occur in the higher intertidal zone and lithify sediments soon after deposition (beach rock; Cooper 1991).

In addition to broken and complete remains of internal and external skeletons of marine organisms, ooids may locally form the bulk of all sandy sediment. Ooids consist of concentrically lamellar, cryptocrystalline aragonite. They are formed in agitated environments along warm- and dry-climate coasts with sea water highly saturated with dissolved carbonate, and strong wave and current activity. It has been advocated that they form inorganically from supersaturated solutions, but there are many indications that their formation is intermediated by algae and/or bacteria (Bathurst 1971). Peloids, early cemented fecal pellets, may externally resemble ooids, but they lack the internal concentric texture. In environments in which ooids and peloids are formed, early cementation of sandy sediment close to the sediment–water interface may occur, and upon erosion of such early lithified sandy sediment, aggregates (grapestones) may be transported and redeposited nearby.

Apart from those parts of the intertidal zone where the proportion of **organic matter** in the sediment may be considerable (in salt marshes and areas with mangrove vegetation; see Sections 9.2 and 9.3), organic matter is found only in minor proportions, as admixture in the sediment. Living organic matter (algae, bacteria) may influence the erosion resistance of the sediment surface (Figure 7.6, and see Section 9.1). Its presence in surficial sediments induces abundant biological and

FIGURE 7.6 Algal cover protecting part of the sandy intertidal flat surface. Prolonged (various tidal cycles) protection of the sediment surface leads to the blurring of ripple structures (central right). Once the algal cover has been eroded, sand transport and the formation of small-scale ripples restarts (left). Drowned Land of Saeftinghe, Zeeland, The Netherlands.

biochemical activity. In the upper millimeters to centimeters of the sediment, oxygen is available, and organic matter is oxidized by burrowing and grazing organisms and by bacterial activity. In all marine depositional environments sulfate-reducing bacteria transfer sulfate into sulfide for the consumption of organic matter in the shallow (cm to dm) anoxic zone below the sediment surface. In this process they produce sulfide which precipitates with Fe as monosulfides and other precursors of pyrite. In special cases, repeated vertical fluctuations of the redox boundary thus may lead to concentration of iron-oxides and the formation of iron crusts. Burrowing organisms may again introduce oxygen into the anaerobic zone, by which the sulfides are reoxidized (Figure 7.7).

7.5. SEDIMENTARY STRUCTURES

The formation of physical sedimentary structures in intertidal environments is basically subject to hydrodynamic processes in a similar way as in all other aqueous environments. However, the intertidal environment distinguishes from other environments as current strength varies greatly, and current reversals commonly occur twice a day, while water depth varies from zero to at maximum the value of the local tidal amplitude. Especially during ebb runoff, upper plane bed conditions (Leeder 1982; Allen 1984) may occur. Thus, evenly laminated sands (with parting lineation) and laminated silts and muds, are much more common bedforms in intertidal environments than in others.

FIGURE 7.7   Redox boundary at some few cm below the sediment surface. The dark color is the result of monosulfides and fine-grained early diagenetically formed pyrite precursors. Note that through the burrow (*Arenicola marina*) oxygen is introduced, reoxidizing the reduced sediment.

The waxing and waning of current velocity and the standstill of currents at high-tide slack water contributes to the formation of the typical lenticular, flaser-linsen bedding (Reineck and Singh 1973) with alternating isolated or continuous clay intervals and continuous or isolated sandy, often rippled intervals. Of course, the latter form during the ebb and/or flood. Note that double mud drapes separating foresets in large-scale crossbedding (Visser 1980), which are often considered to be diagnostic for tidal environments, *only* form in the subtidal domain where there is a water cover during both high-tide and low-tide slack water. In the intertidal area the sediment surface is, by definition, drained during low-tide slack water, so that only one clay drape can be deposited between megaripple foresets formed during the dominant tide (Figure 7.3). This feature, single mud drapes in a series of megaripple foresets with astronomical periodicity (neap-spring cycle and thick-thin alternations due to the diurnal inequality of the tide; de Boer et al. 1989), seems a diagnostic feature of intertidal bedforms. As seen above, in the subtidal environment, such mud drapes are generally double, i.e., between the two mud layers deposited during the two slack-waters generally a (sometimes only microscopically visible) thin sand interval occurs, deposited by the currents between the slack periods.

The variation of current strength and the limited duration of ebb and flood often provide nice examples of the fact that small-scale ripples pass through an evolution from straight-crested to sinusoidal to linguoid ripples through time (Baas et al. 1993).

FIGURE 7.8 Immature, straight-crested small-scale ripples developing into sinuous and linguoid forms (cf. Oost and Baas, 1994). Gradations of immature to mature small-scale ripples may be observed in shallow tidal creeks, where current velocity in the channel axis is sufficient for the development of mature, linguoid ripples during an ebb or flood period, whereas this is not the case along the margins.

The velocity with which this morphological evolution occurs depends on current velocity, and thus it may be seen that in the central, deepest part of intertidal channels, where currents are strongest, mature, linguoid ripples can develop in the few available hours, while toward the margins of the channel current velocities are less, and immature, straight-crested or sinusoidal ripples are found at low tide (Oost and Baas 1996; Figure 7.8).

The small water depth on intertidal flats when they are covered with water, often leads to the production of interference ripples due to the interaction of current and waves. Similarly, a change of wind direction may also produce interference ripples, and a change of current direction may do as well ((Reineck and Singh 1973, p. 366). Although the often very photogenic interference ripples may cover extensive intertidal-flat areas, their preservation potential is low because subsequent waves and currents and laterally migrating tidal channels rework much of the shallow intertidal deposits. Examples of fossilized interference ripples seem to be especially reported from carbonate settings, where algal protection and early cementation protect the sediment surface against erosion (Thorez et al. 1988).

Large-scale crossbedding in intertidal flats deposits often shows a dominance of the flood current. This is due to the common asymmetry of the tidal currents in inshore areas. Maximum ebb current velocities then tend to occur in the second half of the ebb period, when the water level has already fallen below mean sea level, and maximum flood current velocity occurs late in the flood period (van Straaten, 1964; Figure 7.9a,b).

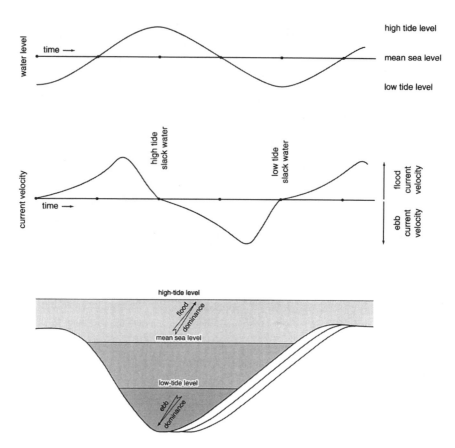

FIGURE 7.9 The asymmetrical character of the tide in most inshore tidal environments leads to a dominance of the ebb current at deeper levels, i.e., in tidal channels, and a dominance of the flood current at higher levels, i.e., on the intertidal flats. As a result, sub- to intertidal sedimentary successions often show a current reversal (ebb → flood) in the vertical direction.

This implies that the larger part of the intertidal flats has already been drained when ebb current velocity is maximal, so that large-scale ebb-oriented ripples are formed in the lower-intertidal and subtidal parts of the channels, while the shoals are not extensively affected at that time. During the rising tide, however, flood currents are maximal in the later stage of the flood period, when much of the shoals has been flooded, so that flood-oriented structures occur there. In the fossil record this phenomenon can often be recognized: many descriptions of subtidal-intertidal successions throughout the geological record indeed show an ebb dominance in the lower parts and an increasing flood dominance higher in the successions.

Current directions are reflected in the orientation of foresets in small-scale and large-scale ripples and in features such as current lineation, scour marks, imbrication of clay flakes, etc. In the open sea, tidal current directions may rotate 360 degrees during the tidal cycle. Toward intertidal areas, currents are forced more and more

FIGURE 7.10   Scour pit in algal-protected sandy intertidal sediment surface. At places where, during high tide, the algal protection is disturbed (e.g., by flatfish or birds), the sediment-starved current erodes the sediment. Freshly eroded and redeposited sand at the down-flow side of the scour indicates the current direction.

into a bidirectional current pattern with ~180 degrees different direction, because of flow through tidal channels and gullies to and from the tidal flats.

7.6.   DEFORMATION STRUCTURES

Physical deformation structures are the product of differences in density in the sediment column (reverse density stratification), of instabilities along slopes, or of strong currents deforming the upper sediment layer (Allen and Banks 1972). Air-filled, cavernous sand (Emery 1945; de Boer 1979) is a typical product of regular drainage and flooding of fine- to medium-grained sandy sediment surfaces (e.g., beach). Slopes along intertidal channels may induce small-scale faults and slumps.

Scouring of the sediment surface may occur when locally erosive processes are relatively strong, or where locally the erosion resistance of the sediment surface is low. The latter often occurs when a protective algal cover is locally destroyed, e.g., by animal activity (Figure 7.10).

Tool marks are formed when some object moves over the sediment surface. The motion of ice and ice-rafted debris may produce deep and extensive scourings in intertidal flats during the winter (Dionne 1988). Also pieces of wood and peat, dead and living animals, etc., may produce tool marks when they are dragged over the sediment surface (Figure 7.11). This may happen especially at a thin water cover during the ebb.

Bioturbation is a major agent in disturbing physical sedimentary structures in intertidal sediments. For example, so-called U burrows (Figure 7.7), produced by,

FIGURE 7.11   Tool mark produced by jellyfish having been dragged over the sediment surface by the ebb current.

among others, worms and shrimps, are typical but not diagnostic for intertidal settings. Echinoids and shells may produce pressure and imprint structures (van den Berg 1981; Figure 7.12; see Section 9.1, and e.g., Bromley 1990 for overviews of the effects of burrowing organisms).

7.7.   FACIES

Detailed descriptions of the various sedimentary facies with intertidal deposition are given in various parts of this book, as well as elsewhere (see e.g., Reineck and Singh 1973; Ginsburg 1975; Terwindt 1988; Reading 1996; Gaily and Hobday 1996, and many others).

7.8.   SEDIMENTARY SEQUENCES

The sedimentary environment, sediment supply, and relative sea level change define the position of intertidal deposits in the vertical succession relative to that of adjacent sedimentary facies (e.g., Ginsburg 1975; Terwindt 1988; Pratt et al. 1992; Gaily and

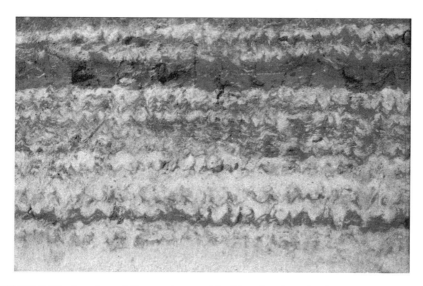

FIGURE 7.12   Imprints of *Cerastoderma edule*. Note that the animals grew during successive migration to higher levels during sedimentation. Subrecent Oosterschelde Estuary, The Netherlands.

Hobday 1996). A common phenomenon in tidal systems is the tendency to build up sedimentary successions from the subtidal zone toward the supratidal zone. In **siliciclastic environments** such shallowing upward sequences are the result of hydrodynamic processes which tend to transport sediments toward the coast. In **carbonate systems,** the *in situ* and/or nearby production of carbonate is of major importance for supplying the sediment for shallowing-upward sequences (Bathurst 1971; Purser 1973). Biogenic frameworks (reefs, carbonate barriers) may protect and enhance carbonate production in back barrier settings (Figure 7.2).

7.9.   PRESERVATION

Preservation of intertidal sediments in the sedimentary record depends on 1) the amount of accommodation space which is created by tectonic subsidence and the rate of eustatic sea level rise, 2) sufficient sediment supply, deposition, and burial preventing later erosion by e.g., subaereal erosion during a fall of sea level.

Upon a transgression, during a rise of the relative sea level, intertidal deposits will be preserved if the vertical accretion rate and the rate of sea level rise are such that the system builds up sufficiently in the vertical direction, to avoid scouring and erosion during landward migration of the coast. Thus, for example, back barrier intertidal deposits have a greater chance to be preserved than intertidal beach sediment along the seashore (Figure 7.13).

Intertidal areas tend to be dissected by tidal channels. In a quantitative sense, such channels occupy a minor part of recent intertidal areas, but because of their (geologically) commonly very rapid lateral migration relative to the vertical accretion rates of intertidal facies, they represent a much greater part of fossil intertidal deposits.

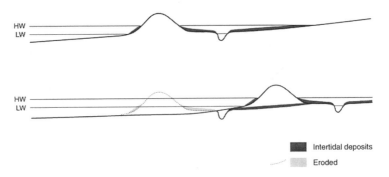

FIGURE 7.13   Although intertidal sediments may occupy vast areas along the shore and in back-barrier basins, they have a limited preservation potential. For example, upon the landward migration of barrier systems, the barrier may initially cover and protect the back-barrier intertidal deposits, but scouring and incision of shoreface processes (waves and tides) may eventually erode them.

7.10. EVOLUTIONARY TRENDS AND VARIATIONS OF CONTROLLING MECHANISMS THROUGH GEOLOGICAL HISTORY

Evolution of flora and fauna define the way in which biological processes have influenced and controlled intertidal facies in the geological past. Organisms which influence intertidal facies in one way or another have appeared and vanished throughout the stratigraphic record. Certain ecological niches may be occupied by different organisms in succession or shared at the same time.

A significant evolutionary trend has been the evolution of filter feeders. Ancient, early Paleozoic and Precambrian (inter)tidal deposits have a marked lack of fine-grained sediment. Apparently a lack of filter feeders in those days meant that suspended mud was not compacted into fecal and pseudofecal pellets, and thus was carried toward the open sea, leaving very sandy, tidal deposits devoid of mud.

We live, at present, in an ice-house period with waxing and waning polar icecaps. This has a twofold effect on intertidal sedimentary facies. First, temperate climate zones extend to much lower latitudes than during greenhouse periods without icecaps. Second, ice-house periods with polar icecaps and extensive glaciers are characterized by very rapid (up to $1 \text{ cm} \cdot \text{yr}^{-1}$) variations in sea level, which have a great effect on the character of coastal areas that have a much greater extension of noncarbonate environments in such periods.

Extension of carbonate environments to much higher latitudes during the Jurassic and Cretaceous was due to the combined effects of 1) a warm global climate in which subtropical areas extended farther north and south than at present, 2) a high sea level leading to broad shallow shelf seas with extensive carbonate production, and 3) relatively few orogenic source areas supplying siliciclastic weathering products. In addition, the warm and humid Mesozoic climate led to relatively strong chemical weathering, so that weatherable source lithologies (feldspar, rock fragments) were degraded to fine-grained mud and clay, and the remaining sand fraction consisted of weathering-resistant minerals, quartz being the dominant one (the Amazon system). As a result, terrigenous sediments had a large proportion of fine-grained sediment as seen today, for example, in the outflow of tropical rivers like the Amazon.

8 Sediment Transport in Intertidal Areas

with H. Ridderinkhof

8.1. INTRODUCTION

The transport, sedimentation, and erosion of sediments in intertidal areas is the result of the interaction between advection and dispersion (i.e., mixing) due to the current field and the bottom sediment processes. Tidal flow usually dominates in intertidal areas, but in exposed shallow areas, the wind can profoundly alter the tidal circulation. During (onshore) storms, sea level is temporarily raised and waves become more important. Most intertidal areas also receive an influx of freshwater from rivers and streams, local runoff, precipitation, and from canals that drain adjacent land areas. This produces vertical as well as horizontal density gradients, that can have a large effect on the water circulation and the transport and deposition of sediment. In the following section a short review of the current field and its driving mechanisms in intertidal areas is presented, followed by a review of the consequences for the sediment transport in different types of intertidal areas. For more general information on the current field in tidal basins, the reader is referred to the textbooks by Bowden (1983), Officer (1976), and Dyer (1986), in which most of the material presented here is discussed in much more detail.

8.2. FLOW IN TIDAL AREAS

The flow field in the tidal channels of a tidal basin can be considered to consist of a large-scale part, that varies depending on the scale of the entire basin, and a small-scale part that varies more locally.

The large-scale current field is forced by the tides at the entrance by large-scale density variations, by river inflow and/or meteorological forcing, and by the large-scale geometry of the tidal basin. The small-scale part arises mainly from the interaction between the large-scale current field and the small-scale bathymetry.

8.3. TIDES

The tides in intertidal areas are co-oscillating with the tides on the shelf and the ocean. The tides are generated by astronomical forces in the deep oceans, where they have a typical amplitude of 10 cm. Subsequently the tidal wave propagates over the shelf where the geometry of the shelf determines the resulting tidal curve at the entrance of intertidal areas. Along the coastlines of a specific shelf the amplitude of the tide can vary considerably, but generally there is a significant increase in the tidal amplitude compared to the oceanic tides. Thus the tides at the entrance of a tidal basin are determined mainly by the geometry of the adjacent shelf and by its position along the coastline. The different constituents that make up a complete tidal signal (M2, S2, K1, O1, etc.) all have their own characteristic propagation over the shelf. This means that different tidal basins along the same shelf can have (completely) different tidal conditions at their entrances.

Within a tidal basin the tidal wave propagates from the entrance toward the end of the basin and can propagate farther into rivers that discharge into the tidal basin. The amplitude and phase of the tidal wave within the tidal basin are determined by the large-scale geometry of the basin. The most important aspects are the overall length and depth of the basin and the relative area of tidal flats.

The overall length and depth in the main channel determine whether or not the basin is close to resonance which occurs if the length of the basin is close to a quarter of the tidal wave length. If these conditions are fulfilled, the interaction of the incoming and the reflected tidal wave can increase the tidal amplitude drastically from the entrance toward the landward end of a basin. In this respect the depth of the basin is important in that it determines the travel speed of the tidal wave and, thereby, its length. The depth of a basin is also important in that it determines the relative influence of bottom friction. Bottom friction dissipates the energy of the incoming wave and can reduce its amplitude. This effect can become important especially in relatively shallow basins.

8.4. TIDAL ASYMMETRIES

Tidal asymmetries are very important for the net, tidally averaged transport of sediments. These asymmetries cause a net transport of sediments although there is no net transport of water. Two types of tidal asymmetries can be distinguished: 1) the average magnitude of the flood currents differs from the average magnitude of the ebb currents, accompanied by a difference in the duration of the flood and ebb periods; and 2) there is a difference in the rate at which velocity and direction of the current change near high-water slack as compared to low-water slack. The first type is most important for bedload transport (coarse-grained material), whereas the second type is important for suspended matter transport (fine-grained material). The first type of tidal asymmetry is discussed extensively by Friedrichs and Aubry (1988), the second type by Postma (1967) and Dronkers (1985).

Both types of tidal asymmetries are caused by the nonlinear distortion of the tide. Nonlinearities, such as a dependence on the square of the oceanic tide, produce compound constituents and higher harmonics of the principal oceanic tide. Because the dominant astronomical constituent is M2, the semidiurnal lunar tide, the most significant overtide formed is M4, the first harmonic of M2. Therefore, the asymmetry in tidal currents is mostly quantified by the ratio between the M2 and M4 tidal current amplitude and the phase difference between both. In combination, these determine the direction and magnitude of the net transport of sediments due to tidal asymmetries.

The M4 overtide is produced by frictional interaction between the tide and the bottom and by the interaction between the tide and the geometry of the tidal basin. In general the influence of these nonlinear interactions on tidal distortion is relatively small on the shelf as compared to their influence in relatively shallow tidal basins with large intertidal areas. Thus tidal distortion is mainly produced within a tidal basin by a combination of frictional effects and the interaction of the tide with the bathymetry.

Nonlinear friction results in greater frictional resistance in shallow water which slows down the propagation of water level changes around low tide relative to high tide. Thus the time delay between low water in the inlet and low water in the inner tidal basin is greater than the time delay between high water. The result is a shorter duration of the flood and higher flood currents (due to mass conservation). The magnitude of this effect depends on the ratio between the amplitude of the water level variations and the water depth.

In tidal basins where the volume of water storage above intertidal areas is large compared to the volume of the channels, the duration of the ebb can be shorter than the duration of the flood, resulting in higher velocities during ebb. In such basins the propagation of the tide at high water is slower than the propagation at low tide. At high water, low velocities above intertidal flats cause high tide to propagate slower than low tide, which propagates mainly in the deeper tidal channels. This results in a shorter duration of the ebb and higher ebb currents. Extreme cases are small tidal basins along the U.S. east coast, where the water depth in the inlet is less than the tidal amplitude; because of bottom friction the flood currents dominate, while the ebb current velocities are considerably lower, and the tides are truncated in the lower part. The result is a relatively strong residual inward transport of sediment, a short rise in water level during the flood and an elongated fall during the ebb (Lincoln and Fitzgerald 1988).

The opposing effect of frictional and geometrical influences on large-scale tidal asymmetries in tidal basins with intertidal areas makes it difficult to predict ebb- or flood dominance in a specific system. This is even further complicated because of the variability in the depth and geometry within a tidal basin. This variability can even cause an ebb-dominance and a flood dominance in some parts of a basin in other parts.

Another type of tidal asymmetry occurs when a water column travels through an area where the local amplitude of the tidal current decreases significantly. This occurs near the landward end of a tidal basin and in intertidal areas. As a result, the change in the magnitude of the velocity that a water column experiences is much slower near high-water slack than near low-water slack. This type of asymmetry is related to the relative increase in wet cross-section with increasing water level at rising tide (Dronkers 1986a, b): when the relative increase in wet cross-section is smaller than the relative increase in storage area, a slack-water period is favored that is longer before the flood than before the ebb; when it is larger, a long slack-water period at low tide before the ebb is favored. In the first case, the tidal channels are deep, and the flats relative to mean sea level; in the second case, the channels are smaller, and the flats low, relative to mean sea level. This asymmetry causes a net transport of fine-grained sediments in these areas.

8.5. DENSITY DRIVEN CIRCULATIONS

In most tidal basins a large-scale longitudinal salinity gradient is present due to the river inflow which causes a decrease in salinity from the entrance toward the land-ward end where the river discharges freshwater into the basin. Depending on the magnitude of the tidal current relative to the river inflow, two typical situations can be distinguished which are classified as partially mixed to well mixed, and salt-wedge type basins.

In partially mixed and well mixed tidal basins, the magnitude of the tidal currents is such that the intensity of the small-scale turbulence generated at the bottom is high enough to keep the water column partially or completely vertically mixed. These basins are most likely to be shallow with a high tidal range. In these tidal basins there is no sharp interface between the upper and lower layer, but a more or less gradual vertical salinity gradient. This gradual vertical gradient indicates that there is vertical mixing over the entire column in which the intensity of vertical mixing is reflected in the magnitude of the vertical salinity gradient. The large-scale longitudinal density gradient drives a residual seaward flow in the upper layer and a landward flow in the lower layer. This vertical gravitational circulation or estuarine circulation is common in most tidal basins with a freshwater inflow. The residual velocities associated with this type of circulation are of the order of 1 to 10 cm · s^{-1}. A turbidity maximum is usually found at the bottom where the inward residual bottom current becomes almost zero.

When the magnitude of the tidal currents is so low that the tidal basin is not vertically mixed, a so-called salt-wedge basin occurs. Here a relatively sharp inter-face (halocline) exists between the fresh, and light, water which flows seaward over the more saline, dense water. Because there is little tidal flow, the saline bottom layer is relatively stationary, and the seaward velocity due to the river inflow in the thin upper layer is relatively high. Although the interface between both layers is rather stable, small amounts of the saline water can be entrained by the upper layer due to waves breaking at the halocline. The salt-wedge tidal basins usually have a pronounced turbidity maximum near the area where the halocline reaches the bottom of the basin.

8.6. METEOROLOGICAL FORCING

Meteorological forcing can also induce a large-scale circulation within an estuary. Surface stress caused by the wind induces a circulation with a current near the surface in the direction of the wind, compensated by a return flow at depth. Apart from a circulation by local forcing, a wind-induced higher level at the entrance of a tidal basin, due to forcing on the adjacent shelf, can have a large temporary effect on the current field: the wind-induced currents associated with the filling and emptying of a basin can have a magnitude of the same order as the tidal currents. Winds can completely reverse the normal tidal flow pattern: in the Potomac river estuary reversal of the flow pattern because of winds occurs approximately 20% of the time, the normal circulation 43% of the time, the same flow direction in the surface water as well as in the bottom water about 20% of the time, and complex flow in the remaining 17% (Elliott 1978).

Another type of meteorological forcing can be attributed to solar heating. In tropical tidal basins, heating can be so intense that evaporation in the shallow, landward part of a basin can cause an increase in the salinity in this part. Then the density increases toward the landward end and a so-called "negative" estuarine circulation, with seaward currents in the lower part of the water column and landward currents in the upper part, can be present (Wolanski 1988).

8.7. THE SMALL-SCALE CURRENT FIELD

The discussions above on the large-scale current field in a tidal basin do not take into account the effect of the small-scale irregularities in the topography and/or density field. These cause the current field to be nonuniform both in a horizontal and vertical direction, which can be very important for the transport of sediment.

Vertical and horizontal shear in the current field is caused by friction along the bottom and side walls of the tidal channels causing a decrease in the magnitude of the current near the channel boundaries. Through interaction with the small-scale turbulent motion, this current shear is very effective in increasing the mixing intensity within a tidal basin, as compared to the mixing intensity due to turbulence alone (Zimmerman 1986).

The interaction between the oscillating tidal current and the irregularities in the topography, like sand banks and channel bends, causes a complex pattern of horizontal residual currents. A dominant feature of this current field is the presence of horizontal residual eddies. A typical value for the length scale of these eddies is in the order of 5 to 10 km, and the magnitude of the velocity in these residual eddies is typically 1 to 10 cm $\cdot$ s^{-1} (Ridderinkhof 1995). Large-scale vertical eddies are induced by the bends in meandering channels. In these eddies, also referred to as secondary circulation cells, the current is directed toward the inner bend near the bottom and toward the outer bend near the surface. One of the consequences of such a complex residual current field is that it contributes to a high mixing intensity in a tidal basin.

Apart from the irregularities in the current field due to the interaction between the tidal current and the topography, irregular patterns in the density field can cause irregular current patterns. Examples are currents induced by lateral salinity gradients when a river outflow is concentrated along one side of a basin. Another example is the occurrence of fronts within a basin, often visible as foam lines in channels where water masses of different origin meet. These water masses can have different densities due to temperature and/or salinity differences, caused for instance by a higher temperature of the water originating from tidal flats, or the phase difference in the currents between shallow and deep regions.

8.8. FLOW OVER TIDAL FLATS

The magnitude of the currents above tidal flats is generally much smaller than the magnitude above tidal channels. In a first approximation (Friedrichs and Aubry 1988), it is often assumed that the shallow tidal flats act only as storage and that all momentum in a tidal basin is transported in the channels, but even though the currents above tidal flats are relatively slow, their presence is of great importance for the large-scale current field and sediment transport in a basin, because the tidal flats have a strong influence on the tidal asymmetries in the channels (see above). Where the flats are covered with vegetation (salt marsh, mangrove), the increased resistance from the vegetation decreases the current velocities compared to the current veloc-ities above bare flats, and the effect on the flow in the channels is enhanced.

8.9. WIND WAVES

Wind waves in a tidal basin are due either to incoming waves from the adjacent shelf or those generated locally by the wind. In most basins the damping of surface waves because of bottom friction is relatively high, so that swell from the adjacent shelf is damped at or near the entrance of the basin. Wind waves in tidal basins are therefore dominantly caused by the local wind blowing over the surface.

These short waves are important because of their ability to bring sediments from the bottom into suspension. The intensity of this process depends on the magnitude of the near bottom orbital velocity. This orbital velocity strongly depends on the wave characteristics and on the local water depth. Thus the effect of waves is much more important above shallow tidal flats than above deep tidal channels. The depen-dency on the wave height and period means that sheltered areas, such as those behind spits or barriers or behind vegetation, are much less directly influenced by wind waves.

8.10. TIDAL FLOW AND SEDIMENT TRANSPORT

The tidal asymmetry with long periods of low current velocities around slack tide favors sediment deposition around high or low tide; the asymmetry with peak velocities during either flood or ebb results in a high sediment transport in the direction of the peak velocities. Long periods of low current velocities around high tide before the ebb and higher flood velocities result in an inward transport (import) of sediment; low current velocities around low tide before the flood, and stronger ebb current velocities result in outward transport (export) of sediment. To this are added other factors (besides freshwater outflow, winds, and water depth, as men-tioned above) that influence the direction of sediment transport, usually favoring inward transport (van Straaten and Kuenen 1958, Postma 1961):

- Around high tide, when the water covers the flats, water depths are lower than around low tide when the water is concentrated in the channels: more suspended material that is settling when current velocities are low will

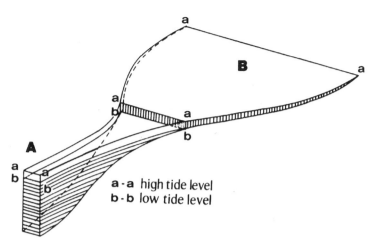

FIGURE 8.1 Water concentrated in the channels at low tide (A) and covering the flats during high tide (B). (Adapted from van Straaten and Kuenen 1957.)

reach the bottom around high-water slack tide than during low-water slack tide (Figure 8.1).

- Around low tide more material will remain in suspension than during high tide because of the greater water depth: more material, not having reached the bottom, is transported inward again with the flood after low tide than is transported outward after high tide (settling lag).

- It takes energy to resuspend particles from the bottom, which implies that the particles are resuspended at a higher current velocity than the velocity at which they were deposited. Therefore, also when the tidal velocity curve is symmetric, there is at high tide level an area where deposited sediment is not resuspended because the flow velocity, sufficiently low for deposition of the particles, is not high enough to resuspend them (van Straaten and Kuenen 1957; Figure 8.2). The amount of additional energy needed for resuspension depends on the strength of adhesion of the particles, to the other bottom sediment particles and resuspension is therefore dependent on the degree of consolidation of the sediment: very little additional energy is needed to resuspend freshly deposited material with a high content of water. Tank experiments by Creutzberg and Postma (1979) with bottom mud from the North Sea have shown that consolidation of the sediment begins to have an effect on the resuspension velocity after about one hour. Many tidal flats fall dry around high tide for a much longer period. On bare flats in temperate areas the salinity in the pore water of the sediment and the concentration of the pore water itself may change when the flats are exposed around low tide, in particular during the summer months. On flats in Adams Cove in New Hampshire (east coast U.S.), Anderson and Howell (1984) found that the salinity can increase by 2.2‰ per hour during low tide, while the pore water concentration is

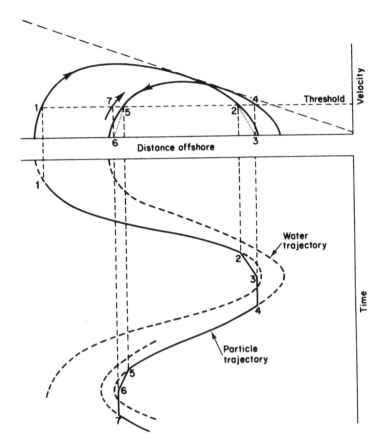

FIGURE 8.2   Schematic representation of the tidal transport of suspended matter shoreward by an asymmetric tidal current. (From Dyer 1986. With permission; adapted from van Straaten and Kuenen 1958, and from Postma 1961.)

reduced by evaporation as well as by drainage. This accelerates consolidation of the sediment, but in dry hot climates (Andros Island, Shark Bay; see Sections 4.16 and 4.26), the exposed sediment can dry out completely and higher up on the flats crusts can be formed that have to be broken up before erosion. Lindsay et al. (1996) measured suspended sediment deposition at velocities above 0.3 m · s^{-1} in the Forth estuary, but resuspension at velocities above 0.6 m · s^{-1}. It is difficult, however, to define a critical threshold for deposition of cohesive sediment, because of *in situ* suspended sediment floc size and turbulence characteristics. Sanford and Halka (1993) therefore suggested that it is better to assume a continuous process of deposition and erosion and to use the relation between the suspended-matter concentration and the rate of deposition, because there is a continuous deposition proportional to the suspended-matter concentration.

- Fixation of sediment particles by benthic fauna and flora occurs on bare flats with mucus in the form of feces, pseudofeces, tubes, algal mats, etc., and in salt marshes and mangrove, with roots, stems, and trunks. This has the same effect as consolidation, but other faunal activities may have the opposite effect by loosening the sediment, increasing the water content, and increasing the roughness of the sediment surface (see Section 9.1), which makes resuspension easier.
- An inward decrease in current velocity causes coarser material to be gradually deposited and finer material to be transported farther inward. This is subordinate to the effects of settling lag and scour lag, which primarily affect fine-grained cohesive material in suspension (Postma 1961): coarser material (sand, coarse silt) is mostly transported on or near the bottom, settles out completely during slack tide, and is also much less affected by a scour lag than the (mostly flocculated) material transported in suspension. Fine sand and silt may also be incorporated in flocs and then behave as the much finer material that forms the bulk of the material in suspension.
- Surface waves, in particular storm waves, predominantly affect areas that are not sheltered from waves, so that mud deposits are usually found in sheltered areas (Figure 8.3). In the more exposed areas, waves primarily affect the flats because the water depth is less than in the channels. This favors resuspension around high tide and seaward removal during the

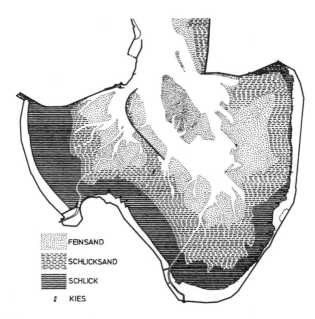

FEINSAND

SCHLICKSAND

SCHLICK

KIES

FIGURE 8.3   Sediment distribution in the Jade (German Wadden Sea). Feinsand = fine sand; Schlicksand = muddy sand; Schlick = mud; Kies = gravel. (From Reineck, ed., 1982. With permission.)

following ebb. The degree of resuspension by waves also depends on sediment consolidation and the particle binding, or loosening, effect of fauna and flora, but in the Wadden Sea fixation of mud by consolidation and flora and fauna is usually not strong enough to prevent erosion by storm waves. Even in a rather sheltered embayment like the Dollard only a few percent of the mud deposited during the summer remains after the winter storms (de Haas and Eisma 1990). Storms enhance a strong ebb flow, because the water is piled up inward (the water level can be raised by several meters) and a strong return flow develops. Waves stir up sediment so that in several intertidal areas (the Wadden Sea, the Eden river mouth, west Korea) high suspended-matter concentrations were found to be correlated with strong winds (Kamps 1962, Jarvis and Riley 1987, Wells et al. 1990). On the Wadden Sea flats, where a strong flood is dominant, suspended-matter concentrations during flood increase approximately quadratically with wave height. During the flood the suspended-matter concentrations also increased where on the flats a turbulent and turbid front is formed that moves inward. This is caused by the friction an accelerating wave experiences when moving upslope.

The inward transport of suspended material is primarily induced by the settling and scour lag, but also by the difference between the inward and outward mean and peak current velocities, which affect the resultant transport (flux) of sand, silt, and mud, as well as the total quantities of the material that is transported (in suspension and at the bottom). Although the lower mean current velocities cover a period of longer duration than the higher mean current velocities (there is no resultant transport of water inward or outward, except temporarily by wind force or locally as a result of density gradients or small-scale residual eddies), this does not compensate for the larger sediment transport in the direction of the stronger currents, because the amount of sediment that is transported is related to the second up to the fifth power of the mean current velocity.

The tidal asymmetry and the settling and scour lags result mostly in a suspended-matter concentration gradient with the highest concentrations inward. Once such a gradient has developed, tidal and turbulent mixing cause a down-gradient exchange of suspended material which counteracts the inward transport (Groen 1967). Measurements in Chignecto Bay (an embayment of the Bay of Fundy; see Section 2.22) during a two-year period indicated the presence of a suspended-matter concentration gradient that varies seasonally, with the total amount of suspended matter varying around 10^5 metric tons, while the concentration varied between 0.6 and 1.5 g · l^{-1} (Amos and Tee 1989). Similar gradients were found in other embayments of the Bay of Fundy with similar tidal ranges. As Amos (1995) points out, this indicates an equilibrium suspended-matter concentration gradient that is independent of the local variations in sediment supply and water flow. It also indicates an equilibrium capacity of the tidal water masses that are involved; both equilibrium gradient and capacity are continuously being upset and re-established based on a dynamic equilibrium between landward transport by tidal asymmetry and seaward dispersal. This system is influenced by the capacity of the tidal flats to accommodate suspended material that is settling out (Evans 1965), and a distinction can be made between a system limited by the accommodation space (essentially the relative area of the tidal flats) and a system limited by sediment supply. Amos and Tee (1989) estimated that variations in the suspended-matter concentration have a much larger effect on the residual suspended-matter flux than variations in the thresholds of erosion and deposition or the particle settling rate. A further differentiation occurs when mixtures of sand and mud are being transported, whereby the gradual inward decrease in tidal current velocity causes a reduction in both capacity and competency of the current so that a gradual differentiation of the sediment load (and of the resultant deposits) occurs (Evans 1965).

Erosion (resuspension) of mud flats occurs in two ways: by breakdown of the weaker bonds between organic aggregates and pellets (what is assumed to cause the scour-lag) and bulk erosion. The first type occurs at low velocities under conditions of hydrodynamically turbulent smooth flow; it ends when more consolidated sediment becomes exposed after removal of the surface layer. The second type occurs under hydrodynamically rough flow conditions: the sediment is broken up along microfractures, which may result in much higher suspended-matter concentrations than the first type.

Sand reaches the flats in suspension; as bed load it would have to go upward from the channel along the steep flanks of the flats. Such a transport is corroborated by the observations of Collins et al. (1982) that also during very quiet periods, measurable amounts of sand were in suspension over the flats. Van Straaten (1951) pointed out that the superposition of tidal currents and waves would result in high transport rates, where transport by currents or waves alone would not amount to very much. The interaction between unidirectional currents and waves, however, is not well understood, and the velocities cannot be simply added up to estimate the effect on sediment transport. However, these considerations make it plausible that the sand is transported toward the flats in suspension. During storms, much sand can be removed from the flats and the flat surface lowered considerably (in the Dutch Wadden Sea in the order of 10 cm or more during the winter storms), but afterward they are built up again, and over longer periods can keep up with the rise in sea level. This indicates that under normal conditions there is a residual inward flux for the fine-grained material in suspension. Because reliable direct measurements of sand transport are lacking, modeling of sand fluxes over tidal flats can give an indication: net-deposition is relatively steady with time and not sensitive to peak tidal current velocities, while erosion is very sensitive to tidal current velocity and is absent during neap tides. Waves are probably an important factor, but their effect is largely unpredictable (Amos 1995).

Large flats induce a maximum ebb current: because of the greater water depth in the channels, the tidal wave usually propagates faster in the channels than on the flats, and the decrease of the water level during ebb is slower on the flats than in the channels. This results in a large difference in water level between flats and channels, in particular during the late ebb and in strong ebb currents along the outer parts of the flats with formation of gullies that often meander. In mangrove swamps and salt marshes this storage of water is enhanced by the vegetation, which increases the bottom roughness. In mangroves the current velocity between the trees can drop to almost zero, while in the nearby channels the tide goes at full velocity. Because of the increased friction by vegetation, a time difference (time lag) develops in channels and creeks between the head and the mouth: when high tide is reached at the head, the water is already falling at the mouth (Wolanski et al. 1992), which enforces the ebb. As sediment transport takes place only at velocities above the critical erosion velocity of the sediment involved, the higher peak velocities as compared with the flood, induce a surplus of sediment transport in the ebb direction. In the more open mangroves along the edges of the mangrove swamp, during periods of strong currents flowing between the trees, resuspension of bottom sediment can create a fluid mud layer (Wolanski and Ridd 1986).

Wind effects can completely change the current velocity pattern during the tidal cycle and reverse the direction of residual transport: Postma's (1961) data for the Amelander Gat in the Dutch Wadden Sea show an asymmetry in the flood direction during northerly winds, and during easterly winds, in the ebb direction. The inlet is open toward the north, while the channel extends inward from the inlet in a south-easterly to easterly direction. Also whether or not the water remains confined to the channels and creeks, influences the asymmetry. In the North Norfolk marshes the asymmetry was not found to be constant (Bayliss-Smith et al. 1979; Healy et al. 1981) and becomes different when during the tidal cycle the water level is able to rise above bank level and the marsh is flooded. Velocity pulses develop when during the flood the water is no longer confined to the channels and creeks and can flow freely over the flats and marshes, while during the ebb the water converges in the channels and creeks and produces a higher discharge than when this does not occur.

8.11. TIDAL AREAS WITH NO OR WEAK DENSITY GRADIENTS

Generally the resulting sediment transport in tidal areas without wind effects and without marked density gradients is inward because of the tidal asymmetry. Density gradients induce an estuarine circulation with a resultant outflow of surface water and inflow of bottom water, when not disturbed by wind effects and density variations. Pejrup (1988) found that during the summer the net influx of suspended matter over a tidal flat in the Danish part of the Wadden Sea is seaward when concentrations are low and inward when concentrations are high. The overall net transport is inward, which corresponds to measured rates of suspended-matter deposition on the inner flats. The net inward flux of suspended matter is opposed by a net outflow through the main channel. These data indicate that scour-lag and settling-lag effects are only pronounced when the suspended-matter concentrations are high, which was attributed to increased flocculation because of higher concentrations, with formation of larger flocs that settle more rapidly to the bottom during slack tide. A seasonal variation in the direction (and quantities) of net sediment transport has been found in other tidal areas as well (as will be seen below), but the residual sediment transport remains problematical where no adequate temporal and spatial measurements of current velocity and suspended-matter concentrations have been made (Boon III,1975; Meade 1969; Boon 1978). At a low (meso- to micro-) tidal range, where the salt marsh channels form intricate meander patterns (usually with interconnecting channels, as along the Georgia coast in the U.S. or in the Niger delta), the flood waters during only the highest tide levels flow directly over the levees. The water mostly follows a long circuitous pathway before reaching the marsh surface, which complicates the effects of a tidal asymmetry. Because of storage of water on the marshes during high tide where the flow is slowed down by the vegetation, the ebb currents usually dominate in the channels (Sapelo Island, Georgia, salt marsh channels: maximum ebb velocity 1.3 m $\cdot$ s^{-1}, maximum flood velocity 1.0 m $\cdot$ s^{-1}, Howard and Frey 1985).

Usually the concentration and deposition of suspended matter in tidal areas is a delicate balance between slow net inward transport and deposition, erosion and net (usually rapid) outward transport during storms, consolidation of the deposited mud, the degree of protection against erosion offered by organisms on the flats, and the force needed for resuspension. Taking these various factors into account, Dronkers (1986a, b) could estimate in the Ameland area (where the net sediment flux is inward) and in the Ooster Schelde (where the net sediment flux is outward) the residual flux as well as its direction to within about one order of magnitude. Pejrup (1986), in a simple model, could describe about 80% of the variance of the suspended-matter concentrations at the observation site on a tidal flat by using wind speed and direction, the salinity, and the tidal current velocity. Sediment is deposited on the flats (and on salt marshes) primarily along the channels and creeks and dispersed over the flats. The bulk deposition may occur on the uppermost parts of the flats, as in San Francisco Bay (Pestrong 1972), or be more equally distributed over the flats as in the Wadden Sea. The formation of levees depends, among other factors, on the overflow frequency of the water in the channels and creeks; flooding gives a rapid decrease of the current velocity directly along the channel, and sediment is deposited. The levees are somewhat higher than the adjacent flats and therefore have a somewhat longer period of exposure and desiccation. Somewhat higher shear stresses are needed for their erosion, which favors their preservation and an increase in relative elevation over the flats. Erosion of bare flats can be severe, but also under normal conditions sediment can be easily resuspended during the flood when a turbid zone develops where the water flows over the flats (Wang and Eisma 1990; de Haas and Eisma 1990). Resuspension is enhanced by the presence on the flats of protruding worm tubes where small local vortices are formed (Carey 1983).

Salt marshes receive suspended matter and some sand during high floods but, in particular, when the water level is raised during storms. Besides the tidal regime and the sediment supply, the wind and wave climate and, over a longer period, the relative sea level movements are important for sediment deposition and erosion on the marsh (Allen and Pye 1992). Waves are damped by the vegetation on the marshes so that suspended matter can settle. It is left behind when the water returns to its normal level, because the vegetation protects the sediment against erosion. This is opposite to what occurs on bare flats, which are usually eroded during storms. In the Graadyb area in the Danish Wadden Sea about 50% of the mud deposited in that area is retained on the salt marshes; the other about 50% is retained on the muddy tidal flats, although the salt marshes cover a relatively small area (in the order of 30%; Bartholdy and Pheiffer Madsen 1985). Erosion occurs on salt marshes along the edges (where a low cliff may be formed), on the marsh itself by enlarging pans or cracks where the vegetation has been damaged, and over the entire marsh surface when a widespread deterioration of the marsh vegetation takes place, e.g., because of a rise of relative sea level (Pethick 1992.) On the exposed (macrotidal) Essex coast the marshes are eroded by episodic storms whereby not only the edge of the marsh is eroded, but also the elevation of a large part of the marsh is lowered. The eroded sediment is deposited on the bare flats directly seaward of the marshes.

The relative extent of salt marshes shows large regional variations from about 3% of the total intertidal area in the Dutch Wadden Sea to almost the entire intertidal area along the Maryland and South Carolina coast in the U.S. Some salt marshes are in the process of being reduced in size (mostly because the area is subsiding), whereas most marshes are being built up. Channels and creeks in the salt marshes may be ebb or flood dominated, but they are mostly ebb-dominated because the water that flows over the flats during flood partly returns through the channels during the ebb. The ebb-dominated channels have a tidal asymmetry with a long period of low current velocities around low tide and a short period around high tide (Boon 1976, 1978; Ward 1981). In the Wadden Sea the larger channels are flood dominated, the smaller channels and gullies ebb dominated, while in some wide channels ebb and flood follow different sides of the channel, so that within the same channel there is a flood-dominated side and an ebb-dominated side. Additionally, the sediment dispersal can be complex because of lateral nonsymmetric mixing (Dronkers and Zimmerman 1982). The differences between the various salt marsh areas arise from differences in the degree and nature of the vegetation, the effects of storms, the occurrence of seasonal changes in sea level, the degree of infill changing (gradually) the tidal characteristics, and, probably most important, the differences in sediment supply.

Net import or export of sediment in tidal areas can vary with the season. In South Carolina marsh channels, suspended-matter concentrations are higher during the summer than during the winter, with maximum concentrations during strong winds and heavy rains in combination with spring tide (Ward 1981). In the marshes there is a net export of organic matter in the summer because of organic matter production, and a balance between import and export of inorganic sediment. On the whole the peak ebb velocities are higher than the peak flood velocities. In Lowe's Cove in Maine the suspended-matter concentrations in the water are seasonally much influenced by ice, bioturbation, and low temperatures during the winter (Anderson and Mayer 1984). Stumpf (1983) found in Holland Glade Marsh, Delaware, that during the summer, normal tidal sediment supply was not sufficient to compensate for the local sea level rise, but that the balance was probably maintained by episodic deposition during winter storms when the marsh is completely flooded. Erosion and deposition vary spatially as well as with time, but on the whole a (short-term) equilibrium is maintained; for long periods no data are available. Generally along the U.S. east coast there is in the lagoons an equilibrium between deposition and the local rate of relative sea level rise. Near deltas there is a surplus of deposition; in subsiding areas, there is a deficit where sediment supply is inadequate (Nichols

1989). In the southern U.S., tidal marshes have been more susceptible to export of sediment, which is probably related to a reduction of sediment supply because of dam construction in many rivers (Court Stevenson 1988).

Most of the estimates of net export or import are based on estimates of net deposition (accretion) rates, which reflect long-term net deposition. This is done with markers (aluminum powder, ground brick) which remain visible in the sediment profile that is built up, or by using short-lived radioisotopes that are adsorbed onto the mud particles and deposited with them; the most used are a natural lead isotope, ^{210}Pb with a half-life of 22.3 years, and fallout isotopes, ^{137}Cs primarily. The fallout isotopes were first deposited in the sediment around 1954 and show a concentration peak in 1963, when a large number of bomb tests were carried out before the test-ban. In western Europe, the fallout from the Chernobyl accident in 1986 is a useful marker. Marshes in western Europe and along the east coast of the U.S. generally have a deposition rate that keeps pace with a local relative sea level rise of 1 to 4 mm · y^{-1} (Pheiffer Madsen and Sorensen 1979; Bartholdy and Pheiffer Madsen 1985; Eisma et al. 1989; Court Stevenson et al. 1988; Wolaver et al. 1988). These estimates are based on a vertical increase in sediment thickness, which introduces uncertainties because of differences in compaction. Estimates therefore are better based on the amount of sediment deposited per unit area and time (g · cm − 2 · s^{-1}), which involves measurements of water content. As the variability from year to year may be large and is obscured in the long-term measurements based on vertical profiles, short-term variations in deposition and erosion, as well as the export–import balance and the processes involved, can best be followed by comparing the amounts of sediment deposited on the flats during a short period, with the suspended-sediment concentrations in the water. Estimates based on flux measurements in channels tend to be misleading, depending on the extent to which storm conditions, spring and neap tides, and transport over the flats can be included in the measurements. In a New Zealand mangrove, export or import of suspended matter was found to depend on the method of estimation (Woodroffe 1985 a, b).

8.12. TIDAL AREAS WITH MARKED DENSITY GRADIENTS: ESTUARIES

Outflow of freshwater from land enhances the ebb flow and creates density differences with a residual seaward flow of low-density water along the surface and a residual inward bottom flow of high-density water. In this way an estuarine circulation develops, with suspended matter being transported seaward with the surface flow and settling toward the bottom where the flow velocity is reduced in contact with the more saline water. Near the bottom, the suspended matter is returned landward with the bottom flow, either in low concentrations (less than about 5 g · l^{-1}) or as a high-concentration near-bottom fluid mud layer of more than 5 g · l^{-1}. In estuaries, with a river-supply of both water and suspended matter, the estuarine circulation as well as lateral processes of sediment supply and deposition along the estuary can result in very complex situations, as is schematically indicated in Figure 8.4 (Eisma 1993). Sediment transport toward a flood plain and toward tidal flats, as

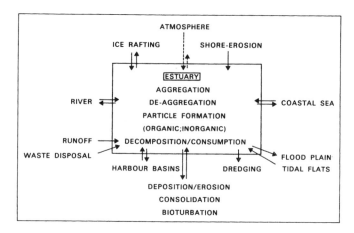

FIGURE 8.4 Diagram of supply and removal of suspended matter and principal particle processes in estuaries. (From Eisma 1986. With permission.)

well as supply from the flats and the floodplain toward the main estuary channel, occur through a complex set of processes that are related to the overall sediment supply and removal in the estuary, as well as to what happens in separate sections of the estuary: what occurs in one area may be very much influenced by what happens in other areas, and vice versa.

Estuarine processes in tidally mixed estuaries primarily affect the sediment in the water. Two aspects of this have a direct bearing on the intertidal deposition along the estuary: the formation of a turbidity maximum and the sediment balance. A turbidity maximum (Figure 8.5) is formed in a tidal estuary where a residual inward bottom flow meets the outward flow of the river, which usually is also the most landward point reached by the inward flow, and where the water is shallower than farther seaward. The residual flow turns upward and the bottom water, mixed with surface water, is moved outward. The actual current velocities in the bottom water here are low so that suspended matter easily settles. The suspended matter at the turbidity maximum, comes from both the river and from the inward flowing bottom water in the estuary. For the formation of a turbidity maximum, however, a relatively strong freshwater outflow that keeps the estuarine circulation going is not necessary. As was demonstrated by Allen et al. (1980) in the Gironde and the Aulne river estuaries, an upstream convergence of the channel, as well as (to a lesser extent) tidal damping by bottom friction because of a shallowing water depth, can cause the tidal amplitude and power dissipation to attain a maximum within the estuary, which contributes considerably to the formation of a turbidity maximum (Figure 8.6). In many estuarine areas, however, a turbidity maximum is not formed. Where river outflow is absent, a salinity gradient can be caused by freshwater runoff, precipitation, and by freshwater discharge from small streams and canals, which causes a weak seaward surface flow and a weak landward bottom flow. In this way, suspended matter is concentrated inward and near shore, as can be seen in many tidal areas and along many coasts, but the suspended-matter concentrations remain limited unless there is supply from another source, and a turbidity maximum is not formed.

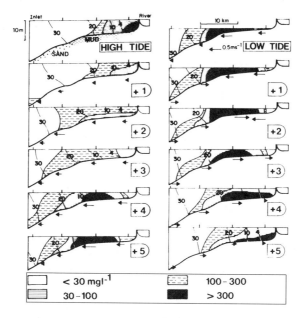

FIGURE 8.5   Turbidity maximum in the Aulne estuary during a spring tide cycle in May 1977, showing the distribution of salinity (numbers indicate ‰ S), and suspended matter concentration distribution. (From Allen et al. 1980. With permission.)

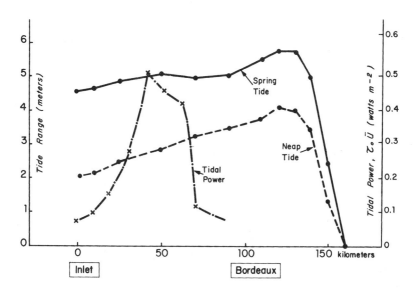

FIGURE 8.6   Longitudinal variation of the tidal range in the Gironde estuary during low river flow, and total tidal power dissipated on the bottom during a tidal cycle. (From Allen et al. 1980. With permission.)

The turbidity maximum is much influenced by variations in river discharge and in the tides (daily at semidiurnal tides and fortnightly at neap and spring tides): changes in the river flow and tidal flow result in a change in the position of the region with low current velocities, to resuspension of the mud deposited there before, and to a shift in the position of the turbidity maximum. In the mesotidal Columbia river estuary, the turbidity maximum shifts daily, fortnightly, and seasonally over a distance in the order of 20 km (Gelfenbaum 1983). Similar shifts occur in the Gironde (Allen et al. 1980), where during spring tides, when the tidal range is about 7 m, the turbidity maximum shifts inward, and the estuary is relatively well mixed. During neap tides, when the tidal range is about 4 m, the turbidity maximum shifts seaward, but is less developed, while the estuary is well stratified. The neap–spring tidal cycle controls, through erosion, deposition and lateral dispersal, the sedimentation along the estuary sides and the seaward escape of sediment into the coastal sea. In the Tejo estuary near Lisbon the turbidity maximum is absent during neap tides (when the tidal range is 1.3 m) and fully developed with concentrations up to $1 \text{ g} \cdot \text{l}^{-1}$ during spring tides (when the tidal range is above 3 m; Vale and Sundby 1987). In the Forth estuary (Scotland) a fluvial-induced stratification is present during neap tides, while the estuary is well mixed during spring tides when there is a strong residual inward transport of sediment (Lindsay et al. 1996). The turbidity maximum can also vary strongly with river discharge and can be completely washed out of the estuary during very high discharges. At the Orinoco river mouth a turbidity maximum is present when river discharge is low but virtually absent during periods of high river discharge.

The high concentrations of suspended matter in the turbidity maximum cause increased flocculation, the formation of larger flocs, more rapid settling, and formation of very high suspended-matter concentrations near the bottom in the order of a gram per liter or more. When the concentration reaches 5 to $10 \text{ g} \cdot \text{l}^{-1}$, the particles (flocs) are in each other's way when settling (hindered settling; Figure 8.7), so that a stable fluid mud develops with concentrations that can rise to $100 \text{ g} \cdot \text{l}^{-1}$. The high suspended-matter concentrations at very low salinities are a source of mud that can be deposited on adjacent intertidal areas. In the Loire estuary this is demonstrated by the content of montmorillonite in the mud deposits at very low salinities along the estuary, which reflects a high montmorillonite content in the muds of the turbidity maximum (probably because of a lower settling velocity of the montmorillonite relative to the settling velocity of the other clay minerals; Gallenne 1974). Also the muds on the low-salinity flats in the Seine estuary have the same composition as the suspended matter in the turbidity maximum, which for about 10% is supplied by the river and the remainder from the coastal sea (Dupont et al. 1994). In other estuaries, e.g., in the estuary of the St. Lawrence in eastern Canada (Dionne 1972a), very muddy deposits are found on the tidal flats along the estuary at low salinities.

Sediment balance estimates are based on a consideration of the quantities of supply and removal, on the use of natural or artificial tracers to characterize sediment of known origin, and/or on the fluxes along the input and output directions, or through transects in the estuary (summarized in Eisma 1993). Usually a time-consuming series of measurements have to be made to estimate residual fluxes that have to be based on small differences between large numbers. Often large variations

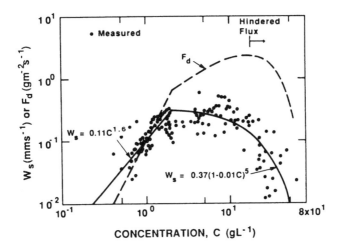

FIGURE 8.7 Sedimentation velocity Ws and sediment flux variations Fd with particle concentration for Tampa Bay mud. (From Mehta 1989. With permission; and adapted from Ross 1988.)

in sediment concentration and current velocity, episodic conditions following large storms when large amounts of sediment are moved, and the effects from factors that are not well known (waves and other wind effects, degree of consolidation of the sediment, bottom transport of suspended matter in large concentrations) make it difficult to estimate reliably the transport of sediment through an estuary. Similar difficulties occur when estimating reliably the fluxes of sediment to and from inter-tidal areas, because sediment transport in tidal channels and creeks is essentially no different from transport in estuaries except, perhaps, in scale and in the greater importance of density gradients in estuaries. Measurements of deposition rates (preferably in $g \cdot m^{-2}$ per year, day, or an even shorter period) make balance estimates more reliable, and estimates can be farther improved by differentiating between organic matter and inorganic material, and between suspended matter (clay + silt) and bedload (sand).

9 Intertidal Fauna and Vegetation

Part 9.1 Influence of Benthic Fauna and Microflora

Gerhard C. Cadée

LIFE ON THE TIDAL FLATS

Activities of organisms influence sedimentation and erosion as well as the physical and chemical nature of sediments. Tidal flats occur mainly in areas where saline and freshwater mix. Benthic organisms occur here usually in high densities because estuaries are among the most productive regions in the sea. Nutrient input by freshwater discharges sustains a relatively high primary production by phytoplankton and micro- and macroflora living on the tidal flats, providing the food for this abundant animal life. Moreover, there is a high input of organic matter (food) from rivers, but, as the organisms must be able to tolerate rapid tidal and seasonal changes in salinity, the number of benthic species is usually lower here than in the open sea and in the freshwater. The Remane curve (Remane 1934) relates the number of species to the gradient in salinity in an estuary: the number of freshwater species rapidly drops with increasing salinity; few survive in salinities in excess of 5‰. Marine species rapidly drop with decreasing salinities. Consequently, a minimum of species occurs around 5‰ because of the decrease of both freshwater and marine species toward 5‰, and relatively few real brackish water species have evolved to populate this brackish part of estuaries. This has been related to the short life span (only thousands of years) of most tidal areas. For plankton also the residence time of the water in the brackish water zone may limit the building up of a brackish-water species assemblage.

It has proved very fruitful to distinguish a number of ecological or functional groups among the intertidal fauna and flora. This enables a much better comparison of intertidal areas even if the species composition is quite different. A differentiation in plankton (free-floating), nekton (free-swimming), and benthos (bottom-living) relates to where the organisms are found. Some organisms stay their whole life in the plankton (holoplankton), but more switch during life: a flatfish such as the plaice starts as a planktonic egg, stays in the water column as a nektonic larva, and finally adopts an (epi)benthic life. Some phytoplankton algae such as *Biddulphia aurita* spend part of their lifecycle on the bottom (tychoplankton), while others, such as most planktonic dinoflagellates, have benthic resting stages, or cysts. Many benthic invertebrates have planktonic larvae. This illustrates a tight coupling between sediment and the overlying water. Because of these tight links, Boero et al. (1996) even question the existence of real plankton and benthos and advocate a growing separation between plankton and benthos research. A full appreciation of these links will force marine ecology toward a more integrated approach.

Another classification of the benthic fauna places them into size classes: micro-, meio-, and macrofauna, with boundaries usually at 0.1 and 1 mm. It is a pity that few biologists study the whole range of sizes; most study only macro- or meiofauna (compare the similar separation in plankton and benthos research mentioned before). A large number of meiobenthic organisms never attain macrofauna size; they live as an interstitial community between the sand grains and include nematodes, copepods, tartigrades, turbellarians, etc. Juveniles of most macrobenthic organisms pass through a meiobenthic stage. Microzoobenthos includes mainly protozoa, and *Foraminifera* are often included although they may be larger than 0.1 mm. Microphytobenthos on tidal flats comprises all microscopic autotrophs such as cyanobacteria, diatoms, (dino)flagellates, and euglenids. Some live interstitially between the sand grains and may show (tidal or diurnal) migration. Others live attached to sand grains or as epiflora on other living or dead objects.

Study of the food chains, or more appropriately trophic relations in food webs, in tidal areas gives an insight in the interdependence of all living organisms. Primary producers are phytoplanktic and benthic micro algae, macrophytes (seagrasses, mangroves, marsh vegetation, see Sections 9.2 and 9.3), and macro algae (see Section 9.4). Input of organic matter from freshwater and trapping of organic matter from the adjacent sea (Postma 1980) gives additional food to the secondary producers, the grazers, in the tidal area. Predators on these secondary producers, such as fish and birds, comprise the tertiary producers. Due to the fact that only some 10% of the food consumed results in growth, the number of steps in a food chain is limited to four or five. Top predators (birds of prey, fish-eating seals, whales, man) form the end of these food chains. Not all primary produced organic matter is directly consumed but may become available later as dead particulate organic matter (detritus, for biologists) after lysis and attack by bacteria and fungi. Such "detritus feeders" may feed in fact mainly on these bacteria and fungi. For practical reasons Mann (1972) and later Day et al. (1989) include these associated microbiota in their definition of detritus: for both the consumers and the scientist it is difficult to distinguish between the nonliving and the living part of detritus. A very small amount of the total produced and imported organic matter may become buried and escape destruction. Part of the primary produced organic matter in the dissolved form (DOM, defined pragmatically as passing through a filter with .45 μm pore size). DOM also comes free during lysis or consumption of phytoplankton. The DOM pool may be equal to or larger than the POM pool in tidal areas (e.g., Dutch Wadden Sea, 1982). Part of this DOM will be consumed by bacteria (the microbial loop). Many other organisms are also able to take up dissolved organic substances. Part is transported to the open sea and added to the oceanic DOM pool.

FIGURE 9.1   Fecal pellets (From *Heteromastus*, a worm) on a tidal flat surface (Wadden Sea). Length of a pellet about 1 mm. (Photograph courtesy of NIOZ.)

BIOLOGICAL EFFECTS ON THE SEDIMENT

In a number of ways biota affect intertidal sediments and sedimentation.

1. Biodeposition: Suspension-feeders catch suspended matter in collecting their food and produce fecal and pseudofecal material containing the inedible fraction of the suspended matter (Figure 9.1). The pseudofeces are formed out of particles that are rejected before ingestion. Both feces and pseudofeces are added to the sediment.
2. Bioturbation: Burrowing organisms disrupt original sedimentary structures by bioturbation (sediment reworking): new structures may result due to biogenic sorting.
3. Traces: Organisms may leave dwelling, feeding, resting, and escape traces in the sediment and on its surface. In ancient sediments these may be the only marks of biological activity (and may give important paleoecological information). They are often difficult to relate to their producer except in Recent environments.
4. Shear strength (internal cohesion) of the sediment: Many intertidal organisms produce mucus (mucopolysaccharides). In phytoplankton this traps suspended particles and may enhance floc formation and sedimentation; mucus produced by diatoms, meio- or macrofauna living on soft bottoms makes sediments sticky. This influences shear strength.
5. Erosion/sedimentation: Mucus production also increases the critical erosion velocity of the sediment; vegetation and tube structures above the sediment surface reduce the water velocity above the sediment, thus reducing erosion and enhancing sedimentation.
6. Permeability: Burrowing organisms and their burrows increase the permeability of the sediment. This enlarges the water content. By pumping respiratory water through their burrows these organisms aerate the sediment and enhance sediment-water exchange, e.g., of nutrients (P, N).

Bioirrigation also increases meio- and microfauna: around burrows more oxygen is available.

7. Biogenic particles: Some consider only particulate organic matter as bio-logic particles (Heip et al. 1995), but organisms also produce other types of particles, e.g., siliceous or carbonate exoskeletons. Sediments may consist partly or entirely of such biogenic exoskeletons. In tidal areas mollusks are particularly important. In addition, mussels, oysters, and the polychaete *Sabellaria* may form reef-like structures on soft sediments. Coral reefs are mainly restricted to hard substrates and will not be dealt with here. Predators may crush shells, producing shell fragments that are added to the sediment.

9.1.1. BIODEPOSITION

EFFECT OF SUSPENSION-FEEDING ON SEDIMENTATION

Suspension-feeding (collection of particles from the water column) is widely distrib-uted among marine organisms among both plankton and benthos. Filter-feeding is often used as a synonym for suspension-feeding, suggesting that sieving is the trap-ping mechanism, but a wide array of methods for collecting particles from suspensions occurs (see e.g., Jorgensen 1966; Levinton 1982; Riisgård and Larsen 1995).

Active or passive collection of particles on a mucus sheet or bag followed by ingestion of the mucus with particles is done by the tube dwelling polychaete *Chaetopterus*, for example, and by ascidians. Also small particles of about 1 μm are trapped.

A wide variety of benthic suspension-feeders, including bivalves, gastropods, and polychaetes, use rows of different types of cilia on arms or gills to collect particles. High densities of particles in suspension induce mucus secretion used to entrap unwanted particles and these are rejected as loose flocs, or pseudofeces. Bivalves have cilia on their mantle, gills, and mouth palps to collect, sort, and transport particles.

Setose suspension-feeding occurs in barnacles, among others. Beating of cirri creates a water current, and successive extension and withdrawal of the cirri enables the collection of food particles. Crinoids, brittle stars, zooplankton-feeding sea anemones, and corals capture food particles (zooplankton) by tentacle-tube feet, often facilitated by mucus production. In sponges a water current is set up by flagellate chambers which cause particles to enter through holes in the surface of the sponge. Inside the sponge, flagellated cells collect particles down to 0.2 μm (Jorgensen 1966; Riisgård and Larsen 1995) including free-living bacteria. The particle-cleared water is vigorously expelled through other, larger openings in the sponge to ensure separation from the incoming water.

Mucus is used in many suspension-feeders to collect food. Sometimes there is a selection for a narrow range of particles of certain diameter, but mucus-bag suspension feeding is unspecialized with regard to particle size. This implies that

FIGURE 9.2   Sand flat with mussel banks (*Mytilus edulis*). (Photograph courtesy of J. van de Kam.)

most suspension feeders compete for the same food source. Kamermans (1994) even indicated that food competition with deposit-feeders may occur when both groups feed on the same fluffy layer at the sediment–water interface.

Verweij (1952) was among the first to realize and estimate the quantitative role of suspension-feeders in biodeposition. He estimated that mussels and cockles in the Wadden Sea together could filter all the water present in two weeks (Figure 9.2). Later research (Dankers et al. 1989; Dame et al. 1991; Oost 1996) has proved this to be an underestimate. Mussels alone can filter all the water of the western Wadden Sea within one week and produce more feces and pseudofeces than estimated by Verweij, amounting to some 40,000 ton/day for the entire western Wadden Sea (1560 km^2). As there will also be resuspension and thus recycling of this material by mussels, such data cannot simply be extrapolated to annual biodeposition figures by multiplication, but it indicates that suspension-feeders are even more important than stressed by Verweij (1952). Taking into account that other suspension-feeders also occur in the Wadden Sea (e.g., bivalves *Mya arenaria* and polychaetes such as *Lanice conchilega*) and that some versatile ("opportunistic;" Cadée 1984) organisms can switch to suspension-feeding when this is more profitable (*Macoma balthica*, Hummel 1985; *Nereis diversicolor*, Riisgård 1991), one realizes that suspension-feeding plays a dominant role in sedimentation here: the suspended particles are packed into larger pseudofeces and into (often more resistant) fecal pellets that also, due to their size, will settle much more easily than the original suspended particles. Smaal and Prins (1993) compared the residence time of the water in a tidal area with the clearance time (the time needed to filter the total water volume of a tidal area) by the most important suspension-feeders and found in some cases that the clearance time even exceeded the residence time. This makes it understandable that Cloern (1982) observed in South San Francisco Bay that the seasonal cycle of tidal phytoplankton was mainly regulated by benthic suspension-feeders.

THE ROLE OF DEPOSIT-FEEDERS IN BIODEPOSITION

Benthic deposit-feeders feed on organic matter in and on the sediment. Strictly following this definition (feeding on sediment) they do not contribute to deposition. However, benthic suspension-feeders and deposit-feeders in a tidal area rely partly on the same food source (Kamermans 1994): the particles present at the sediment–water interface. Like suspension-feeders, deposit-feeders produce fecal pellets, which makes resuspension of the original fine particles more difficult. In some cases fecal pellets are very resistant (e.g., for *Heteromastus;* see Cadée 1979), also have a fossilization potential (Abel 1935), and are often very species specific (Häntzschel 1975). These fecal pellets still contain organic matter not digested by the producer. They can become rapidly colonized by bacteria and become food particles for their producers or other coprophagous organisms. Newell (1965) could feed starved *Hydrobia ulvae* (a small gastropod) on its own feces after keeping the feces for three days in sea water, during which period the nitrogen content increased 40 times due to an increase in bacterial protein. Carbon decreased by only 1%. The bivalve *Macoma balthica*, however, could not recycle its own feces. Frankenberg and Smith (1967) studied coprophagy in some 20 benthic organisms.

9.1.2. BIOTURBATION (SEDIMENT REWORKING)

Bioturbation comprises all processes of sediment reworking by organisms. The main bioturbators are found among deposit-feeders that feed on organic particles in soft sediments. Like suspension-feeders, deposit-feeders have developed several ways to collect these particles (see e.g., Levinton 1972). Swallowers (polychaetes such as *Arenicola*, sipunculids, spatangoid sea urchins such as *Echinocardium*) ingest all particles (except those too large to ingest). They need to process large amounts of sediment because the organic matter content of intertidal sediments is usually low (a few percent at most, but usually far less than 1%), and most of this is indigestible. Tentacle feeders (some tube-dwelling polychaetes, nuculoid bivalves, sea cucumbers) use tentacles to gather particles and transport them to their mouth; they are more selective than swallowers. Surface siphon-feeders occur among tellinid bivalves. The inhalant siphon is used like a vacuum cleaner to ingest surface sediment. Microalgae living on and in the sediment surface layer are the main food. Most deposit-feeding tellinids can switch to suspension-feeding (for *Macoma balthica*, see Olafson 1988; Kamermans 1994; Figure 9.3). Setose deposit-feeders occur among crustaceae (e.g., *Corophium*; Figure 9.4). They move appendages equipped with setae to trap sediment particles. *Corophium* scrapes surface material with its long gnathopods toward its U-shaped burrow, through which it creates a current with its pleopods. With its pleopods, provided with long plumose setae, it collects suitable particles from this current (Meadows and Reid 1966). This once more indicates that the transition from deposit-feeders to suspension-feeders is gradual.

Many deposit-feeders feed on surface sediments. This is related to the higher food content because of the presence of microalgae thriving in the euphotic zone of these sediments (only a few millimeters thick) and freshly deposited organic matter

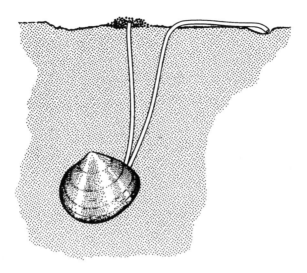

FIGURE 9.3 *Macoma balthica* in living position with siphon searching food particles on the tidal flat surface.

with its microbial flora. These surface deposit-feeders usually deposit their feces on the sediment surface. Surface deposit feeding crabs dominate in (sub)tropical tidal flats and, among these, fiddler crabs (*Uca*) are best known (see Section 9.3). Early on, Verweij (1930) described in detail the way they scrape the sediment surface for digestible particles, discarding the larger sand grains in characteristic round balls, often deposited in rows radiating from their dwelling hole. These sand grains are scooped from the sediment surface with small chelae and transported to the mouth parts where sand grains are scrubbed free of the microbiota used as food (see also Miller 1961). Fiddler crabs living in muddy sediments are able to sort edible particles from inedible ones by a kind of flotation method (Altevogt 1955,1957). Montague (1980) stressed the key role fiddler crabs play in intertidal ecology and gives numerous references.

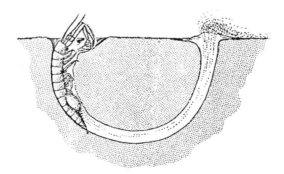

FIGURE 9.4 *Corophium sp.* (a crustacean) in its burrow. Depth of the burrow about 5 cm.

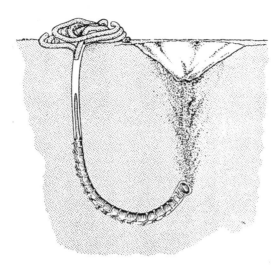

FIGURE 9.5 *Arenicola marina* (a worm) in its burrow. Depth of burrow about 15 cm.

A special group of surface deposit-feeders are funnel-feeders. They feed at depth in the sediment but collect surface sediment that sinks to the feeding depth in a funnel-like structure. The best-studied example is *Arenicola marina* (Figures 9.5 and 9.6; see Cadée 1976 and references given there). *Balanoglossus* is an example of a funnel-feeding enteropneust (Stiasny 1910). Some deposit-feeders such as *Pectinaria*, maldanid and capitellid polychaetes, and the holothurians (sea cucumbers *Leptosynapta, Caudina, Molpadia*) feed on deeper sediments and deposit their feces on the sediment surface. Rhoads (1974) called such deposit-feeders conveyer-belt species because they function as a vertical transport tube transporting subsurface sediment (as feces) to the sediment surface. A few surface deposit-feeders deposit feces below the sediment surface, such as the intertidal polychaetes *Scolecolepis squamata* (see Wohlenberg 1937) and *Pygospio elegans* (see Reise 1981). These can be called inverted conveyer-belt feeders.

FIGURE 9.6 Sandy tidal flat (Wadden Sea) with *Arenicola* (feces heaps). (Photograph courtesy of NIOZ.)

FIGURE 9.7 *Hydrobia* sp. living on mud flats. Length of shell about 3 mm.

Funnel-feeders and conveyer-belt deposit-feeders produce a biogenic layering of the sediment. They feed at a certain depth below the sediment and ingest here only those particles they can swallow; later these are deposited at the sediment surface. As a result, the coarser particles are left behind at the feeding depth, forming there a layer that is rich in coarse particles. Such a layer from the Wadden Sea was first described by van Straaten (1952) as a *Hydrobia* layer because it consisted mainly of shells of *Hydrobia* (Figure 9.7). It occurs at some 30 cm below the surface of many tidal flats. Van Straaten (1952) could relate this layer to the feeding activity of the polychaete *Arenicola marina*, and since then similar layers produced by *Arenicola* have been observed in many places (Trewin and Welsh 1976; Cadée 1976). Cadée (1979) found that locally in muddy sediments another conveyer-belt poly-chaete, *Heteromastus*, may help in the formation of this coarse shell layer. Rhoads and Stanley (1965) reported a comparable shelly layer produced by the polychaete *Clymenella torquata* from Barnstable Harbor. Warme (1967) reported a similar layer produced by *Callianassa* from Mugu Lagoon, as did Myrick and Flessa (1996) for Cholla Bay. Meldahl (1987) named this process "biogenic stratification."

Intertidal areas generally have sedimentation rates in the order of a few mm/yr (see Chapter 8). Sediment reworking rates by deposit-feeders may be one or two orders of magnitude higher (see tables in Cadée, 1976 and Lee and Swartz 1980). This implies that surface sediments are passed several times per year through the deposit-feeding benthos. As a result, primary sediment structures may be completely disturbed by bioturbation. At higher sedimentation rates the number of organisms that can live under these circumstances decreases and bioturbation decreases, as well. Higher sedimentation rates on Chinese tidal flats (Hangzhou Bay) of tens of centimeters per year were also used by Cadée et al. (1994) to explain the absence of bioturbation structures (except short-lived traces on the surface layer). Also in the more exposed intertidal areas of the Wadden Sea, sediment reworking by physical processes (e.g., during formation of megaripples, migrating tidal gullies) causes the disappearance of bioturbate structures, which is, moreover, enhanced by the lower number of benthic organisms that live there (Reineck 1977; Ma et al. 1995).

Bioturbation is not confined to deposit-feeders. Some predators among birds and fish disturb the sediment to dig for prey. Best known are the deep pits made by rays (Howard et al. 1977; Gregory et al. 1979; Myrick and Flessa 1996). Cadée (1990) described comparable deep feeding holes (diameter up to 60 cm, depth up to 20 cm) made by shelduck in the Wadden Sea. Smaller feeding pits are made by benthic flatfish such as flounders. Some gulls also make characteristic feeding traces. Cadée (1990) quantified the role of shelduck and gulls in the sediment reworking of the surface layer of intertidal Wadden Sea sediments. This was not negligible and amounted to about 2.5 cm annually, but it was much less than that of the deposit-feeders at the same site (equaling a sediment layer of 35 cm · y^{-1}).

Many benthic organisms make (semi)permanent dwelling holes in intertidal sediments. Excavation of these burrows also results in sediment reworking. Good examples are crustaceans such as fiddler crabs and callianassids (see Bromley 1990 for numerous pictures of their often intricate burrowing systems). Verweij (1930) was one of the first to study their tidal zonation and often species-specific holes in Indonesian intertidal areas. The introduction of the use of polyester resins to make casts of these burrows has greatly facilitated their study (Farrow 1975).

Many epibenthic organisms rest temporarily on or in the sediment surface layer. Usually they dig in the sediment and cover themselves with sediment to hide from predators. The habit is well known for flatfish, shrimps, and crabs (Schäfer 1962), but occurs also in horseshoe crabs (*Limulus*; Rhoads 1967) and the gastropod *Bullacta exarata* (Cadée et al. 1994). The quantitative role in sediment reworking has never been estimated, but it will cause a continual mixing of the upper centimeters of the sediment. Carnivorous animals, burrowing through the upper sediment layer in search of prey, also play a role in mixing sediment. Examples are the gastropod *Polinices* (Trueman 1968), priapulids such as *Priapulus caudatus* (Elder and Hunter 1980), and the polychaete *Polyphysia crassa* (Elder 1973).

Meiofauna lives predominantly in the uppermost few centimeters of the sediment. The constant activity of these ostracods, nematodes, copepods, juvenile polychaetes, mollusks, etc., tends to homogenize this surface layer, thereby also destroying traces made by the macrofauna (Bromley 1990).

9.1.3. TRACES

Traces left by organisms in Recent intertidal sediments have been studied for a long time (see Osgood 1975), but few biologists have studied these traces (Ennion and Tinbergen 1967; Falkus 1978); most trace specialists are geologists. This emphasizes the special role of trace fossils in the fossil record: they represent fossil behavior (Seilacher 1967). However, unlike body fossils, trace fossils are usually anonymous; the producing organism is unknown. Many students of traces (ichnologists) therefore have studied Recent environments to see the trace producers at work. Much early work in intertidal areas has been done in the Wadden Sea (compiled in Schäfer 1962, 1972, and Hertweck 1994), whereas Cadée (1976, 1979, 1990) quantified sediment reworking activities of organisms in this area. Traces in many other intertidal areas have been studied since (Bromley 1990). However, Curran (1994) still mentions the overall lack of knowledge about modern trace-making organisms in tropical carbonate environments.

Seilacher proposed an ethological classification of animal traces, which has been amended and illustrated by Bromley (1990) and is widely in use now. Resting traces (cubichnia) are left by epibenthic organisms (fish, crabs, shrimps) that rest tempo-

FIGURE 9.8   Feeding traces of a crab (*Macrophthalmus*) on Hangzhou Bay flats. (Photograph courtesy of G. C. Cadée.)

rarily on the sediment surface and conceal themselves from the eyes of potential predators by burrowing temporarily in the superficial sediment layers. Crawling traces (repichnia) are produced by organisms moving on or in the sediment, usually in one direction. Feeding activity may occur during this movement, but this is not clearly visible from the traces. Grazing traces (pascichnia) are often meandering or spiraling traces that result from the exploitation of a sediment surface area for food (Figure 9.8). Feeding traces (fodinichnia) are related to deposit-feeders and combine feeding and dwelling functions. Gulls and shelducks leave characteristic feeding traces consisting of several-cm-deep troughs that are up to 60 cm wide and may be up to 3 m long (Cadée 1990; Figures 9.9 and 9.10). Dwelling traces (domichnia) are structures used as semipermanent dwelling places. Traps and gardening traces (agrichnia) are complicated grazing traces that suggest they are regularly visited by its maker. Predation traces (praedichnia) are confined to hard substrates and relate to, for example, holes drilled in molluskan shells or shells crushed by predators. Escape traces (fugichnia) are produced by organisms buried suddenly under a sediment layer. Organisms that constantly adjust their burrows to an aggrading or degrading sediment layer because they need to maintain contact with the surface make equilibrium traces (equilibrichnia; Figure 9.11).

Many traces have a low fossilization potential. Hertweck (1972), studying high-energy environments on the Georgia coasts, observed no uninhabited burrows: physical reworking of the upper sediment layers destroyed structures as fast as they were produced. Only the deepest parts of burrows of *Callianassa major* were below the depth of physical reworking. Meiofauna homogenizes the upper centimeters of sediment, and highly mobile infaunal predators destroy deeper traces. Due to this

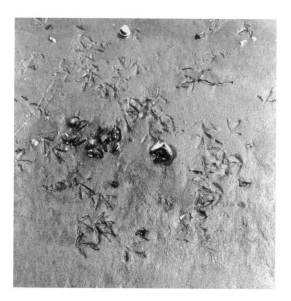

FIGURE 9.9   Footprints of the oyster catcher (*Haematopus*) on a muddy flat (Wadden Sea). (Photograph courtesy of NIOZ.)

physical and biological destruction, the traces that are preserved give no complete picture of the trace-producing fauna. A similar discrepancy due to differences in fossilization potential is observed in body fossils (Cadée 1968; Schopf 1978).

9.1.4.   EROSION

The erodibility of sediments can be assessed by measuring their critical erosion velocity, which is the water velocity (at a certain level above the bottom) at which the sediment surface starts to be eroded. Critical erosion velocities were established

FIGURE 9.10   Feeding traces (trampling holes) of shelducks (*Tadorna tadorna*) in the Wadden Sea. Diameter 30 to 50 cm. (Photograph courtesy of G. C. Cadée.)

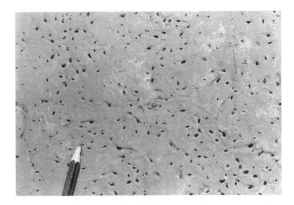

FIGURE 9.11   Siphon openings of burrowed clams (*Glauconome chinense*) on tidal flats along north Hangzhou Bay, China. (From Cadée et al. 1994. With permission.) (Photograph courtesy of G. C. Cadée.)

long ago for sediments in the laboratory and without biota (Hjulström 1935). Hjulström found a curvilinear relation with particle size, with at small particle sizes a critical erosion velocity that is higher for consolidated than for unconsolidated sediments.

The role of biota on the erodibility of sediments was realized later. The three most important stabilizing mechanisms are mucus binding, tube construction by bottom fauna, and the growth of vegetation (for the latter see Sections 9.2, 9.3, and 9.4). Such stabilizing processes counteract sediment reworking activities because they decrease erosion and increase deposition of fine-grained material.

Any clean surface exposed in an aqueous habitat will become covered within seconds with a film of organic molecules. Microorganisms will adsorb to this layer and metabolize, grow, form colonies, and produce mucus (also called exopolymers or extracellular substances, EPS). In this way a so-called biofilm is formed (Neu 1994), and EPS are involved in a wide range of interactions. Microbial biofilms develop at interfaces. In intertidal sediments they may evolve into true microbial mats at the sediment–water interface, dominated by *Cyanobacteria*, colorless and purple sulfur bacteria, and sulfate-reducing bacteria, all living in a complex interaction. A clear layering of these differently colored organisms leads to the "Farbstreifensandwatt" of German researchers (Krumbein 1987). These mats trap sediment and bind sediment particles. They now occur only where, due to adverse conditions, grazing and bioturbation are low: on high intertidal or supratidal flats and in hypersaline coastal lagoons. Early in the geologic record, when bioturbation was negligible to absent, algal mats were far more abundant, and they fossilized as stromatolites. Bioturbation increased during the Cambrian revolution in benthic organisms; deposit-feeders increased and mat communities became much more rare (Seilacher and Pfleger 1994).

No microbial mats are formed in most intertidal sediments but nevertheless, microbiota are present in high numbers (Fenchel 1969; Figure 9.12). Diatoms, both free-living and attached to particles, are more abundant. Free-living diatoms secrete

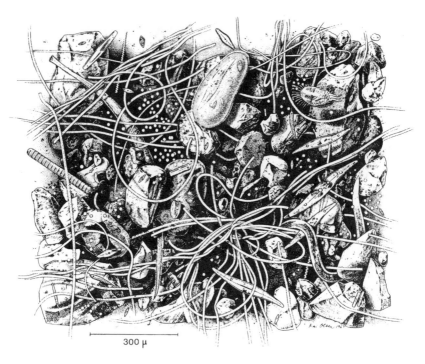

300 μ

FIGURE 9.12 Microflora and fauna in tidal flat sand in Nivao Bay (northern Denmark). Sand grains mixed with ciliates, diatoms, nematodes, blue-green algae, bacteria. Density is about 2000 to 3000 organisms per cm^2. (From Fenchel 1969. With permission.)

EPS as part of their locomotive mechanism (Edgar and Picket-Heaps 1984). Attached living diatoms use EPS for adhesion to sediment grains (Daniel et al. 1987). The role of diatoms in sediment stabilization had already been mentioned by Brockmann (1935) and van der Werff (1960) and was studied later more quantitatively (Holland et al. 1974; Frostick and McCave 1979; De Boer 1981; Vos et al. 1988). Paterson (1989) found that EPS released for diatom locomotion forms an extensive extracellular matrix throughout the surface sediment. SEM pictures of this mucus were given by Gouleau (1975) among others. Decho (1990) indicates that EPS secretions of both algae and bacteria are linked in surface sediments. This holds particularly for microbial mats and stromatolites (see Section 9.4). In conclusion, microbiota bind and stabilize intertidal sediments by mucus (EPS) secretion which enhances cohesion of particles. Mucus is, however, produced by all kinds of organisms (see Decho 1990 for a review). The combined mucus production of all biota will help in stabilizing intertidal sediments. Filamentous *Cyanobacteria*, moreover, also stabilize sediments by the formation of a network of filaments (Krumbein et al. 1994). Only the uppermost sediment layers are stabilized because the *Cyanobacteria* involved depend on light which is rapidly absorbed in the top millimeters. Grant et al. (1986) found that in intertidal sands on the Nova Scotia coast near Halifax, Canada, diatom mucus films were patchy on a scale of centimeters, corresponding to the structure of the sand ripples. After a storm such films were absent.

Many deposit-feeders produce a fecal pile at the sediment surface. Meadows and Tufail (1986) observed that casts of *Arenicola marina*, although protruding from the sediment surface, had a higher critical erosion velocity than the surrounding sediments. This was caused, again, by mucus secretions binding the fecal coils.

Dense stands of seagrasses or protruding tubes of benthic organisms increase the bottom roughness, reduce the near-bottom current velocity, and thus reduce the rate of sediment erosion (Featherstone and Risk 1977). Tube-building organisms, moreover, bind sediment in their tubes. In temperate and boreal areas polychaetes such as *Lanice conchilega* and *Pygospio elegans* are particularly important. A special example is the reef-building polychaete *Sabellaria*. This polychaete builds 50- to 70-cm-long tubes of 6 to 7 mm diameter, which consist of sand grains cemented in an organic matrix that are densely packed and firmly connected. They form rigid reef-like structures that may grow up to 30 cm above the sediment surface (Gruet 1972; Caline 1982). Such reefs withstand erosion. Schäfer (1962) mentions reefs of several square kilometers in the Wadden Sea involving large quantities of sand. They have vanished now from the Wadden Sea due to dredging and trawling (Reise and Schubert 1987), but some still occur along the French coast in the Baie du Mont-St. Michel (see Section 2.2, and Caline 1982), and they are also observed along subtropical and tropical coasts (along the Florida coast and near Sao Paulo (Remane 1954).

9.1.5. SHEAR STRENGTH, INTERNAL SEDIMENT STABILITY

The transition of surface cohesion to internal cohesion in sediments is a gradual one, and mucus also plays a dominant role here. The internal cohesion of a sediment is assessed by its shear strength, the lateral pressure needed to break a block of sediment or disrupt its structure. Lee and Swartz (1980) and Meadows and Meadows (1991) reviewed biological effects on sediment cohesion. Although it is widely accepted that biota influence shear strength, few have studied these relations (Deans et al. 1982; Meadows and Tait 1985; Tufail et al. 1989). The effects of cohesion caused by microbes, for example, in algal mats, have been studied best (Krumbein et al. 1994).

Mucopolysaccharide binding material is used by many infaunal invertebrates to construct the walls of their tubes. This causes an increase in the shear strength of the sediment (Meadows and Tait 1985), as was also found for *Corophium* and *Nereis diversicolor* burrows in intertidal sediments in the laboratory (Meadows and Tait 1989). Meadows and Meadows (1991) suggest that byssus thread complexes of mussels will be highly efficient in stabilizing muddy inshore sediments. This, together with increased deposition of feces and pseudofeces around mussel beds causes mussel beds to grow above the sediment surface (see also Oost 1996).

Sediment reworking activities may destabilize the sediment (Rhoads and Young 1970). De Deckere et al. (1996) observed a decrease in sediment cohesion due to the feeding activity of *Corophium*. Consumption of the diatom layer that stabilizes the sediment results in destabilization of the sediment. This indicates that conflicting influences of different biota on sediment cohesion exist, which in the case studied by De Deckere et al. (1996), results in a seasonal variation of sediment cohesion: high when the feeding activity of *Corophium* is low (in winter and spring), low in summer and autumn when the feeding activity of *Corophium* is high.

9.1.6. PERMEABILITY (EXCHANGE SEDIMENT/WATER)

The flow of water through sediments is related to grain size: large grains leave more voids between the grains, hence there is more volume for water flow. This volume for water flow can be enlarged by organisms. Burrows left by the organisms will form channels along which water can flow. Meadows and Tait (1989) observed an increase in permeability in sediments with *Nereis diversicolor* burrows, which were 30 to 40 cm long and oriented perpendicular to the sediment surface, but with *Corophium volutator* burrows they found no increase in permeability: the U-shaped burrows probably form a physical barrier. Mucus produced by many different organisms will clog sediment, thereby decreasing the sediment permeability.

As long as organisms are still living in their burrows, pumping of respiratory water through the burrows (bioirrigation) considerably enhances the water exchange between the sediment and the overlying water. Dworschak (1981) measured up to 2 liters of water pumped per gram dry weight in the burrowing crustacean *Upogebia pusilla*. The total population in the Lagoon of Grado pumped 1/4 to 1/1 of all the water entering this lagoon each tide from the open Adriatic Sea through their burrows. Aller (1978) demonstrated that bioirrigation increases the thickness of the oxygen-rich zone in sediments from 1 cm (without infauna) to 3 to 6 cm. Around burrows locally oxygenated zones occurred even deeper (Aller 1982). The oxygen-rich zones around the burrows enhance the growth of meio- and microfauna (Reise 1981; Meyers et al. 1987, 1988). Bioirrigation also effects biogeochemical processes in the sediment, while the intense decomposition of organic matter around the burrow walls causes mobilization of metals like Fe, Mn, and Zn (Aller and Yingst 1978).

9.1.7. BIOGENIC PARTICLE PRODUCTION

Particles in estuaries and coastal waters formed by organisms consist largely of organic matter (a review is given in Heip et al. 1995). Biogenic production of calcium carbonate occurs as bacterial-induced precipitation of calcium carbonate (e.g., Chafets 1994) or it is used in skeletons (e.g., mollusks, foraminifera). Opal is produced in diatom frustules. More quantitative studies exist on organic-C production than on production of other biogenic particulate material.

Organic matter is produced by the primary producers: the autotrophic plants. For the intertidal areas a wide array of autotrophs are important: the phytoplankton in the water column, the microphytobenthos on the sediment surface, and macrophytes growing on tidal marshes replaced in tropical areas by mangroves. In their reviews Smith and Hollibaugh (1993) and Heip et al. (1995) conclude that in many tidal areas the respiration of organic matter exceeds the primary production. This can be balanced only by an import of organic matter from outside the area. This import may stem from two sources: the rivers that discharge into the tidal area and the coastal sea. In tidal areas suspended matter is concentrated (see Chapter 8), including suspended load, containing organic matter, that is supplied from drainage areas far exceeding the size of the tidal area. Considerable effort has been put into estimating the size of the riverine input into tidal areas on a global scale as part of estimating the global balance of organic carbon (Degens et al. 1991). Postma (1980) indicated how (organic) particles may become trapped in tidal areas by a combination of three mechanisms: wave transport, tidal transport, and estuarine circulation. When the respiration exceeds the primary production, this indicates that losses of organic particles to the sediment must be low and occur only in those cases where input of allochthonous organic particles is high.

Few studies relate to opal production and the addition of opal to intertidal sediments. Van Bennekom et al. (1974) estimated for the western Dutch Wadden Sea an annual input of 140,000 tons of SiO_2: about 20,000 tons is deposited annually as diatom frustules.

Many organisms use calcium carbonate to build skeletons, and (parts of) these skeletons may become part of intertidal sediments or even be the only constituent. Shells of mollusks are most conspicuous (Van Straaten 1954; Frey and Dörjes 1988). Few quantitative studies exist on shell carbonate production (e.g., Beukema 1980, 1982). In tropical areas near coral reefs calcareous algae and corals also add to bioclastics. Microscopic inspection reveals skeletons of foraminifera, ostracods, skeletons and spines of echinoderms, and spiculae of holothurians and *Octocorallia*, etc. (e.g., Van Straaten 1954; Newell et al. 1959).

Large amounts of mollusk shells become fragmented due to the activities of predators that consume them. In intertidal areas like the Wadden Sea this biologic fragmentation is far more important than physical fragmentation (Cadée, 1994, 1995). This may hold for all rather sheltered areas. Only on exposed beaches may physical fragmentation be more important. Boring organisms (polychaetes, sponges, algae, fungi, lichens) living in mollusk shells also contribute to their destruction (Golubic et al. 1975). Deterioration of mollusk shells starts during the life of the animal (Cadée 1968) and is most intense when after death these shells are exposed on the sediment surface, as burial under a sediment cover will stop the activity of most borers (Flessa et al. 1993).

9.1.8. GEOGRAPHICAL DISTRIBUTION OF THE INFLUENCE OF BENTHIC FAUNA AND MICROFLORA

Benthic fauna and microflora occur virtually everywhere in tidal areas, but may be present only in very low numbers or temporarily absent 1) in very cold areas that are frozen during a long period in the winter, 2) in very mobile sediment that is easily reworked or resuspended, and 3) where deposition rates are very high so that benthic organisms are quickly buried. Reise (1991) compared three intertidal areas (in southern Chile, in the Wadden Sea, and along the Andaman Sea coast of Thailand; Table 9.1).

TABLE 9.1

Climate and Hydrographical Characteristics of a Cold, a Temperature, and a Tropical Intertidal Area

	Bahia Quillaipe Chile	Königshafen North Sea	Ao Nam Bor Thailand
Average summer temperature	16	15	28
Average winter temperature	11	4	29
Annual rainfall (mm)	2300	731	2266
Salinity ‰	9–24	26–32	32–35
Neap tidal range (m)	2.0	1.7	1.3
Spring tidal range (m)	6.3	1.9	2.5
Median grain size (mm)	0.50	0.46	0.09

From Reise 1991. With permission.

Figure 9.13 gives the benthic species assemblages in mud (above) and in sand (below) for each area. In the two areas with a temperate climate the number of species is low in relation to the number of individuals per unit area, but in the tropical intertidal area the number of species is high. In general the faunas are rather similar: closely related species occupy similar ecological niches, and unrelated species have analogous ecological functions. Individual species and species groups follow an often worldwide geographical distribution pattern, related to environmental factors and to their distributional history, which in most cases extends over geological time periods, but also can be related to a very recent history of migration, e.g., to dispersal by ships. Within a tidal area there can be, and usually is, a species zonation, but, as was seen above, biodeposition, bioerosion, and bioturbation can be the outcome of very different species-specific processes, which all end up in a change of the original sediment structure or composition. The degree of change can be related to the time needed to change the sediment, and to the activities, the presence or absence of specific benthic organisms, which depends on a large number of local and regional factors. As the presence or absence of benthic organisms does not show a basic large-scale geographical pattern and consists of a multitude of species-related distribution patterns, it can be concluded that bioturbation, biodeposition, and bioerosion occur everywhere in intertidal areas where organisms can live. Microflora need light for photosynthesis and therefore are restricted in salt marshes and mangroves, where leaves reduce the light that reaches the bottom and competition for space on the sediment surface can be prohibitive for a microflora to develop well. On the more open areas and in particular on the soft mud between mangrove trees, a microflora is present.

Conclusion

Biota do influence intertidal sediments and sedimentation. Biodeposition enhances sedimentation. Mucus, produced by many kinds of organisms, influences sediment properties and sedimentation. Organic particles and skeletons of organisms may

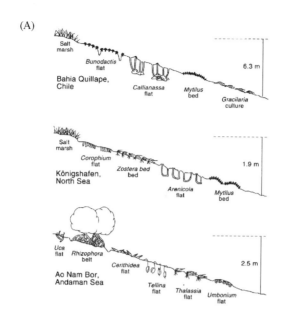

FIGURE 9.13 (A) Zonation of macrobenthos in Bahia Quillaipe (Chile), Koenigshaven (Wadden Sea), and Ao Nam Bor (Thailand). Spring tidal range to the right. (B) Species assemblages in mud (above) and sand (below) in Bahia Quillaipe (Chile), Koenigshaven (Wadden Sea), and Ao Nam Bor (Thailand). I = *Bunodactis*, a sea anemone; 2 = *Perinereis*, a worm; 3 = small worms (*Scoloplos, Capitella, Naididae*); 4 = *Corophium insidiosum*, a burrowing crustaceae; 5 = *Hemigraptus*, a crab; 6 = *Hydrobia ulvae*, a small gastropod; 7 = *Nereis diversicolor*, a worm; 8 = small worms (*Pygospio, Tharyx, Tubificidae*); 9 = *Corophium volutator*, a borrowing crustaceae; 10 = *Macoma balthica*, a bivalve; 11 = *Heteromastus*, a worm; 12 = *Carcinus*, a crab; 13 = *Tylonereis*, a worm; 14 = *Diogenes*, a hermit-crab; 15 = *Cerithidea*, a gastropod; 16 = *Ancistrosyllis*, a worm; 17 = *Enteropneusta*, a worm-like *Hemichordata*; 18 = *Poecilochaetus* sp. a worm; 19 = *Solen*, a bivalve; 20 = *Dotilla* sp., a crab; 21 = *Heteromastus* sp., a worm; 22 = *Scoloplos*, a worm; 23 = *Polygordius*, a worm; 24 = *Caecum*, a small gastropod; 25 = *Callianassa*, a crab; 26 = *Hemipodus*, a worm; 27 = *Phoxocephalidae*, an amphipod; 28 = *Exosphaeroma*, a small crustaceae; 29 = *Gracillaria*, a seaweed; 30 = *Tagelus*, a bivalve; 31 = *Diplodonta*, a bivalve; 32 = *Lumbrinereis*, a worm; 33 = *Scoloplos*, a worm; 34 = *Microphthalamus*, a small worm; 35 = *Arenicola marina*; 36 = *Eteone*, a worm; 37 = *Cerastoderma*, a bivalve; 38 = *Zostera*, a seagrass; 39 = *Pygospio*, a worm; 40 = *Lanice*, a worm; 41 = *Scoloplos*, a worm; 42 = *Tellina*, a bivalve; 43 = *Glycera*, a worm; 44 = *Aricidea*, a bivalve; 45 = *Tellinidae*, a bivalve; 46 = *Lumbrinereis*, a worm; 47 = *Diopatra*, a worm; 48 = *Thalassia*, a seagrass; 49 = *Nassarius*, a gastropod. (*Source*: Reise 1991.) (Continued on following page)

form an important component of the sediment. Vice versa, sediments (grain size composition, water content, and related mechanical properties) also influence the biota: muddy sediments have a fauna quite different from clean sands; water content determines the ability of organisms to burrow in the sediment. Large gaps, however, still exist in our understanding of organism–sediment relationships. We hardly know which sediment properties are most relevant to organisms: grain size, water content, organic matter content, or others?

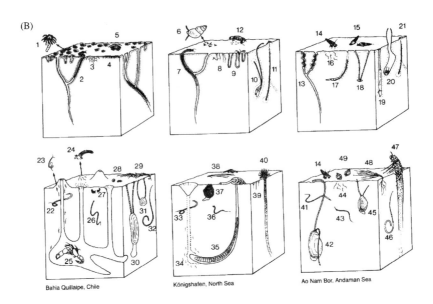

(B)

Bahia Quillaipe, Chile

Königshafen, North Sea

Ao Nam Bor, Andaman Sea

FIGURE 9.13 *Continued.*

Part 9.2 The Influence of Salt Marsh Vegetation on Sedimentation

with K. S. Dijkema

9.2.1. INTRODUCTION

Coastal salt marshes may be defined as areas vegetated by herbs, grasses, and low shrubs, bordering saline water bodies. Although such areas are exposed to the air most of the time, they are subjected to periodic flooding as a result of fluctuations in the level of the adjacent water body (Adam 1990). Salt marshes are intertidal with regular sedimentation and erosion; the sediments and soils are wet, saline, and poorly aerated. These conditions are unfavorable for most higher plants, so that the plants, salt marsh species as well as mangrove species, are adapted to the special conditions in the tidal zone. Salt marshes occur mainly around high tide level in the temperate zone, with extensions to higher (colder) and lower (warmer) latitudes (Figures 9.14, 9.15, and 9.16). The flats below the salt marsh are usually bare sediment, with locally a vegetation of seagrass (*Zostera*; Figure 9.37) or, more rarely, macroalgae (Figure 9.17). At lower latitudes, along tropical and subtropical coasts, the salt marsh is replaced by mangrove which covers the flats from below mean tide level upward. In the transition zones (north coast Gulf of Mexico, southeast Australia), salt marshes cover the upper parts of the flats, while mangrove forms a fringe on the lower parts.

The most diverse salt marshes grow on muddy, meso- to macrotidal flats in humid-temperate areas. Where the temperature is the limiting factor (at higher

FIGURE 9.14 Salt marsh with low cliff edge showing horizontally layered sediment (Wadden Sea). (Photograph courtesy of J. van de Kam.)

FIGURE 9.15   Mud flats along salt marsh with flowering *Armeria* (Dollard, Wadden Sea). (Photograph courtesy of J. van de Kam.)

FIGURE 9.16   Low salt marsh with *Salicornia* (Wadden Sea). (Photograph courtesy of J. van de Kam.)

FIGURE 9.17   Sand flats with gullies and macroalgae (Ooster Schelde). (Photograph courtesy of J. van de Kam.)

latitudes), grasses and sedges dominate; where extreme salinity is the limiting factor (at lower latitudes), succulent shrubs dominate; where a low salinity is the limiting factor, reeds (*Phragmites communis*) dominate (Barson 1982). The vegetation of salt marshes has a similar appearance throughout the world, caused by the abiotic factors salinity and waterlogging. The vegetation is characterized by its tolerance to these conditions and consists of halophytic (terrestrial) plants. Such a worldwide vegetation is called azonal. Other examples of azonal vegetation are found in mangroves, sand dunes, and submerged aquatic environments. The resemblance in the salt marsh vegetation can be seen in the physiognomy (succulence, dull foliage, small-leaf or grass type, etc.) and in the occurrence of the same genera and even species of plants (Chapman 1974). The halophytic flora is poor in species and mainly consists of perennial grasses, rushes, (dwarf-)shrub, and some annuals (*Salicornia, Suaeda*), which are exclusively found in these saline environments.

Favorable conditions for the development of salt marshes are a flat interface between sea and land, relatively low energy from waves and currents, and the availability of a suitable substrate (Dijkema 1987). Coastal salt marshes occur on a substrate of alluvial sediment or peat within the reach of tidally fluctuating sea water (e.g., Atlantic Ocean, North Sea) or seasonally fluctuating sea water (e.g., Baltic, Mediterranean). The substrate originates from deposition of allochthonous sediment through waves and currents (the common salt marsh types on sedimentary shores and in bays of rocky shores) or from autochthonous material (by peat formation, or by an isostatic rise of marine sediment, as, for example, on skerry coasts).

9.2.2. ZONATION OF SALT MARSHES

The distribution of intertidal plant species and vegetation shows a zonation related to the frequency of flooding or to the elevation above mean sea level; flooding can be the result of a temporary rise in sea level because of a storm as well as of the tides. A pioneer vegetation of *Spartina anglica* and *Salicornia* can begin 20 to 40 cm below high tide level on the west European flats; a similar vegetation of *Spartina alterniflora* on the North American flats starts at mean sea level. The elevation at which salt marshes start to grow on the flats is determined, besides by the elevation or frequency of flooding, by the size (energy) of the incoming waves and the degree of consolidation of the sediment, which needs to reach a certain firmness before vegetation can establish itself.

Around mean high tide level the coverage of the sediment with vegetation increases and other species appear, such as (along the Dutch–German–Danish Wadden Sea and other west European salt marshes) *Suaeda, Aster* and, finally, the perennial grass *Puccinellia*. This species is the principal species growing at the edge of the salt marshes in western Europe; elsewhere other plant species and genera occupy the same position along salt marshes: *Puccinellia phryganoides* along the arctic coasts of Europe, North America, and Siberia (Thannheiser 1975), *Spartina alterniflora* with *Juncus* and *Distichlis* along the Atlantic coast of the U.S. and Canada (Reimold 1977), *Spartina* and *Salicornia* along the northern Pacific coast of Canada and the U.S., and *Puccinellia* more to the south (MacDonald 1977), *Suaeda* along the northern coasts of China and Japan, *Triglochin, Salicornia,* and

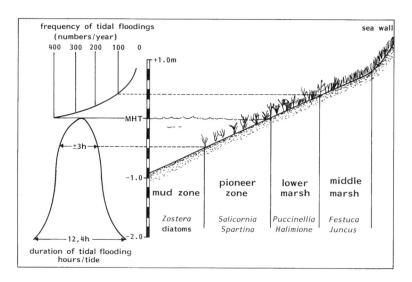

FIGURE 9.18   Frequency and average duration of tidal flooding of the principal zones in west European salt marshes. (From Erchinger 1985. With permission.)

Puccinellia farther south, *Salicornia*, *Suaeda*, and *Triglochin* in southern Australia, and *Spartina* in southern South America (Chapman 1977). The pioneer vegetation is almost daily flooded, the lower salt marshes several hundred times a year and increasingly less at higher elevation (Figure 9.18). The duration of submergence not only strongly influences the species composition of the marsh vegetation, but also the plant growth in general (usually negative) and sediment deposition (usually positive; Reed 1990). On very flat coasts, where the salt marshes also have a low gradient because of a small tidal range (as along parts of the U.S. east coast) flooding is more irregular and related to storm surges, which, however, over the years may show a regular pattern. In these, the frequency of flooding, and therefore also a zonation, is more related to irregularities in the surface relief of the marsh surface than to the general elevation of the marsh above sea level (Craft et al. 1993). At higher latitudes on most salt marshes at higher elevation more different species are present with an increasing number of species that are less tolerant to salt.

The type and density of the vegetation determine its effect on sedimentation, which is enhanced when the water flow is more reduced by the leaves and stems of the plants and the deposited sediment is held by the roots. A *Salicornia* vegetation, being open, without leaves and absent in winter, is considered to have a very small effect on sediment deposition, but can initiate the growth of *Puccinellia* and thus indirectly also affect sediment deposition (Kamps 1962). *Spartina* is often considered valuable in protecting a coast against erosion, but *Spartina* can only prosper in areas that are sheltered against large waves (König 1948). The perennial grass *Puccinellia*, growing at the edge of the salt marsh, induces the fastest rate of sediment deposition during the entire lifetime of a marsh, induces the development of a natural creek system and counteracts erosion of the newly formed salt marsh by protecting the sediment against resuspension and being washed away. The salt marsh vegetation is very sensitive to even short time — year to year — changes in mean high tide level and will shift parallel to the variations or more permanent changes in sea level, as has been observed along coasts in the Baltic where the land is rising (Cramer and Hytteborn 1987). Even a change in level of 5 to 10 cm can result in a shift in the occurrence of some species (Beeftink 1987). The results of a change in level can be seen within a year and will slow down in the following years. In man-made salt marshes changes in mean high tide level have shown major shifts of the salt marsh within a short time. For the long-term development of a salt marsh the balance of sediment deposition, sea level, and soil subsidence is the determining factor.

The salt marsh vegetation strongly influences the accumulation of sediment. During the first stages of the development of a salt marsh the aeration and stability of the sediment (soil) are essential for plant growth; with increasing elevation of the marsh, the soil salinity and the competition between species become more important. Armstrong et al. (1985) distinguished the following zones of soil aeration:

1. The pioneer zone, which is characterized by a predominantly reduced sediment; oxidized sediment occurs only during neap tide and then only at the surface to a depth of at most 5 cm in the sediment.
2. The lower salt marsh with predominantly oxidized sediment with a lower oxygen content around spring tide (for a few days).
3. The mid salt marsh with the longest period of aeration without fluctuations in the oxygen content, except during the highest tides.

Not all salt marsh areas may show this sequence, as their development depends on the inward increase in elevation of the marsh; in particular in some low-lying, flat marshes along the U.S. east coast, it may not be found.

The development of a creek system, usually with levees, is an essential element in the development of a salt marsh because it distributes the sediment supply and stimulates the growth of salt marsh species. It lowers the water content in the sediment and enhances aeration. In this way the succession to other types of salt marsh vegetation is stimulated. The relation between marsh accretion and sediment flow in the channels and creeks, however, and the processes on the marsh surface are not well known. One of the reasons is that usually the vertical accretion (in mm · y^{-1}) is measured: Reed (1989) indicates that collection of the sediment deposited during a certain time interval will give more significant results, because the effect of compaction is avoided.

Along the outer edge of a salt marsh often a zone is present which has a somewhat higher elevation than the more inward parts of the salt marsh. This is caused not by a stronger compaction of the older sediment more inward, and not by a higher deposition because of the plant species growing in that zone, but is the result of higher deposition when sea level is raised temporarily during storms. Then more sandy material is transported toward the salt marshes and is deposited on the outer edge as a "salt marsh levee" where the water flow is reduced, while the finer material is dispersed over the salt marsh. Waves during normal high tides can erode the outer edge of the salt marsh so that a low cliff is formed where the sediment is sufficiently consolidated (see Chapter 6).

9.2.3. SALT MARSH DEVELOPMENT

The development of salt marshes follows a chain of interactions between physical and biological processes. Salt marsh formation begins with the settling of a pioneer vegetation near mean high tide level where the coast is flat and sheltered against waves: the water has to be sufficiently quiet for the seeds to germinate. As soon as plants are there, the conditions for sediment deposition improve, in particular when the *Salicornia* is replaced by plants with broader leaves and roots. While salt marsh growth continues, creek systems develop: this may begin during the pioneer stage when the water can flow off between the plants, or the salt marsh creeks may develop out of creeks or gullies that already existed on the (bare) flats, with the vegetation developing between the creeks. Usually a system of levees along the creeks (with basins between them) also develops at an early stage, as well as a salt marsh levee along the outer edge of the marsh (see Chapter 6). With the development of a perennial vegetation, the rate of sediment deposition increases. In man-made marshes, the deposition increased after the development of a perennial vegetation to an average rate of about 15 mm · y^{-1}, while the deposition rate on flats with a nonperennial pioneer vegetation was, on the average, 3 mm · y^{-1}. Measurements during four years at the barrier island Ameland in the Dutch Wadden Sea show an increase in average elevation on low marshes (less than 40 cm above mean high tide level), while on the higher marshes (more than 40 cm above mean high tide level) the average elevation remained constant (van Duin personal communication). Both where average elevation increases and where it remains constant, a low net sedimentation rate is found during the summer (when the rate is measured in mm and not in cm^3) because of compaction. With increasing elevation the frequency of flooding decreases and with it the deposition of sediment.

The creek system is not only important for the transport of water and dewatering of the marshes, but also for the transport of sediment and nutrients. Mobilization and transport of sediment within the creek system is important for the supply of sediment to the salt marshes and hence for salt marsh growth. The creek systems grow with the marsh and both erosion and deposition take place, with deposition dominating. Older creeks can become so deeply incosed that during normal high tides the water does not flow over the adjacent marshes, which are flooded only during exceptionally high tide or storm surges. During storms, sediment is mobilized and deposited on the marsh while they are flooded, as has been observed on many marshes (summarized in Reed 1990). Tsunamis have a similar effect, as was observed in Willapa Bay (Reinhart and Bourgeois 1987) and on the Oregon coast (Grant 1987).

Deposition of sediment in salt marshes results in characteristic features in the marsh: levees between the vegetation along the channels formed by the tides and a parallel layering in the sediment with root marks, marginal levees formed by waves, usually along a low erosional scarp, sand sheets formed during exceptionally high tides and storm floods, irregular deposits formed by ice-rafting, which may contain coarse sand and gravel, and blocks of consolidated sediment with vegetation. Erosion occurs along channel banks and in channels (mainly by tidal currents), along the salt marsh margin (by waves) resulting in a scarp or slope, on the marsh surface by the wind, waves, and ice, resulting in ponds that can be further eroded by the wind, and by grazing of geese and cattle.

A larger tidal range can have a positive effect on sediment deposition: Stevenson et al. (1986) found for marshes along the U.S. coast a correlation between tidal range and the vertical accretion of the marsh minus the local relative sea level rise. Marshes with a tidal range larger than 3 m were not included, but this relationship suggests that mesotidal and probably also macrotidal marshes, with a higher accretion rate, have a better chance of surviving an increase in sea level rise than microtidal marshes.

Erosion of the marsh by water flow was found to be negligible in Cedar Creek, Florida (Leonard et al. 1995). During the entire tidal cycle, flow velocities over the marsh remained below 10 cm · s⁻¹ and were inversely related to the distance from a creek edge and with vegetation (stem) density. As is also found in other marshes (Louisiana, Delaware, Wadden Sea; DeLaune et al. 1978, Baumann et al. 1984, Stumpf 1983, Stoddart et al. 1989, Esselink et al. in press), sediment deposition was related to the proximity of a tidal channel or creek. During winter storms, deposition increased through an increase in the concentrations of suspended material, which was related to an increase in supply, not to an increase of erosion of the marsh: also during storms there was no evidence for erosion. Because storms are incidental, the regular deposition during the rest of the year, and even the summer deposition rate, may exceed the winter rates with only the nonstorm deposition being important. Along the coast of western Florida, the marshes are protected by a broad shallow shelf, which reduces the effects of a storm. In other, more exposed salt marshes, storms may have more effect along the marsh margin or on the marsh itself, but since marshes can only develop in relatively sheltered areas, the effects of storms will be limited. Where major storms do erode the salt marsh, as on the Dengie Peninsula, Essex, England, measurements from 1987 to 1990 have shown that the normal increase in marsh elevation because of sediment deposition was interrupted by storm erosion, which lowered the marsh surface by about 6 mm. It took more than a year to return to the level from before the storm, and it was estimated that it would take about 5 years for the marsh to recover completely and reach the level it would have reached without the storm erosion (Pethick 1992; Figure 9.19).

Ponds are very common on salt marshes and can remain behind during the first stage of marsh growth, when areas with bare sediment are completely closed off by the vegetation so that the water cannot flow off. Pools near a salt marsh levee or a former salt marsh cliff can remain when slump material forms a hollow area, while creek pools can develop in a creek that is blocked, for example, because of bank erosion or by sediment deposition.

9.2.4. THE INFLUENCE OF VEGETATION ON SEDIMENTATION

For the preservation of a salt marsh the balance between accretion and erosion, relative to the local relative sea level rise, is essential. The salt marsh vegetation creates conditions favorable for sedimentation: reduction of the flow (in particular between the stems) when the marsh is submerged, so that suspended matter is deposited, and protection of the sediment against erosion by both flow reduction near the bottom (Pethick et al. 1992) and damping of waves, and by consolidating

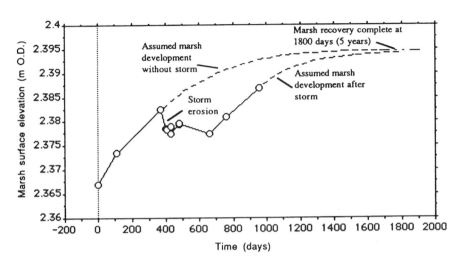

FIGURE 9.19   Theoretical recovery of a salt marsh after storm erosion, based on data from the Dengie Peninsula, Essex. (From Pethick 1992. With permission.)

the sediment with roots. The actual flow velocities within and just above a *Spartina* canopy are quite complex and include flow reversal in the lowest part of the canopy, while above the canopy the distribution of flow velocity with height is semi-logarithmic (Shi et al. 1995). Conditions critical for the formation and development of a salt marsh are those during its pioneer phase: the dynamic conditions during that phase must allow sediment deposition and stabilization of the sediment, as well as the establishment of vegetation. Pioneer vegetation is very sensitive to mechanisms that induce or prevent sedimentation or erosion. In the Dutch Wadden Sea an important pioneer species is *Salicornia dolichostachya*: its survival depends on biological factors such as seed production, germination rate, and growth of seedlings, and on hydrodynamic factors such as sediment deposition rate, disturbance of the sediment by currents and waves, and washing away of seeds and seedlings (Houwing et al. in press). The density of *Salicornia* at the end of the growing season depends mainly on the number of seeds that remain in the upper layers of the bottom sediment and the growth of seedlings in spring and summer. At low shear strength the bed is easily disturbed and 90% of the seeds and seedlings is washed away during storms. At high shear strength this is reduced to 70%. The higher shear strengths are frequently found in the transition zone between vegetated and unvegetated flats.

Accretion depends on the supply of inorganic as well as organic material by the tides, waves, episodic high tides and storm tides, and by ice-rafting. Inorganic sediment is mainly supplied, together with fine-grained organic material in suspension and in the form of flocs, but sand can be also deposited on the marsh during periods with large waves and high tides. This sand can be eroded from the tidal flats and channels seaward of the marsh, but also during high tide, from supratidal sandy deposits landward of the marsh, and is deposited on the marsh as a sheet-like layer. Organic material can be supplied with inorganic material in the form of fine plant detritus but is also produced on the marsh itself. This supply depends on the conditions for plant growth on the marsh, as well as on the conditions for preservation of the organic material that is produced. In the Dollard and in the Ooster Schelde, Dankers et al. (1984) and Oenema and DeLaune (1988) have shown that up to 90% of the surface-deposited plant material is not incorporated into the sediment but is removed by tidal flow or by microbial degradation. West European salt marsh sediments usually have low organic matter contents (in the order of 5 to 10%), but in salt marshes along the U.S. east coast (Louisiana, Georgia, Virginia, Massachusetts, Maine) the organic content of marsh sediments is much higher (up to more than 90%). This is probably not related to a larger plant growth on the marshes, but to different conditions of preservation and in particular to the degree of preservation of roots. Gardner (1990) found that the turnover rate and the mean diameter of the roots has a strong influence on the accumulation of organic matter. This is species-dependent: different *Spartina* species have a live mass in the ground that varies from 1.7 to 2.4 times the live mass above ground (Schubauer and Hopkinson 1984), and the turnover time of the rhizomes varies from $1.42 \ y^{-1}$ to $3.22 \ y^{-1}$.

The tidal range may be a decisive factor: the west European salt marshes occur along macro- or mesotidal coasts and the upper layers of the marsh sediment tend to be well-aerated, whereas the marshes along the U.S. east coast occur along coasts that are microtidal or low mesotidal (Georgia; Figure 9.20) and have low wave energy, where conditions are favorable for the formation of a high groundwater table in the marshes, development of anaerobic conditions near the sediment surface, and preservation of the root systems and other plant material. When the supply of sediment or organic matter to the marsh is reduced, the marsh deteriorates, as can be seen in Louisiana and at the Nile delta marshes, where river supply of sediment has been reduced because of dam construction. Deterioration of marshes is enhanced by subsidence, but all salt marshes, even those along stable coasts, need a supply of inorganic sediment and/or organic matter to compensate for the general sea level rise, which at present is in the order of 1 to 2 $mm \cdot y^{-1}$ (Gornitz et al. 1982 give an average of 1.2 $mm \cdot y^{-1}$). At the present salt marshes along the U.S. east coast, the yearly supply of organic matter and inorganic sediment is sufficient to balance the (local) sea level rise (Stevenson et al. 1986; Figure 1 in Reed 1990), except in Louisiana where river supply and marsh growth can only maintain a vertical accretion rate of 8 $mm \cdot y^{-1}$, while the local submergence rate is 1.2 $cm \cdot y^{-1}$ because of subsidence (DeLaune et al. 1983). This is partly related to a decrease in river sediment supply since the mid-19th century because of dam construction, but also because deposition during major storm events with extensive flooding of the marsh (considered to be the central factor in the sediment budget of microtidal marshes along the U.S. east coast; Leonard et al. 1995) is not sufficient. In the Dutch Wadden Sea, which is subsiding at a rate of 1 to 2 $mm \cdot y^{-1}$, land reclamation since 1930 has changed an erosional situation into an accretional situation, where salt marshes were formed in sheltered areas along sand dikes and along the inner margins of the Wadden Sea (Houwing et al. in press; Figure 9.21). The question of the response of a salt marsh to sea level variations, and in particular to a sea level rise, is important when a worldwide increase in sea level is considered to be likely in the near future (see Chapter 11), but at present generalizations on the response of salt marshes are apt to be inadequate. A promising attempt to model a sediment budget to predict

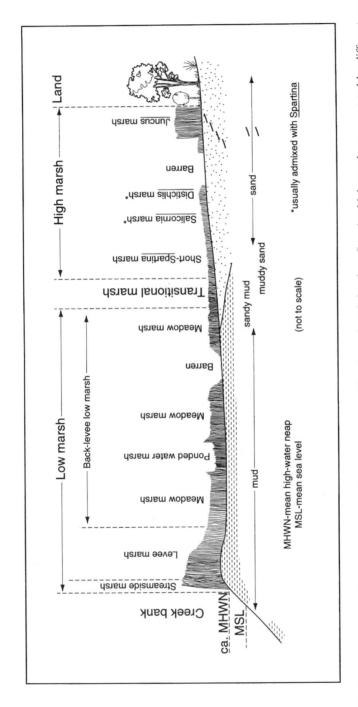

FIGURE 9.20 Georgia (U.S.) salt marsh habitats. The low transitional marsh and the short-*Spartina* high marsh are vegetated by different growth forms of *Spartina alterniflora*. (Adapted from Edwards and Frey 1977.)

FIGURE 9.21 Artificial salt marsh accretion along the Wadden Sea. (Photograph courtesy of J. Abrahamse.)

the change in marsh elevation has been made by Allen (1990), but Luternauer et al. (1995) pointed out that it predicts surface elevations for one single coast. Although this may be quite sufficient for local purposes, for more generally applicable predictions the complexities of sediment transport in relation to tidal height have to be better known.

The effect of grazing on salt marshes and salt marsh sedimentation has been shown for marshes in the Wadden Sea (Leybucht; Andresen et al. 1990). Grazing considerably reduced the sedimentation rate, which probably is related to a lowering of the height of the vegetation, damage to the turf, and reduction of litter production. The influence of the vegetation on the deposition of sediment along the channels and creeks and on the formation of levees can indirectly be seen from the effect of grazing in the Dollard (Esselink et al. 1996, in press). After the artificial drainage system was abandoned in 1982, the number of levees as well as their elevation relative to the marsh behind them, increased in a less pronounced manner in the areas where grazing by cattle and geese was more intensive. Grazing resulted in a different vegetational structure and a reduction in accretion of the marsh. Geese not only graze the surface, but also grub for the tubers of *Spartina* in the sediment.

The type of vegetation can have a decisive influence on sediment deposition: some marsh species, such as *Spartina alterniflora*, collect suspended matter as a film on stems and leaves, whereas at other species, such as *Juncus* sp., such films are absent (Leonard et al. 1995). Stumpf (1983) found in Delaware Bay marshes that up to 50% of the material deposited on the marshes was collected on the *Spartina* vegetation, whereas in the Louisiana marshes only 4 to 5% was collected by the *Juncus* vegetation. Alizai and McManus (1980) found in the Tay estuary (Scotland) a high retention of sediment by broken reed stems (*Phragmites communis*), which may account for an annual accretion of 0.6 mm (French and Spencer 1993). In a North Norfolk marsh French and Spencer (1993), however, found that retention by

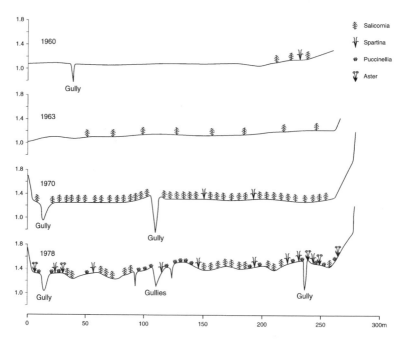

FIGURE 9.22   Development of relief, vegetation, and elevation of the Cappeler Tief tidal flat. (From Grotjahn et al. 1983. With permission.)

the plant surfaces was extremely low (within the probable experimental error). The marsh vegetation was a mixture of salt marsh plants (with *Salicornia* and *Puccinellia*) dominated by *Aster tripolium.*

The general influence of the marsh vegetation on the development of tidal marsh relief can be seen from the profiles of tidal flats and marsh vegetation made at the Cappeler Tief in the German Wadden Sea (Figure 9.22; Grotjahn et al. 1983). A small creek was present on the bare flats in 1960 at the onset of the development of a marsh vegetation and had disappeared several years later. Formation of a creek system started with the development of a vegetation of *Salicornia* and *Spartina* and resulted in an irregular relief six years later when a vegetation with *Puccinellia* and *Aster* had developed.

Reversely, creeks are very important for the distribution of sediment across the marshes (Stoddard et al. 1989): individual creeks greatly influence local patterns of accretion, and sedimentation in the North Norfolk marshes (Hut marsh) is determined by drainage density, channel (creek) size, and the velocity regime. Sediment, deposited during neap tides in the creeks, is mobilized by velocity pulses and during spring tide is reworked and partly dispersed, so that not only sediment availability, but also the opportunities for mobilization and transport within the marsh system as well as the interaction between the creeks and the vegetated surface determine the development of both the salt marshes and the creek systems.

Part 9.3 Mangrove

Mangroves are trees and bushes living along sheltered shores between spring tide level and mean sea level (or mid-tide level) in tropical, subtropical, and some warm-temperate areas (Macnae 1968); in the tropics they form about 25% of the coastline (Hatcher et al. 1989). "Mangal" indicates a mangrove vegetation or plant community which consists mainly of mangrove trees and (wooden) shrub with some herbs, ferns, and algae. They have in common that they are living in warm tidal waters of varying salinity. Most of them have root adaptations to live in regularly submerged sediment. "Mangrove" is a collective name for about 19 genera (or about 60 species) among which twelve genera, all trees, dominate (*Aegiceras, Avicennia, Brugueira, Ceriops, Conocarpus, Kandelia, Laguncularia, Lumnitzera, Nypa, Rhizophora, Sonneratia,* and *Xylocarpus*). They mainly belong to four closely related families. Also an often specialized fauna of fish, mollusks, crustaceans, reptiles, amphibians, and insects, adapted to intertidal conditions, lives in mangroves on or in the sediment and the trees. Geographically mangroves live between 32° N (the southern tip of Japan) and 38° S (the northern end of New Zealand) but these limits are the extremes where only one mangrove species is able to live (a *Kandelia* species in Japan, an *Avicennia* species in New Zealand). Generally mangroves flourish where the air temperatures in the coldest month do not fall below 20° C and the temperature range is about 10° C at most (Chapman 1977). In Florida, the northern Red Sea, Brazil, and New Zealand some *Avicennia* species are capable of tolerating minimum temperatures between 10° and 15° C.

There is a major division between genera that occur around the world (*Avicennia, Rhizophora*), a few genera that only live in the Americas and along the west coast of Africa (*Laguncularia, Conocarpus, Pelliciera*) and genera (the majority) that live in the Indo-West Pacific area from East Africa to East Asia, Australia, and Micronesia. The largest species diversity is found in southeast Asia (Malaysia, Indonesia) and from there decreases in all directions, not only to the north and south but also to the west toward Africa, and toward the east (there are few mangroves in Polynesia). Some of the major genera were already present in the Cretaceous in Borneo (*Nypa,* a low palm) and the early Tertiary (Eocene) in Brazil, Borneo, India, southeast Asia, west Australia, western Europe (*Nypa, Sonneratia, Avicennia, Rhyzophoraceae*), in the Oligocene in Borneo, Venezuela, Nigeria, New Guinea (*Rhizophora*), in the Miocene in east Asia (*Avicennia, Sonneratia, Lumnitzera, Ceriops, Bruguiera;* Churchill 1973). In Borneo an uninterrupted succession of *Nypa* was found from the Eocene up to the present. The distribution in the Cretaceous and Eocene, ranging from Australia to South America, suggests a dispersal through the Tethys Sea and an early development of *Nypa, Rhizophora,* and *Avicennia,* which could migrate westward before the Tethys closed in the early Tertiary (Figure 9.23). After that, endemic species developed and a number of other genera became mangroves, prob-

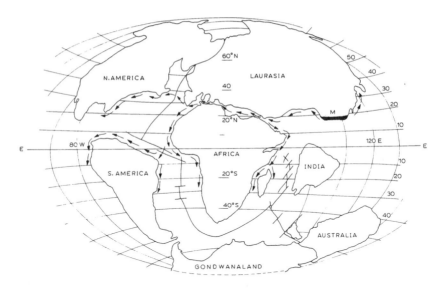

FIGURE 9.23 Sketch map of the world at the end of the Cretaceous with the probable area of origin of mangroves; probable dispersal indicated by arrows. (From Chapman 1977. With permission.)

ably out of terrestrial plants. At present the highest number of mangrove species occurs exclusively in Asia and the adjacent parts of Irian Jaya/Papua-New Guinea and Australia, with a much lower number of species that occur also outside this region. The number of mangrove species living exclusively in Africa and the Americas is much lower (Saenger et al. 1983). Plaziat (1995) pointed out that modern mangroves are rarely and then often imperfectly fossilized; older mangrove deposits can be recognized by their mollusk fauna (encrusted oysters), pollen, fruits, wood, peat, and the branching roots of *Avicennia*.

The dispersal from east Asia was primarily determined by climate, by the presence or absence of sheltered coasts (bays, estuaries) and suitable intertidal sediments, and by the distribution of cold and warm currents, so that today along the east coasts of Australia, Africa, and South America mangroves are present more to the south (down to 27° to 30° S) than along the west coasts (down to 20° S; in South America down to 7° S). Along the South America west coast this distribution is also influenced by the absence of sheltered coasts south of 7° S. In the Pacific the dispersal eastward was restricted because the principal currents in the tropics and subtropics go westward. Samoa is the natural eastern limit; mangroves occurring farther east were introduced by man. In the tropics and subtropics mangrove is only absent along coasts that are exposed to waves and wind, and generally is less well developed away from the wet-tropical areas with soft mud. Mangroves can live, however, on all types of sediment ranging from very stiff or soft mud to even gravel or rock. They can also live in a wide range of climates ranging from very wet and humid lowlands to very dry desert coasts (Red Sea, Persian Gulf, Gulf of California). Both the sediment type and the air humidity can, to a large extent, determine the species composition of a mangrove or mangal. The water content and the frequency of flooding are also important as some species cannot tolerate very wet or water-logged soils. Many mangrove species have adaptations to an environment with often strong tidal currents, waves, occasional strong winds, and mud deposition. Special root systems developed for a better fixation of the trees, and a better aeration in waterlogged muddy soil where sediment is regularly being deposited (Tomlinson 1986). The viviparous seedlings that occur in *Rhizophora*, *Avicennia*, and *Aegiceras* species favor settling in freshly deposited mud where normal seeds are not likely to germinate. In Panama, mangroves growing near open water or a channel have large and heavy propagules; those living more inward produce smaller ones that in some cases need some resting time before they can start to grow (Rabinowitz 1978).

A mangrove vegetation usually, but not necessarily, consists of zones of different species composition, although in the Americas and along west Africa a zonation is often less clear or absent, probably because there are relatively few genera present. Zonation is primarily related to the frequency and duration of flooding by the tides and to the salinity of the water (Watson 1928, De Haan 1931) and follows the coastline, the tidal channel banks, and the borders of estuaries. At river mouths, where mangrove is best developed, the zonation is related to the tidal amplitude (Balzer 1969): where the tidal range is small, the mangrove belt tends to be narrow and to consist of only one species. Where a zonation is less well developed (southeast Mexico, Colombia, Orinoco delta), the distribution of species is, more than to flooding, related to land forms and types of substrate, to the presence of areas of active sedimentation, and to the frequency of shifts in the channels. A supply of fine sediment is favorable for mangrove growth: in areas dominated by carbonate (coral, rock), mangroves cover only limited areas and the trees are usually smaller (or occur only as shrubs). Only a few genera can settle on recently deposited sediment and form pioneer vegetations: *Avicennia*, in the Indo-West-Pacific area, also *Sonneratia*, and in some areas *Rhizophora* (Senegal, lagoons from Liberia to Nigeria, coral islands). On rapidly accumulating mud flats *Spartina* can form a pioneer vegetation, with *Avicennia* more inward. A little more inland *Rhizophora* forms the outer fringe along the lower river courses and the tidal channels. The *Avicennia* zone can be very small — only one to three trees deep — and lies around mean sea level, but on accreting coasts can be up to 800 m wide (Malaysia, Indonesia). *Avicennia* cannot grow in the shade: they die also when they are older, when other trees (*Rhizophora*, *Bruguiera*) start to grow over them.

Sonneratia can replace, or grow seaward of *Avicennia*, or it can mix with it. It grows around mean tidal level and can stand being wetted daily with sea water. *Avicennia*, *Sonneratia*, and several other mangrove species have a shallow root system with cable roots that form the main network which is kept in place by anchor roots and with pneumatophores that grow upward into the air and with thin nutritive roots (Figures 9.24, 9.25, and 9.26). *Rhizophora* mangrove is usually present from mid-tide level to about halfway to high-tide level and is characterized by arched prop roots, which support the trunk and make passage very difficult (Figures 9.27, 9.28, 9.29, and 9.30). Different species of *Rhizophora* live in areas dominated by sea water, and others in areas dominated by freshwater. Along the Senegal coast different *Rhizophora* species form zones: *Rh. racemosa* along the seaward margin and along the lower tidal channels, *Rh. harrisonii* at the middle level, and *Rh. mangle*

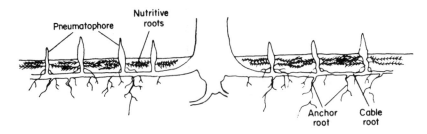

FIGURE 9.24   Root system of *Avicennia, Sonneratia, Lumnitzera* sp., and *Xylocarpus* sp. (From Macnae 1968. With permission.)

FIGURE 9.25   *Avicennia* sp. with pneumatophores (Kenya). (Photograph courtesy of M. Hemminga.)

FIGURE 9.26   *Avicennia* sp. with pneumatophores (Isla Margarita, Venezuela). (Photograph courtesy of D. Eisma.)

FIGURE 9.27   Prop roots of *Rhizophora*. (From Macnae 1968. With permission.)

FIGURE 9.28   *Rhizophora* (Kenya). (Photograph courtesy of M. Hemminga.)

FIGURE 9.29   *Rhizophora* with prop roots (Isla Margarita, Venezuela). (Photograph courtesy of D. Eisma.)

FIGURE 9.30   Mud flats with mangrove fringe (*Rhizophora*; Guinea Bissau). (Photograph courtesy of J. van de Kam.)

along the drier inner margins, whereas undergrowth can be rare. Higher up in the tidal zone there are forests of *Brugueira* (with also *Xylocarpus* and *Lumnitzera*) on the more consolidated sediments with incipient soil formation and often a well-marked surface drainage. In humid areas these forests reach up to high tide level: sediment trapping and deposition of organic material results in a higher elevation, which also results in a change in vegetation (e.g., more *Ceriops*). *Ceriops* and *Brugueira* forests can grade toward the supratidal zone into a *Nypa* mangal, which is typical for the brackish-freshwater transition and is dominated by a palm (*Nypa*) that can live in waterlogged soil. On the drier parts it can grade into a *Ceriops* or a *Barringtonia* vegetation which prefers relatively well-drained soils and grows behind mangals as well as behind sandy beaches. *Nypa* mangals can be very dense with hardly any other plants.

A full range of mangrove vegetation develops where the rainfall is more than 2000 mm per year. *Avicennia* trees can reach 30 m where they grow at optimum conditions, (Amapà–Guyana coast), *Rhizophora* more than 40 m (Colombia-Ecuador, Nigeria, Cameroon, Zaire, estuaries in east Africa, Indonesia) and *Heritiera* up to 30 m (Bangladesh, Burma). Most mangals are of medium height and form rather thin forests, because even where the yearly rainfall is more than 2000 mm · y^{-1} very dry seasons can occur. During such periods the salinity of the water in the soil can rise to 50 to 100‰. *Avicennia* is the most adaptable and can stand soil salinities above 90‰. *Sonneratia* prefers salinities below 35‰, and *Bruguiera* prefers salinities below 25‰. At extreme salinities (strong evaporation), poor aeration (stagnant water), or a very poor substratum such as coral rock, mangrove shrub or dwarfed mangrove is found. Where evaporation and transpiration are far in excess of the freshwater supply, a salt vegetation of bushes and low shrubs develops, and where it becomes very dry, the soil becomes bare without vegetation. Along the west African coast such bare salt flats are formed landward of the mangrove at high tide level or just above it, and are called "tannes" (Figure 9.31). Extensive supratidal bare salt flats generally occur where occasional flooding with sea water is combined with long periods of drought.

FIGURE 9.31   Tanne with mangrove (Guinea Bissau). (Photograph courtesy of J. van de Kam.)

Mangrove forests can be divided into five types (Lugo and Snedaker 1974):

- River mangroves: along river channels flooded during most high tides and during the wet season
- Basin mangrove: in a local depression, flooded by few high tides during the dry season, by most high tides during the wet season
- Fringe mangroves: along shorelines with a steep elevation gradient, flooded during all high tides
- Overwash mangroves: completely flooded (overwashed) during all high tides (low islands, peninsulas)
- Dwarf mangrove: on flats above mean high tide level, only flooded during the wet season and dry during most of the year, and generally present in areas where growth conditions for mangroves are marginal.

Woodroffe (1992) distinguished river-dominated, tide-dominated, wave-dominated, and composite river- and wave-dominated mangroves, mangroves in drowned bedrock valleys, mangroves in a carbonate setting with mangroves in embayments or lagoons on the Pacific islands, reef flat mangroves, mangroves in depressions, and inland mangroves.

Mangrove normally occurs along sheltered parts of the coast: waves are not favorable for mangrove growth, in particular not for the rooting of propagules. Once there, however, mangroves can protect the coast against waves and create quieter areas as long as the waves are not too large. Zenkovich (1967) observed on Hainan Island, China, that waves were attenuated in the outer margin of the mangrove, with the water flow remaining strong enough to resuspend bottom sediment in the mangrove. In northern Australia waves were observed to break within the outer margin, but can uproot mangrove, especially young specimens and especially during storms. In channels, continuing wave erosion can result in regression of the mangrove through bank erosion (Augustinus 1985). Wave attenuation by mangrove not only reduces the waves but also their reflection so that, as observed by Baltzer (1975), a quiet area is created in front of the mangrove, where mud can be deposited. Large waves during tropical storms can erode the mangrove, but usually the high current velocities that develop because of heavy rainfall and flooding, as well as the strong winds, can erode the mangrove. A storm surge, however, protects the mangrove against the wind, and during destruction of the mangrove the roots and trunks protect the sediment, so that actually little is eroded (Spenseley 1977). Along estuaries and large channels, waves can erode the banks, which finally results in slumping of the mangrove.

RELATION OF MANGROVES TO SEDIMENTS

Mangrove sediments in general have a relatively high salt and water content, are low in oxygen or completely anoxic, and in the latter case contain abundant hydrogen sulfide. They are often little consolidated and can contain a high percentage of humus or mangrove peat, which consists of fibrous soil containing the remains of roots and other wood. The organic matter content of such peaty deposits can be up to 95%.

Calcareous material comes mostly from foraminifera and mollusk shells. Sulfur bacteria decompose the sulfur-containing organic compounds in the shells, which makes the latter more fragile and the calcium more available to other organisms. Because of this, shells are usually not well preserved in mangrove deposits.

Sediment has to be deposited first before a mangrove vegetation can develop. Mud banks are barren when they are freshly deposited, and algae, probably also bacteria, are necessary to prepare the deposited mud for utilization by higher plants (Schuster 1952, Emerit 1960). The water depth at which *Avicennia* seeds can develop and grow roots is not known, but *Rhizophora* species need very shallow water before they can settle. The seeds grow roots while still on the trees (they are viviparous); they are up to several decimeters long and fall down in a vertical position. When they fall into mud, they stick upright in the bottom and start to grow immediately. When they fall into water, they float horizontally for 10 to 30 days. Then they turn vertical (in the sunlight; in the shade it can take several months longer), and where the roots touch the bottom they form seedlings, rooting in the sediment. The first leaves are formed after 40 to 50 days; in the shade this takes much longer and can take well over three months (Banus et al. 1975). In freshwater the seeds float vertically almost immediately after reaching the water surface and will settle relatively quickly. Seeds washed ashore between mean sea level and high tide level will die when they become covered with mud.

Avicennia settles at all salinities and is able to tolerate a wide salinity range. All three genera that settle on fresh mud (*Avicennia, Sonneratia, Rhizophora*) grow (or grow best) without shade. It is not clear why a certain species acts as a pioneer and others do not: it is probably important who is first. *Rhizophora* is often the pioneer in the West Indies and usually on coral debris (east Africa, Madagascar, Malaysia, Indonesia, west Pacific). The other two genera settle more on soft mud.

Once mangroves have settled, accretion of the coast is enhanced. This is mainly caused by the root system. Almost all mangroves have shallow roots, which can extend over considerable distances beyond the area covered by the branches. *Rhizophora* has prop roots with bundles of thick, air-filled roots about 30 cm underground (Figure 9.27). From there, nutritive roots penetrate the upper soil. *Bruguiera* has air-filled roots that elbow upward out of the soil. *Avicennia, Sonneratia, Lumnitzera,* and *Xylocarpus* have an extensive system of cable roots at 20 to 50 cm below the surface, with anchoring roots going downward and aerial roots (pneumatophores) going upward and sticking out of the soil. Nutritive roots grow out of the pneumato-phores in the top soil and form a firm layer at 10 to 25 cm below the sediment surface (Figure 9.24). *Xylocarpus*, which grows in anoxic soil, has plank roots that protrude above the soil. The root systems hold the sediment together by protecting it against erosion, while the tree trunks and pneumatophores, prop roots and other roots that stick out of the sediment obstruct the water flow, reduce erosion, and induce sediment deposition. The seaward zone, with *Avicennia, Sonneratia,* and *Rhizophora* is the zone of the largest deposition. The rapid accretion of some muddy deposits, particularly in Indonesia, has been ascribed to the influence of the man-groves on the settling of mud (100 to 200 m · y^{-1} of coastal deposits along the north coasts of Sumatra and Java) but sedimentation has also been enhanced by a large supply of fine-grained material from easily erodible volcanic soils, and by the presence of a very shallow sea, which can rapidly be filled in. Junghuhn in 1853 had already indicated that the mangrove follows the siltation rather than precedes or initiates it (Van Steenis 1958).

Mud is also settling, although in smaller quantities, at the higher levels in the mangroves up to high tide level. Litter is usually decomposed quickly, but near high tide level the production of organic material is in excess of decomposition: peaty deposits are formed that gradually raise the sediment surface to above high tide level. This is counteracted by compaction and consolidation of both mud and peaty sediment. Where the climate is dry and a dry vegetation with halophytes grows around high tide level, there is often a low "salting" cliff which is eroded during the normal high tides. The salt vegetation at higher levels is only episodically flooded during spring tides or storm surges (which are rare along mangrove coasts).

The relation between mangrove genera (and species) and sediment types, ground water characteristics, degree of flooding, salinity, and the content of oxygen and/or of organic matter is not clear. *Avicennia*, which can live in a very broad range of conditions, is found on freshly deposited mud as a pioneer plant (along many coasts), but also on older, compacted mud (Guyana, southeast Mexico), on sandy mud or sand (west and east Africa, India, Indonesia), on calcareous mud (Bimini), in oxi-dized sediment (in many areas, as far as known), in waterlogged sediment (west Africa), or temporarily submerged (India). It forms high trees (up to 30 m), medium sized trees, low shrub, is stunted or dwarfed, and occurs at a very broad range of temperatures up to the limits of the mangrove distribution, and near to human habitation, resulting in disturbance and (partial) destruction. *Rhizophora* also is a pioneer in freshly deposited mud, but generally tends to grow somewhat more inland, more in brackish water, in oxygenated as well as deeply reduced mud (Guyana, southeast Mexico) but also in sandy clays, in waterlogged soils (India), usually in low grounds (Indonesia), as tall trees (>40 m) as well as low trees, shrub, stunted or dwarfed, and on calcareous mud flats (Bimini). *Laguncularia* is found (in west Africa and the Americas) on compacted consolidated clay (Guyana), and on sandy soils (southeast Mexico). A local differentiation in sediment can give a differentiation in species: in Tabasco, Mexico, *Avicennia* grows on point bar ridges with a slightly coarser sediment, while *Rhizophora* and *Laguncularia* grow in the adjacent muddy swales (Thom 1967). *Sonneratia* (from east Africa to the Pacific) can be a pioneer plant and grows on muddy sediment (many areas), on sandy muds (east Africa, Indonesia), on waterlogged soils (India), on sandy clay soils inland (Bombay area), and on coral limestone (Java). *Bruguiera* is usually found on higher grounds, on sandy mud, at anoxic conditions and high organic matter content (Red Sea), and at the transition from salt to freshwater (Indonesia). *Xylocarpus* occurs also on higher grounds and on well-drained soils that are still regularly flooded. Together with *Lumnitzera* it occurs at the transition from salt to brackish water (east Africa, Seychelles, Comores, Malaysia, Indonesia), and with *Ceriops* on the highest parts, which are usually better drained and less saline (southeast Asia). *Exoecaria* occurs on stiff clays with poor drainage and reduced salinity; *Heritiera* at low salinity with regular river flooding but less saltwater flooding (India, Burma, Indonesia). It forms tall trees (more than 30 m) but also, at high salinity, saltwater mixed forests (India). *Nypa*, as was seen above, is typical for the transition of brackish to freshwater. Where fresh groundwater and freshwater floods occur, there is hardly any mangrove. Mangrove species can grow in freshwater, but freshwater favors the growth of many species that cannot grow in saline water, which leaves the saline waters to the mangroves. The highest salinities at which mangrove (*Avicennia*) can grow, are about 90‰ (Augustinus 1995).

For some genera different species are known to live on different sediment types: *Avicennia marina* occurs on sand, *A. alba* on mud (Indonesia), *Rhizophora apiculata* on mud, *Rh. stylosa* on more sandy mud, *Rh. racemosa* more on soft mud, *Rh. harrisonii* and *Rh. mangle* more on compacted mud (west Africa), and *Rh. mucronata* on sand (Malaysia-Indonesia, Papua New Guinea); *Brugueira parviflora* lives more on wetter soil, *Br. cylindrica* on dryer soil, and *Br. sexaplana* at the transition to freshwater; *Lumnitzera littorea* on loamy soils, *L. racemosa* on clays. Probably there is also a differentiation within species, in analogy to the ability to tolerate low temperatures: *Avicennia germinans* and *A. marina* collected at high latitudes (27° to 37°) can endure night chilling down to 3 to 4°C better than those collected at low latitudes (12° to 29°), while for *Laguncularia* species there is no difference (McMillan 1975). To what extent such differences between genera and species influence sedimentation is not known. *Rhizophora* sediments are probably more easily eroded than sediments with *Avicennia* and *Sonneratia* because of the less compact structure of the root system in *Rhizophora*.

It should be realized that in many areas mangroves are no longer in their natural state. Cutting, either to use the wood, bark, and leaves, or to use the soil (e.g., for conversion into fish ponds), has been and still is widespread. This is evident particularly in India where *Rhizophora* has practically disappeared from the west coast and is rare in many parts along the east coast. *Avicennia*, which is very adaptable and reproduces easily, and *Exoecaria* have become the dominant mangrove genera. In Vietnam mangroves were intensively sprayed with herbicides during the Vietnam war, which has resulted in defoliation and death of trees on a large scale. Sprayed areas remained without trees for at least 6 years.

Bottom Fauna

The benthic fauna and flora in mangroves consists of bacteria, cyanobacteria, microalgae, macroalgae, protozoa, foraminifera, fungi, meiofauna and macrofauna. Micro- and meiofauna in general loosen the sediment and redistribute the particles so that after some time the original structure is lost. Microalgae (diatoms) have been observed on mangrove sediment (they live at the surface because they need light for photosynthesis) down to a depth of 6 cm in the sediment. It is not known to what extent they increase sediment cohesion and thus affect its erodibility, as they do on mud flats at higher latitudes (see Section 9.1). Macroalgae can be abundant at the bases of trees and on pneumatophores and their presence can increase, or reduce, the bottom roughness. Nothing is known about the effect of the other groups (except macrobenthos) on sediment deposition and erosion. It is also not known to what extent birds disturb the bottom, as they do on tidal flats in more temperate climates. Reworking of bottom sediment in mangroves is done primarily by the benthic macrofauna, which consists of burrowing and crawling animals (crustaceans, mollusks, fish, worms). The burrowers also crawl when they are out of their burrows. The distribution of the bottom fauna in mangroves is not very much related to the type of vegetation but primarily to the nature of the substrate: particle size, degree of consolidation, humidity, salt content. The following genera are usually present (Macnae 1968):

- Crabs (Crustaceae):
  - *Thalassina anomala* (mud lobster; Figure 9.32): it makes deep burrows that are mostly U-shaped with a long horizontal or nearly horizontal shaft with several side arms that end blindly. The horizontal shaft goes abruptly downward and ends in a circular pool at ground water level, which is at most 75 cm deep below the sediment surface. The upper end is a mud hill, consisting of mud that was discarded when the burrow was made, and which can reach up to 70 cm above ground level.

FIGURE 9.32 *Thalassina anomala*, side view and front of a female. (From Macnae 1968. With permission.)

FIGURE 9.33 *Sesarma smithii.* (From Macnae 1968. With permission.)

Thalassina occurs in southeast Asia and around the Indian Ocean and is abundant at higher levels in the mangrove (with *Bruguiera*, *Ceriops*) up to above the high water mark.

- *Sesarma sp.* (Figure 9.33) makes a straight burrow with a circular entrance, which ends in a pool at ground water level. It burrows in all types of sediment ranging from mud to sand and usually dominates the mangrove fauna at the landward fringes. It also makes burrows in the sides of *Thalassina* mounds. It feeds on diatoms living on the sediment surface (which results in a 2-cm-wide track where the brown surface film has been removed), on feces, and on litter.

- *Uca sp.* (fiddler crab; Figure 9.34) makes vertical burrows both in mud and in sand and is common along the edges of clearings at all levels in the mangrove, and along the banks of channels and creeks. Most species prefer shade and avoid dry air, but some live on bare open flats tolerating high salinities and dry conditions. They occur at all levels in the mangroves.

- *Upogebia sp.* makes U-shaped burrows in very wet mud along the edges of tidal creeks and channels.

- *Scylla serrata* (Figure 9.35) is a large crab (carapace-width up to more than 20 cm) that makes wide, shallow surface burrows up to 3 to 4 m long at all levels in the mangroves.

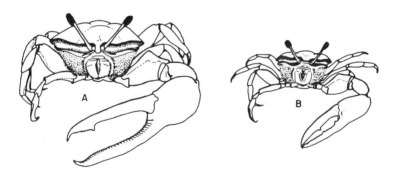

FIGURE 9.34 Fiddler crabs. A = *Uca dussunieri;* B = *Uca lactea.* (From Macnae 1968. With permission.)

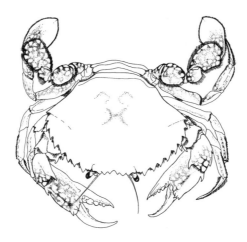

FIGURE 9.35 *Scylla serrata.* (From Macnae 1968. With permission.)

- *Alpheus sp.* makes anastomosing burrow systems (labyrinths) in soft mud, mostly at the upper and middle levels in the mangrove. Dominates in certain *Rhizophora* forests.
- Mollusks: mainly gastropods, only a few bivalves. The gastropods graze the sediment surface for microorganisms; the different species live at different levels and in different sediment types. *Terebralia*, the most common bivalve, occurs in fine mud at the seaward fringe and in *Bruguiera* forest. *Crassostrea* (an oyster), *Littorina,* and *Balanus* (a barnacle) live on tree trunks and exposed roots (in particular the prop roots of *Rhizophora*).
- Fish: *Periophthalmus sp.* (mudskipper) can be up to 25 cm long and moves on its frontal fins and tail over the sediment surface, making characteristic tracks (Figure 9.36). Its burrows are saucer or bowl-shaped hollows in the mud surface with a wide opening and the rim raised 2.5 to 3 cm above the sediment level; they are formed by pushing the mud outward (bull-dozing). Also Y-shaped burrows are formed which go vertically downward and end at the surface in two turrets that protrude up to 15 to 20 cm above the ground. They are built up with small pellets and have their top (with the opening) well below high tide level. *P. spec.* is abundant on the banks and along the channels.
- *Polychaetes* (burrowing worms) are usually scarce and only live in oxygen-depleted muds, but locally can be present in appreciable numbers (Malaysia).

Densities of macrobenthos in mangroves and adjacent bare flats have been listed by Alongi and Sasekumar (1992) and range from only a few specimens per m^2 (in very low and very high tidal flats) to 1000 or more on mid-tidal mangroves and in *Avicennia* forest, and to almost 6000 specimens per m^2 on bare mud flats. This variation can be related to a number of factors including physical conditions (rainfall,

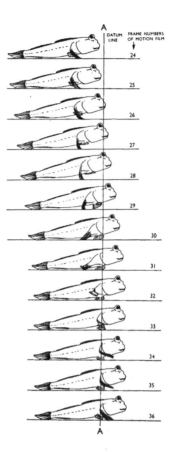

FIGURE 9.36 *Periophtalmus sp.*, a mud skipper. Cycle of fin movements during locomotion on land by crutching. From a motion film: each frame represents a time lapse of 0.03 sec. (From Macnae 1968. With permission.)

temperature, desiccation), to competition with epifauna, predation, food quality, and chemical defense (through production of soluble tannine) by mangrove trees.

There is a general predominance of burrowing crustaceae and gastropods in mangrove bottom fauna communities. Burrows provide a refuge, but are often also a center of territorial behavior, or an accessory to feeding (in particular the U-shaped burrows). It is difficult to assess the influence of the benthic organisms on sedimentation and erosion. Disturbance of the top sediment on a large scale as occurs in many mangroves, will loosen the sediment and make it more erodible. Groundwater comes up through crab-burrows during the incoming tide, but it is not clear to what extent this enhances erosion. Burrows and mounds, however, survive flooding during high tide: the normal currents are not able to remove them. Many organisms (crustaceae, mollusks, mudskippers) fix mud in the form of feces or pellets, which makes the sediment less susceptible to erosion. The reduction of flow and damping of waves by roots and trunks, and the relative rarity of storms along the sheltered coasts where mangroves grow, reduce the chances for erosion of mangrove sediment.

THE INFLUENCE OF MANGROVES ON SEDIMENTATION AND EROSION

The role of mangroves in sedimentary processes has been discussed primarily in relation to geomorphology and, in the earlier years, mainly in terms of two opposing views: mangroves initiate deposition of coastal sediment (landbuilding), or they follow the depositional patterns already established. This discussion was already settled before 1970: mangroves follow the existing patterns of sedimentation and their distribution is determined by the formation of banks, deltas, channels, levees, lagoons and bays (Thom 1967, see also Egler 1950, West 1956, Vann 1959). Mangroves settle on already deposited sediment: their development is favored by a large supply of fine-grained sediment, a low coastal relief, a large tidal range and low wave energy (Scholl 1968). Once established, mangrove influences sedimentation and erosion processes, and high in the intertidal zone they may form peat layers (Davis 1940). This leaves the question to what extent, and how, sedimentation and erosion in the mangroves are influenced by the vegetation and the associated (benthic) fauna.

In the Niger delta, which has been extensively described by Allen (1965), in Cameroon and in New Caledonia (Baltzer 1975), mangrove swamps have channels bordered by levees which grade into backswamps that form the core of the inter-channel flats. The mangrove trees are highest on the levees and the point bars, where they can reach a height of 30 to 45 m on the extreme outer edges. Toward the backswamps the tree-height decreases. This indicates a rather stable vegetation that is in balance with the slowly changing geomorphological conditions. Lateral accretion of the interchannel flats occurs through deposition of mud banks along the fringes. Mangroves start to grow on these banks when they emerge above low-tide level. The young mangrove vegetation enhances mud deposition, and this provides a basis for further plant growth. At the same time there is mud deposition on the interchannel flats, but in much lower amounts.

Along other coasts mangroves occur as a fringe along a straight coast, or are present on mud flats without channels, or occur in bays or lagoons (Thom 1984); there is no typical mangrove morphology, which strengthens the view, already expressed more than a century ago by Junghuhn, that mangroves follow the siltation rather than precede or initiate it. Augustinus (1995), however, indicated features that can be regarded as typical mangrove morphology:

- In Queensland (Australia) along the Wenlock river, narrow mangrove less than 200 m wide usually has a smooth gentle slope toward open water or a channel; the wider mangrove has a surface that is dissected by narrow, shallow creeks about 1000 m apart, that only drain the mangrove swamp (Wolanski and Ridd 1986).
- In lagoons the mangrove provides shelter against waves, which can result in a rapidly prograding mud flat. Here an *Avicennia* species is always the pioneer, presumably because of fluid mud formation during rapid siltation, which prevents *Rhizophora* seedlings from settling (Augustinus 1978). There is clearly a relation between mangrove ecology and the geomorphology of estuaries (Thom 1967, Thom et al. 1975, Semeniuk 1980). In

the Cambridge Gulf–Ord River area in northwestern Australia bands of mangrove occur along river channels with a relatively large input of freshwater, a rapid rate of erosion and deposition and of mud-bank formation, as well as prograding shorelines influenced by waves and tides. Where the direct influence of river discharge and waves is reduced or negligible, tidal sedimentation results in extensive bare flats at approximately high spring tide level. The flats as well as the channels and creeks are fringed with mangrove. About 6000 to 7000 BP the flats were covered with mangrove that developed at the end of the Holocene transgression. Since then 3 to 4 m of sediment have been deposited raising the flats to present level and reducing the area of mangrove growth (Thom et al. 1975). This indicates that the presence of mangroves is controlled to a large extent by the geomorphology, but this is partly induced by the mangrove itself through enhancing sediment deposition, which finally reduces the mangroves to areas that are frequently flooded. So finally there is also a strong influence of the mangrove on the geomorphology, which also applies to the interaction between the effects of sea level change and mangroves; the relation between mangroves and geomorphological conditions is of potential use for paleogeographical reconstructions.

A mangrove vegetation increases the bottom roughness with an often dense above-ground root network with prop roots, pneumatophores, and trunks. Additional relief is provided by crab burrows or mounds. The roughness is a factor of the vegetation density (Bunt and Wolanski 1980, Wolanski et al. 1980), but the data of Spenceley (1977) on *Avicennia* pneumatophores suggest that at relatively low initial velocities the roots reduce the flow, but at medium-high velocities eddies are formed behind obstacles with local scour. Strong flow reduction has been observed in mangrove: in Bimini lagoon a *Rhizophora* prop root network with approximately 15 cm distance between the roots reduces a 40 cm · s^{-1} flow to almost zero (Scoffin 1970); in Coral Creek (Queensland) current velocities in mangrove creeks were above 1 m · s^{-1}, while in the adjacent mangrove swamps they were never above 7 cm · s^{-1} (Wolanski 1992). Suspended matter settles in the mangrove during high tide, while the ebb currents are too low to remove the deposited sediment. Mangroves in this way actively contribute to the formation of mud banks (Furukawa and Wolanski 1996). By reducing the flow, mangroves also have a strong influence on the intertidal hydrography: water, temporarily stored in the mangrove around high tide, is released during the ebb, which gives increased peak velocities during the ebb. In Coral Creek (Queensland) this results in an asymmetric tide with peak current velocities during the ebb that are 20% to 50% higher than the peak flood velocities, while the flood has a longer duration than the ebb (Bunt and Wolanski 1980). The degree of storage is related to the relative elevation of the mangrove flats (with a stronger ebb at higher elevations (Larcombe and Ridd 1996), and to the ratio of swamp area to creek/channel area: a higher ratio increases the tidal prism of the estuary (Wolanski et al. 1980, 1990; Wolanski, 1992). When the area with mangrove swamps is reduced, the storage capacity decreases as well as the ebb flow, and the creeks and channels are silted up. Trapping also increases the residence time of mineral-rich water in the mangrove (Francis 1992) and controls the lateral mixing along a channel. This effect is enhanced during high freshwater discharge, when the water level is raised and freshwater tends to remain in the mangrove swamp because of its buoyancy. Similar trapping of water occurs on bare tidal flats, where it results in strong ebb flow through the tidal creeks (Wadden Sea; van Straaten 1954) and on salt marshes (North Norfolk marshes; Healy et al. 1981, Bayliss-Smith et al. 1979), where storage around high tide results in strong velocity pulses in the ebb flow.

The erosion that follows removal of mangrove (e.g., Report, Forest Administration, Malaysia 1956) points to protection of deposited sediment by the mangrove. This can occur by reduction of the water flow within the mangrove and probably also by the extensive root systems at, or just below the sediment surface, which hold plants and sediment together. Once the vegetation is there, sedimentation, when the mangrove is flooded, is induced by trapping suspended sediment behind trunks, roots, and the pneumatophore network (Young and Harvey 1996), as well as through reducing the flow velocity. As was seen above, this probably only takes place at medium-high current velocities, with deposition of sediment at lower velocities. Benthic diatoms, that live on the sediment surface, can have an important effect on the consolidation of bottom sediment by gluing the particles together with mucus and by moving upward through freshly deposited sediment to reach sunlight again. In this way mud can become, at least temporarily, more resistant to resuspension. On open tidal flats without vegetation this effect is limited because of the strong wave turbulence that occurs during storms, but in an environment like mangroves, where storm waves are less common and the sediment is protected, the effect of benthic diatoms may be more lasting. Reduction of sunlight below the mangrove vegetation, however, may limit the growth of diatoms. The author is not aware of any work that may have been done on this, as was already indicated in the review paper on mangrove diatoms by Cooksey (1984).

Mangroves enhance sediment deposition: an approximately tenfold increase has been observed in mangroves in Senegal and Guinea (Emerit 1960). Spenceley (1977, 1987) argued that stabilization of initial sediment by roots and pneumatophores was more important than the creation of quiet flow favorable for sedimentation. Water dripping from leaves and flow around *Ceriops* roots would result in (local) erosion, and erosion is induced by increased turbulence around roots and trunks when the flow is strong: there probably is a point at which flow reduction because of the network of obstacles turns into enhanced erosion when the flow velocity increases and a stronger turbulence develops around the obstacles. When strong currents move over fine colloidal mud between the mangrove, resuspension may result in local high suspended-matter concentrations and even in a fluid mud layer, but it is not necessarily removed. The overall effect of trapping mud and in this way is its influence on the dispersal of material along a channel or river (Wolanski and Ridd 1986).

Part of the mangrove sediment is produced within the mangrove itself. Near high tide level, i.e., more inland in the mangrove, litter is not entirely removed by the flooding and a fibrous peat may be formed that can be a large contribution to sedimentation within the mangrove. This is extensively described by Davis (1940) and Scholl (1964) in Florida, where peat layers were formed in relation to a local relative sea level rise. Mangrove mud deposits at lower levels have a much lower content of organic matter and may contain only little peaty material or plant remains, but usually remains of mollusk shells are present that in spite of corrosion can be recognized as belonging to species that live in mangroves (*Ostrea sp.*, gastropods). This makes identification of mangrove deposits feasible. Identification can also be based on the presence of pollen and seeds, but these may have been supplied from elsewhere by wind (pollen) and/or surface waves and currents.

Bird (1986) describes how in Westernport Bay (Victoria, Australia) trapping of sediment results in a depositional terrace reaching up to high spring tide level, followed by the development of a salt marsh. Where such a terrace is degraded because the mangrove died or has been cleared away, the intertidal platform was cut by wave-action. After revival of the mangrove the terrace was formed again. Although at present mangroves are degraded mainly by human activity, natural processes can also cause mangrove degradation. Thus a *Bruguiera* mangrove died near Cilacap (Java, Indonesia) after flooding by fresh river water and rapid siltation (Soerianegara 1968). Hurricanes, which defoliate or completely destroy the trees, are also a natural cause of mangrove degradation. During a storm, seedlings can be dispersed over a wide area including areas that are hardly suitable for mangroves so that recovery after a hurricane can be slow (Thom, 1967). Mangroves can therefore be regarded as being a stable vegetation, able to respond to slow geomorphological changes but disrupted by sudden changes in siltation or by damage.

The influence of the bottom fauna in the mangrove on sedimentation and erosion processes is not clear. Burrowing loosens the soil, and water can flow in and out the burrows. Most burrows have a rather fixed position (as far as is known) and at most are widened during the animal's life span so that the sediment between the burrows remains undisturbed. Mud brought to the surface is glued together with mucus into pellets or mud balls, which reduces resuspension; surface structures (turrets, mounds) can withstand the tidal flow through the mangroves, probably also because of flow reduction by the vegetation. There are no clear indications for either a sedimentary or eroding effect of the bottom fauna, but fauna, and particularly crabs, can be important in removing litter. In the Indo–West Pacific region they can remove up to 30% of the annual leaf fall, while up to 80% is removed by the tides (Robertson 1986), but in the Caribbean, where they are predominantly carnivorous, crabs are less important for litter removal (McIvor and Smith 1995). Amphipods can also remove litter but only in the subtidal zone. High in the intertidal zone, bacterial decay takes over the removal of the litter but often removal is not complete so that peat formation results.

Accretion rates in mangroves in Mexico and Florida, determined with ^{137}Cs and ^{210}Pb, were found to be between 1.0 and 4.4 mm $\cdot$ y^{-1} (corrected for sediment consolidation; Lynch et al. 1989): the rates in fringe mangroves was generally higher than in basin mangroves, which Lynch et al. (1989) attributed to a greater subsidence rate in the basins, The rates are generally of the same order as the local relative sea level rise (1.4 to 1.6 mm $\cdot$ y^{-1}).

Part 9.4 Algal Mats

Four kinds of organic layering are called "mats": the networks formed by the filaments of blue-green algae, the thin diatom-rich layer at the surface of mud flats, the mm-thick layers with high concentrations of bacteria or green microalgae, and dense covers of macroalgae. The diatom and bacteria mats have some cohesion because of mucopolysaccharides that are produced by the organisms to attach themselves to sediment particles. Diatoms have therefore some influence on the erosion of fine-grained sediment, as discussed in Section 9.1. Bacterial mats and mats of green algae have no known influence on sediment behavior (but the concentration of organic matter in the mats may attract benthic animals who use it as food). Macroalgae usually occur scattered over intertidal flats, but dense covers of macroalgae (*Enteromorpha, Ulva*) are found where sewage effluents are dispersed on or near tidal flats (Soulsby et al. 1982; see also Chapter 11). Mats of blue-green algae occur on low-energy tidal flats and supratidal flats in rather arid areas. They have been studied extensively on Andros Island in the Bahamas and at Shark Bay, west Australia, and have been described from Mauritania, Sierra Leone, Langebaan Lagoon, the Red Sea, the Trucial Coast in the Persian Gulf, Coorong Lagoon and Spencer Gulf in southern Australia, Baja California (Mexico), Bermuda, the Bahamas, Caicos Island, Florida, the Mississippi delta coast, and the Texas coast (Black 1933, Monty 1967, 1972, Davies 1970, Hardie 1977, Logan et al. 1964, 1970, 1974, von der Borch 1976, Einsele et al. 1974, Friedman et al. 1973, Horodyski et al. 1977, Phleger 1965, Ginsburg 1955, Sorensen and Conover 1962, Dalrymple 1965, Gebelein 1969, Sections 4.5, 4.8, 4.10, 4.16, 4.17, 4.23, 4.26 of this book). Fossilized mats are called stromatolites; a broad introduction to algal mats, recent and fossil, is given in Walter (1976).

Algal mats consist of intertwined filaments and/or gelatinous material (Golubic 1976). They spread over bare sediment and bind the particles in what may become a hard crust. Like all green plants, they need light, nutrients, and CO_2 as well as oxygen for other microorganisms incorporated in the mats. The metabolic activity of the mat community itself regulates to a large extent the immediate environment of the mats and stabilizes it. Light penetrates only a few millimeters into the mat. The different microorganisms in the mat are adapted to changes in light intensity and light spectra inside the mats. There is often a layering within the mats: an example from the Persian Gulf (in Golubic 1976) shows a brown filamentous surface layer (brown to protect the algae against intense sunlight), a blue-green layer with algae, a salmon-pink layer of pigmented filamentous bacteria, which marks the transition from the aerobic surface layer to an anaerobic deeper layer, a purple-pink layer of anaerobic sulfur bacteria, and a black bottom layer with anaerobic bacteria stained by iron sulfide. The algae rapidly colonize newly deposited sediment and grow upward when new sediment is being deposited on top of them.

The microorganisms living in the mats with time become buried and encrusted when they have no mechanism to move upward or sideways. Different mechanisms have been developed that are reflected in the mat structure (Golubic 1976): 1) formation of an unidirectional gel (by the coccoid algae), 2) formation (by filamentous algae and bacteria) of filaments to glide along, 3) growth in a vertical (upright) or radial direction, and 4) growth of horizontal filaments and networks. The following types of mat-building (or stromatolite formation) can be listed (Logan et al. 1974, Golubic 1976; freshwater types are not included):

- Nonmotile coccoid single cells that form gelatinous, pustular, or mammillate mats or domes; the sediment is trapped in the depressions and then overgrown. When the sedimentation rate increases and the mats become covered, cells are released that recolonize the sediment surface through passive transport by currents.
- Single cells that form a gelatinous stalk or filament and use it to elevate themselves out of the sediment.
- Gliding filamentous bacteria or algae that do not produce cohesive sheaths and, although often present in substantial numbers, do not form a large contribution to the mat structure.
- Sheathed filamentous algae that form extensive mats (the common mat-forming algae: *Microleus sp., Schizotrix sp.*): they produce a more permanent cohesive structure in a variety of environments. Some (like *Lyngbya sp.*) are stained yellow to brown by sunlight and thus provide a protective surface layer.
- Alternating vertical/horizontal filament units and sediment, together forming finely laminated "biscuit."
- Bundled and ramifying filaments that grow upward and radially. Sediment is trapped between them and conical mats up to 80 cm high can be formed.
- A smooth leathery coating on developed mats which protects the mats from wind-blown sand by deflecting the sand grains.
- Calcified vertical and horizontal structures.

Calcium carbonate precipitation is the result of CO_2 extraction by the algae during photosynthesis, whereby the content of CO_2 and HCO_3 in the vicinity of the algae is reduced. Friedman et al. (1973) describe the formation of cryptocrystalline laminae in the algal mats along the Red Sea near Aqaba that consist of high-magnesium calcite or aragonite. They are formed by algal precipitation and lithified in the mats. During decomposition of dead algae the production of CO_2 can result in dissolution of the carbonate. The large variety of algal mats is related to the variety in environmental conditions, ranging from subtidal to intertidal and supratidal, from marine to freshwater conditions, from (relatively) low temperature to warm springs and geysers, and from hard to soft substrate. Intertidal mats are mostly of the mammillate type with, at a higher intertidal level in sheltered locations, unlithified flat low mats, that are often cracked into polygons, pinnacle-type mats, and blistered or wrinkled mats. The last three types of mats are built up by multilayered communities of blue-green algae and bacteria.

Desiccation is an important factor in shaping the mats: flat-topped mounds and horizontal mats broken by desiccation cracks occur where the mats have grown up to the water surface. Mats with concave and upward saucers develop under conditions of extreme desiccation (Horodyski et al. 1977). In very arid environments induration crusts, forming extensive brecchiated pavements 2 to 3 cm thick, occur on intertidal and supratidal flats as a result of extreme desiccation. Simultaneously cryptocrystalline aragonite is formed and a fenestral structure parallel to the bedding plane of the crust. The fenestrae are unsupported voids that develop by the interaction between the algal mat and the sediment, and by desiccation, oxidation, and lithification (Logan et al. 1974). The indurated crusts along tidal channels have been called "wall rock" (Davies 1970). On the mats on intertidal flats gas blisters can be formed by concentration of gas below an indurated crust; they are most common on flats with a high organic content and fine-grained sediment. Formation of brecchias can be initiated by thermal expansion and contraction, and by volume changes during induration. Along the joints the crust wedges upward by continuation of the joint-forming processes, by volume increase under the crust, and/or by crystallization of gypsum which is commonly present below the crusts. Also (dolomite) crusts have been observed that were tilted by mangrove roots and pneumatophores. Finally the brecchia blocks are overturned during storms or by tidal scour.

Intertidal mats are grazed by mollusks and fish, and by burrowing crabs, which destroy the mats by stripping the algae from the surface (Garrett 1970, Davies 1970, Friedman et al. 1973): the cohesion of the mats is reduced and the rate of algal regrowth retarded. The damaged mats can then be easily eroded by the tidal currents and during storms.

Part 9.5 Seagrasses

with C. J. M. Philippart

Seagrasses (Figure 9.37) occur worldwide with about 16 intertidal species, six of which are found only near low tide level. Three species that normally are subtidal are also marginally intertidal and occur at low spring tide level. Other seagrass species are only subtidal (den Hartog 1970). One of the most common species around the North Atlantic, *Zostera marina*, known as eelgrass, is mainly subtidal but penetrates to some extent on the tidal flats. It is the only species that occurs in the Arctic. Seagrasses grow on all kinds of substrate, from hard coarse sand, gravel and coral debris to soft mud (Figure 9.38). Besides on tidal flats, seagrasses grow in shallow pools on coral reefs and along small creeks in mangroves. They settle only where some organic matter is present in the sediment. Leaves and fertile shoots become detached and float for some time, as do seeds. On the sediment they form shoots that extend horizontally, so that a bed or meadow is formed (Hatcher et al. 1989).

The influence of seagrass on sedimentation is similar to the influence of salt marshes: reduction of the flow velocity near the bottom and, in particular where the vegetation canopy reaches to the water surface, by damping of waves. Seagrasses strongly influence the flow over the bed, mostly with a strong stratification at the top of the canopy and reduction of turbulence within the canopy. This effect is reduced by deflection and compression of the plant canopy with increasing flow velocity, which reduces the friction. The regular bending and unbending (waving) in the flow also has a physiological function for the plants (pumping material through the vascular system) and contributes to avoiding desiccation during emergence. A high density of shoots has little effect on the flow velocity (Fonseca 1996). At low

FIGURE 9.37   Seagrass at low tide (outer Wadden Sea). (Photograph courtesy of J. van de Kam.)

FIGURE 9.38 Tidal flat with seagrass beds at low tide (Banc d'Arguin, Mauritania). (Photograph courtesy of J. van de Kam.)

water levels, when the top of the canopy emerges, water is retarded in the seagrass beds; the outflow from the beds then reaches a high suspended-matter concentration (up to five times the concentrations over unvegetated areas).

Reduction of turbulence by a seagrass bed results in deposition of suspended matter. To this is added the supply of organic matter that is produced by the seagrass itself. Not all species have the same effect. Those that have vertical rhizomes are able to fix the sediment by growing upward and forming shoots at lateral buds with new horizontal branches. In this way a stable sediment layer of up to 50 cm thickness may be formed by *Halodule uninervis*, which lives around the Indian Ocean and along the western Pacific and penetrates in mangrove forests up to the pneumatophores of *Avicennia*. Species like *Halophila sp.*, which are small and do not form a dense network, do not stabilize the sediment, but grow easily in very sheltered places and are often pioneers on freshly formed sediment. Also other species, like *Halodule uninervis*, can be a pioneer. Subtidal seagrasses that induce sedimentation, like *Thalassina* in Florida (Biscayne Bay; Craighead 1964, Zieman 1972), may affect intertidal deposition by raising the sediment surface to a water depth at which mangrove (*Rhizophora*) can settle; then mangrove-induced sedimentation takes over. Usually periods of low current velocities are necessary for seagrasses to settle, but once there, they can withstand velocities up to more than $1.5 \text{ m} \cdot \text{s}^{-1}$, and may spread to areas with high current velocities (Fonseca 1996). They may colonize sediment accumulation formed at rather high flow velocities (banks or large ripples), but also can cause them to be formed by reducing the turbulence. An often patchy distribution may result from local disturbance of the seagrass bed and the formation of blow-outs in analogy to the blow-outs formed in dunes. Such blow-outs can migrate by up to 2 m per year. Some distribution patterns, however, seem to have their origin in the formation of horizontal shoots forming rows and patches on a scale of tens of meters (Williams 1990; Fonseca 1996). Seagrasses in Arcachon Bay, France, were found to enhance the abundance of benthic macrofauna: the seagrasses produce organic detritus that serves as food and provide protection against predators (Castel et al. 1989). Shifting margins of the seagrass beds can be related to bioturbation by shrimp, echinoderms, and crabs that disturb the root systems (Philippart 1994, Fonseca 1996). Also the duration of emergence during ebb can be a limiting factor (den Hartog 1971).

Subtidal seagrass beds influence intertidal deposition and erosion, where they occur in the adjacent shallow water and dampen waves, capture sediment, produce and export large amounts of organic material, and cause finer sediment to be transported landward and deposited (Fonseca 1996). Loss of the seagrass beds results in a shift from fine mud-sand sediment to coarse sands and even pebbles (where available). It also can lead to increased sedimentation of coarse material on the coast and higher current velocities over the intertidal areas. In general, a loss of seagrass beds results in an increased sediment transport, as the seagrass no longer stabilizes the sediment with roots and canopy. Dead seagrass is deposited often on the coast and has in one case (in Mauritania) even resulted in the formation of sand dunes over masses of dead seagrass (wrach; Hemminga and Nieuwenhuize 1990).

Erosion of the sediment accumulated by seagrass can occur during storms, or as a result of grazing by geese or swans, or of human activity. During storms the sediment may be washed away and the plants sometimes are uprooted. Grazing by geese or swans can result in damage to the vegetation, which then can be more easily eroded during storms. Ferguson Wood (1959; in den Hartog 1970) and den Hartog (1970), however, describe damage to *Zostera capricorni* beds in Australia by swans and *Zostera noltii* in western Europe without causing any damage to the community as a whole. There are indications that for seagrasses an optimum current velocity lies between 20 and 40 cm · s^{-1} (Fonseca and Kenworthy 1987). At lower velocities the metabolism of the plants is limited by restricted diffusion rates; at higher velocities the plants are disrupted. *Zostera marina*, however, occurs at flow velocities of 120 to 150 cm · s^{-1}. At such speeds the plants can survive, but there is no reduction of turbulence (deGroodt 1992). Very likely, the optimum velocity range will differ for different species. Erosion can also be induced by human activities such as nearby dredging resulting in a higher turbidity in the water, or increased sediment supply to the seagrass beds so that the plants do not receive sufficient light any more or are even choked by an increased sediment deposition. Mortality becomes significant when more than 50% of the plants are buried below sediment. Other changes in the environment as well, such as disease, may result in death of the vegetation: during the first half of this century many seagrass beds along the northern Atlantic have disappeared because of a microbial disease (Nienhuis 1994). Recent changes in the environment of the Dutch Wadden Sea have led to both a local increase and a local decrease in seagrass abundance (Philippart and Dijkema 1995).

The eroded intertidal seagrass is probably mainly deposited along the shore where it is washed together by waves. Hemminga and Nieuwenhuize (1991) observed at the Banc d'Arguin (Mauritania) that seagrass accumulated along the shore and not on the subtidal sea floor or the tidal flats. Fossil or subrecent seagrass-induced sediment deposits are not known; in analogy to the muddy sediment induced by mussel beds, which accumulate where, without mussels, sandy deposits are formed, the muddy seagrass-induced deposits are not stable after the seagrass has died and been eroded.

10 Holocene Development of Intertidal Deposits

During the Holocene, two (related) processes occurred worldwide along coasts: the postglacial sea level rise, which lasted to approximately 6000 BP, and an increase in temperature, which affected coastal areas predominantly at higher latitudes. After the postglacial Holocene sea level rise, sea level changes continued because of regional or local tectonics and with more short-term changes, because of meteorological conditions (winds, air pressure) and geological developments (coral reef growth, formation of spits and barriers). The effect of these processes on intertidal sedimentation are different at a tide-dominated and at a river-dominated coast. At the latter, the formation of intertidal deposits is very much linked to the depositional conditions and history of the river mouth (discharge, sediment supply, development of an estuary or a delta). River mouths generally have followed the Holocene sea level rise: depending on the relation between river flow, the tides, sediment supply and water depth, a delta has been built up or an estuary has been formed. Delta formation at some river mouths has resulted in large intertidal areas (Niger, Irrawaddy), whereas along estuaries the tidal areas are usually small (Gironde, Severn). In tide-dominated areas, the first intertidal deposits tended to be formed at the end of the Holocene sea level rise between 6000 BP and 5000 BP, when sea level became more stable, but deposition started earlier in the Baie du Mont-St. Michel (around 8000 BP), the Thames estuary (8900 BP), in King Sound (7850 BP), at the Ord-Victoria river flats (8000 BP), San Francisco Bay (about 10,000 BP), the Bay of Fundy (8500 BP), San Sebastian Bay (about 7000 BP) and Shark Bay (7400 BP; see Sections 2.2, 2.16, 2.22, 2.26, 3.18, 4.16, for the Thames estuary see Greensmith and Tucker 1973). In the southern North Sea intertidal deposits were first formed around 8000 BP and with the rise of sea level, increasingly extended toward the edges (the present Wadden Sea area, where intertidal deposition started around 5000 BP, the North Norfolk flats that started about 6600 BP, and the Wash, where intertidal deposition started around 5000 BP; see Sections 2.1, 2.4, 3.1, and Eisma et al. 1981), while at the same time the deeper parts of the present southern North Sea were flooded. The start of intertidal deposition seems to be related to the moment of flooding of the lowest areas. Sea level rose around 9000 BP by about 1 m per 100 years, and decreased between 7000 and 6000 BP to about 50 cm per 100 years, and between 6000 and 5000 BP became so low that other factors causing sea level changes, started to dominate. Intertidal deposition started in the Bassin d'Arcachon in 5000 BP, along the China coast in 6600 BP, in Broad Sound in 6000 BP, in the Gulf of Carpentaria in 5600 BP, in Bahia Blanca before 4600 BP, in the Dyfi estuary around 6000 BP, in Florida around 5500 BP and along the New Jersey coast around 6000 BP (see Sections 2.3, 2.11, 2.12, 2.18, 3.14, 3.23, 4.24, 4.25, 4.27).

The Holocene rise in relative sea level, however, was highly variable. A gradual decrease in sea level rise velocity is reflected in the Holocene history of the Dyfi river estuary infill: around 10,000 BP this started with deep water subtidal deposits when the area was flooded. Around 6000 BP deposition changed to shallow water tidal and intertidal deposits. Around 3500 BP salt marsh deposition started; the present salt marshes probably date from about 2000 BP (see Section 4.1). In the Baie du Mont-St. Michel intertidal deposition had its greatest extension around 7500 BP, when sea level was at −10 m (Section 2.2). The flats along the coast between the Ord and Victoria rivers in northwest Australia were around 7500 BP only little smaller than the recent flats, but at a lower level (see Section 2.16). Around 6000 BP sea level remained temporarily about 2.5 to 3 m below present level, then rose to approximately present level ±1 m, while the flats were brought up to present high tide level. In the Pacific on Kosrae Island (East Carolines), during a period of rapid sea level rise (about 10 mm · y^{-1}) between 4100 and 3700 BP, mangrove forests retreated, but developed into the present forests after 2000 BP when the sea level rise became 1 to 2 mm · y^{-1}, while before 3500 BP they formed narrow bands in the inlets (Fujimoto et al. 1996). In Fiji mangrove peats indicate (through ^{14}C-dating) a decreasing sea level rise up to 2000 to 1500 years BP and since then a gentle fall in sea level. More to the west, in New Caledonia, mangrove peat ages indicate a fairly stable sea level during the past 6000 years (Woodroffe 1987). At Langebaan Lagoon, San Sebastian Bay, and Mar Chiquita at the end of the postglacial sea level rise between 6000 and 5000 BP, sea level was about 2.5 to 3.5 m higher and subsequently gradually dropped to present level (see Sections 2.26, 4.8, and 4.21). In Langebaan Lagoon intertidal sediments were deposited only during a short period between 2000 and 700 BP, and since that time have only been somewhat redistributed. In Shark Bay the recent evaporite deposits and algal mats form a thin veneer (up to 1.5 m thick) over Pleistocene deposits. The hypersaline conditions that prevail at present developed during the Holocene from oceanic conditions with salinities of 35‰ S to 40‰ S at the end of the Pleistocene. The intertidal deposits probably originated at the end of the Holocene sea level rise when the sea water reached the inner parts of Shark Bay (Hagan and Logan 1974).

This short survey shows that the Holocene sea level rise and its effects on intertidal deposition strongly varied along the world's coasts. A complex relation between the Holocene eustatic sea level rise, caused by the melting of the glacial ice sheets and the increase in sea water temperatures, and the rise of the land that had been covered and pressed down by these ice sheets, was recognized in northern Europe, Canada/Alaska, and New Zealand (e.g., Mörner 1971), while other coasts were known to have experienced during the Holocene only a rising sea level, often with some temporary standstills. Clark et al. (1978), assuming a eustatically rising sea level up to 5000 BP, estimated the way sea level moved in relation to the land, based on 1) crustal rebound: upward where the ice cap had melted away, downward where an ice cap had not been formed, and 2) a waterload in the oceans which increases with rising sea level. The decrease in sea level outside the formerly glaciated areas was related to the distance from the former ice limit and turned into a (relatively small) upward movement when after the initial dip the crust started to reach equilibrium. The increased water load in the oceans resulted in a tilting of the coast and to uplifted (emerging) shores. This approach in large part can explain, at least qualitatively, the observed relative sea level movements during the Holocene along many coasts, with the largest errors near the former ice caps. In particular for intertidal deposits, which are formed within a few meters of local mean sea level, the relative sea level curves are of great interest, but the relation between the deposits and sea level is complicated by meteorological effects that result in temporary changes in sea level, as well as by the possibility that deposition and/or erosion occur mainly during abnormal periods and are not primarily related to the normal local sea level. Also regional/local tectonics interfere with a more zonal variation in sea level rise. In southwest Australia, in the Perth region, relative sea level during the Holocene varied strongly over only 170 km along the coast: in the northern region present sea level was reached about 5100 BP, in the central part sea level reached +2.5 m around 6400 BP and then was gradually lowered to present level, while in the southern part sea level remained at –2 to –3 m between 7000 BP and 5500 BP, then rose to +3 or +4 m between 4800 BP and 3800 BP, and then was lowered to present level around 2800 BP. Semeniuk and Searle (1986) related this to local tectonics. Generally, however, the observed sea level changes fit reasonably within the different Holocene sea level change patterns indicated by Clark et al. (1978; the numbers correspond to the areas given in Figure 10.1):

I) In the formerly glaciated areas a continuous rise of the land after the ice had been removed. Formerly glaciated areas often have valleys with steep slopes (fjords) and few shallow coasts. Intertidal flats therefore have been formed mainly along shallow flat basins such as the Hudson Bay and, on a smaller scale, Leaf Basin and the local bights along Labrador and Baffin Island (Sections 2.24, 3.20). In northern Europe a shallow basin is partly filled in by the White Sea (Section 4.2). The Baltic would be a favorable area for intertidal deposition if the tidal range was not too small. Northern Canada during the Weichselian was covered by a several-km-thick ice cap until about 8000 BP and since then has risen by 120 to 150 km. Around Ungava Bay the rate of uplift declined gradually between 7000 and 6000 BP and rapidly after 6000 BP, interrupted by periods when the land remained more or less stationary around 6000, 5600, and 5300 BP. During these interruptions boulder ridges were formed approximately at the high tide level of that time and remained behind when the coastline receded farther. After about 5000 BP intertidal deposits accumulated, probably during a period of about 3000 years, but only a thin layer of up to several meters thick was formed, which now is partly supratidal. Similar thin deposits, containing stones and boulders, were formed around Hudson Bay and at Baffin Island: conditions for the formation of intertidal deposits

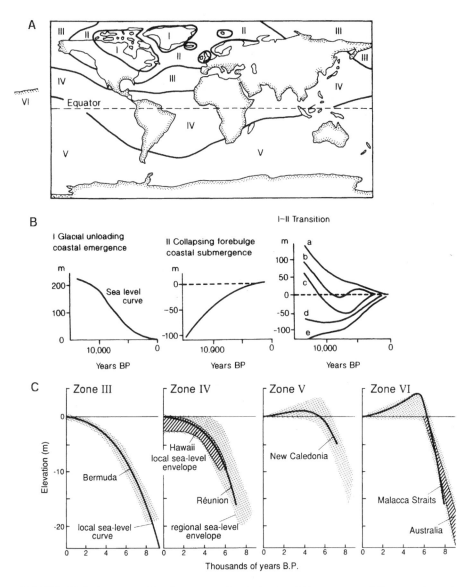

FIGURE 10.1 Patterns of Holocene sea level rise in different zones (VI = continental shorelines). (From Woodroffe 1992; Viles and Spencer 1995. With permission; and adapted from Clark et al. 1978.)

have been poor because of a very limited sediment supply, in the absence of deep weathering and large streams, and because the area is covered by ice and snow for about six months.

II) In the areas surrounding the formerly glaciated areas a general rise in sea level. This is found along western Europe and along the east coast of the U.S. and the adjacent parts of Canada (Bay of Fundy). There was a rapid rise up to about 6000 BP, then the rate of sea level rise slowed down and at about 2000 BP sea level was very near present level (at most 1 to 2 m lower). By that time in the Wash, along the North Norfolk coast, the Baie du Mont-St. Michel, and the Dyfi estuary, the present salt marshes started to form (Sections 2.1, 2.2, 2.4, and 4.1). In the Wadden Sea and along the Norfolk coast the flooding by the rising sea level was interrupted by periods when most of the area was not flooded and (fresh water) peat was formed (Sections 2.4, 3.1). This can have been related to a possibly lower subsidence rate or to a lower frequency of storms in the southern North Sea. Irregularities in the rate of sea level rise (or the subsidence rate) have been observed along the coast of Maine (Section 3.19) and in Chesapeake Bay (Section 4.27).

III) In a relatively small zone, the Holocene sea level rise reached to about 0.5 m (at most) above present level a few thousand years ago and then lowered gradually to present level. The higher level was so little above present level that an emerged coast hardly developed. Also in this zone, however, raised beaches in some areas indicate regional irregularities in the relative sea level movements (in this case by a regionally stronger uplift; Clark et al. 1978). The intertidal/supratidal deposits along Florida, the Bahamas, Belize, and Grand Cayman Island do not necessarily indicate a higher sea level during the past several thousand years: the supratidal deposits (in particular in Belize) can also be considered storm deposits (Sections 4.24, 4.25, 4.26).

IV) In a broad tropical–subtropical zone reaching along eastern Asia up to Siberia, sea level rose a little above present level around 5000 BP, followed by submergence that started around 3000 to 2000 BP. This sequence is seen along the Persian Gulf (Section 4.10) where sea level about 4000 BP was about 0.5 m above present level, but in other intertidal areas in this zone (China, Bahia Blanca; Sections 2.12, 3.23, 4.15) the available data do not indicate more than that about 5000 to 4000 BP sea level was near or at present level.

V) In this large southern area sea level rose about 5000 years ago 1.5 to 2 m above present level and then was lowered to present level. Along the continental shorelines, in this zone as well as in the other zones, sea level may have risen even higher (probably up to +5 m), because of the tilting caused by the increased ocean water load. This can be seen around Australia (Ord river–Victoria river coast, Broad Sound, the Gulf of Carpentaria, northern Queensland, and Shark Bay; Sections 2.16, 2.18, 3.14, 4.16), although deposits indicating a higher sea level have also been explained by deposition during storms (Broad Sound, northern Queensland). It has also been observed at San Sebastian Bay and Langebaan Lagoon (Sections 2.26 and 4.8).

Although there is a certain zonation in the way the Holocene sea level rise proceeded, every area with intertidal deposits seems to have its own Holocene history, as far as is known: of about 60% of the intertidal areas discussed here the

Holocene history is not known and of about half of the remaining 40% the Holocene history is only known in a general way, which makes generalizations doubtful. But almost all intertidal areas have remained intertidal from the moment the first intertidal sediments were deposited. The most notable exception is the southern North Sea area, where a large Holocene intertidal area was completely flooded after about 6000 BP and fresh water peat areas developed in the Wadden Sea and Norfolk flats areas during periods when no intertidal sediments were being deposited there. One may speculate how many other intertidal areas were drowned during the Holocene sea level rise, in particular on inner shelves, but it is likely that before about 8000 BP, when sea level rose relatively fast, most intertidal deposits were thin and eroded within a short time.

In some coastal areas during part of the Holocene the intertidal areas had a much wider extension than at present, but this is not necessarily related to the relative sea level movements. In northern Australia at the (macrotidal) South Alligator river estuary, the Holocene sea level rise ended with a transgression period between 8000 BP and 6800 BP and the formation of a large mangrove swamp that remained in existence until about 5300 BP while sea level remained stable (Figure 10.2). By

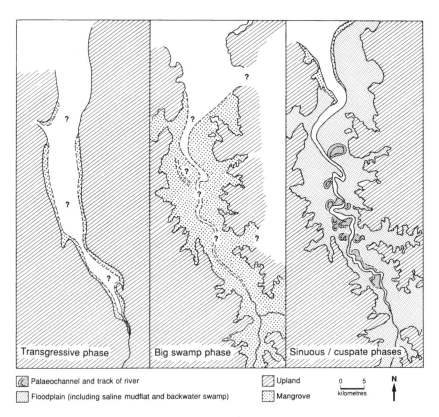

FIGURE 10.2 Holocene development of the South Alligator river mouth plains (From Woodroffe (1992). Transgressive phase: 8000 to 6000 BP; Big (mangrove) swamp phase: 6000 to 5300 BP; Sinuous phase: 5300 to 2500 BP; Cuspate phase: 2500 BP to Recent.

natural vegetational succession, the original *Rhizophora* mangrove swamp changed into an *Avicennia* swamp, which was then gradually replaced by flat plains covered with sedges and grasses when the area reached supratidal level by continued (tidal) deposition. This turned into a river floodplain that only changed its character around 2500 BP by channel adjustments under conditions of stable sea level (Woodroffe et al. 1989). Data from numerous river systems have shown that mangrove forests were widespread around 6000 BP. In semiarid western Australia they changed during the Holocene into broad saline mud flats, whereas in seasonally wet northern Australia they were replaced by clayey flood plains with sedges and grass (Woodroffe 1993).

An important question regarding the Holocene development of intertidal areas is whether the tides have remained the same. This can be assumed for the astronomical tides, but the actual tides along a coast are much influenced by the coastline configuration and the distribution of water depth. Both, in most areas, changed during the Holocene. For the North Sea during the (mid-Holocene) period when the southern part of it was finally flooded, both dissipation of tidal energy resulting in lower tidal ranges, and resonance, resulting in higher tides, may have occurred in the shallow but deepening waters of that time. A reconstruction of the tides in a Holocene tidal embayment in The Netherlands, the Bergen inlet, with a relatively well-known depositional history, has been made by Van der Spek (1994; Figure 10.3) but was restricted by uncertainties in the assumptions that have to be made about the depth distribution, bottom roughness, and the tides in the adjacent North Sea at the time the embayment existed (about 7200 BP to about 3300 BP). The interpretation of vertical salt marsh accretion in the Severn estuary since 150 AD is influenced by a possible change in tidal range because of increasing water depth during the sea level rise, and by a possible rise of tidal level because of wetland reclamation (Allen and Rae (1988). Temperatures, and climate in general, around 6000 to 5000 BP were very similar to present temperatures and climate. Recent changes in intertidal flora and fauna can be related to a multitude of local or regional conditions, but there are no large-scale changes related to Holocene changes in climate.

For a few intertidal areas detailed profiles through the Holocene deposits are available. All profiles reflect the Holocene sea level rise, but apart from that they are very different. A profile from the (mesotidal) Wadden Sea (Figure 10.4, after van der Spek 1994) shows a basal (fresh water) peat layer that was partly eroded during the transgression, and sandy and clayey sediments on top with intercalations of (fresh water) peat formed when the sea temporarily retreated, or the flood could not reach that part of the Wadden Sea. Most of the deposits are channel deposits formed by lateral migrating channels. The channel deposits and tidal flat sands are partly covered (on the seaward side) by barrier island deposits (beach sands, beach flats, aeolian dunes), salt marshes, and a dike (usually a series of dikes). A (mesotidal) Gulf of Carpentaria profile (after Rhodes 1982; Figure 10.5) shows a basal layer of lateritic paleosols with subtidal and low-intertidal muds on top, which are covered by high tidal muds and cheniers. A profile of the (macrotidal) Colorado river delta plain along the northern Gulf of California (from Thompson 1968; Figure 10.6) shows older transgressive sands, interbedded with mollusk shells and grey muds, on top of grey laminated sand and clay, which is interpreted as subtidal and transgressive. These deposits are covered by grey burrowed clay with a brown mottled mud

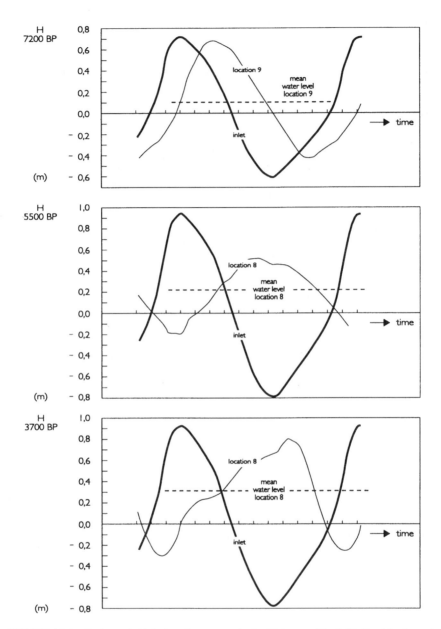

FIGURE 10.3   Estimated tidal elevation curves in the Holocene North Holland-basin at the tidal inlet (heavy lines) and near the inner margin (location 8 and 9; thin lines) in 7200 BP, 5500 BP, and 3700 BP. (From van der Spek 1994. With permission.)

on top (that is similar to muds that are at present burrowed by crabs), by brown laminated silts and by so-called chaotic muds, which are a mixture of gypsum, halite and silt/clay, in combination with beach ridges that probably are cheniers and have been formed over approximately the past 1600 years.

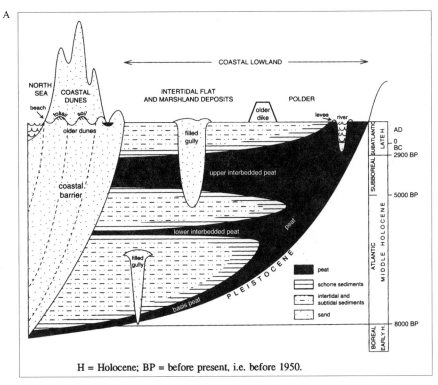

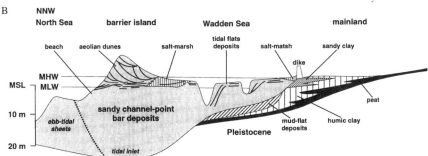

FIGURE 10.4 Schematic profiles through the Holocene along the Belgian coast (A) and in the western Wadden Sea (B). (From Houthuys et al. 1990 and Postma 1996. With permission; and adapted from van der Spek 1994.)

A profile from (macrotidal) King Sound (from Semeniuk 1980; Figure 10.7) shows a basal layer of nodular sand grading landward into mottled muddy sand with muddy mangrove deposits on top, laminated muds, and bioturbated muds, that are overgrown by the present mangrove forest. On the seaward side the profile is cut off because of the erosion that now dominates this area. On top is a thin veneer of modern ephemeral muds and sands. A profile across the (microtidal) chenier plain along the Firth of Thames (New Zealand), the Miranda coastal plain, (after Woodroffe et al.

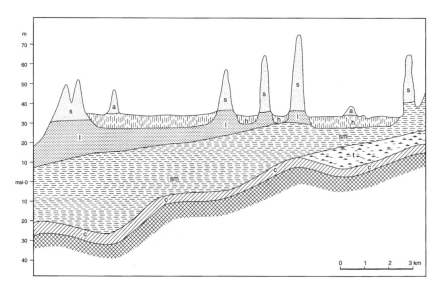

FIGURE 10.5 Holocene profile across a chenier plain along the Gulf of Carpentaria. a = aeolian silts and clays; s = chenier sands and shells; h = high tide muds; l = low tide muds; sm = subtidal muds; t = transgressive sands; c = paleosol. (Adapted from Rhodes 1982.)

1983; Figure 10.8), shows a basal layer of marine sands and muds covering peat and pre-Holocene sediments. On top of this lies a bluish mud that has been deposited in an intertidal embayment of uncertain age (about 4000 to 1000 BP ?), which is covered by the present layer of cheniers, marshes, and interridge muds, which has a thickness of less than 1 to more than 2 m. A more complex area is the intertidal area along (microtidal) west Florida, where a variety of interpretations has been given for the Holocene development of the present intertidal mangrove swamp (Figure 10.9; from Woodroffe 1992). The series of profiles given here shows a wide variation in the Holocene development of intertidal areas, that reflects the present diversity of intertidal deposits and intertidal areas. Local and regional conditions and climate determine, to a large extent, the formation of intertidal deposits with only a limited influence of larger (global) zonations and general trends.

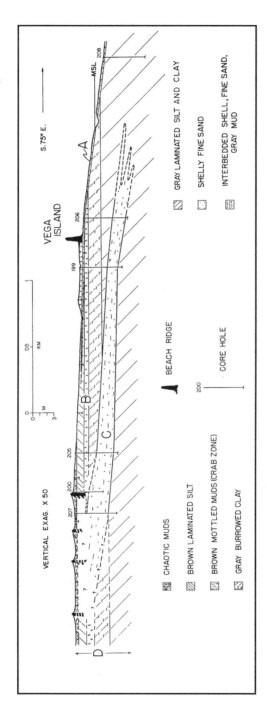

FIGURE 10.6 Holocene profile across the flats south of the Colorado river delta. A = modern transgressive, reworked sand-shell layer; B = mud flat deposits with beach ridges (probably cheniers); C = old transgressive sands with grey sandy and silty clays; D = mud flat deposits underlying the high flats. (From Thompson 1968. With permission.)

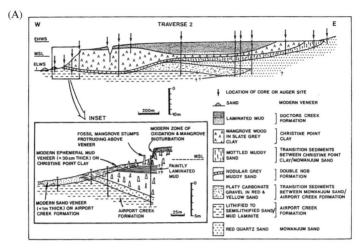

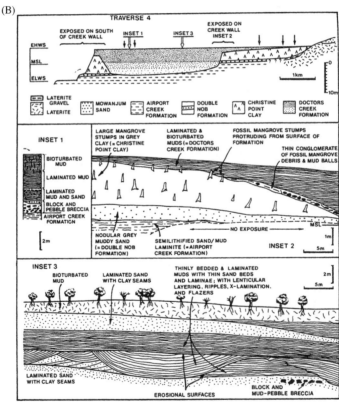

FIGURE 10.7   Holocene profiles across flats at King Sound, northwestern Australia. (From Semeniuk 1980. With permission.)

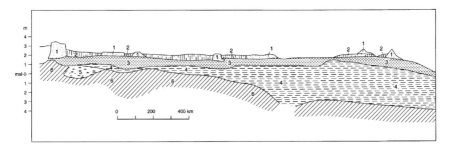

FIGURE 10.8 Holocene profile across a chenier plain along the Firth of Thames, New Zealand. 1 = sand and shells; 2 = marsh and interridge deposits; 3 = blue mud; 4 = sandy-muddy marine sediments; 5 = peat; 6 = clay-tephra and pre-Holocene sediments. (Adapted from Woodroffe et al. 1983.)

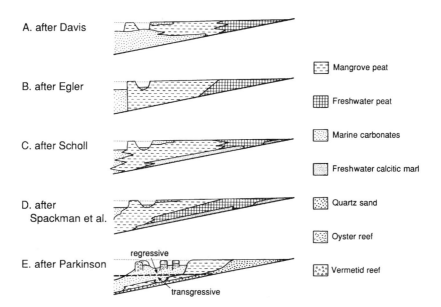

FIGURE 10.9 Profiles across the southwestern Florida coast according to different stratigraphic interpretations: A) adapted from Davis 1940; B) adapted from Egler 1952; C) adapted from Scholl 1964a, 1964b; D) adapted from Spackman et al. 1966, 1969; E) adapted from Parkinson 1989. (From Woodroffe 1992. With permission.)

11 Influence of Man

Intertidal areas — and in particular the higher parts that are rarely flooded, including salt marshes — were reclaimed a thousand years ago in several parts of the world: Vietnam, China, the Low Countries in western Europe. Low dikes and embankments along rivers and lakes enclosing and protecting land that was used for agriculture, husbandry, and human occupation had already been constructed before the 4th century BC, when an attempt was made to drain a lake in Boeotia (Smith 1976). Intertidal areas along the Dutch Wadden Sea coast had been occupied since the 7th century BC by raising small artificial hills with sediment collected from the nearby mud flats (Bos 1995), but reclamation of intertidal areas bordering the sea involved larger constructions, because the attack from the sea could be severe during storms, and because large areas were involved. For this a larger organization was needed, including more than the local farmers alone. Reclamations along the sea were therefore initiated by royalty or the church. Later reclamations passed, at least in western Europe, more and more into the hands of local or regional communities, or became a financial operation. At present large former intertidal areas in east Asia (Korea, China, Vietnam) and in western Europe (mainly along the southern North Sea) have been reclaimed and the inner, landward border of the intertidal areas is a dike. Reclamations were carried out to use the land as farmland and for coastal protection. More recently reclamations were also made for harbor developments and industry, and in some cases (as in northern Holland) for the construction and protection of a canal dug through the tidal flats. In Britain large intertidal areas were reclaimed in the 17th to 19th centuries (in the Severn since Roman times), but the remaining salt marshes tended to deteriorate: in the Medway area 55% of the salt marshes existing in 1800 was reclaimed (enclosed), followed by a loss of 25% through erosion, while only 20% now remains, mostly reduced by dissection, with only a few percent in its original state (Doody 1992). In the Dutch Wadden Sea reclamation has led to a complete disappearance of the natural salt marshes along the mainland coast: the reclamations were much more rapid than the formation of new marsh land by natural sedimentation. All marsh lands now existing are artificial, in which sedimentation is enhanced by a system of ditches and low embankments (Oost and Dijkema 1993). Although reclamation is not needed any more, the system is still maintained to prevent erosion of the coast.

Tidal flats have not only been used for reclamation, but also for the extraction of sediment (sand, mud), for fishing (bottom fish, shellfish), and for collecting mollusks or worms for food (worms also for fish bait). Salt marshes were, and are, used for grazing cattle, sheep and geese, for the cutting of turf, for raising salt-tolerant crops (also for raising halophytes that can be planted on bare flats to induce sediment deposition and the formation of new marshes), and for the cutting of salt marsh vegetation as forage. Minor use of salt marshes includes cutting and collecting plants for ornamental use or as potherbs and for horticulture, collecting grasses for hay, *Juncus* for fibers and paper making, and several plant species for vegetable medicine, food flavoring, and perfumes (Mudie 1974). Mangroves have been and are being cut for the wood to use as timber, for burning, or for charcoal fabrication. Mangroves have also been exploited for collecting fibers and for extraction of tannin (before the chemical industry took over). *Nypa* palm leaves have been used for roof cover and fruits for distillation of sugar and alcohol. Large areas of mangrove have been cleared for rice fields and for fish or shrimp ponds. Mangrove species have also been collected for medicine, perfume, or flavoring. Salt extraction since the Middle Ages from mangrove soils along the coast of west Africa (from Casamance and Gambia to Benin) has resulted in wide bare flat areas (salt flats). In the areas with a dry climate, (as in Casamance and Gambia) such flats are formed around and above high tide level through natural desiccation and exposure (tanns: see Sections 3.3, 4.4, 4.5); salt extraction has led to a greater extension of these flats (Paradis 1980).

Indirectly man has influenced intertidal areas by changing the sediment supply through the construction of river dams and through changes at river mouths, whereby sediment supply or erosion were reduced, or increased, and/or sheltered areas were formed, where sedimentation led to formation of tidal flats, salt marshes, or mangroves. More recently intertidal areas have increasingly been used, and changed, as recreational areas for tourists (e.g. crossing mud flats on foot during low tide) and as nature reserves, or as combinations of both.

Reclamation

Reclamation of intertidal areas, including cutting mangroves, is the most radical change but not completely irreversible because reclaimed land can be flooded again, and mangrove can be replanted (as in Bangladesh where this was done to protect the coast against waves; Saenger and Siddiqi 1993). Reflooding has happened during many storms in the past and in the near future may be done intentionally in The Netherlands to create new lakes. Modern techniques, however, make unintentional reflooding less likely to happen and the overall result of about a thousand years of reclamation is an enormous increase in flat coastal land. No statistics are available but large areas along the coasts of south and east Asia from India to Korea, western Europe (including about half The Netherlands), and smaller areas in the Guyanas and the U.S. are reclaimed land with little loss after reclamation. To this come large areas of cleared mangrove: 75% of the original mangrove in Puerto Rico, 20% in peninsular Malaya, 50% in the Philippines, 40% in Sabah, as well as large areas in Malaysia, India, Bangladesh, Venezuela, Benin, and Ghana. In some areas (in particular in India) the original mangrove has largely been replaced by other species that are economically less valuable (Hatcher et al. 1989). Reclamations as well as the clearing of mangroves continue, but large projects tend to be limited by the costs involved and by the opposition of other, conflicting claims involving nature conservancy, tourism or recreation, and fisheries. Many intertidal areas are nursing areas for young fish, crustaceae, and shellfish, or produce nutrients that are dispersed into the coastal sea and lead to a high productivity. Reduction or removal of such intertidal areas results in severe damage and economic loss of commercial fisheries in the adjacent coastal sea. This applies to areas like the Wadden Sea and probably also to mangroves, although this is not evident, as some mangroves probably are nursery areas, while others are not (Hatcher et al. 1989, Vance et al. 1990, Chong et al. 1990, John and Lawson 1990). Also the dispersal of dissolved nutrient and organic carbon from mangroves into the coastal sea is disputed (Hatcher et al. 1989, Rivera-Monroy et al. 1995, Boto and Bunt 1981). Reclamation and cutting can be partial, leaving the intertidal area as a whole more or less as it is with only a small area used for a road or highway, a small harbor, an industrial activity (including a canal or a drill site for oil or gas), or even an airfield, as near Vancouver, Canada. This usually means that the remaining flats are affected by pollution, noise, changes in water level and are disturbed, which results in changes in flora and fauna and a reduction of its suitability for recreation, although the presence of a road or a canal can also have the opposite effect.

Sustainable Use

Opposite to reclamation is the use of intertidal areas that leaves them largely undamaged, if not completely unchanged. This includes fishing (mainly for benthos and shrimp), the use of salt marshes for collecting grass and cutting turf, the collection of wood, leaves, seeds, grasses, and other plants and fauna in mangroves as well as salt marshes. To this can be added the extraction of sediment (sand, gravel, mud) from bare flats for a variety of uses from dike construction to the construction of roads and industrial sites, and the destruction of vegetation (seagrasses) and benthic fauna through a high man-induced turbidity of the water, which reduces the light in the water and results in clogging animal gills with sediment. Vegetation and also benthic fauna can be replanted or re-established, but while replanting of mangrove has been successful, replanting of seagrasses has met with variable success (Hatcher et al. 1989). Thus a variety of uses is sustainable or repairable as long as it remains within limits so that no over-exploitation develops. Estimating when these limits are reached, with legal consequences including the establishment of quotas and penalties, is usually very difficult, may have to be based as much on experience and intuition as well as on scientific knowledge, and may involve political decisions that reach much farther than the exploitation of an intertidal area. An absence of good estimates, laws, or regulations usually results in serious over-exploitation and lasting, irreparable damage; where necessity, greed, or overpopulation dominate, irreplaceable losses are the final result.

Impact from Outside Activities

Reclamation and human use of intertidal areas, sustainable or not, involve local or regional *in situ* activities with local or regional effects. But also activities outside the intertidal area, in some cases located at quite a large distance, can have an influence: the most prominent are water and air pollution, dam construction, and a global change in climate, or more specifically, a global rise in sea level.

Pollution, in particular water pollution, can easily affect intertidal areas: they are flooded daily and the pollution can come from far away, from polluted rivers, harbors, city waste discharge points, industrial dumps. More diffuse sources are ships (oil, waste dumps, ships' paints). In addition to the oil, predominantly the paints, meant to reduce the growth of algae, barnacles, and mollusks on the ship's hull, negatively affect other organisms (fish, mollusks) through dissolution of their toxic substances in the water. Air pollution also can affect intertidal areas far away from the source, as is the case with DDT and related substances that are found in intertidal areas the world over. Intertidal areas are particularly sensitive to bulk pollutants in coastal waters, predominantly oil and masses of dead plankton after a bloom. This is deposited on the flats and becomes incorporated in the sediment. Dead plankton leads to (local) reduction of the oxygen concentration, and finally anaerobic conditions. Oil covers flora and fauna, mostly resulting in death, and gives a persistent presence of oil and oil compounds, as well as of oil-cleaning compounds, in the tidal flat sediments, with the possibility that the cleaning compounds finally are more poisonous than the oil compounds.

An oil spill at Bahia las Minas, Panama, led to defoliation, loss of branches, and eventually dead trees in a fringing mangrove, with cleaning operations hampered by the inaccessibility of the mangrove. Five years after the spill, the mangrove affected by the oil was still reduced: the surviving trees had fewer and shorter prop roots and a higher percentage of dead roots, which reduced the surface area of the submerged roots 33% along the open coast to 74% in the streams leading out of the mangrove. This not only affected the trees but also the organisms living on the roots (Garrity et al. 1994). A full range of aromatic hydrocarbon residues from earlier oil spills was present at least 20 years after the event and led to an increased proportion of dead mangrove roots (Burns et al. 1994). Toxic hydrocarbon residues were still released from the sediment after five years and were stored in bivalve tissue. It resulted in a persistent reduction of epibiota ranging from 33% along the open coast to 99% in streams (Levings et al. 1994). Although initially the biodegradation of the oil was fast, so that most of the volatile hydrocarbons had been removed from the surface sediment within six months, this rate was not maintained and for at least five years, oil still leached out of the sediment. Accumulation of oil compounds in organisms reflected this release (Burns and Yelle-Simmons 1994). Discharges of organic waste and nutrients can result in a larger productivity in the water and an increase in the growth of pelagic algae, macroalgae, and benthos, which afterward leave large amounts of dead and decaying organic matter: where this is deposited, anaerobic conditions can develop at the bottom, and in the bottom sediment (as was recently found in the Wadden Sea and the adjacent shallow parts of the North Sea floor; Neira and Rackemann 1996). An increased growth of macroalgae and the formation of macroalgal mats because of sewage effluent discharges has been described by Soulsby et al. (1982). This results also in a change in the fauna, both in the benthic fauna and in the predators that live on it (fish, birds), and in the deposition of fine-grained sediment.

RELATIVE SEA LEVEL CHANGE

A change in relative sea level caused by human impact can be the result of local or regional subsidence or by a global rise in sea level. Local or regional subsidence can be the result of groundwater extraction for a large city (Thailand: Bangkok; Taiwan: Taipei; China: Shanghai, Tianjin; Ren 1994, Nutalaya et al. 1996, Jelgersma 1996). Subsidence may also be caused by extraction of hydrocarbons (oil and gas along the U.S. Gulf coast; gas in the Dutch Wadden Sea), but except perhaps in Louisiana, this has not affected intertidal areas very much. The subsidence is a local phenomenon that tends to be compensated by sediment deposition.

In Louisiana subsidence occurs in the Mississippi delta over a wide area, including the intertidal area and freshwater marshes. Deltas subside because the weight of the sediment bends the local crust downward, but normally this is roughly compensated by deposition. In Louisiana, however, the relative sea level rise increased in this century from a net 1 to 2 mm · y^{-1} to a rate of 1 cm · y^{-1} or more, which resulted in increased flooding and a loss of up to 73 km^2 of wetland each year. This loss was relatively modest until 1940, increased dramatically up to 1968, and then declined somewhat (Figure 11.1; Boesch et al. 1994). Because of the

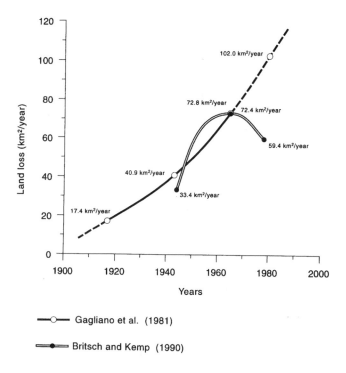

FIGURE 11.1   Coastal land loss curves for the Mississippi delta plain. (Adapted from Gagliano et al. 1981 and Britsch and Kemp 1990.)

(naturally) shifting deposition of the river, areas where no deposition takes place for some time tend to be gradually drowned by subsidence, but this has been enhanced by the reduction of sediment supply from the Mississippi river by about 50% between 1963 and 1982 because of dam construction and diversion of river water, as well as by the construction of levees, which started at the end of the 19th century, and by restriction of the river overflow within a narrow flood plain. The virtual elimination of sediment deposition on the marshes gave an accretion deficit of 4.1 to 8.1 mm · y^{-1} depending on the type of marsh land. The natural subsidence was locally accelerated by hydrocarbon extraction and sulfur production (by subterraneous dissolution) and by soil dewatering at the surface. Canals and navigation channels contributed by the formation of spoil banks that inhibit overbank flooding, and indirectly by erosion and interference with the marsh hydrology. Boesch et al. (1994) indicate that at least 26% of the Louisiana wetland losses between 1955 and 1978 are caused by direct human impact, to which should be added an unknown percentage of loss by indirect human impact effects for a total of human impact losses between 30% and 59%.

The possibility of global warming because of the release of fossil CO_2 into the atmosphere through the burning of coal, oil, and natural gas, was put forward in 1896 (by S. Arrhenius). The first detailed measurements, which show a rising CO_2 concentration in the atmosphere (between 1957 and 1960, were published in 1960 (Keeling 1960, in Weart 1997). To what extent this will result in a global rise of sea level depends on the degree of melting of the ice caps in Antarctica and Greenland, the melting of glaciers, and the degree of expansion of the ocean water volume because of the higher temperature. Opinions on what will happen vary widely, and the extent of a future global sea level rise varies from several decimeters to a meter within the next century (Mörner 1995, Raper et al. 1996). Intertidal areas will be very sensitive to even a small rise in global sea level but react primarily to the actual local change in sea level, which are related, as seen above, to regional tectonics and local movements. Tidal flats, salt marshes, and mangroves are expected to respond differently to a rise in relative sea level (see discussions in Eisma ed. 1995). Mangroves may collapse at a rise of over 12 cm within the next century (Ellison and Stoddart 1991), and it looks as if even man-made salt marshes will not be able to withstand a future sea level rise of more than 1 to 2 cm · y^{-1}, natural salt marshes probably less (Dijkema et al. 1990). Both in salt marshes and mangroves the effect of sea level rise will depend on the extent to which sediment is supplied (Bird 1993): when this is not sufficient, they will be eroded and cut back on the seaward side, and migrate landward when a low-lying area is available there. Sieffert and Lassen (1987) indicate that during the past 120 years sea level rose 20 cm, while the tidal flats in the German Wadden Sea remained at the same level as 120 years before. For the Dutch Wadden Sea, Louters and Gerritsen (1991) assume that deposition of sand supplied from the North Sea coast will raise the level of the flats in accordance with the rise in sea level as long as the local sea level rise does not exceed 60 cm in the next 100 years.

A climate change because of global warming produces other changes besides a change in temperature and sea level, such as changes in precipitation and in the frequency and force of the wind. Precipitation changes affect the water balance and water table and the flux of sediments and nutrients, while changes in the wind, and particularly in the frequency and intensity of storms, will result in increased erosion. The change in CO_2 concentration itself induces an increased plant growth and an increased litter production, but a higher temperature has an opposite effect, with reduced plant growth and reduced litter production. In dryer climates an increased desiccation of bare sediment surfaces is likely (Pernetta 1993).

References

Abel, O., 1935. Vorzeitliche Lebensspuren. G. Fischer, Jena. 644 pp.

Abeywickrama, B. A., 1965. The estuarine vegetation of Ceylon, in *Scientific Problems of the Humid Tropical Zone. Proc. Dacca Symp.* Feb.-March 1964, 207-210.

Adam, P., 1990. Saltmarsh ecology. Cambridge University Press, Cambridge, U.K., 461 p.

Ahmad, E., 1972. *Coastal Geomorphology of India.* Orient Longman, 222 p.

Ahmad, N., and F. K. Khan, 1959. The Sundarban forests of East Pakistan. *The Oriental Geographer,* July 1959, 13-32.

Alejo, I., M. I. de Ramon, M. A. Nombela, M. J. Reigosa and F. Vilas, 1990. Complejo intermareal de la Ramallosa (bahia de Baiona, Pontevedra). I. Ecologia y evolucion. *Thalassas* 8, 45-56.

Alestalo, J., 1985. Finland, in E. C. F. Bird and M. L. Schwartz eds.: *The World's Coastline.* Van Nostrand Reinhold. New York, 295-302.

Alexander, C. R., Ch. A. Nittrouer, D. J. DeMaster, Y.-A. Park and S.-Ch. Park, 1991. Macrotidal mudflats of the southwestern Korean coast: a model for interpretation of intertidal deposits. *J. Sed. Petrol.* 61, 805-824.

Alexander, C. S., 1985. Tanzania, in E. C. F. Bird and M. L. Schwartz eds.: *The World's Coastline.* Van Nostrand Reinhold. New York, 691-695.

Aliotta, S., and E. Farinati, 1990. Stratigraphy of Holocene sand-shell ridges in the Bahia Blanca estuary, Argentina. *Mar. Geol.* 94, 353-360.

Aliotta, S., and G. M. E. Perillo, 1987. A sand wave field in the entrance to Bahia Blanca estuary, Argentina. *Mar. Geol.* 76, 1-14.

Alizai, S. A. K., and J. McManus, 1980. The significance of reed beds on siltation in the Tay estuary. *Proc. Roy. Soc. Edinburgh* 78B, s1-s13.

Allen, G. P., P. Castaing and J. M. Jouanneau, 1977. Mecanismes de remise en suspension et de dispersion des sédiments fins dans l'estuaire de la Gironde. *Bull. Soc. Geol. de France,* (7) XIX (2), 167-176.

Allen, G. P., D. Laurier, and J. Thouvenin, 1979. Etude sedimentologique du delta de la Mahakam. Notes et Memoires Total, Cie. Franc. Petrol., Paris, 156 p.

Allen, G. P., J. C. Salomon, P. Bassoulet, Y. du Penhoat, and C. de Grandpré, 1980. Effects of tides on mixing and suspended sediment transport in macrotidal estuaries. *Sed. Geology* 26, 69-90.

Allen, J. R. L., 1964. Sedimentation in the modern delta of the River Niger, West Africa, in L. M. J. U. van Straaten ed.: *Deltaic and Shallow Marine Deposits,* Elsevier, Amsterdam, 26-34.

Allen, J. R. L., 1965a. Coastal Geomorphology of eastern Nigeria: beach-ridge barrier islands and vegetated tidal flats. *Geol. Mijnb.* 44, 1-24.

Allen, J. R. L., 1965b. Late Quaternary Niger delta and adjacent areas. *Assoc. Petr. Geol. Bull.* 49, 547-600.

Allen, J. R. L., 1965c. A review of the origin and characteristics of recent alluvial sediments. *Sedimentology* 5 (2), 89-191 (Special Issue).

Allen, J. R. L., 1984. Sedimentary structures, their character and physical basis. *Dev. in Sedimentology* 30A/B. Elsevier, Amsterdam, 593/663 pp.

Allen, J. R. L., 1990. Salt-marsh growth and stratification: A numerical model with special reference to the Severn estuary, southwest Britain. *Mar. Geol.* 95, 77-96.

Allen, J. R. L., 1990a. The Severn estuary in southwest Britain: its retreat under marine transgression. *Sed. Geol.* 96, 13-28.

Allen, J. R. L. and Banks, N. L., 1972. An interpretation and analysis of recumbent-folded deformed cross-bedding. *Sedimentology*, 19, 257-283.

Allen, J. R. L., and J. E. Rae, 1988. Vertical salt-marsh accretion since the Roman period in the Severn estuary, southwest Britain. *Mar. Geol.* 83, 225-235.

Allen, J. R. L., and K. Pye, 1992. Coastal saltmarshes: their nature and importance, in J. R. L. Allen and K. Pye eds.: *Saltmarshes*. Cambridge University Press. 1-18.

Aller, R. C. 1978. Experimental studies of changes produced by deposit feeders on pore water, sediment, and overlying water chemistry. *Am. J. Sci.* 278, 1185-1234.

Aller, R. C. 1982. The effects of macrobenthos on chemical properties of marine sediment and overlying water chemistry, in K. R. Tenore and B. C. Coull eds. *Marine Benthic Dynamics*. University of North Carolina Press, Chapel Hill, 285-308.

Aller, R. C. and J. Y. Yingst, 1978. Biogeochemistry of tube dwellings: a study of the sedentary polychaete *Amphitrite ornata* (Leidy). *J. Mar. Res.* 36, 201-254.

Allersma, E., 1971. Mud on the oceanic shelf off Guiana. *Symp. Invest. Res. Caribbean Sea and Adjacent Regions.* UNESCO (Paris), 193-203.

Allison, M. A., C. A. Nittroner and G. C. Kineke, 1995a. Seasonal sediment storage on mud flats adjacent to the Amazon river. *Mar. Geol.* 125, 303-328

Allison, M. A., C. A. Nittroner and L. E. C. Faria Jr., 1995b. Rates and mechanisms of shoreface progradation and retreat down drift of the Amazon river mouth. *Mar. Geol.* 125, 373-392.

Allison, M. A., C. A. Nittrouer, L. E. C. Faria, Jr., O. M. Silveira, and A. C. Mendes, 1996. Sources and sinks of sediment to the Amazon margin: the Amapa coast. *Geo-Marine Letters* 16, 36-40.

Alongi, D. M., and A. Sasekumar, 1992. Benthic communities, in A. I. Robertson and D. M. Alongi, eds.: *Tropical Mangrove Ecosystems. Coastal Estuarine Studies* 41, Amer. Geophys. Union, Washington D.C., 137-171.

Altenburg, W., 1982. Description of the area. In W. Altenburg, M. Engelmoer, R. Mes and Th. Piersma: *Wintering Waders on the Banc d'Arguin, Mauretania*. Report NOME, Wadden Sea Working Group Comm. 6, 34-48.

Altevogt, R. 1955. Beobachtungen und Untersuchungen an indischen Winkerkrabben. *Z. Morph. Ökol. Tiere* 43, 501- 522.

Altevogt, R. 1957. Untersuchungen zur Biologie, Ökologie und Physiologie indischer Winkerkrabben. *Z. Morph. Ökol. Tiere* 46, 1-110.

Amos, C. L., 1978. The postglacial evolution of the Minas Basin, N. S. A sedimentological interpretation. *J. Sed. Petrol.* 48(3), 965-982.

Amos, C., 1995. Siliclastic tidal flats, in G. M. E. Perillo ed.: *Geomorphology and Sedimentology of Estuaries. Developm. Sedimentol.* 53, Elsevier, Amsterdam, 10, 273-306.

Amos, C. L. and B. F. N. Long, 1980. The sedimentary character of Minas Basin, Bay of Fundy, in S. B. McCann ed.: *The Coastline of Canada*. Canada Geol. Survey Paper 80-10, 153-180.

Amos, C. L., and K. T. Tee, 1989. Suspended sediment transport processes in Cumberland Basin. *J. Geophys. Res.* 94, 14407-14417.

Amos, C. L., J. V. Barrie, and J. T. Judge, 1995. Storm-enhanced sand transport in a macrotidal setting, Queen Charlotte Islands, British Columbia, Canada. Spec. Publ. Int. Ass. Sediment. 24, 53-68.

Anderson, F. E., 1973. Observations of some sedimentary processes acting on a tidal flat. Mar. Geol. 14, 101-116.

Anderson, F. E., and B. A. Howell, 1984. Dewatering of an unvegetated muddy tidal flat during exposure — desiccation or drainage? Estuaries 7, 225-232.

Anderson, F. E., and L. M. Mayer, 1984. Seasonal and spatial variability of particulate matter of a muddy intertidal flood front. Sedimentology 31, 383-394.

Anderson, F. E., L. Black, L. E. Watling, W. Mook and L. M. Mayer, 1981. A temporal and spatial study of mudflat erosion and deposition. J. Sed. Petrol. 51, 729-736.

Andresen, H., J. P. Bakker, M. Brongers, B. Heydemann, and U. Irmler, 1990. Long-term changes of salt marsh communities by cattle grazing. Vegetatio 89, 137-148.

Armstrong, W., E. J. Wright, S. Lythe, and J. K. Gaynard, 1985. Plant zonation and the effects of the spring-neap tidal cycle on soil aeration in a Humber saltmarsh. J. Ecol. 73, 323-339.

Aronow, S., and R. T. Kaczorowski, 1985. Texas, in E. C. F. Bird and M. L. Schwartz eds.: The World's Coastline. Van Nostrand Reinhold. New York, 129-145.

Ashley, G. M., 1980. Channel morphology and sediment movement in a tidal river, Pitt River, British Columbia. Earth Surf. Proc. 5, 347-368.

Ashley, G. M., 1988. Tidal channel classification for a low mesotidal saltmarsh. Mar. Geol. 82, 17-32.

Ashley, G. M., ed., 1988. The hydrodynamics and sedimentology of a back-barrier lagoon-salt marsh system, Great Sound, New Jersey. Mar. Geol. 82, V-XV, 1-132.

Augustinus, P. G. E. F., 1978. The changing shoreline of Surinam (South America). Thesis Utrecht University, 232 p.

Augustinus, P. G. E. F., 1980. Actual development of the chenier coast of Suriname (South America). Sed. Geol. 26, 91-114.

Augustinus, P. G. E. F., 1987. The geomorphologic development of the coast of Guiana between the Corentyne river and the Essequibo river, in V. Gardiner, ed.: International Geomorphology. part. I, 1281-1292.

Augustinus, P. G. E. F., 1995. Geomorphology and sedimentology of mangroves, in G. M. E. Perillo ed.: Geomorphology of Estuaries. Devel. Sedimentol. 53, Elsevier, 233-357.

Augustinus, P. G. E. F., ed., 1989. Cheniers and Chenier Plains. Mar. Geol. 90 (4) (Spec. Issue), 219-354.

Augustinus, P. G. E. F., L. Hazelhoff and A. Kroon, 1989. The chenier coast of Suriname: modern and geological development. Mar. Geol. 90, 269-281.

Avoine, J., 1981. L'estuaire de la Seine: sédiments et dynamique sédimentaire. Thesis, Universite de Caen, 236 p.

Baas, J. H., A. P. Oost, O. Sztanó, P. L. de Boer, and G. Postma, 1993. Time as an independent variable for current ripples developing towards linguoid equilibrium morphology. Terra Nova 5, 29-35.

Bagnold, R. A., 1960. Some aspects of the shape of river meanders. U.S. Geol. Surv. Prof. Papers 282E, 135-144.

Bakker, J. P., J. de Leeuw, K. S. Dijkema, P. C. Leendertse, H. H. T. Prins, and J. Rozema, 1993. Salt marshes along the coast of the Netherlands. Hydrobiologia. 265, 73-95.

Baltzer, F., 1975. Mapping of mangroves in the study of sedimentological processes and the concept of swamp structure, in G. Walsh, S. Snedaker and H. Teas eds.: Proc. Int. Symp. Biol. Management of Mangroves, Honolulu, Oct. 1974: 499-512.

Banus, M. D., E. Seppo and S. E. Kolehmainen, 1975. Floating, rooting and growth of red mangrove (*Rhizophora mangle L.*) seedlings: effect on expansion of mangroves in southwestern Puerto Rico, in G. Walsh, S. Snedaker and H. Teas, eds.: *Proc. Int. Symp. Biol. Management of Mangroves.* Honolulu, Oct. 1974: 370-334.

Barson, M. M., 1982. Vegetation coasts, in M. L. Schwartz ed.: *Encycl. Beaches Coast. Env.*, Hutchinson Ross Publ., Stroudsburg, 847-848.

Bartholdy, J., and P. Pheiffer Madsen, 1985. Accumulation of fine-grained material in a Danish tidal area. *Mar. Geol.* 67, 121-137.

Barwis, J. H., 1978. Sedimentology of some South Carolina tidal-creek point bars, and a comparison with their fluvial counterparts, in A. D. Miall ed.: *Fluvial Sedimentology.* Canad. Soc. Petr. Geol., Calgary, 129-160.

Basan, P. B., 1973. Aspects of sedimentation and development of a carbonate bank in the Barracuda Keys, South Florida. *J. Sed. Petrol.* 43, 42-53.

Bathurst, R. G. C., 1971. Carbonate sediments and their diagenesis. *Developments in Sedimentology* 12. Elsevier, Amsterdam, 620 pp.

Baumann, R. H., J. W. Day, and C. A. Miller, 1984. Mississippi deltaic wetland survival: sedimentation vs. coastal submergence. *Science* 224, 1093-1095.

Bayliss-Smith, T. P., R. Healey, R. Lailey, T. Spencer, and D. R. Stoddart, 1979. Tidal flows in salt marsh creeks. *Est. Coast. Mar. Sc.* 9, 235-255.

Beeftink, W. G., 1977. The coastal salt marshes of western and northern Europe: an ecological and phytosociological approach, in V. J. Chapman ed.: *Wet Coastal Ecosystems.* Elsevier, Amsterdam, 109-155.

Beeftink, W. G., 1987. Vegetation responses to changes in tidal inundations of salt marshes, in J. van Andel et al. eds.: *Disturbance in grasslands.* Junk, Dordrecht, 97-117.

Bejan, A., 1982. Theoretical explanation for the incipient formation of meanders in straight rivers. *Geophys. Res. Letters* 9, 831-834.

Beukema, J. J. 1980. Calcimass and carbonate production by molluscs on the tidal flats in the Dutch Wadden Sea. I. The tellinid bivalve *Macoma balthica. Neth. J. Sea Res.* 14, 323-338.

Beukema, J. J. 1982. Calcimass and carbonate production by molluscs on the tidal flats in the Dutch Wadden Sea. II. The edible cockle *Cerastoderma edule. Neth. J. Sea Res.* 15, 391-405.

Bidet, J. C., and C. Carruesco, 1982. Etude sédimentologique de la lagune de Oualidia (Maroc). *Oceanol. Acta Symp. Bordeaux* Sept. 1981, 29-37.

Bird, E. C. F., 1970. *Coasts.* M. I. T. Press, Cambridge, MA, 310 p.

Bird, E. C. F., 1970. The deltaic shoreline near Cairns, Queensland. *Australian Geographer* 11, 138-147.

Bird, E. C. F., 1984. *Coasts.* Blackwell, Oxford, 320 p.

Bird, E. C. F., 1985a. England and Wales, in E. C. F. Bird and M. L. Schwartz eds.: *The World's Coastline.* Van Nostrand Reinhold. New York, 359-369.

Bird, E. C. F., 1985b. Indonesia, in E. C. F. Bird and M. L. Schwartz eds.: *The World's Coastline.* Van Nostrand Reinhold. New York, 879-888.

Bird, E. C. F., 1985c. Papua New Guinea, in E. C. F. Bird and M. L. Schwartz eds.: *The World's Coastline.* Van Nostrand Reinhold. New York, 889-897.

Bird, E. C. F., 1985d. Fiji, in E. C. F. Bird and M. L. Schwartz eds.: *The World's Coastline.* Van Nostrand Reinhold. New York, 1003-1010.

Bird, E. C. F., 1986. Mangroves and intertidal morphology in Westernport Bay, Victoria, Australia. *Mar. Geol.* 69, 251-271.

Bird, E. C. F., 1993. *Submerging Coasts.* John Wiley and Sons, Chichester, 169 p.

Bird, E. C. F. and M. L. Schwartz, 1985. *The World's Coastline*. Van Nostrand Reinhold. New York, 1071 p.

Bird, E. C. F., and J. Iltis, 1985. New Caledonia and the Loyalty Islands, in E. C. F. Bird and M. L. Schwartz eds.: *The World's Coastline*. Van Nostrand Reinhold. New York, 995-1002.

Black, M., 1933. The algal sediments of Andros Island, Bahamas. *Phil. Trans. Royal Soc. London,* ser. B, 222, 165-192.

Bodéré, J.-C., 1985. Iceland, in E. F. C. Bird and M. L. Schwartz eds.: *The World's Coastline*. Van Nostrand Reinhold. New York, 267-271.

Boero, F., G. Belmonte, G. Fanelli, S. Piraino and F. Rubino, 1996. The continuity of living matter and the discontinuities of its constituents: do plankton and benthos really exist? *Trends Ecol. Evol.* 11, 177-180.

Boersma, J. R., 1969. Internal structure of some tidal mega-ripples on a shoal in the Westerschelde estuary, the Netherlands. *Geol. Mijnb.* 48, 409-414.

Boersma, J. R., and J. H. J. Terwindt, 1981. Neap-spring tide sequences of intertidal shoal deposits in a mesotidal estuary. *Sedimentology* 28, 151-170.

Boesch, D. F., M. N. Josselyn, A. J. Mehta, J. T. Morris, W. K. Nuttle, Ch. A. Simenstadt, and D. J. P. Swift, 1994. Scientific assessment of coastal wetland loss, restoration and management in Louisiana. *J. Coast. Res. Spec. Issue* no. 20, 103+v p.

Boon III, J. D., 1975. Tidal discharge asymmetry in a salt marsh drainage system. *Limnol. Oceanogr.* 20, 71-80.

Boon III, J. D., 1978. Suspended-solids transport in a saltmarsh creek — an analysis of errors, in B. J. Kjerve ed.: *Estuarine Transport Processes*. University South Carolina Press, Columbia, 147-160.

Borowka, R. K., 1985. Poland, in E. C. F. Bird and M. L. Schwartz eds.: *The World's Coastline*. Van Nostrand Reinhold. New York, 311-314.

Bos, J. M., 1995. Archeologie van Friesland. Uitgeverij Matrijs, Utrecht, 229 p. (in Dutch)

Boto, K. G., and J. S. Bunt, 1981. Tidal export of particulate organic matter from a northern Australian mangrove system. *Est. Coast. Shelf Sci.* 13, 247-255.

Bowden, K. F., 1983. *Physical Oceanography of Coastal Waters*. John Wiley and Sons, Chichester, 302 p.

Boyd, R., R. Dalrymple, and B. A. Zaitlin, 1992. Classification of clastic coastal depositional environments. *Sed. Geol.* 80, 139-150.

Breitner, H. J., 1953. Ueber die Geschwindigkeit des Dickenwachstums von Eisdecken, *Deutsch. Hydr. Zeitschr.* 6, 34-38.

Bressolier-Bousquet, C., 1991. Geomorphological effects of land reclamation in the eighteenth century at the mouth of the Leyre river, Arcachon Bay, France. *J. Coast. Res.* 7, 113-126.

Bricker-Urso, S., S. W. Nixon, J. K. Cochran, D. J. Hirschberg, and C. Hunt, 1989. Accretion rates and sediment accumulation in Rhode Island salt marshes. *Estuaries* 12, 300-317.

Bridges, P. H., and M. R. Leeder, 1976. Sedimentary model for intertidal mudflat channels, with examples from the Solway Firth, Scotland. *Sedimentology* 23, 533-552.

Brierley, G. J., 1989. River planform facies models: the sedimentology of braided, wandering and meandering reaches of the Squamish River, British Columbia. *Sed. Geol.* 61, 17-93.

Brinkman, R. and L. J. Pons, 1968. A pedo-geomorphological classification and map of the Holocene sediments in the coastal plain of three Guianas. *Soil Survey Papers,* Soil Survey Inst. Wageningen, 4, 1-40.

Britsch, L. D., and E. B. Kemp, 1990. Land loss rates: Mississippi deltaic plain. US Army C. Eng. Techn. Rep. GL/90/2.

Brockmann, C. 1935. Diatomeen und Schlick im Jade Gebiet. Abh. Senckenberg. *Naturforsch. Ges.* 430, 1-64.

Bromley, R. G., 1990. *Trace Fossils: Biology and Taxonomy.* Unwin Hyman. 280 pp.

Brown, R. G., and P. J. Woods, 1974. Sedimentation and tidal-flat development, Nilemah Embayment, Shark Bay, Western Australia, in B. W. Logan et al.: *Evolution and Diagenesis of Quaternary Carbonate Sequences,* Shark Bay, Western Australia, A. A. P. G. Memoir 22, 316-340.

Burjis, M. J., and J. J. Symoens, 1987. *African Wetlands and Shallow Water Bodies.* ORSTOM, Paris, publ. 211.

Burns, K. A., and L. Yelle-Simmons, 1994. The Galeta oil spill. IV. Relationship between sediment and organism hydrocarbon loads. *Est. Coast. Shelf Sci.* 38, 397-412.

Burns, K. A., S. D. Garrity, D. Jorissen, J. MacPherson, M. Stoelting, J. Tierney, and L. Yelle-Simmons, 1994. The Galeta oil spill. II. Unexpected persistence of oil trapped in mangrove sediments. *Est. Coast. Shelf Sci.* 38, 349-364.

Byrne, J. V., D. O. LeRoy, and Ch. M. Riley, 1959. The chenier plain and its stratigraphy, southwestern Louisiana. *Trans. Gulf Coast Assoc. Geol. Soc.* 9, 237-259.

Cadée, G. C. 1979. Sediment reworking by the polychaete Heteromastus filiformis on a tidal flat in the Dutch Wadden Sea. *Neth. J. Sea Res.* 13, 441-456.

Cadée, G. C. 1982. Tidal and seasonal variation in particulate and dissolved organic carbon in the western Dutch Wadden Sea and Marsdiep tidal inlet. *Neth. J. Sea Res.* 15, 228- 249.

Cadée, G. C. 1984. 'Opportunistic feeding', a serious pitfall in trophic structure analysis of (paleo)faunas. *Lethaia* 17, 289-292.

Cadée, G. C. 1990. Feeding traces and bioturbation by birds on a tidal flat, Dutch Wadden Sea. *Ichnos* 1, 23-30.

Cadée, G. C. 1994. Eider, shelduck, and other predators, the main producers of shell fragments in the Wadden Sea: palaeoecological implications. *Palaeontology* 37, 181- 202.

Cadée, G. C. 1995. Birds as producers of shell fragments in the Wadden Sea, in particular the role of the Herring gull. *Geobios M. S.* 18, 77-85.

Cadée, G. C., Ma Hong Guang, and Wang Bao Can, 1994. Animal traces on a tidal flat in Hangzhou Bay, China. *Neth. J. Sea Res.* 32, 73-80.

Cadée, G. C. 1968. Molluscan biocoenoses and thanatocoenoses in the Ria de Arosa, Galicia, Spain. *Zool. Verh. Leiden* 95, 1-121.

Cadée, G. C. 1976. Sediment reworking by Arenicola marina on tidal flats in the Dutch Wadden Sea. *Neth. J. Sea Res.* 10, 440-460.

Caline, B. 1982. Le secteur occidental de la Baie du Mont Saint-Michel. Morphologie, sédimentologie et cartographie de l'Estran. Documents BRGM 42, 1-250.

Cao, K., Y.F. Dong, S. Zh. Yan and G. Ch. Gu, 1989. Basic characteristics of the tidal flat on the north coast of Hangzhou Bay. *Oceanol. Limnol. Sinica* 20, 412-421.

Cao, P. K., F. X. Hu, G. Ch. Gu and Y. Q. Zhou, 1989. Relationship between suspended sediments from the Changjiang estuary and the evolution of the embayed muddy coast of Zhejiang province. *Acta Oceanol. Sinica* 8, 273-283.

Carey, D. A., 1983. Particle resuspension in the benthic boundary layer induced by flow around Polychaete tubes. *Can. J. Fish. Aquat. Sci.* 40 (Suppl. 1), 301-308.

Carruesco, C., 1989. Evolution sedimentologique a l'Holocene de quelques systemes lagunaires du littoral atlantique (Oualidia, Moulay Bou Salham, Maroc — Arcachon, France). Thesis University Bordeaux, 308 p.

Castaing, P., and A. Guilcher, 1995. Geomorphology and sedimentology of rias, in G. M. E. Perillo ed.: Geomorphology and Sedimentology of Estuaries. *Devel. in Sediment.* 53, Elsevier, 69-111.

Castel, J., P.-J. Labourg, V. Escaravage, I. Auby, and M. E. Garcia, 1989. Influence of seagrass beds and oyster parks on the abundance and biomass patterns of meio- and macrobenthos in tidal flats. *Est. Coast. Shelf Sci.* 28, 71-85.

Chafets, H. S. 1994. Bacterially induced precipitates of calcium carbonate and lithification of microbial mats, in W. E. Krumbein, D. M. Paterson and L. J. Stal eds. *Biostabilization of Sediments.* BIS-Verlag. University Oldenburg, 149-163.

Chang, K. S. T., and F. L. W. Tang, 1970. Studies on the shore processes and wave features of the western coast of Taiwan. Proc. 12th. Coast. Eng. Conf. Wash., II, 729-737.

Chang, J. H., S. S. Chun, S. J. Kwon, D.-H. Shin, S.-J. Han and Y. A. Park, 1993. Sedimentary characteristics and evolution history of chenier, Gomso Bay tidal flat, western coast of Korea. *J. Oceanol. Soc. Korea* 28, 212-228.

Chapman, V. J., 1974. *Salt Marshes and Salt Deserts of the World.* Cramer, Lehre, 2nd ed., 392 p.

Chapman, V. J. ed., 1977. *Wet Coastal Ecosystems. Ecosystems of the World 1.* Elsevier, Amsterdam: 428 p.

Chapman, V. J., 1977. Introduction, in V. J. Chapman ed.: *Wet Coastal Ecosystems.* Elsevier, 1-29.

Chappell, J., and J. Grindrod, 1984. Chenier plain formation in northern Australia, in *Coastal Geomorphology in Australia.* Academic Press Australia, 197-231.

Chen, C. J., 1991. Development of depositional tidal flat in Jiangsu province. *Oceanol. Limnol. Sinica* 22, 360-368.

Chen, Ji-Yu, Yun Caixing, and Xu Haigen, 1982. The model of development of the Chang Jiang estuary during the last 2000 years, in V. S. Kennedy ed.: *Estuarine Comparisons,* Academic Press. 655-666.

Chen Jiyu, Liu Cangzi, Zhang Chongle, and H. J. Walker, 1990. Geomorphological development and sedimentation in Qiantang estuary and Hangzhou Bay. *J. Coastal Res.* 6, 559-572.

Chong, V. C., A. Sasekumar, M. U. C. Leh, and R. D'Cruz, 1990. The fish and prawn communities of a Malaysian coastal mangrove system, with comparisons to adjacent mud flats and inshore waters. *Est. Coast. Shelf Sci.* 31, 703-722.

Chung, G. S., and Y. A. Park, 1978. Sedimentological properties of the recent intertidal flat environment, southern Nam Yang Bay, west coast of Korea. *J. Oceanol. Soc. Korea* 13, 9-18.

Churchill, D. M., 1973. The ecological significance of tropical mangroves in the early Tertiary floras of Southern Australia. *Geol. Soc. Austr. Spec.* Publ. 4: 79-86.

Clague, J. J., and B. D. Bornhold, 1980. Morphology and littoral processes of the Pacific coast of Canada, in S. B. McCann ed.: *The Coastline of Canada,* Canada Geol. Survey Paper 80-10, 339-380.

Clark, J. A., W. E. Farrell, and W. R. Peltier, 1978. Global change in post glacial sea level: a numerical calculation. *Quaternary Res.* 9, 265-289.

Cloern, J. E. 1982. Does the benthos control phytoplankton biomass in south San Francisco Bay? *Mar. Ecol. Prog. Ser.* 9, 191-202.

Cook, P. J., and H. A. Polach, 1973. A chenier sequence at Broad Sound, Queensland, and evidence against a Holocene high sea level. *Mar. Geol.* 14, 253-268.

Cook, P. J., and W. Mayo, 1977. Sedimentology and Holocene history of a tropical estuary (Broad Sound, Queensland). *Bureau Min. Res., Geol. and Geophys. Dept. Natl. Development Bull.* 170, Canberra, 206 p.

Cooksey, K. E., 1984. Role of diatoms in the mangrove habitat, in S. C. and J. G. Snedaker eds. *The Mangrove Ecosystem: Research Methods,* UNESCO, ch. 11: 175-180.

Cooper, J. A. G., 1991. Beach rock formation in low latitudes; implications for coastal evolutionary models. *Mar. Geol.* 98, 145-154.

Cooper, J. A. G., 1993. Sedimentation in the cliff-bound, microtidal Mtamvuna estuary, South Africa. *Mar. Geol.* 112, 237-256.

Court Stevenson, J., L. Ward and M. S. Kearney, 1988. Sediment transport and trapping in marsh systems: implications of tidal flux studies. *Mar. Geol.* 80, 37-59.

Craft, C. B., E. D. Seneca and S. W. Broome, 1993. Vertical accretion in microtidal regularly and irregularly flooded estuarine marshes. *Est. Coast. Shelf Sci.* 37, 371-386.

Craighead, F. C., 1964. Land, mangroves and hurricanes. *Bull. Fairchild Trop. Gdn.,* 19 (4), 1-2 G.

Cramer, W., and H. Hytteborn, 1987. The separation of fluctuation and long-term change in vegetation dynamics of a rising seashore. *Vegetatio* 69, 157-167.

Creutzberg, F., and H. Postma, 1979. An experimental approach to the distribution of mud in the southern North Sea. *Neth. J. Sea. Res.* 13, 211-261.

Curran H. A. 1994. The palaeobiology of ichnocoenoses in Quaternary, Bahamian-style carbonate environments: the modern to fossil transition, in S. K. Donovan ed. *The Palaeobiology of Trace Fossils.* J. Wiley, New York, 83-104.

Curray, J. R. and D. G. Moore, 1971. Growth of the Bengal Deep-Sea fan and denudation in the Himalayas. *Bull. Geol. Soc. Amer.* 82, 563-572.

Dacey, M. F., and W. C. Krumbein, 1976. Three growth models for stream channel networks. *J. Geol.* 84, 153-163.

Dalrymple, D. W., 1965. Calcium carbonate deposition associated with blue-green algal mats, Baffin Bay, Texas. *Publ. Inst. Mar. Science,* 10, 187-200.

Dalrymple, R. W., R. J. Knight and G. V. Middleton, 1975. Intertidal sand bars in Cobequid Bay (Bay of Fundy), in L. E. Cronin ed.: *Estuarine Research* vol. II, 293-307.

Dalrymple, R. W., R. J. Knight and J. J. Lambiase, 1978. Bedforms and their hydraulic stability relationships in a tidal environment, Bay of Fundy, Canada. *Nature* 275, 100-104.

Dalrymple, R. W., B. A. Zaitlin, and R. Boyd, 1992. Estuarine facies models: conceptual basis and stratigraphic implications. *J. Sed. Petrol.* 62, 1130-1146.

Dame, R. F., N. Dankers, T. Prins, H. Jongsma and A. Smaal, 1991. The influence of mussel beds on nutrients in the western Wadden Sea and the Eastern Scheldt estuaries. *Estuaries* 14, 130-138.

d'Anglejean, B., 1980. Effects of seasonal changes on the sedimentary regime of a subarctic estuary, Rupert Bay, Canada. *Sed. Geol.* 26, 51-68.

Daniel, G. F., A. H. L Chamberlain and E. B. S. Jones, 1987. Cytochemical and electron microscopical observations on the adhesive materials of marine fouling diatoms. *Brit. Phycol. J.* 22, 101-118.

Daniel, J. R. K., 1989. The chenier plain coastal system of Guyana. *Mar. Geol.* 90, 283-287.

Dankers, N., K. Koelemaij and J. Zegers. 1989. De rol van de mossel en de mosselcultuur in het ecosysteem van de Waddenzee. RIN-rapport 89/9, Texel 66 pp.

Dankers, N., M. Binsbergen, K. Zegers, R. Laane, and M. R. van der Loeff, 1984. Trans-portation of water, particulate and dissolved organic and inorganic matter between a salt marsh and the Ems-Dollard estuary, The Netherlands. *Est. Coast. Shelf Sci.* 19, 143-165.

Davies, G. R., 1970. Algal-laminated sediments, Gladstone Embayment, Shark Bay, Western Australia, in Logan, B. W., et al.: *Carbonate Sedimentation and Environments, Shark Bay, Western Australia,* A. A. P. G. Memoir 13, 169-205.

Davies, J. L., 1964. A morphogenic approach to world shorelines. *Zeitschr. Geomorph. N. F. Bd.* 8, Sonderheft, 127-142.

Davies, J. L., 1972. *Geographical Variation in Coastal Development.* Oliver and Boyd, Edinburgh, 204 p.

Davis, J. H. Jr., 1940. The ecology and geologic role of mangroves in Florida. *Pap. Tortugas Lab.* 32 (Publ. Carnegie Inst. 517): 303-412.

Day, J. W., C. A. S. Hall, W. M. Kemp and A. Yanez-Arancibia, 1989. *Estuarine Ecology.* J. Wiley, New York. 558 pp.

de Boer, P. L. 1981. Mechanical effects of microorganisms on intertidal bedform migration. *Sedimentology* 28: 129-132.

de Boer, P. L., 1979. Convolute lamination in modern sands of the estuary of the Oosterschelde, the Netherlands, formed as a result of entrapped air. *Sedimentology* 26, 283-294.

de Haas, H., and D. Eisma, 1993. Suspended-sediment transport in the Dollard estuary. *Neth. J. Sea Res.* 31, 37-42.

Deans, E. A., P. S. Meadows and J. G. Anderson, 1982. Physical, chemical and microbiological properties of intertidal sediments and sediment selection by Corophium volutator. *Int. Rev. Ges. Hydrobiol.* 67: 261-269.

DeDeckere Decho, A. W. 1990. Microbial exopolymer secretions in ocean environments: their role(s) in food webs and marine processes. *Oceanogr. Mar. Biol. Annu. Rev.* 28, 73- 153.

Defant, A., 1961. *Physical Oceanography,* II. Pergamon Press, Oxford, 598 p.

Degens, E. T., S. Kempe and J. E. Richey 1991. *Biogeochemistry of Major World Rivers.* SCOPE 42. John Wiley and Sons, Chichester, 356 pp.

DeGroodt, E. G., 1992. *Ecological Profile of Macrophytes and Macroalgae.* Res. Rep. Delft Hydraulics, 17 p.

DeLaune, R. D., 1983. Relationships among vertical accretion, coastal submergence, and erosion in a Louisiana Gulf Coast marsh. *J. Sed. Petrol.* 53, 147-157.

DeLaune, R. D., S. R. Pezeshki, and W. H. Patrick, Jr., 1987. Response of coastal plants to increase in submergence and salinity. *J. Coastal Res.* 3, 535-546.

DeLaune, R. D., W. H. Patrick and R. J. Buresh, 1978. Sedimentation rates determined by 137-Cs dating in a rapidly accreting salt marsh. *Nature* 275, 532-533.

Delft Hydraulics, 1962. *Demerara Coastal Investigation.* Delft, Netherlands, 232 p.

den Hartog, C., 1970. The sea-grasses of the world. Verh. Kon. Ned. Akad. Wet. afd. Nat. Tweede reeks, 59 (1), 275 p.

den Hartog, C., 1971. The dynamic aspect in the ecology of seagrass communities. *Thalassia Jugoslavica* 7, 101-112.

Depuydt, F., 1972. De Belgische strand- en duinformaties. Verh. Kon. Academic Wet., Lett. Schone Kunsten van Belgie, Kl. Wet., 34, nr. 122, 228 p.

deVries Klein, G., 1963. Bay of Fundy intertidal zone sediments. *J. Sed. Petrol.* 33(4), 844-854.

deVries Klein, G. 1964. Sedimentary facies in Bay of Fundy intertidal zone, Nova Scotia, Canada, in L. M. J. U. van Straaten ed.: *Deltaic and Shallow Marine Deposits.* Elsevier, Amsterdam 193-199.

deVries Klein, G., 1967. Comparison of Recent and Ancient tidal flat and estuarine sediments, in G. H. Lauff ed.: *Estuaries.* Publ. 83, Amer. Assoc. Adv. Sci., Washington, 207-218.

deVries Klein, G., and J. E. Sanders, 1964. Comparison of sediments from Bay of Fundy and Dutch Wadden Sea tidal flats. *J. Sed. Petrol.* 34, 19-24.

Dijkema, K. S., 1987. Geography of salt marshes in Europe. *Z. Geomorph. N. F.* 31, 489-499.

Dijkema, K. S., 1989. *Habitats of The Netherlands, German and Danish Wadden Sea.* Texel, Leiden, 30 pp.

Dijkema, K. S., in press. Impact prognosis for salt marshes from subsidence by gas extraction. *J. Coastal Res.*

Dijkema, K. S., J. H. Bossinade, P. Bouwsema, and R. J. de Glopper, 1990. Salt marshes in the Netherlands Wadden Sea: rising high-tide levels and accretion enhancement, in J. J. Beukema et al. eds.: *Expected Effects of Climatic Change on Marine Coastal Ecosystems*. Kluwer Academic Publ., Dordrecht, 173-188.

Dionne, J. C., 1971. Polygonal patterns in muddy tidal flats. *J. Sed. Petrol.* 41, 838-839.

Dionne, J. C., 1972a. Caracteristiques des schorres des regions froides en particulier de l'estuaire du Saint-Laurent. *Z. Geomorph. N. F. Suppl. Bd.* 13, 131-162.

Dionne, J. C., 1972b. Ribbed grooves and tracks on mud tidal flats of cold regions. *J. Sed. Petrol.* 42(4), 848-851.

Dionne. J. C., 1973. Monroes: a type of so-called mud volcanoes in tidal flats. *J. Sed. Petrol.* 43(3), 848-856.

Dionne J. C., 1980. An outline of the eastern James Bay coastal environments, in S. B. McCann eds.: *The Coastline of Canada*. Canada Geol. Survey Paper 80-10, 311-338.

Dionne, J. C., 1985. Tidal marsh erosion by geese, St. Laurence Estuary, Québec. *Geogr. Phys. et Quatern.* 39(1), 99-105.

Dionne, J. C., 1988. Characteristic features of modern tidal flats in cold regions, in P. L. de Boer ed.: *Tide-Influenced Sedimentary Environments and Facies*. Reidel Publ. Co. 301-332.

Dionne, J. C., 1988. Characteristic features of modern tidal flats in cold regions (1988) In.: P. L. de Boer, A. van Gelder, and S. D. Nio, eds.: *Tide-Influenced Sedimentary Environments and Facies*. Reidel, Dordrecht, 301-332.

Diop, E. S., 1985. Guinea Bissau and Republic of Guinea, in E. C. F. Bird and M. L. Schwartz eds.: *The World's Coastline*. Van Nostrand Reinhold, New York, 561-567.

Doody, J. P., 1992. The conservation of British saltmarshes, in J. R. L. Allen and K. Pye eds.: *Saltmarshes*. Cambridge University Press, 80-114.

Drew, K. S., 1985. United Arab Emirates, in E. C. F. Bird and M. L. Schwartz eds.: *The World's Coastline*. Van Nostrand Reinhold. New York, 723-727.

Dronkers, J., 1986a. Tidal asymmetry and estuarine morphology. *Neth. J. Sea Res.* 20, 117-131.

Dronkers, J., 1986b. Tide-induced residual transport of fine sediment, in J. v. d. Kreeke ed.: *Physics of Shallow Estuaries and Bays*. Springer Verlag, 228-244.

Dronkers, J., and J. T. F. Zimmerman, 1982. Some principles of mixing in tidal lagoons. *Oceanol. Acta,* Proc. Int. Symp. Coastal Lagoons, Bordeaux, Sept. 1981, 107-117.

Dupont, J.-P., R. Lafitte, M.-F. Huault, P. Hommeril, and R. Meyer, 1994. Continental/marine ratio changes in suspended and settled matter across a macrotidal estuary (the Seine estuary, northwestern France). *Mar. Geol.* 120, 27-40.

Dworschak, P. C. 1981. The pumping rates of the burrowing shrimp *Upogebia pusilla* (Petagna) (Decapoda: Thalassinidea). *J. Exp. Mar. Biol. Ecol.* 52, 25-35.

Dyer, K. R., 1986. *Coastal and Estuarine Sediment Dynamics*. John Wiley and Sons, Chichester, 342 p.

Edgar, L. A. and J. D. Pickett-Heaps 1984. Diatom locomotion. *Progress Phycol. Res.* 3, 447-88.

Edwards, J. M., and R. Frey, 1977. Substrate characteristics within a Holocene salt marsh, Sapelo Island, Georgia. *Senckenb. Marit.* 9, 215-259.

Egler, F., 1950. Southeast saline Everglades vegetation, Florida, and its management. *Vegetatio* 3: 213-265.

Ehlers, J., 1988. *The Morphodynamics of the Wadden Sea*. Balkema, Rotterdam, 397 p.

Einsele, G., D. Herm, and H. Schwartz, 1974. Sea level fluctuations during the past 6000 years at the coast of Mauritania. *Quaternary Res.* 4, 282-289.

Eisma, D., 1981. Supply and deposition of suspended matter in the North Sea. *Int. Assoc. Sedimentol. Spec. publ.* 5, 415-428.

Eisma, D., 1985. Vietnam, in E. C. F. Bird and M. L. Schwartz eds.: *The World's Coastline.* Van Nostrand Reinhold. New York, 805-811.

Eisma, D., 1985. North Korea and South Korea, in E. C. F. Bird and M. L. Schwartz eds.: *The World's Coastline.* Van Nostrand Reinhold, New York, 833-841.

Eisma, D, 1990. Transport and deposition of suspended matter in the North Sea and the relation to coastal siltation, pollution and bottom fauna distribution. *Rev. Aquat. Sci.* 3, 181-216.

Eisma, D., 1993. *Suspended Matter in the Aquatic Environment.* Springer Verlag, Berlin-Heidelberg, 315 p.

Eisma, D., ed. 1995. *Climate Change: Impact on Coastal Habitation.* Lewis Publ. Boca Raton, 260 p.

Eisma, D., P. G. E. F. Augustinus and C. Alexander, 1991. Recent and subrecent changes in the dispersal of Amazon mud. *Neth. J. Sea. Res.* 28 (3), 181-192.

Eisma, D., and A. J. van Bennekom, 1978. The Zaire river and estuary and the Zaire outflow in the Atlantic Ocean. *Neth. J. Sea Res.* 12, 255-272.

Eisma, D., G. W. Berger, W-Y. Chen, and J. Shen, 1989. Pb-210 as a tracer for sediment transport and deposition in the Dutch-German Waddensea, in *Proc. Coastal Lowlands Symp.* Den Haag, May 1987, 237-253.

Eisma, D., S. J. van der Gaast, J. M. Martin and A. J. Thomas, 1978. Suspended matter and bottom deposits of the Orinoco delta: turbidity, mineralogy and elementary composition. *Neth. J. Sea Res.* 12, 224-251.

Eisma, D., and G. Irion, 1988. Suspended matter and sediment transport, in W. Salomons, B. L. Bayne, E. K. Duursma and U. Förstner eds.: *Pollution of the North Sea, An Assessment.* Springer Verlag Berlin Heidelberg, 20-35.

Eisma, D., and H. W. van der Marel, 1971. Marine muds along the Guiana coast and their origin from the Amazon basin. *Contrib. Mineral. Petrol.* 31, 321-334.

Eisma, D., W. G. Mook and C. Laban, 1981. An early Holocene tidal flat in the Southern Bight. Spec. Publ. Int. Ass. Sediment 5, 229-237.

Eisma, D., and W. J. Wolff, 1980. The development of the westernmost part of the Wadden Sea in historical times, in K. S. Dijkema, H.-E. Reineck and W. J. Wolff, eds.: *Geomorphology of the Wadden Sea Area.* Rep. 1. Wadden Sea Working Group, Leiden, 95-103.

Elder, H. Y. 1973. Direct peristaltic progression and the functional significance of the dermal connective tissues during burrowing in the polychaete Polyphysia crassa (Oersted). *J. Exper. Biol.* 58, 637-655.

Elder, H. Y. and R. D. Hunter, 1980. Burrowing of Priapulus caudatus (vermes) and the significance of the direct peristaltic wave. *J. Zool. London* 191, 333-351.

Elliott, A. J., 1978. Observations of the meteorologically induced circulation in the Potomac estuary. *Est. Coast. Mar. Sci.* 6, 285-290.

Ellison, J. C., and D. R. Stoddart, 1991. Mangrove ecosystem collapse during predicted sea-level rise: Holocene analogues and implications. *J. Coast. Res.* 7, 151-165.

Emerit, M., 1960. Etude granulométrique de la mangrove de Joal, Senegal. *Ann. Fac. Sci.* University Dakar, 5: 107-115.

Emery, K. O., 1945. Entrapment of air in beach sand. *J. Sedim. Petrol.* 15, 39-49.

Emery, K. O., 1968. Relict sediments on continental shelves of world. *Am. Assoc. Petrol. Geol. Bull.* 52, 445-464.

Ennion, E. A. R. and N. Tinbergen, 1967. *Tracks.* Oxford University Press, London, 63 pp.

Enos, P., and R. D. Perkins, 1979. Evolution of Florida Bay from island stratigraphy. *Geol. Soc. Amer. Bull.* 90, 59-83.

Erchinger, H. F., 1985. Dünen, *Watt und Salzwiesen. Niedersächsischer Minister fuer Ernährung, Landwirtschaft und Forsten.* Hannover, 1-59.

Esselink, P., K. S. Dijkema, S. Reents, and G. Hageman, submitted. Vertical accretion and profile changes in abandoned man-made tidal marshes in the Ems Dollard estuary, the Netherlands. *J. Coastal Res.*

Evans, G., 1965. Intertidal flat sediments and their environments of deposition in the Wash. *Quart. J. Geol. Soc. London,* 121, 209-245.

Evans, G., 1975. Intertidal flat deposits of the Wash, western margin of the North Sea, in R. N. Ginsburg, ed: *Tidal Deposits,* Ch. 2, 13-20.

Evans, G., V. Schmidt, P. Bush, and H. Nelson, 1969. Stratigraphy and geologic history of the Sabkha, Abu Dhabi, Persian Gulf. *Sedimentology* 12, 145-159.

Evans, G., and M. B. Collins, 1975. The transportation and deposition of suspended sediment over the intertidal flats of The Wash, in J. Hails and A. Carr, eds., *Nearshore Sediment Dynamics and Sedimentation.* London, Wiley-Interscience, 273-304.

Faas, R. W. and J. T. Wells, 1990. Rheological control of fine-sediment suspension, Cape Lookout Bight, North Carolina. *J. Coast. Res.* 6, 503-515.

Faas, R. W., 1995. Mudbanks of the southwest coast of India III: Role of non-Newtonian flow properties in the generation and maintenance of mudbanks. *J. Coast. Res.* 11, 911-917.

Fairbridge, R. W., 1980. The Estuary: Its definition and geodynamic cycle, in E. Olausson and I. Cato eds.: *Chemistry and Biogeochemistry of Estuaries.* Wiley and Sons, 1-35.

Falkus, H. 1978. *Nature Detective.* Victor Gollancz Ltd. London. 254 pp.

Farrow, G. E. 1975. Techniques for the study of fossil and Recent traces, in R. W. Frey (ed.) *The Study of Trace Fossils.* Springer Verlag, Berlin. 562, 537-554.

Fasano, J. L., M. A. Hernandez, F. I. Isla and E. J. Schnack, 1982. Aspectos evolutivos y ambientales de la laguna Mar Chiquita (provincia de Buenos Aires, Argentina). *Oceanologica Acta Spec. Vol. Symp.* Bordeaux 1981, 285-292.

Faugères, J. C., R. Cuignon, H. Ferries and J. Gayen, 1986. Caractères et facteurs d'un comblement littoral a l'Holocene supérieur: passage d'un domaine estuarien à un domaine lagunaire (Bassin d'Arcachon, France). *Bull. Inst. Geol. Bassin d'Aquitaine,* 39, 95-116.

Featherstone, R. P. and M. J. Risk, 1977. Effect of tubebuilding polychaetes on intertidal sediments of the Minas Basin, Bay of Fundy. *J. Sed. Petrol.* 47, 446-450.

Fenchel, T., 1969. The ecology of marine microbenthos. IV. Structure and function of the benthic ecosystem, its chemical and physical factors and the microfauna communities with special reference to the ciliated Protozoa. *Ophelia* 6, 1-182.

Feng, H. Z. and Wang Z. T., 1989. Holocene sea level changes and coastline shifts in Zhejiang province, *China Acta Oceanol. Sinica* 8, 101-111.

Feng, Y. J., Li, Y., and Li, B. G., 1993. Relationship between the characteristics of landforms and sediments as well as the plume front in the Hangzhou Bay. *Acta Oceanol. Sin.* 12, 133-144.

Finley, R. J., 1975. Hydrodynamics and tidal deltas of North Inlet, South Carolina, in L. E. Cronin ed.: *Estuarine Research* vol. II. Academic Press, New York. 277-291.

Fisher, J. J., 1985. Atlantic USA — North, in E. C. F. Bird and M. L. Schwartz eds.: *The World's Coastline.* Van Nostrand Reinhold, New York, 223-234.

Fisk, H., 1959. Padre Island and the Laguna Madre flats, coastal south Texas, in R. J. Russell ed.: *2nd Coastal Geography Conference,* Louisiana State University April 1959, Washington D. C., 103-151.

Flemming, B. W., 1977. Langebaan Lagoon: a mixed carbonate-siliclastic tidal environment in a semi-arid climate. *Sed. Geol.* 18, 61-95.

Flessa, K. W., A. H. Cutler and K. H. Meldahl, 1993. Time and taphonomy: quantitative estimates of time-averaging and stratigraphic disorder in a shallow marine habitat. *Paleobiology* 19, 266-286.

Fletcher III, Ch. H., H. J. Knebel and J. C. Kraft, 1992. Holocene depocenter migration and sediment accumulation in Delaware Bay: a submerging marginal marine sedimentary basin. *Mar. Geol.* 103, 165-183.

Fonseca, M. S., 1996. The role of seagrasses in nearshore sedimentary processes: a review, in K. F. Nordstrom and C. T. Roman eds.: *Estuarine Shores.* Wiley and Sons Ltd., Chichester, 261-286.

Fonseca, M. S., and W. J. Kenworthy, 1987. Effects of current on photosynthesis and distribution of seagrasses. *Aquat. Bot.* 27, 59-78.

Francis, J., 1992. Physical processes in the Rufiji delta and their possible implications on the mangrove ecosystem, in V. Jaccarini and E. Martens, eds.: *The Ecology of Mangrove and Related Ecosystems. Hydrobiologia.* 247, 173-179.

Frankenberg, D., S. L. Coles and R. E. Johannes, 1967. *Limnol. Oceanogr.* 12, 113-120.

French, J. R., and T. Spencer, 1993. Dynamics of sedimentation in a tide-dominated back-barrier salt marsh, Norfolk, UK. *Mar. Geol.* 110, 315-331.

Frey, R. W., and J. D. Howard, 1969. A profile of biogenic sedimentary structures in a Holocene barrier island — salt marsh complex, Georgia. *Gulf Coast Assoc. Geol. Soc. Trans.* 19, 427-444.

Frey, R. W., and P. B. Basan, 1985. Coastal salt marshes, in R. A. Davis ed.: *Coastal Sedimentary Environments*, 2nd ed., Springer Verlag, 225-303.

Frey R. W. and Dörjes, J. 1988. Carbonate skeletal remains in beach-to-offshore sediments, Pensacola, Florida. *Senckenb. Marit.* 20: 31-57.

Frey, R. W., J. D. Howard, S.-J. Han and B.-K. Park, 1989. Sediments and sedimentary sequences on a modern macrotidal flat, Inchon, Korea. *J. Sed. Petrol.* 59, 28-44.

Friedman, G. M., A. J. Amiel, M. Braun, and D. S. Miller, 1973. Generation of carbonate particles and laminites in algal mats — example from sea-marginal hypersaline pool, Gulf of Aqaba, Red Sea. *Bull. A. A. P. G.,* 57, 541-557.

Friedrichs, C. T., and D. G. Aubrey, 1988. Non-linear tidal distortion in shallow well-mixed estuaries: a synthesis. *Est. Coast. Shelf Sci.* 27, 521-545.

Froidefond, J. M., R. Griboulard, R. Prud'homme and M. Pujos, 1987. Deplacement des bancs de vase et variation du littoral de la Guyane francaise. *Bull. Inst. Geol. Bassin d'Aquitaine,* 42, 67-83.

Froidefond, J. M., M. Pujos and X. André, 1988. Migration of mud banks and changing coastline in French Guyana. *Mar. Geol.* 84, 19-30.

Frostick, L. E. and I. N. McCave, 1979. Seasonal shifts of sediment within an estuary mediated by algal growth. *Est. Coastal Mar. Sci.* 9: 569-576.

Fujimoto, K., T. Miyagi, T. Kikuchi and T. Kawana, 1996. Mangrove habitat formation and response to Holocene sea-level changes on Kosrae Island, *Micronesia. Mangroves and Saltmarshes,* 1, 47-57.

Furukawa, K., and E. Wolanski, 1996. Sedimentation in mangrove forests. *Mangroves and Saltmarshes* 1, 3-10.

Gagliano, S. M., K. J. Meyer-Arendt, and K. M. Wickerm, 1981. Land loss in the Mississippi river deltaic plain. *Trans. Gulf Coast Assoc. Geol. Soc.* 31, 295-300.

Gaily, W. E. and D. K. Hobday, 1996. *Terrigenous Clastic Depositional Systems.* Springer. 489 pp.

Gallenne, B., 1974. Study of fine material in suspension in the estuary of the Loire and its dynamic grading. *Est. Coast. Mar. Sci.* 2, 261-272.

Galloway, R. W., 1981. An inventory of Australia's coastal lands. *Austr. Geogr. Studies* 19, 107-116.

Galloway, R. W., 1985. Northern Territory, in E. C. F. Bird and M. L. Schwartz eds.: *The World's Coastline*. Van Nostrand Reinhold. New York, 949-956.

Gardner, L. R., 1990. Simulation of the diagenesis of carbon, sulfur and dissolved oxygen in salt marsh sediments. *Ecol. Monogr.* 60, 91-111.

Gardner, L. R., and M. Bohn, 1980. Geomorphic and hydraulic evolution of tidal creeks on a subsiding beach ridge plain, North Inlet, S. C. *Mar. Geol.* 34, M91-M97.

Gardner, L. R., L. Thomas, D. Edwards and D. Nelson, 1989. Time series analyses of suspended sediment concentrations at North Inlet, South Carolina. *Estuaries* 12, 211-221.

Garofalo, D., 1980. The influence of wetland vegetation on tidal stream channel migration and morphology. *Estuaries* 3, 258-270.

Garrett, P., 1970. Phanerozoic stromatolites: noncompetitive ecologic restriction by grazing and burrowing animals. *Science* 169 (3941), 171-173

Garrity, S. D., S. C. Levings, and K. A. Burns, 1994. The Galeta oil spill. I. Long-term effects on the physical structure of the mangrove fringe. *Est. Coast. Shelf Sci.* 38, 327-348.

Gebelein, C. D., 1969. Distribution, morphology and accretion rate of recent subtidal algal stromatolites — Bermuda. *J. Sed. Petrol.* 39, 32-49.

Gebelein, C. D., 1976. Open marine subtidal and intertidal stromatolites (Florida, the Bahamas and Bermuda). In M. R. Walter ed.: *Stromatolites. Developments in Sedimentology* 20, Elsevier, ch. 8. 1, 381-388.

Gelfenbaum, G., 1983. Suspended-sediment response to semidiurnal and fortnightly tidal variations in a mesotidal estuary: Columbia river, U.S.A. *Mar. Geol.* 52, 39-52.

Gellatly, D. C., 1970. Cross-bedded tidal megaripples from King Sound (northwestern Australia). *Sed. Geol.* 4, 185-191.

Gibb and partners and NEDECO, 1976. Rangoon Sea Access Channel and associated port improvement study. Final Report vol. II March 1976, 310 p.

Gierloff-Emden, H. G., 1985a. East Germany (German Democratic Republic), in E. C. F. Bird and M. L. Schwartz eds.: *The World's Coastline*. Van Nostrand Reinhold. New York, 315-319.

Gierloff-Emden, H. G., 1985b. Baltic West Germany (Federal Republic of Germany), in E. C. F. Bird and M. L. Schwartz eds.: *The World's Coastline*. Van Nostrand Reinhold. New York, 321-323.

Ginsberg, S. S., and G. M. E. Perillo, 1990. Channel bank recession in the Bahia Blanca estuary, Argentina. *J. Coastal Res.* 6, 999-1009.

Ginsburg, R. N., 1955. Recent stromatolitic sediments from south Florida. *J. Paleont.* 29 (4), 723-724.

Ginsburg, R. N., 1975. *Tidal Deposits: A Casebook of Recent Examples of Fossil Counterparts*. Springer. 428 pp.

Ginsburg, R. N., L. A. Hardie, O. P. Bricker, P. Garrett and H. Wanless, 1977. Exposure index: a quantitative approach to defining position within the tidal zone, in L. A. Hardie ed. *Sedimentation on the Modern Carbonate Tidal Flats of Northwest Andros Island, Bahamas*. John Hopkins University Studies in Geology No. 22, 7-11.

Giusti, E. V., and W. J. Schneider, 1965. The distribution of branches in river networks. *U.S. Geol. Surv. Prof. Paper* 422-G, 10 p.

Glaeser, J. D., 1978. Global distribution of barrier islands in terms of tectonic setting. *J. Geol.* 86, 283-297.

Godfrey, P. J., and M. M. Godfrey, 1972. Comparison of ecological and geomorphic interactions between altered and unaltered barrier island systems in North Carolina, in D. R. Coates ed.: *Coastal Geomorphology*. State University New York, Binghampton.

Golubic, S., 1976. Organisms that build stromatolites. In M. R. Walter ed.: *Stromatolites. Developments in Sedimentology* 20, Elsevier, 113-126.

Golubic, S., R. D. Perkins and K. J. Lukas, 1975. Boring microorganisms and microborings in carbonate substrates, in R. W. Frey ed.: *The Study of Trace Fossils.* Springer Verlag, Berlin, 229-259.

Gonzalez Lastra, J., and J. R. Gonzalez Lastra, 1984. Zonacion ambiental de la Ria de San Vicente de la Barquera (Cantabria). *Thalassas* 2, 43-48.

Gornitz, V., S. Lebedeff, and J. Hansen, 1982. Global sea level trend in the past century. *Science* 215, 1611-1614.

Gostin, V. A., J. R. Hails and A. P. Belperio, 1984. The sedimentary framework of northern Spencer Gulf, South Australia. *Mar. Geol.* 61, 111-138.

Gottschalk, L. C., 1945. Effects of soil erosion on navigation in upper Chesapeake Bay. *Geogr. Rev.* XXXV, 219-238.

Gould, H. R., and E. McFarlan, Jr., 1959. Geologic history of the chenier plain, southwestern Louisiana. *Trans. Gulf Coast Assoc. Geol. Soc.* 9, 261-270.

Gouleau, D. 1975. Les premiers stades de la sédimentation sur les vasières littorales Atlantiques, role de l', émersion. Thèse University Nantes. 241 + 123 pp.

Grant, J., U. V. Bathman and E. L. Mills, 1986. The interaction between diatom films and sediment transport. *Est. Coast. Shelf Sci.* 23, 225-238.

Grant, W. C., 1987. Evidence for Holocene subduction earthquakes along the northern Oregon coast. *Eos* 68, 1239.

Grant, J., 1988. Intertidal bedforms, sediment transport, and stabilization by benthic microalgae, P. L. de Boer, A. van Gelder and S. D. Nio, eds.: in *Environments and Facies*, Reidel Publ., Dordrecht, 499-510.

Greenberg, D. A., and C. L. Amos, 1983. Suspended sediment transport and deposition modeling in the Bay of Fundy, Nova Scotia — a region of potential tidal power development. *Can. J. Fish. Aquat. Sci.* 40 (Suppl. 1), 20-34.

Gregory, M. R., P. F. Ballance, G. W. Gibson and A. M. Ayling, 1979. On how some rays (Elasmobranchia) excavate feeding depressions by jetting water. *J. Sed. Petrol.* 49, 1125-130.

Grindrod, J., and E. G. Rhodes, 1984. Holocene sea-level history of a tropical estuary: Mission Bay, North Queensland, in *Coastal Geomorphology in Australia.* Academic Press Australia, 150-178.

Grinham, D. F., and I. P. Martini, 1983/1984. Sedimentology of the Ekwan shoal, Akimiski Strait, James Bay, Canada. *Sed. Geol.* 37, 273-294.

Groen, P., 1967. On the residual transport of suspended matter by an alternating tidal current. *Neth. J. Sea Res.* 3, 564-574.

Grotjahn, M., H. Michaelis, B. Obert, and H.-J. Stephan, 1983. Höhenentwicklung, Sediment, Vegetation und Bodenfauna in den Landgewinnungsfeldern beiderseits des Cappeler Tiefs (1957 bis 1978). Jahr. ber. Forsch.-Stelle Norderney 34, 94 + 16 p.

Gruet, Y. 1972. Aspects morphologiques et dynamiques de constructions de l'annelide polychète Sabellaria alveolata (Linn,). *Rev. Trav. Inst. Pêches Marit.* 36, 131-161.

Gudelis, V., 1967. Morphogenetic types of the Baltic Sea coasts. *Baltica* 3, 123-145.

Gudelis, V., 1985. Baltic USSR, in E. C. F. Bird and M. L. Schwartz eds.: *The World's Coastline.* Van Nostrand Reinhold. New York, 303-310.

Guilcher, A., 1959. Coastal sand ridges and marshes and their continental environment near Grand Popo and Ouidah, Dahomey, in R. J. Russell ed.: *Second Coastal Geography Conference,* Louisiana State University, Washington D. C., 189-212.

Guilcher, A., 1979. Marshes and estuaries in different latitudes. *Interdisc. Sci. Rev.* 4, 158-168.

Guilcher, A., 1985. Senegal and Gambia, in E. C. F. Bird and M. L. Schwartz eds.: *The World's Coastline.* Van Nostrand Reinhold. New York, 555-560.

Guilcher, A., and J. P. Nicolas, 1954. Observations sur la Langue de Barbarie et les bras du Senegal aux environs de Saint Louis. Bull. Comité d'Océanogr. et d'Etudes des Côtes, Paris, 6, 227-242.

Guilcher, A., and L. Berthois, 1957. Cinq annees d'observations sedimentologiques dans quatre estuaires temoins de l'Ouest de la Bretagne. Revue de Geomorphologie dynamique, VIII, (1-2), 67-86.

Guilcher, A., B. Andrade, and M. M. Dantec, 1982. Diversite morpho-sedimentologique des estuaires de Finistere. Norois, Poitiers, 114 (29), 205-228.

Guillaumont, B., 1991. Utilisation de l'imagerie satellitaire pour les comparaisons spatiales et temporelles en zone intertidale, in M. Elliott and J.-P. Ducrotoy eds.: *Estuaries and Coasts,* ECSA 19 Symp., Olsen and Olsen, 63-68.

Gunther, H.-J, 1963. Untersuchungen zur Verbreitung und Okologie von Uca Tangeri an der SW-Iberischen Kueste. *Z. Morph. Okol. Tiere,* 53, 242-310.

Häntzschel, W. 1975. Coprolites, in C. Teichert ed. Treatise on Invertebrate paleontology Part. W. Miscellanea suppl. 1. Trace fossils and problermatica. *Geol Soc. Amer.* Boulder and University Kansas, Lawrence, 139-143.

Haan, J. H. de, 1931. On the tidal forests of Tjilatjap (in Dutch). *Tectona* 24: 39-76.

Hack, J. T., 1957. Studies of longitudinal stream profiles in Virginia and Maryland. *U.S. Geol. Surv. Prof. Papers* 294B, 45-97.

Hagan, G. M., and B. W. Logan, 1974. History of Hutchinson Embayment tidal flat, Shark Bay, western Australia, in B. W. Logan et al. eds.: *Evolution and Diagenesis of Quaternary Carbonate Sequences, Shark Bay, Western Australia.* Amer. Assoc. Petrol. Geol. Mem. 22, Tulsa, Oklahoma U.S.A., 283-315.

Hahn, S. D., 1980. Variability of tidal range at Inchon. *J. Oceanol. Soc. Korea* 15, 123-128.

Harbison, P., 1984. Regional variation in the distribution of trace metals in modern intertidal sediments of northern Spencer Gulf, south Australia. *Mar. Geol.* 61, 221-247.

Hardie, L. A. ed., 1977. *Sedimentation on the Modern Carbonate Tidal Flats of Northwest Andros Island, Bahamas.* John Hopkins University Press, Baltimore, 196 p.

Hardie, L. A., and P. Garrett, 1977. General environmental setting, in L. A. Hardie, ed.: *Sedimentation on the Modern Carbonate Tidal Flats of Northwest Andros Island, Bahamas.* The John Hopkins University Studies in Geology No. 22, The John Hopkins University Press, Baltimore/London, 12-49.

Hardie, L. A., and R. N. Ginsburg, 1977. Layering: the origin and environmental significance of lamination and thin bedding, in L. A. Hardie ed.: *Sedimentation on the Modern Carbonate Tidal Flats of Northwest Andros Island, Bahamas.* John Hopkins University Studies in Geology No. 22, 50-123.

Harris, P. T., and M. B. Collins, 1985. Bedform distribution and sediment transport paths in the Bristol Channel and Severn estuary, U.K. *Mar. Geol.* 62, 153-166.

Harrison, S. C., 1975. Tidal-flat complex, Delmarva Peninsula, Virginia, in R. N. Ginsburg ed.: *Tidal Deposits.* Springer, 31-38.

Hatcher, B. G., R. E. Johannes and A. I. Robertson, 1989. Review of research relevant to the conservation of shallow tropical marine ecosystems, *Oceanogr. Mar. Biol. Annu. Rev.* 27, 337-414.

Hatton, R. S., R. D. deLaune, and W. H. Patrick, Jr., 1983. Sedimentation, accretion, and subsidence in marshes of the Barataria Basin, Louisiana. *Limnol. Oceanogr.* 28, 494-502.

Hayden, B. P., and R. T. Dolan, 1979. Barrier islands, lagoons and marshes. *J. Sed. Petrol.* 49, 1061-1072.

Hayes, M. O., 1975. Morphology of sand accumulation in estuaries: an introduction to the symposium, in L. E. Cronin ed.: *Estuarine Research,* vol. II, Academic Press. 3-22.

Hayes, M. O., 1985. Atlantic USA — south, in E. C. F. Bird and M. L. Schwartz eds.: *The World's Coastline.* Van Nostrand Reinhold. New York, 207-211.

Hayes, M. O. and J. Michel, 1982. Shoreline sedimentation within a forearc embayment, lower Cook Inlet, Alaska. *J. Sed Petrol.* 52(1), 251-263.

Healey, R. G., K. Pye, D. R. Stoddart, and T. P. Bayliss-Smith, 1981. Velocity variations in salt marsh creeks, Norfolk, England. *Est. Coast. Shelf Sci.* 13, 535-545.

Heip, C., N. K. Goosen, P. M. J. Herman, J. Kromkamp, J. J. Middelburg and K. Soetaert, 1995. Production and consumption of biological particles in temperate tidal estuaries. *Ocean. Mar. Biol. Ann. Rev.* 33, 1-149.

Hemminga, M. A., A. Cattrijsse, and A. Wielemaker, 1996. Bedload and nearbed detritus transport in a tidal saltmarsh creek. *Est. Coast. Shelf Sci.* 42, 55-62.

Hemminga, M. A., and J. Nieuwenhuize, 1990. Seagrass wrack-induced dune formation on a tropical coast (Banc d'Arguin, Mauritania). *Est. Coast. Shelf Sci.* 31, 499-502.

Hemminga, M. A., and J. Nieuwenhuize, 1991. Transport, deposition and in situ decay of seagrasses in a tropical mudflat area (Banc d'Arguin, Mauritania). *Neth. J. Sea Res.* 27 (2), 183-190.

Herd, D. G., T. L. Youd, H. Meyer, J. L. Arango C., W. J. Person, and C. Mendoza, 1981. The great Tumaco, Colombia earthquake of 12 December 1979. *Science* 211, 4481, 441-445.

Hertweck, G. 1972. Georgia coast region, Sapelo Island, USA: sedimentology and biology 5. Distribution and environmental significance of Lebensspuren and in-situ skeletal remains. *Senckenb. Marit.* 4, 125-167.

Hertweck, G. 1994. Zonation of benthos and Lebensspuren on tidal flats of the Jade Bay, southern North Sea. *Senckenberg. Marit.* 24, 157-170.

Heydorn, A. E. F., and B. W. Flemming, 1985. South Africa, in E. C. F. Bird and M. L. Schwartz eds.: *The World's Coastline.* Van Nostrand Reinhold, New York, 653-667.

High Jr., L. R., 1975. Geomorphology and sedimentology of Holocene coastal deposits, Belize, in K. F. Wantland and W. C. Pusey III, eds.: Belize Shelf — *Carbonate Sediments, Clastic Sediments, and Ecology.* A. A. P. G. Studies in Geology 2, 53-96.

Hine, A. C., 1975. Bedform distribution and migration patterns on tidal deltas in the Chatham Harbor estuary, Cape Cod, Massachusetts, in L. E. Cronin ed.: *Estuarine Research,* II, Academic Press Inc. New York. 235-252.

Hine, A. C., D. F. Belknap, J. G. Hutton, E. B. Osking, and M. W. Evans, 1988. Recent geological history and modern sedimentary processes along an incipient, low-energy, epicontinental-sea coastline: northwest Florida. *J. Sed. Petrol.,* 58, 567-579.

Hjulström, F. 1935. Studies of the morphological activity of rivers as illustrated by the river Fyris. Bull. Geol. Inst. Uppsala 25, 221-527.

Hobbs, III, C. H., J. P. Halka, R. T. Kerhin, and M. J. Carron, 1992. Chesapeake Bay sediment budget. *J. Coast. Res.* 8, 292-300.

Hoekstra, P., 1989. River outflow, depositional processes and coastal morphodynamics in a monsoon-dominated deltaic environment, East Java, Indonesia. Thesis Utrecht University, 215 p.

Holland, A. F., R. G. Zingmark and J. M. Dean, 1974. Quantitative evidence concerning the stabilization of sediments by marine benthic diatoms. *Mar. Biol.* 27, 191-196.

Hopley, D., 1982. *The Great Barrier Reef.* Wiley Interscience, 453 p.

Hopley, D., 1982. *Geomorphology of the Great Barrier Reef.* Wiley Interscience.

Hopley, D., 1985. Queensland, in E. C. F. Bird and M. L. Schwartz eds.: *The World's Coastline.* Van Nostrand Reinhold. New York, 957-967.

Horodyski, R., B. Bloeser and S. P. Vonder Haar, 1977. Laminated algal mats from a coastal lagoon, Laguna Mormona, Baja California, Mexico. *J. Sed. Petrol.* 47, 680-696.

Horton, R. E., 1945. Erosional development of streams and their drainage basins; hydrophysical approach to quantitative morphology. *Bull. Geol. Soc. Amer.* 56, 275-370.

Houthuys, R., G. de Moor, and J. Somme, 1993. The shaping of the French-Belgian North Sea coast throughout Recent geology and history, in R. Hillen and H. J. Verhagen, eds.: *Coastlines of the Southern North Sea.* ASCE, New York, 27-40.

Houwing, E.-J., W. E. van Duin, Y. Smit-van der Waaij, K. Dijkema and J. H. J. Terwindt, submitted. The bio- and hydrodynamics of the intertidal pioneer zone along the Dutch Wadden Sea. *Neth. J. Sea Res.*

Howard, J. D., and R. W. Frey, 1985. Physical and biogenic aspects of backbarrier and sedimentary sequences, Georgia coast, U.S.A. *Mar. Geol.* 63,77-127.

Howard, J. D., T. V. Mayou and R. W. Heard 1977. Biogenic sedimentary structures formed by rays. *J. Sed. Petrol.* 47, 339-346.

Hsu, T. L., 1962. A study on the coastal geomorphology of Taiwan. *Proc. Geol. Soc. China* 5, 29-45.

Hsu, T. L., 1965. The tidal flat of the Chiayi area. *Bull. Geol. Survey Taiwan,* 16, 18-54.

Hubbard, D. K., G. Oertel and D. Nummedal, 1979. The role of waves and tidal currents in the development of tidal inlet sedimentary structures and sand body geometry: examples from North Carolina, South Carolina and Georgia. *J. Sed. Petrol.* 49, 1073-1092.

Hummel, H. 1985. Food intake of *Macoma balthica* (Mollusca) in relation to seasonal changes in its potential food on a tidal flat in the Dutch Wadden Sea. *Neth. J. Sea Res.* 19, 52-76.

Hutchinson, I., 1982. Vegetation-environment relations in a brackish marsh, Lulu Island, Richmond, B. C. *Can. J. Bot.* 60, 452-462.

Hutchinson, S. M., and D. Prandle, 1994. Siltation in the saltmarsh of the Dee estuary derived from 37-Cs analysis of shallow cores. *Est. Coast. Shelf Sci.* 38, 471-478.

Huy, D. V., and T. D. Thanh, 1994. Shoreline development in Haiphong — Quangyen area during Holocence by the investigation of old beach ridge systems. Haiphong Inst. of Oceanol.: *Marine Environment and Resources* (1991-1993), Hanoi, Science Techn. Publ. House, 61-65. (in Vietnamese).

Imberger, J., T. Berman, R. R. Christian, E. B. Sherr, D. E. Whitney, L. R. Pomeroy, R. G. Wiegert and W. J. Wiebe, 1983. The influence of water motion on the distribution and transport of materials in a salt marsh estuary. *Limnol. Oceanogr.* 28, 201-214.

Intasen, W. and P. S. Roy, 1996. Large-scale coastal behaviour: the geological measurement of long-term coastal change. *Intern. Report Univ. New South Wales,* 164 pp.

Isla F. I., F. E. Vilas, G. G. Bujalesky, M. Ferrero, G. Gonzalez Bonorino and A. Arche Miralles, 1991. Gravel drift and wind effects on the macrotidal San Sebastian Bay, Tierra del Fuego, Argentina. *Mar. Geol.* 97 (1/2), 211-224.

Isla, F. I., 1995. Coastal Lagoons, in G. M. E. Perillo ed.: *Geomorphology and Sedimentology of Estuaries.* Developm. in Sediment. 53, Elsevier, 241-273.

Islam, M. A. 1978. The Ganges-Brahmaputra river delta. *J. University Sheffield Geol. Soc.* 7(3), 116-122.

Jacobson, H. A., 1988. Historical development of the saltmarsh at Wells, Maine. *Earth Surf. Process. Landf.* 13, 475-486.

Jacobson, H. A., G. L. Jacobson, and J. T. Kelley, 1987. Distribution and abundance of tidal marshes along the coast of Maine. *Estuaries* 10, 126-131.

Jago, C. F., 1980. Contemporary accumulation of marine sand in a macrotidal estuary, southwest Wales. *Sed. Geol.* 26, 21-49.

James, W. R., and W. C. Krumbein, 1969. Frequency distributions of stream link lengths. *J. Geol.* 77, 544-565.

Jarvis, J., and C. Riley, 1987. Sediment transport in the mouth of the Eden estuary. *Est. Coast. Shelf Sci.* 24, 463-481.

Jelgersma, S., 1996. Land subsidence in coastal lowlands, in J. D. Milliman and B. U. Haq eds.: *Sea-Level Rise and Coastal Subsidence.* Kluwer Academic Publ., Dordrecht, 47-62.

Jennings, J. N., and E. C. F. Bird, 1967. Regional geomorphological characteristics of some australian estuaries, in G. H. Lauff ed.: *Estuaries.* A. A. A. S. Publ. 83, Washington D. C., 121-128.

Jennings, J. N., and R. J. Coventry, 1973. Structure and texture of a gravelly barrier island in the Fitzroy estuary, western Australia, and the role of mangroves in the shore dynamics. *Mar. Geol.* 15, 145-167.

John, D. M., and G. W. Lawson, 1990. A review of mangrove and coastal ecosystems in West-Africa and their possible relationships. *Est. Coast. Shelf Sci.* 31, 505-518.

Johnson, H. D. and C. T. Baldwin, 1996. Shallow siliciclastic seas, in Reading, H. G., ed.: *Sedimentary Environments; Processes, Facies and Stratigraphy.* Blackwell Science, Oxford, p. 232-280.

Johnston, W. A., 1921. *Sedimentation of the Fraser River Delta.* Geol. Surv. Can. Mem. 125, 46 p.

Jorgensen, C. B. 1966. *Biology of Suspension Feeding.* Pergamon Press, Oxford. 357 pp.

Junghuhn, F., 1851-1854. Java, seine Gestalt, Pflanzendecke und innere Bauart.

Kaczorowski, R. T., 1980. The Louisiana chenier system — some preliminary reinterpretations and refinements. *Trans. Gulf Coast Assoc. Geol. Soc.* 30, 427-430.

Kamermans, P. 1994. Similarity in food source and timing of feeding in deposit and suspension- feeding bivalves. *Mar. Ecol. Prog. Ser.* 104, 63-75.

Kamps, L. F., 1962. Mud distribution and land reclamation in the Eastern Wadden Shallows. Rijkswaterstaat Comm. 4, Den Haag, 73 p.

Kaplin, P., 1985. Pacific USSR, in E. C. F. Bird and M. L. Schwartz eds.: *The World's Coastline.* Van Nostrand Reinhold, New York, 857-862.

Ke, X., G. Evans, and M. B. Collins, 1996. Hydrodynamics and sediment dynamics of the Wash embayment, eastern England. *Sedimentology* 43, 157-174.

Kestner, F. J. T., 1975. The loose-boundary regime of The Wash. *Geogr. J.,* 141, 388-414.

Khalaf, F. I., and M. Ala, 1980. Mineralogy of the recent intertidal muddy sediments of Kuwait — Arabian Gulf. *Mar. Geol.* 35, 331-342.

Kim, J.-L., and S.-Ch. Park, 1985. Intertidal flat sediments and characteristic sedimentary structures in the Changgu Bay, west coast of Korea. *J. Oceanol. Soc. Korea* 20, 43-49.

Kingo Jacobsen, N., 1980. Form elements of the Wadden Sea area, in K. S Dijkema, H.-E. Reineck and W. J. Wolff eds., *Geomorphology of the Wadden Sea Area.* Rep. 1 Wadden Sea Working Group, 5071.

Kirby, R., 1992. Effects of sea-level rise on muddy coastal margins, in D. Prandtle ed.: *Dynamics and Exchanges in Estuaries and the Coastal Zone.* AGU, Washington.

Kirby, R., and W. R. Parker, 1982. A suspended sediment front in the Severn estuary. *Nature* 295, 396-399. *Int. Explor. Mer* 181, 59-63.

Kirby, R., and W. R. Parker, 1983. Distribution and behavior of fine sediment in the Severn estuary and inner Bristol Channel, U.K. *Can. J. Fish. Aquat. Sci.* 40 (Suppl. 1), 83-95.

Kjerve, B., and K. E. Magill, 1989. Geographic and hydrodynamic characteristics of shallow coastal lagoons. *Mar. Geol.* 88, 187-199.

Klingebiel, A. and J. Gayet, 1995. Fluvio-lagoonal sedimentary sequences in Leyredelta and Arcachon Bay, and Holocene sea level variations along the aquitaine coast (France). *Quat. Int.* 29/30, 111-117.

Knight, R. J., 1980. Linear sand bar development and tidal current flow in Cobequid Bay, Bay of Fundy, N. S., in S. B. McCann ed.: *The Coastline of Canada*. Canada Survey Paper 80-10, 123-152.

Knight, R. J., and R. W. Dalrymphe, 1975. Intertidal sediments from the south shore of Cobequid Bay, Bay of Fundy, Nova Scotia, Canada, in R. N. Ginsburg ed.: *Tidal Deposits*, Springer Verlag, Berlin-Heidelberg, Ch. 6., 47-55.

Knighton, A. D., K. Mills and C. D. Woodroffe, 1991. Tidal creek extension and saltwater intrusion in northern Australia. *Geology* 19, 831-834.

Knighton, A. D., C. D. Woodroffe and K. Mills, 1992. The evolution of tidal creek networks, Mary river, northern Australia. *Earth Surf. Proc. Landf.* 17, 167-190.

Komar, P. D., 1985. Oregon, in E. C. F. Bird and M. L. Schwartz eds.: *The World's Coastline*. Van Nostrand Reinhold, New York, 23-26.

König, D., 1948. *Spartina townsendii* an der Westküste von Schleswig-Holstein. *Planta* 36, 34-70.

Kolb, Ch. R., and J. R. van Lopik, 1966. Depositional environments of the Mississippi river deltaic plain — southeastern Louisiana, in M. L. Shirley and J. A. Ragsdale eds.: *Deltas in Their Geologic Framework*. Houston Geol. Soc., 17-61.

Kolb, Ch. R., and W. K. Dornbusch, 1975. The Mississippi and Mekong deltas — a comparison, in M. L. Broussard ed.: *Deltas, Models for Exploration*. Houston Geol. Soc., 193-207.

Koopmans, B. N., 1964. Geomorphological and historical data of the lower course of the Perak River (Dindings). JMBRAS 37, 175-191.

Krumbein, W. E. 1987. Das Farbstreifensandwatt: Bau, Struktur und Erdgeschichte von Mikrobenwatten, in G. Gerdes, W. E. Krumbein and H.-E. Reineck eds.: Mellum- Portrait einer Insel. Kramer, Frankfurt am Main, 170-187.

Krumbein W. E., D. M. Paterson and L. J. Stal (eds). 1994. *Biostabilization of Sediments*. BIS-Verlag. University Oldenburg. 529 pp.

Kuehl. S. A., T. M. Hariu and W. S. Moore. 1989. Shelf sedimentation of the Ganges-Brahmaputra river system: evidence for sediment bypassing to the Bengal Fan. *Geology*, 17. 1132-1135.

Kulm, L. D., and J. V. Byrne, 1967. Sediments of Yaquina Bay, Oregon, in G. H. Lauff ed.: *Estuaries*. A. A. A. S. Publ. 83, Washington D. C., 226-238.

Labourg, P. J., E. J. Soriano-Sierra and I. Anby, 1995. Evolution recente de la vegetation intertidale du delta de l'Eyre, in *Le Delta de la Leyre*. Trav. Coll. Sc. 21-23 Okt. 1993. 47-56.

Larcombe, P., and P. V. Ridd, 1996. Dry season hydrodynamics and sediment transport in a mangrove creek, in Ch. Pattiaratchi ed.: *Mixing in Estuaries and Coastal Seas*. Coast. Estuar. Studies 50, Amer. Geophys. Union, Washington D. C., 388-404.

Larsonneur, C., 1975. Tidal deposits, Mont Saint Michel Bay, France, in R. N. Ginsburg ed.: *Tidal Deposits*, Ch. 3, 21-30.

Larsonneur, C., and co-authors, 1994. The Bay of Mont-Saint-Michel: a sedimentation model in a temperate macrotidal environment. *Senckenbergiana maritima* 24, 3-63.

Lauriol, B. and J. T. Gray, 1980 Processes responsible for the concentration of boulders in the intertidal zone of Leaf Basin, Ungava, in S. B. McCann ed.: *The Coastline of Canada*. Canada Geol. Survey Paper 80-10, 281-290

Lee, Ch.-B., Y.-A. Park and Ch.-H. Koh, 1985. Sedimentology and geochemical properties of the Banweol area in the southern part of Kyeonggi Bay, Korea. *J. Oceanol. Soc. Korea* 20, 20-29.

Lee, Ch.-B., H.-Rh. Yoo and Ky.-S. Park, 1992. Distribution and properties of intertidal surface sediments of Kyeonggi Bay, west coast of Korea. *J. Oceanogr. Soc. Korea* 27, 277-289.

Lee, H. and R. C. Swartz. 1980. Biological processes affecting the distribution of pollutants in marine sediments. Part II. Biodeposition and bioturbation, in R. A. Baker ed. *Contaminants in Sediments* 2. Ann Arbor Science Publ. Ann Arbor, 555-606.

Leeder, M. R., 1982. *Sedimentology: Process and Product.* Allen and Unwin. 344 pp.

Lees, A., 1975. Possible influences of salinity and temperature on modern shelf carbonate sedimentation. *Mar. Geol.* 19, 159-198.

Lees. B. G., 1992. The development of a chenier sequence on the Victoria delta, Joseph Bonaparte Gulf, northern Australia. *Mar. Geol.* 103, 215-224.

Leonard, L. A., A. C, Hine, and M. E. Luther, 1995. Surficial sediment transport and deposition processes in a Juncus roemerianus marsh, west-central Florida. *J. Coastal Res.* 11, 322-336.

Leontiev, O. K., 1956. Coasts with windy flats as a special type of coast. *Proc. Ac. Sci. USSR,* 5, 67-78. (in Russian)

Leontiev, O. K., 1985. Caspian USSR, in E. C. F. Bird and M. L. Schwartz eds.: *The World's Coastline.* Van Nostrand Reinhold. New York, 481-486.

Leopold, L. B., and M. G. Wolman, 1960. River meanders. *Bull. Geol. Soc. Am.* 71, 769-794.

Leopold, L. B., M. G. Wolman, and J. P. Miller, 1964. *Fluvial Processes in Geomorphology.* Freeman, San Francisco, 522 p.

Levings, S. C., S. D. Garrity, and K. A. Burns, 1994. The Galeta oil spill. III. Chronic reoiling, long-term toxicity of hydrocarbon residues and effects on epibiota in the mangrove fringe. *Est. Coast. Shelf Sci.* 38, 365-395.

Levinton, J. S. 1982. *Marine Ecology.* Prentice Hall, Englewood Cliffs N.J. 526 pp.

Li Guangtian, Fu Wenxia, and Jia Xijun, 1986. The comprehensive characteristic of the Liaodong peninsular tidal flat. *Acta Geogr. Sinica* 41, 262-273.

Li, B., Z. Yang and Q. Xie., 1993(1992). Suspended sediment transport of the Hangzhou Bay and its related hydronamic analyses. *Acta Oceanol. Sinica,* 12(1), 39-50.

Li, J. F., 1991. The rule of sediment transport on the Nanhui tidal flat in the Changjiang estuary. *Acta Oceanol. Sin.* 10, 117-127.

Li, Y., and Xie, Q. Ch., 1993a. Zonation of sediment and sedimentary rate on Andong tidal flat in Hangzhou Bay, China. *Donghai Mar. Sci.* 10, 21-33. (English abstract)

Li, Y., and Xie, Q. Ch., 1993b. Dynamical development of the Andong tidal flat in Hangzhou Bay, China. *Donghai Mar. Sc.* 11, 25-33. (English abstract)

Lincoln, J. M., and D. M. Fitzgerald, 1988. Tidal distortions and flood dominance at five small tidal inlets in southern Maine. *Mar. Geol.* 82, 133-148.

Lindsay, P., P. W. Balls, and J. R. West, 1996. Influence of tidal range and river discharge on suspended particulate matter fluxes in the Forth estuary (Scotland). *Est. Coast. Shelf Sci.* 42, 63-82.

Lisitzin, E., 1974. *Sea Level Changes.* Elsevier Oceanographic Series 8, 286 pp.

Little-Gadow, S., and H.-E. Reineck, 1974. Diskontinuierliche Sedimentation von Sand und Schlick in Wattensedimenten. *Senckenbergiana maritima* 65, 149-159.

Liu C. Z., Wu, L. Ch., and Cao M., 1987. Sedimentary characteristics of cheniers in southern Changjiang delta and their origin and age determination. *Acta Oceanol. Sin.* 6, 405-412.

Liu Cangzi and H. Jesse Walker, 1989. Sedimentary characteristics of cheniers and the formation of the chenier plains of East China. *J. Coastal Res.* 5, 353-368.

Logan, B. W., R. Rezak and R. N. Ginsburg, 1964. Classification and environmental significance of algal stromatolites. *J. Geol.* 72, 68-83.

Logan, B. W., and D. E. Cebulski, 1970. Sedimentary environments of Shark Bay, Western Australia, in Logan, B. W., et al.: *Carbonate Sedimentation and Environments, Shark Bay, Western Australia,* A. A. P. G. Memoir 13, 1-37.

Logan, B. W., P. Hoffman and C. D. Gebelein, 1974. Algal mats, cryptalgal fabrics, and structures, Hamelin Pool, Western Australia, in B. W. Logan et al.: *Evolution and Diagenesis of Quaternary Carbonate Sequences, Shark Bay, Western Australia,* A. A. P. G. Memoir 22, 140-194.

Lohmann, H. 1896. Die Appendicularien der Plankton Expedition. Ergebn. Plankt. Exp. Humboldt- Stiftung. 2 (E. c.), 1-148.

Louters, T., and F. Gerritsen, 1994. Het mysterie van de wadden. Report RIKZ-94. 040, 69 p. (in Dutch)

Lubowe, J. K., 1964. Stream junction angles in the dendritic drainage pattern. *Amer. J. Sci.* 262, 325-339.

Lüders, K., 1934. Ueber das Wandern der Priele. Abh. Naturwiss. Verein Bremen 19-32.

Lugo, A. E., and S. C. Snedaker, 1974. The ecology of mangroves. *Ann. Rev. Ecol. and System.* 5, 39-64.

Luternauer, J. L., 1980. Genesis of morphological features on the western delta front of the Fraser river, British Columbia — Status of knowledge, in S. B. McCann ed.: *The Coastline of Canada,* Geol. Survey of Canada Paper 80-10, 381-391.

Luternauer, J. L., and J. W. Murray, 1973. Sedimentation on the western Delta-front of the Fraser River, British Columbia. *Can. J. Earth Sci,* 10, 1642

Luternauer, J. L., R. J. Atkins, A. I. Moody, H. F. L. Williams and J. W. Gibson, 1995. Salt marshes, in G. M. E. Perillo ed.: Geomorphology and Sedimentology of Estuaries. *Developm. in Sediment.* 53, Elsevier, 307-332.

L'Yavanc, J., and Ph. Bassoulet, 1991. Nouvelle approche dans l'étude de la dynamique sedimentaire des estuaires macrotidaux a faible debit fluvial. *Oceanol. Acta* 11, 129-136.

Lynch, J. C., J. R. Meriwether, B. A. McKee, F. Vera-Herrera, and R. R. Twilley, 1989. Recent accretion in mangrove ecosystems based on 137-Cs and 210 Pb. *Estuaries* 12, 284-299.

Ma Hongguang, Yu Zhiying and G. C. Cadée, 1995. Macrofauna distribution and bioturbation on tidal confluences of the Dutch Wadden Sea. *Neth. J. Aquat. Ecol.* 29, 167-176.

MacPherson, H., and P. G. Kurup, 1981. Wave damping at the Kerala mud banks. Indian *J. Mar. Sci.* 10, 154-160.

Macdonald, K. B., 1977. Plant and animal communities of Pacific north american salt marshes, in V. J. Chapman ed.: *Wet Coastal Ecosystems,* Elsevier, 167-191.

Macnae, W., 1968. A general account of the fauna and flora of mangrove swamps and forests in the Indo-Westpacific region. *Adv. Mar. Biol.* 6: 72-270.

Magnier, Ph., T. Oki, and L. Witoelar Kartaadiputra, 1975. *The Mahakam Delta, Kalimantan, Indonesia.* Proc. 9th World Petrol. Congress, Tokyo, 239-250.

Mallik, T. K., K. K. Mukherji, and K. K. Ramachandran, 1988. Sedimentology of the Kerala mud banks (fluid muds?). *Mar. Geol.* 80, 99-118.

Mangum, C. P., S. L. Santos, and W. R. Rhodes Jr., 1968. Distribution and feeding in the onuphid polychaete, *Diopatra cuprea* (Bosc). *Mar. Biol.* 2, 33-40.

Mann, K. H. 1972. Macrophyte production and detritus food chains in coastal waters. *Mem. Ist. Ital. Idrobiol.* 29 (Suppl.), 353-383.

Mao, Zh. Ch., 1993. Wave caused by typhoon and its impacts on the erosion-accumulation of the eastern Chongming flat. *Donghai Mar. Sci.* 11, 8-16.

Marker, M. E., 1967. The Dee estuary: its progressive silting and saltmarsh development. *Transact. Inst. Brit. Geogr.* 241, 65-71.

Marshall, J. R., 1964. The morphology of the upper Solway salt marshes. *Scottish Geogr. Mag.* 78, 81-99.

Martini, I. P., 1981. Morphology and sediments of the emergent Ontario coast of James Bay, Canada. *Geogr. Annaler* 63 A(1-2), 81-94.

Martini, I. P., D. W. Cowell, and G. M. Wickware, 1980. Geomorphology in southern James Bay: a low energy, emergent coast, in S. B. McCann ed.: *The Coastline of Canada*. Geol. Survey Canada paper 86 (10), 293-301.

Mathew, J., and M. Baba, 1995. Mudbanks of the southwest coast of India. II. Wave-Mud interactions. *J. Coastal Res.* 11, 179-187.

Mathew, J., M. Baba and N. P. Kurian, 1995. Mudbanks of the southwest coast of India. *J. Coast. Res.* 11, 168-178.

Mathews, W. H., and F. P. Shepard, 1962. Sedimentation of Fraser River delta, British Columbia. *Bull. Amer. Assoc. Petrol. Geolog.*, 46, 1416-1443.

Mazure, J. P. ed., 1974. Rapport van de Waddenzeecommissie. Rijkswaterstaat, Den Haag. 328 p. (in Dutch).

McCann, S. B., 1980. Classification of tidal environments, in S. B. McCann ed.: *Sedimentary Processes and Animal-Sediment Relationships in Tidal Environments*. Geol. Assoc. Can. Short Course Notes 1, 1-24.

McCann, S. B., 1985. Atlantic Canada, in E. C. F. Bird and M. L. Schwartz eds.: *The World's Coastline*. Van Nostrand Reinhold, New York, 235-240.

McCann, S. B., J. E. Dale and P. B. Hale, 1981. Subarctic tidal flats in areas of large tidal range, southern Baffin Island, Eastern Canada. *Géogr. Phys. et Quaternaire* 35(2), 183-204.

McCave, I. N., and A. C. Geiser, 1978. Megaripples, ridges and runnels on intertidal flats of the Wash, England. *Sedimentology* 26, 353-369.

McGowen, J. H., and A. J. Scott, 1975. Hurricanes as geologic agents, in L. E. Cronin ed.: *Estuarine Research,* Academic Press Inc. New York-London, II, 23-46.

McIvor, C. C., and T. J. Smith III, 1995. Differences in the crab fauna of mangrove areas at a southwest Florida and a northeast Australia location: implications for leaf litter processing. *Estuaries* 18, 591-597.

McLean, R. F., 1978. Recent coastal progradation in New Zealand, in J. L. Davies and M. A. J. Williams eds.: *Landform Evolution in Australasia*. Australian Ntl. University Press, Canberra, 168-196.

McLean, R. F., 1985. New Zealand, in E. C. F. Bird and M. L. Schwartz eds.: *The World's Coastline*. Van Nostrand Reinhold, New York, 981-994.

McMillan, C., 1975. Adaptive differentiation to chilling in mangrove populations, in G. Walsh, S. Snedaker and H. Teas eds.: Proc. Int. Symp. Biol. Management of Mangroves, Honolulu Oct. 1974: 62-68.

Meade, R. H., 1969. Landward transport of bottom sediments in estuaries of the Atlantic coastal plain. *J. Sedim. Petrol.* 39, 222-234.

Meadows P. S. and A. Reid, 1966. The behaviour of Corophium volutator. *J. Zool. Lond.* 150, 387-399.

Meadows, P. S. and J. Tait, 1985. Bioturbation, geotechnics and microbiology at the sediment water interface in deepsea sediments, in P. E. Gibbs ed. Proc 19th EMBS Symp. Cambridge University Press. Cambridge, 191-199.

Meadows, P. S. and A. Tufail, 1986. Bioturbation, microbial activity and sediment properties in an estuarine ecosystem. *Proc. R. Soc. Edinb.* (B) 90, 129-142.

Meadows, P. S., J. Tait and S. A. Hussain, 1990. Effects of estuarine infauna on sediment stability and particle sedimentation. *Hyrobiologia* 190, 263-266.

Meadows, P. S. and A. Meadows, 1991. The geotechnical and geochemical implications of bioturbation in marine sedimentary ecosystems. *Symp. Zool. Soc. London* 63, 157- 181.

Meckel, L. D., 1975. Holocene sand bodies in the Colorado delta area, northern Gulf of California, in M. L. Broussard ed.: Deltas, models for exploration. *Houston Geol. Soc.,* 239-265.

Mehta, A. J., 1989. On estuarine cohesive sediment suspension behaviour. *J. Geophys. Res.* 94, C10, 14. 303-14. 314.

Meldahl, K. H. 1987. Sedimentologic, stratigraphic, and taphonomic implications of biogenic stratification. *Palaios* 2, 350-358.

Meulenbergh, J., 1974. La Mangrove Zairoise. Academic Roy. Sci. d'outre-mer, Cl. Sci. Techn. (N. S.) 17 (8), 1-86.

Meyers, M. B., F. Fossing and E. N. Powell, 1987. Microdistribution of interstitial meiofauna, oxygen and sulfide gradients, and the tubes of macro-infauna. *Mar. Ecol. Prog. Ser.* 35, 223-241.

Meyers, M. B., E. N. Powell and H. Fossing. 1988. Movement of oxybiotic and thiobiotic meiofauna in response to changes in pore water oxygen and sulfide gradients around macro- infaunal tubes.- *Mar. Biol.* 98, 395-414.

Miller, D. C. 1961. The feeding mechanism of fiddler crabs, with ecological considerations of feeding adaptations. *Zoologica* 46, 89-100.

Miller, J. A., 1975. Facies characteristics of Laguna Madre wind-tidal flats, in R. N. Ginsburg ed.: *Tidal Deposits.* Springer Verlag, Berlin-Heidelberg, 67-73.

Milliman, J. D., 1980. Sedimentation in the Fraser river and its estuary, southwestern British Columbia (Canada). *Est. Coast. Mar. Sci.* 10, 609-633.

Milliman, J. D., J. Butenko, J. P. Barbot and J. Hedberg, 1982. Depositional patterns of modern Orinoco/Amazon muds on the northern Venezuelan shelf. *J. Mar. Res.* 40, 643-657.

Milliman, J. D. and R. H. Meade, 1983. Worldwide delivery of river sediment to the oceans. *J. Geology* 91, 1-21.

Milliman, J. D. and J. M. P. Syvitski, 1992. Geomorphic and tectonic control of sediment discharge to the ocean: the importance of small mountainous rivers. *J. Geology,* 525-544.

Moerner, N.-A., 1971. The Holocene eustatic sea level problem. *Geol. Mijnb.* 50, 699-702.

Moerner, N.-A., 1995. Recorded sea level variability in the Holocene and expected future changes, in D. Eisma ed.: *Climate Change, Impact on Coastal Habitation,* Lewis Publ., Boca Raton, 17-28.

Moni, N. S., 1970. Study of mudbanks along the southwest coast of India. Proc. 12th Conf. Coastal Eng. ASCE, 2, 739-750.

Montague, C. L. 1980. A natural history of temperate western Atlantic fiddler crabs (genus Uca) with reference to their impact on the salt marsh. *Contr. Mar. Sci.* 23, 25-55.

Monty, C. L. V., 1967. Distribution and structure of recent stromatolitic algal mats, Eastern Andros Island, Bahama. *Ann. Soc. Geol. Belgique* 90 (3), 55-100.

Monty, C. L. V., 1972. Recent algal stromatolitic deposits, Andros Island, Bahamas. Preliminary Report. *Geol. Rundschau* 61, 742-783.

Moore, T. C and P. C. Scrutton, 1957. Minor internal structures of some Recent unconsolidated sediments. *Bull. Am. Assoc. Pet. Geol.* 41, 2723-2751.

Moreira, M. E. Soares de Albergaria, 1992. Recent saltmarsh changes and sedimentation rates in the Sado estuary, Portugal. *J. Coastal Res.* 8, 631-640.

Mosetti, F., 1971. Considerazoni sulle cause dell'acqua alta a Venezia. Boll. Geofisica Teor. ed Appl. 50, 169-184. (in Italian)

Mudie, P. J., 1974. The potential economic uses of halophytes, in R. J. Reimold and W. H. Queen, eds.: *Ecology of Halophytes.* Academic Press, New York, 565-597.

Mukherjee, K. N., 1969. Nature and problems of neo-reclamation in the Sundarbans. *Geogr. Review of India,* 31(4), 1-9.

Murça Pires, J., 1965. The estuaries of the Amazon and Oiapoque rivers and their floras, in *Scientific Problems of the Humid Tropical Zone,* Proc. Dacca Symp. Feb.-March 1964, UNESCO, 211-218.

Myers, A. C., 1972. Tube-worm-sediment relationships of Diopatra cuprea (Polychaeta: Onuphidae). *Mar. Biol.* 17, 350-356.

Myrick, J. L. and K. W. Flessa, 1996. Bioturbation rates in Bahia La Choya, Sonora, Mexico. *Ciencias Marinas* 22, 23-46.

Nagaraja, V. N., 1965. Hydrometeorological and tidal problems of the deltaic areas in India, in *Scientific Problems of the Humid Tropical Zone.* Prod. Dacca Symp. Feb.-March 1964. 115-121.

NEDECO, 1965. A study on the siltation of the Bangkok Port Channel, vol. II, *The Field Investigations.* The Hague, Holland.

NEDECO, 1968. Surinam Transportation Study: Report on Hydraulic Investigation. Den Haag, 293 pp.

Neira, C., and M. Rackemann, 1996. Black spots produced by buried macroalgae in intertidal sandy sediments of the Wadden Sea: effects on the meiobenthos. *J. Sea Res.* 36, 153-170.

Neu, T. R. 1994. Biofilms and microbial mats, in W. E. Krumbein, D. M. Paterson and L. J. Stal eds. *Biostabilization of Sediments.* BIS-Verlag. University Oldenburg, 9-15.

Newell, R. 1965. The role of detritus in the nutrition of two marine deposit feeders, the prosobranch Hydrobia ulvae and bivalve Macoma balthica. *Proc. Zool. Soc. Lond.* 144, 25-45.

Newell, N. D., J. Imbrie, E. G. Purdy and D. L. Thurber, 1959. Organism communities and bottom facies, Great Bahama Bank. *Bull. Am. Mus. Nat. Hist.* 117, 177-228.

Nichols, F. H., and J. Thompson, 1985. Persistence of an introduced mudflat community in south San Francisco Bay, California. *Mar. Ecol. Progr. Ser.* 24, 83-97.

Nichols, M. M., 1989. Sediment accumulation rates and relative sea-level rise in lagoons. *Mar. Geol.* 88, 201-219.

Nienhuis, P. H., 1994. Causes of the eelgrass wasting disease: Van der Werff's changing theories. *Neth. J. Aquat. Ecol.* 28 (1), 55-61.

Nikodic, J., 1981. Dynamique sédimentaire dans la partie occidentale de la Baie du Mont Saint-Michel. Thèse University Nantes, 180 p.

Nio, S, J. H. van den Berg, M. Goesten, and F. Smulders, 1980. Dynamics and sequential analysis of a mesotidal shoal and intershoal channel complex in the Eastern Scheldt (southwestern Netherlands). *Sed. Geol.* 26, 263-279.

Nummedal, D., G. F. Oertel, D. K. Hubbard, and A. C. Hine, 1977. *Tidal Inlet Variability — Cape Hatteras to Cape Canaveral.* Proc. Conf. Coastal Sed. A. S. C. E., Charleston S. C., 543-562.

Nummedal, D., J. M. Coleman, R. Boyd and S. Penland, 1985. Louisiana, in E. C. F. Bird and M. L. Schwartz eds.: *The World's Coastline.* Van Nostrand Reinhold. New York, 147-153.

Nutalaya, P., R. N. Yong, T. Chumnankit, and S. Buapeng, 1996. Land subsidence in Bangkok during 1978-1988, in J. D. Milliman and B. U. Haq eds.: *Sea-Level Rise and Coastal Subsidence.* Kluwer Academic Publ., Dordrecht, 105-130.

Oele, E., 1969. The quaternary geology of the Dutch part of the North Sea, north of the Frisian Isles. Geol. en Mijnb. 48, 467-480.

Oenema, O., and R. D. DeLaune, 1988. Accretion rates in salt marshes in the eastern Scheldt, south-west Netherlands. *Est. Coast. Shelf Sci.* 26, 379-394.

Oertel, G. F., and S. P. Leatherman eds., 1985. Barrier Islands. *Mar. Geol.* 63 Spec. Issue.

Oertel, G. F., M. S. Kearney, S. P. Leatherman and H.-J. Woo, 1989. Anatomy of a barrier platform: outer barrier lagoon, southern Delmarva peninsula, Virginia. *Mar. Geol.* 88, 303-318.

Officer, C. B., 1976. *Physical Oceanography of Estuaries and Associated Coastal Waters.* John Wiley and Sons, New York, 465 p.

Ohya, M., 1965. Comparative study on the geomorphology and flooding in the plains of the Cho-Shui-Chi, Chao-Phya, Irrawaddy and Ganges, in *Scientific Problems of the Humid Tropical Zone*. Proc. Dacca Symp. Feb.-March 1964, 23-28.

Ojany, F. F., 1985. Kenya, in E. C. F. Bird and M. L. Schwartz eds.: *The World's Coastline*. Van Macoma balthica: an experimental study. Thesis, Lund University. 122 pp. Nostrand Reinhold. New York, 697-701.

Oost, A. P., 1995. Dynamics and sedimentary development of the Dutch Wadden Sea with emphasis on the Frision inlet. *Meded. Fac. Aardwetensch.* University Utrecht, 454 p.

Oost, A. P., and K. S. Dijkema, 1993. Effecten van bodemdaling door gaswinning in de Waddenzee. Rapport IBN-025. IBN-DLO/University Utrecht Fac. Aardwetensch., 133 p. (in Dutch)

Oost, A. P. and J. H. Baas, 1994. The development of small-scale bedforms in tidal environments.: an empirical model for unsteady flow and its applications. *Sedimentology*, 41, 883-903.

Oost, A., and P. L. de Boer, 1994. Sedimentology and development of barrier islands, ebb-tidal deltas, inlets and backbarrier areas of the Dutch Wadden Sea. *Senckenbergiana maritima* 24, 63-115.

Orme, A. R., 1972. Barrier and lagoon systems along the Zululand coast, South Africa, in D. R. Coates ed. *Coastal Geomorphology*. State University New York, Binghampton, 181-217.

Osgood, R. G. 1975. The history of invertebrate ichnology, in R. W. Frey ed. *The Study of Trace Fossils*. Springer, New York, 4-12.

Otvos Jr., E. G., and W. Armstrong Price, 1979. Problems of chenier genesis and terminology — an overview. *Mar. Geol.* 31, 251-263.

Ovenshine, A. T., D. E. Lawson, and S. R. Bartsch-Winkler, 1976. The Placer River silt — an intertidal deposit caused by the 1964 Alaska earthquake. *J. Res. U.S. Geol. Survey*, 4(2), 151-162.

Owens, E. H., and J. R. Harper, 1985. British Columbia, in E. C. F. Bird and M. L. Schwartz eds.: *The World's Coastline*. Van Nostrand Reinhold, New York, 11-15.

Paradis, G., 1980. Un cas particulier de zones denudees dans les mangroves d'Afrique de l'Ouest: celles dues a l'extraction de sel. *Bul. Mus. Natn. Hist. Nat.* Paris, 4e serie 2, B (3), 227-261.

Park. Y. A., S. C. Kim, and J. H. Choi, 1986. The distribution and transportation of fine-grained sediments on the inner continental shelf off the Keum river estuary, Korea. *Cont. Shelf. Res.* 5, 499-519.

Parkinson, R. W., 1989. Decelerating Holocene sea-level rise and its influence on south-western Florida coastal evolution. *J. Sed. Petrol.* 59, 960-972.

Parkinson, R. W., 1991. Geologic evidence of net onshore sand transport throughout the Holocene marine transgression, southwest Florida. *Mar. Geol.* 96, 269-277.

Paterson, D. M. 1986. The migratory behaviour of diatom assemblages in a laboratory tidal micro-ecosystem examined by low temperature scanning electron microscopy. *Diatom Research* 1, 227-239.

Paterson, D. M. 1989. Short term changes in the erodibility of intertidal cohesive sediments related to the migratory behaviour of epipelic diatoms. *Limnol Oceanogr.* 34, 223- 234.

Pejrup, M., 1986. Parameters affecting fine-grained suspended sediment concentrations in a shallow micro-tidal estuary. Ho Bugt, Denmark. *Est. Coast. Shelf Sci.* 22, 241-254.

Pejrup, M., 1988. Suspended sediment transport accross a tidal flat. *Mar. Geol.* 82, 187-198.

Perillo, G. M. E., ed., 1995. Geomorphology and sedimentology of estuaries. *Dev. in Sedimentology* 53, 471 pp.

Perillo, G. M. E., 1995. Definitions and geomorphologic classifications of estuaries, in G. M. E. Perillo ed.: Geomorphology and Sedimentology of Estuaries. *Developm. in Sediment.* 53, Elsevier, 17-47.

Perillo, G. M. E., and M. E. Sequeira, 1989. Geomorphologic and sediment transport characteristics of the middle reach of the Bahia Blanca estuary (Argentina). *J. Geophys. Res.* 94, 14, 351-14, 362.

Perillo, G. M. E., M. D. Ripley, M. C. Piccolo, and K. R. Dyer, 1996. The formation of tidal creeks in a salt marsh: new evidence from the Loyola Bay salt marsh, Rio Gallegos estuary, Argentina. *Mangroves and Salt Marshes* 1, 37-46.

Pernetta, J. C., 1993. Mangrove forests, climate change and sea level rise: hydrological influences on community structure and survival, with examples from the Indo-West Pacific. IUCN, Gland, Switzerland, 46 p.

Pestrong, R., 1965. The development of drainage patterns on tidal marshes. *Stanford University Publ. Geol. Sci. X* (2), 87 + VIII p.

Pestrong, R., 1972. Tidal-flat sedimentation at Cooley Landing, southwest San Francisco Bay. *Sed. Geology* 8, 251-288.

Pethick, J. S., 1980a. Salt-marsh initiation during the Holocene transgression: the example of the North Norfolk marshes, England. *J. Biogeogr.* 7, 1-9.

Pethick, J. S., 1980b. Velocity surges and asymmetry in tidal channels. *Est. Coast. Mar. Sci.* 11, 331-345.

Pethick, J. S., 1992. Saltmarsh geomorphology, in J. R. L. Allen and K. Pye eds.: *Saltmarshes.* Cambridge University Press, 41-79.

Pethick, J. S., 1996. The geomorphology of mudflats, in K. F. Nordstrom and C. T. Roman eds.: *Estuarine Shores.* Wiley and Sons Ltd., Chichester, 185-211.

Pethick, J., D. Leggett and L. Husain, 1992. Boundary layers under salt marsh vegetation developed in tidal currents, in J. B. Thorne ed.: *Vegetation and Erosion Processes and Environments.* Wiley and Sons Ltd, London, 113-124.

Pheiffer Madsen, P. and J. Sorensen, 1979. Validation of the lead-210 dating method. *J. Radioanal. Chem.* 54 (1-2), 39-48.

Pheiffer Madsen, P., 1981. Accumulation rates of heavy metals determined by 210-Pb dating: experience from a marsh area in Ho Bay. In H. Postma ed.: *Sediment and Pollution Interchange in Shallow Seas.* Rapp. proc.-verb. reun. ICES, Proc. Workshop Texel spt. 1979, 59-63.

Philippart, C. J. M., 1994. Interactions between Arenicola marina and Zostera noltii on a tidal flat in the Wadden Sea. *Mar. Ecol. Progr. Ser.* 111, 251-257.

Philippart, C. J. M., and K. S. Dijkema, 1994. Wax and wane of Zostera noltii Hornem. in the Dutch Wadden Sea. *Aquat. Bot.* 49, 255-268.

Phleger, F. B., 1965. *Sedimentology of Guerrero Negro Lagoon, Baja California, Mexico.* Proc. 17th Symp. Colston Res. Soc. London, 205-237.

Piccolo, M. C., and G. M. E. Perillo, 1990. Physical characteristics of the Bahia Blanca estuary (Argentina). *Est. Coast. Shelf Sc.* 31, 303-317.

Pitman, J. I., 1985. Thailand, in E. C. F. Bird and M. L. Schwartz eds.: *The World's Coastline.* Van Nostrand Reinhold. New Tork, 771-787.

Pizzuto, J. E., and E. W. Rogers, 1992. The Holocene history and stratigraphy of palustrine and estuarine wetland deposits of Central Delaware. *J. Coastal Res.* 8, 854-867.

Plaziat, J.-C., 1995. Modern and fossil mangroves and mangals: their climate and biogeographic variability, in D. W. J. Bosence and P. A. Allison eds.: *Marine Paleoenvironmental Analysis from Fossils.* Geol. Soc. Spec. Publ. 83, 73-96.

Postma, H., 1954. Hydrography of the Dutch Wadden Sea. *Arch. Neerl. Zool.* 10, 405-511.

Postma, H., 1957. Size frequency distribution of sands in the Dutch Wadden Sea. *Arch. Neerl. Zool.* 12, 319-349.

Postma, H., 1961. Transport and accumulation of suspended matter in the Dutch Wadden Sea. *Neth. J. Sea Res.* 1 (1/2), 148-190.

Postma, H., 1967. Sediment transport and sedimentation in the marine environment, in G. H. Lauff ed.: *Estuaries.* Amer. Assoc. Advanc. Sci. publ. 83, Washington, 158-179.

Postma, H. 1980. Sediment transport and sedimentation. In. E. Olausson and I. Cato eds., *Chemistry and Biogeochemistry of Estuaries.* J. Wiley, New York, 153- 186.

Postma, H., 1996. Sea-level rise and the stability of barrier islands, with special reference to the Wadden Sea, in J. D. Milliman and B. U. Haq eds.: *Sea-Level Rise and Coastal Subsidence.* Kluwer Academic Publ., Dordrecht, 269-280.

Powell, E. N. 1977. Particle size selection and sediment reworking in a funnel feeder, Leptosynapta tenuis (Holothuroida, Synaptidae) *Int Rev. Ges. Hydrobiol.* 62, 385-408.

Pratt, B. R., N. P. James, and C. A. Cowan, 1992. Peritidal carbonates In.: Walker, R. G. and N. P. James, eds.: Facies models; response to climate change. *Geol. Assoc. Can.* 303-322.

Prost, M. T., 1989. Coastal dynamics and chenier sands in French Guiana. *Mar. Geol.* 90, 259-267.

Purser, B. H., ed., 1973. *The Persian Gulf.: Holocene Carbonate Sedimentation and Diagenesis in a Shallow Epicontinental Sea.* Springer, Berlin. 471 pp.

Purser, B. H., and G. Evans, 1973. Regional sedimentation along the Trucial Coast, SE Persian Gulf, in B. H. Purser ed.: *The Persian Gulf.* Springer Berlin Heidelberg, 211-231.

Pye, K., 1992. Saltmarshes on the barrier coastline of North Norfolk, eastern England, in J. R. L. Allen and K. Pye eds.: *Saltmarshes.* Cambridge University Press, 149-181.

Rabinowitz, D., 1978. Early growth of mangrove seedlings in Panama, and an hypothesis concerning the relationship of dispersal and zonation. *J. Biogeogr.* 5, 113-133.

Rao, M. S., and R. Vaidyanadhan, 1979. Morphology and evolution of Godavari delta, India. *Z. Geomorph. N. F.* 23, 243-255.

Raper, S. C. B., T. M. L. Wigley, and R. A. Warrick, 1996. Global sea-level rise: past and future, in J. D. Milliman and B. U. Haq eds.: *Sea-Level Rise and Coastal Subsidence.* Kluwer Academic Publ., Dordrecht, 11-45.

Reading, H. G., ed. 1996. *Sedimentary Environments: Processes, Facies and Stratigraphy.* Blackwell Science, Oxford, 687 pp.

Redfield, A. C., 1967. The ontogeny of a salt marsh estuary, in G. H. Lauff ed.: *Estuaries.* A. A. A. S. Publ. 83, Washington DC, 108-114.

Reed, D. J., 1989a. Patterns of sediment deposition in subsiding coastal salt marshes, Terrebonne Bay, Louisiana: the role of winter storms. *Estuaries* 12, 222-227.

Reed, D. J., 1989b. The role of salt marsh erosion in barrier island evolution and deterioration in coastal Louisiana. *Trans. Gulf Coast Assoc. Geol. Soc.* 39, 501-510.

Reed, D. J., 1990. The impact of sea-level rise on coastal salt marshes. *Progr. in Oceanogr.* 14, 465-481.

Reimold, R. J. 1977. Mangals and saltmarshes of eastern United States In: V. J. Chapman ed: *Wet Coastal Ecosystems.* Elsevier, 157-166.

Reineck, H.-E., 1967. Layered sediments of tidal flats, beaches and shelf bottoms of the North Sea, in G. H. Lauff ed.: *Estuaries.* Amer. Assoc. Adv. Sci. publ. 83, Washington D. C., 191-206.

Reineck, H.-E., 1970. *Das Watt.* Frankfurt am Main, 142 p.

Reineck, H.-E., 1975. German North Sea tidal flats, in R. N. Ginsburg ed.: *Tidal Deposits.* Springer, Berlin Heidelberg, 5-12.

Reineck, H.-E. 1977. Natural indicators of energy level in Recent sediments: application of ichnology to a coastal engineering problem. *Geol. J.* (Spec. Issue) 9, 265-272.

Reineck, H.-E., 1982. *Das Watt*. 3rd ed. Kramer, Frankfurt. 185 p.

Reineck, H.-E. and I. B. Singh, 1973. *Depositional Sedimentary Environments*. Springer, Berlin. 439 pp.

Reineck, H.-E., and Y. M. Cheng, 1978. Sedimentologische und faunistische Untersuchungen an Watten in Taiwan. I. Aktuogeologische Untersuchungen. *Senckenb. Marit.*, 10, 85-115.

Reinhart, M. A., and J. Bourgeois, 1987. Distribution of anomalous sand at Willapa Bay, Washington: evidence for large-scale landward-directed processes. *Eos* 68, 1469.

Reise, K. 1981. High abundance of small zoobenthos around biogenic structures in tidal sediments of the Wadden Sea. *Helgoländer Wiss. Meeresunters.* 3, 413-425.

Reise, K., 1982. Long-term changes in the macrobenthic invertebrate fauna of the Wadden Sea: are polychaetes about to take over? *Neth. Journ. Sea Res.* 16, 29-36.

Reise, K., 1991. Macrofauna in mud and sand of tropical and temperate tidal flats, in M. Elliott and J.-P. Ducrotoy eds.: *Estuaries and Coasts*. ECSA Symposium, Olsen and Olsen, 211-216.

Reise, K. and A. Schubert, 1987. Macrobenthic turnover in the subtidal Wadden Sea: the Norderaue revisited after 60 years. *Helgoländer Wiss. Meeresunters.* 41, 69-82.

Rejmanek, M., C. E. Sasser, and G. W. Peterson, 1988. Hurricane-induced sediment deposition in Gulf coast marsh. *Est. Coast. Shelf Sci.* 21, 217-222.

Remane, A. 1934. Die Brackwasserfauna. *Zool. Anz. Suppl.* 7, 34-74.

Remane, A. 1954. Wurmen-Riffe am Tropenstrand. *Nat. u. Volk* 84, 177-183.

Ren Mei-e, 1992. Human impact on coastal landform and sedimentation — the Yellow river example. *Geojournal* 28. 4, 443-448.

Ren Mei-e, 1993. Relative sea level rise and coastal erosion in the Yellow River delta, China, and management strategies. Proc. "Coastal Zone '93" New Orleans, 42-53.

Ren Mei-e, 1994. Relative sea level rise in China and its socioeconomic implications. *Mar. Geol.* 17, 17-44.

Ren Mei-e, and Zhu Xianmo, 1994. Anthropogenic influences on changes in the sediment load of the Yellow River, China, during the Holocene. *The Holocene* 4, 314-320.

Ren Mei-e, Zhang Ren-shun, and Yang Ju-hai, 1985. Effect of typhoon no. 8114 on coastal morphology and sedimentation of Jiangsu province, People's Republic of China. *J. Coastal Res.* 1, 21-28.

Ren, M.-e., and Y.-L. Shi, 1986. Sediment discharge of the Yellow River (China) and its effect on the sedimentation of the Bohai and the Yellow Sea. *Cont. Shelf. Res.* 6, 785-810.

Ren, Mei-e ed., 1986, *Modern Sedimentation in the Coastal and Nearshore Zones of China*. China Ocean Press Beijing, Springer Verlag Berlin Heidelberg, 466 p.

Rhoads, D. C. 1967. Biogenic reworking of intertidal and subtidal sediments in Barnstable Harbor and Buzzards Bay, *Massachusetts. J. Geol.* 75, 461-476.

Rhoads, D. C. 1974. Organisms-sediment relations on the muddy sea floor. *Oceanogr. Mar. Biol. Ann. Review* 12, 263-300.

Rhoads, D. C., and D. J. Stanley, 1965. Biogenic graded bedding. *J. Sed. Petrol.*, 35, 956-963.

Rhoads, D. C. and D. K. Young, 1970. Yhe influence of deposit- feeding organisms on sediment stability and community trophic structure. *J. Mar. Res.* 28, 150-178.

Rhodes, E. G., 1982. Depositional model for a chenier plain, Gulf of Carpentaria, Australia. *Sedimentology* 29, 201-221.

Richard, G. A., 1978. Seasonal and environmental variations in sediment accretion in a Long Island salt marsh. *Estuaries* 1, 29-35.

Ridderinkhof, H., 1995. Lagrangian flows in complex Eulerian current fields, in D. R. Lynch and A. M. Davies eds.: *Quantitative Skill Assessment for Coastal Ocean Models*. AGU Coastal and Marine Studies 47, 31-48.

Riisgård, H. U. and P. S. Larsen. 1995. Filter-feeding in marine macro-invertebrates: pump characteristics, modelling and energy cost. *Biol. Rev.* 70: 67-106.

Riisgård, H. U. 1991. Suspension feeding in the polychaete Nereis diversicolor. *Mar. Ecol. Prog. Ser.* 70: 29-37.

Rine, J. M., and R. N. Ginsburg, 1985. Depositional facies of a mud shoreface in Suriname, South-America. A mud analogue to sandy, shallow-marine deposits. *J. Sed. Petrol* 55. 633-652.

Rivera-Monroy, V. H., J. W. Day, R. R. Twilley, F. Vera-Herrera, and C. Coronado-Molina, 1995. Flux of nitrogen and sediment in a fringe mangrove forest in Terminos Lagoon, Mexico. *Est. Coast. Shelf Sci.* 40, 139-160.

Robertson, A. I., 1986. Leaf-burying crabs: their influence on energy flow and export from mixed mangrove forests (Rhizophora spp.) in northeastern Australia. *J. Exp. Mar. Biol. Ecol.* 102: 237-248.

Rodolfo, K. S., 1975. The Irrawaddy delta: tertiary setting and modern offshore sedimentation, in M. L. Broussard ed.: *Deltas, Models for Exploration. Houston Geol. Soc.,* 339-356.

Rogers, A., 1870. A few remarks on the geology of the country surrounding the Gulf of Cambay, in western India. *Quart. J. (Proc.). Geol. Soc.* 26, 118-124.

Rosen, P., 1979. Boulder barricades in central Labrador. *J. Sed. Petrol.* 49, 1113-1124.

Rosen, P. S., 1980a. Coastal environments of the Makkovik region, Labrador, in S. B. McCann, ed.: *The Coastline of Canada,* Geol. Survey of Canada Paper 80-10, 267-280.

Rosen, P., 1980b. Erosion susceptibility of the Virginia Chesapeake Bay shoreline. *Mar. Geol.* 34, 45-59.

Ross, M. A., 1988. Vertical structure of estuarine fine sediment suspensions. Thesis University Florida, Gainesville (in Mehta 1989).

Roy, P. S., B. G. Thom, and L. D. Wright, 1980. Holocene sequences on an embayed high-energy coast: an evolutionary model. *Sed. Geol.* 26, 1-19.

Roy, P. S., and B. G. Thom, 1981. Late Quaternary marine deposition in New South Wales and southern Queensland — an evolutionary model. *J. Geol. Soc. Austr.* 28, 471-489.

Ruhe, R. V., 1952. Topographic discontinuities of the Des Moines lobe. *Amer. J. Sci.* 250, 46-56.

Rusnak 1967 in Schäfer, W. 1962. Aktuopaläontologie nach Studien in der Nordsee. W. Kramer Frankfurt/Main. (translated 1972 as Ecology and paleoecology of marine environments. Oliver and Boyd, Edinburgh).

Saenger, P., E. J. Hegerl and J. D. S. Davie, eds., 1983. Global status of mangrove ecosystems. *The Environmentalist* 3, suppl. 3, 1-88.

Salinas, L. M., R. D. deLaune, and W. H. Patrick, Jr., 1986. Changes occurring along a rapidly submerging coastal area: Louisiana, USA. *J. Coast. Res.* 2, 269-284.

Salomon, J. N., 1978. Contribution a l'étude écologique et géographique des mangroves. *Rev. Géomorph. Dynamique,* 27, 63-80

Sanford, L. P., W. Panageotou and J. P. Halka, 1991. Tidal resuspension of sediments in northern Chesapeake Bay. Mar. Geol. 97, 87-103.

Sanford, L. P., and J. P. Halka, 1993. Assessing the paradigm of mutually exclusive erosion and deposition of mud, with examples from upper Chesapeake Bay. *Mar. Geol.* 114, 37-57.

Sanlaville, P., 1985. Arabian Gulf Coasts, in E. C. F. Bird and M. L. Schwartz eds.: *The World's Coastline.* Van Nostrand Reinhold. New York, 729-733.

Schaefer, W., 1962. *Aktuo-Palaeotologie Nach Studien in der Nordsee.* Waldemar Kramer, Frankfurt, 666 p.

Schaefer, W., 1972. *Ecology and Paleoecology of Marine Environments,* Oliver and Boyd, and Univ. of Chicago Press, Edinburgh and Chicago, 568 p.

Scheidegger, A. E., *Theoretical Geomorphology,* 3rd ed. 1991. Springer Verlag, 434 p.

Schnack, E. J., 1985. Argentina, in E. C. F. Bird and M. L. Schwartz eds.: *The World's Coastline.* Van Nostrand Reinhold. New York, 69-78.

Scholl, D. W., 1964a. Recent sedimentary record in mangrove swamps and rise in sea level over the southwestern coast of Florida. *Mar. Geol.* 1, 344-366.

Scholl, D. W., 1964b. Recent sedimentary record in mangrove swamps and rise in sea level over the southwestern coast of Florida: part 2. *Mar. Geol.* 2, 343-364.

Scholl, D. W., 1968. Mangrove swamps, in R. W. Fairbridge ed.: *Encyclopedia of Geomorphology.* Reinhold Book Corp.: 683-688.

Scholl, D. W., and M. Stuiver, 1967. Recent submergence of southern Florida. A comparison with adjacent coasts and other eustatic data. *Geol. Soc. Amer. Bull.* 78, 437-454.

Scholl, D. W., F. C. Craighead Sr. and M. Stuiver, 1969. Florida submergence curve revisited: its relation to coastal sedimentation rates. *Science* 163, 562-564.

Schopf, T. J. M. 1978. Fossilization potential of an intertidal fauna. *Paleobiology* 4, 261-270.

Schubauer, J. P., and C. S. Hopkinson, 1984. Above- and below-ground emergent macrophyte production and turnover in a coastal marsh ecosystem. *Limnol. Oceanogr.* 29, 1052-1065.

Schumm, S. A., 1963. A tentative classification of alluvial river channels. *U.S. Geol. Surv. Circ.* 477, 10 p.

Schuster, W. H., 1952. Fish culture in brackish water ponds of Java. *Spec. Publ. Indo Pac. Fish Counc.,* no. 1.

Schwartz, M. L., 1985a. Cameroon and Equatorial Guinea, in E. D. F. Bird and M. L. Schwartz eds.: *The World's Coastline.* Van Nostrand Reinhold. New York, 621-623.

Schwartz, M. L., 1985b. Libya, in E. C. F. Bird and M. L. Schwartz eds.: *The World's Coastline.* Van Nostrand Reinhold. New York, 519-521.

Schwartz, M. L., 1985c. Pacific Colombia, in E. C. F. Bird amd M. L. Schwartz eds.: *The World's Coastline.* Van Nostrand Reinhold, New York, 45-47.

Scott Warren, R., and W. A. Niering, 1993. Vegetation change on a northeast tidal marsh:interaction of sea-level rise and marsh accretion. *Ecol.* 74, 96-103.

Scoffin, T. P., 1970. The trapping and binding of subtidal carbonate sediment by marine vegetation in Bimini Lagoon, Bahamas. *J. Sed. Petrol.* 40, 249-273.

Scott, P. A., 1985. Sierra Leone, in E. C. F. Bird and M. L. Schwartz eds.: *The World's Coastline.* Van Nostrand Reinhold. New York, 569-573.

Seilacher, A. 1967. Fossil behavior. *Scient. Amer.* 217, 72- 80.

Seilacher, A. and F. Pfleger, 1994. From biomats to benthic agriculture: a biohistoric revolution, in Krumbein W. E., D. M. Paterson and L. J. Stal eds.: *Biostabilization of Sediments.* BIS-Verlag. University Oldenburg, 97-105.

Semeniuk, V., 1980a. Quaternary stratigraphy of the tidal flats, King Sound, western Australia. *J. Royal Soc. Western Australia,* 63(3), 65-78.

Semeniuk, V., 1980b. Mangrove zonation along an eroding coastline in King Sound, northwestern Australia. *J. Ecol.* 68, 789-812.

Semeniuk, V., 1981a. Sedimentology and the stratigraphic sequence of a tropical tidal flat, north-western Australia. *Sed. Geology,* 29, 195-221.

Semeniuk, V., 1981b. Long-term erosion of the tidal flats, King Sound, north-western Australia. *Mar. Geol.* 43, 21-48.

Semeniuk, V., and D. J. Searle, 1986. Variability of Holocene sealevel history along the southwestern coast of Australia — evidence for the effect of significant local tectonism. *Mar. Geol.* 72, 47-58.

Settlemyre, J. L., and R. L. Gardner, 1977. Suspended sediment flux through a salt marsh drainage basin. *Est. Coast. Mar. Sci.* 5, 653-663.

Sha, L. P., 1989. Variation in ebb-delta morphologies along the west and east Frisian Islands, the Netherlands and Germany. *Mar. Geol.* 89, 11-28.

Sha, L. P., and P. L. de Boer, 1991. Ebb-tidal delta deposits along the west Frisian Islands (The Netherlands): processes, facies architecture and preservation, in D. G. Smith et al. eds.: Clastic Tidal Sedimentology. *Can. Soc. Petr. Geol. Mem.* 16, 199-218.

Shi, Z., 1991. Tidal bedding and tidal cyclicities within the intertidal sediments of a microtidal estuary, Dyfi river estuary, west Wales, U.K. *Sed. Geol.* 73, 43-58.

Shi, Z., 1993a. Recent saltmarsh accretion and sea level fluctuations in the Dyfi estuary, central Cardigan Bay, Wales, UK. *Geo-Marine Letters* 13, 182-188.

Shi, Z., 1993b. Application of the 'Pejrup Approach' for the classification of the sediments in the microtidal Dyfi estuary, West Wales, U.K. *J. Coast. Res.* 8, 482-491.

Shi, Z., and H. F. Lamb, 1991. Post-glacial sedimentary evolution of a microtidal estuary, Dyfi estuary, west Wales, U.K. *Sed. Geol.* 73, 227-246.

Shi, Z., H. F. Lamb and R. L. Collin, 1995. Geomorphic change of saltmarsh tidal creek networks in the Dyfi estuary, Wales. *Mar. Geol.* 128, 79-83.

Shi, Z., J. S. Pethick and K. Pye, 1995. Flow structure in and above the various heights of a saltmarsh canopy: a laboratory flume study. *J. Coast. Res.* 11, 1204-1209.

Shinn, E. A., 1968. Burrowing in recent lime sediments of Florida and the Bahamas. *J. Paleont.* 42, 879-894.

Shinn, E. A., 1973. Carbonate coastal accretion in an area of longshore transport, NE Qatar, Persian Gulf, in B. H. Purser ed.: *The Persian Gulf.* Springer Berlin Heidelberg, 179-191.

Shinn, E. A., R. M. Lloyd, and R. N. Ginsburg, 1969. Anatomy of a modern carbonate tidal flat, Andros Island, Bahamas. *J. Sed. Petrol.* 39, 1202-1228.

Shreve, R. L., 1966. Statistical law of stream numbers. *J. Geol.* 74, 17-37.

Shreve, R. L., 1967. Infinite topologically random channel networks. *J. Geol.* 75, 178-186.

Shuisky, Y. D., 1985. Northern Black Sea and Sea of Azov, USSR, in E. C. F. Bird and M. L. Schwartz eds.: *The World's Coastline.* Van Nostrand Reinhold. New York, 467-472.

Siefert, W., and H. Lassen, 1987. Zum sakularen Verhalten der mittleren Watthohen an ausgewahlten Beispielen. *Die Kuste* 45, 59-70.

Sioli, H., 1965. General features of the delta of the Amazon, in *Scientific Problems of the Humid Tropical Zone.* Proc. Dacca Symp. Feb.-March 1964, UNESCO, 381-390.

Smaal A. C. and T. C. Prins, 1993. The uptake of organic matter and the release of inorganic nutrients by bivalve suspension feeder beds, in R. F. Dame ed. *Bivalve Filter Feeders in Estuarine and Coastal Ecosystem Processes.* NATO ASI Series, Series G: Ecological sciences vol 33, 271-298. Berlin, Springer.

Smart, J., 1976. The nature and origin of beach ridges, western Cape York Peninsula, Queensland. BMR *J. Aust. Geol. Geophys.* 1, 211-218.

Smith, N., 1975. *Man and Water.* Peter Davies, London, 239 p.

Smith, N. D., A. C. Phillips and R. D. Powell, 1990. Tidal drawdown: a mechanism for producing cyclic sediment laminations in glaciomarine deltas. *Geology,* 18, 10-13.

Smith, S. V. and J. T. Hollibaugh, 1993. Coastal metabolism and the oceanic organic carbon balance. *Rev. Geophysics* 31, 75-89.

Snead, R. E., 1985. Pakistan, in E. C. F. Bird and M. L. Schwartz eds.: *The World's Coastline.* Van Nostrand Reinhold. New York, 735-740.

Snedaker, S. C. and J. G. Snedaker, 1984. The mangrove ecosystem research methods. UNESCO: 251 pp.

Soerianegara, I., 1968. The causes of mortality of Bruguiera trees in the mangrove forest near Tjilatjap, Central Java. *Rimba Indonesia,* XIII: 1-11.

Sorensen, L. O., and J. T. Conover, 1962. Algal mat communities of Lyngbya confervoides (C. Agardh) Gomont. Publ. Inst. Mar. Sci. University Texas 8, 61-74.

Sornin, J.-M., 1981. Processus sédimentaires et biodéposition liées; differents modes de conchyliculture. Thèse University Nantes, 47 p.

Soulsby, P. G., D. Lowthion and M. Houston, 1982. Effects of macroalgal mats on the ecology of intertidal mudflats. *Mar. Poll. Bull.* 13, 162-166.

Sovershaev, V. A., 1977. Windy sea-level changing as a factor of development of the Chukchee Sea coasts, in B. A. Popov and I. D. Danilov eds.: *Geographical Problems in the Study of Northern Areas,* Moscow University Press, 161-166.

Spackman, W., C. P. Dolsen, and W. Riegel, 1966. Phytogenic organic sediments and sedimentary environments in the Everglades-mangrove complex. Part 1. Evidence of a transgressive sea and its effect on environments of the Shark river area of southwest Florida. *Palaeontographica* B117, 135-152.

Spackman, W., W. Riegel and C. P. Dolsen, 1969. Geological and biological interactions in the swamp-marsh complex of southern Florida, in E. C. Dapples and M. B. Hopkins eds.: *Environments of Coal Deposition,* Geol. Soc. Amer. Spec. Paper, 114, 1-35.

Spenceley, A. P. 1977. The role of pneumatophores in sedimentary processes. *Mar. Geol.* 24: M31-M37.

Spenceley, A. P., 1987. Mangroves and intertidal morphology in Westernport Bay, Victoria, Australia. Comment. *Mar. Geol.* 77: 327-331.

Steenis, C. G. G. J. van, 1958. Rhizophoraceae. Ecology. *Fl. Males, ser.* I, 5:431.

Steers, J. A., 1938. The rate of sedimentation on salt marshes on Scolt Head Island. Norfolk *Geol. Mag.* 75, 26-39.

Steers, J. A., 1967. Geomorphology and Coastal Processes, in G. H. Lauff ed.: *Estuaries.* Amer. Assoc. Adv. Sci. publ. no 83, 100-107.

Stevenson, J. C., M. S. Kearney and E. C. Pendleton, 1985. Sedimentation and erosion in a Chesapeake Bay brackish marsh system. *Mar. Geol.* 67, 213-235.

Stevenson, J. C., L. G. Ward, and M. S. Kearney, 1986. Vertical accretion in marshes with varying rates of sea level rise, in D. A. Wolfe ed.: *Estuarine Variability.* Academic Press, Orlando, 241-260.

Stiasny, G. 1910. Zur Kentniss der Lebensweise von Balanoglossus clavigerus Delle Chiaje. *Zool. Anzeiger* 35, 561-565, 633.

Stoddart, D. R., D. J. Reed, and J. R. French, 1989. Understanding salt-marsh accretion, Scolt Head Island, Norfolk, England. *Estuaries* 12, 228-236.

Strahler, A. N., 1957. Quantitative analysis of watershed geomorphology. *Trans. Amer. Geophys. Union* 38, 913-920.

Stumpf, R. P., 1983. The processes of sedimentation on the surface of a salt marsh. *Est. Coast. Shelf Sci.* 17, 495-508.

Su, J., K. Wang and Y. Li., 1993 (1992). Fronts and transport of suspended matter in the Hangzhou Bay. *Acta Oceanol. Sinica,* 12(1), 1-15.

Sugden, W., 1963. Some aspects of sedimentation in the Persian Gulf. *J. Sed. Petrol.* 33, 355-364.

Swan, B., 1985. Sri Lanka, in E. C. F. Bird and M. L. Schwartz eds.: *The World's Coastline.* Van Nostrand Reinhold. New York, 749-759.

Teh Tiong Sa, 1985. Peninsular Malaysia, in E. C. F. Bird and M. L. Schwartz eds.: *The World's Coastline.* Van Nostrand Reinhold. New York, 789-795.

Terwindt, J. H. J., 1988. Palaeo-tidal reconstructions of inshore tidal depositional environments. In.: P. L. de Boer, A. van Gelder and S. D. Nio, eds.: *Tide-Influenced Sedimentary Environments and Facies.* Reidel Dordrecht. 233-262.

Terwindt, J. H. J. and H. N. C. Breusers, 1972. Experiments on the origin of flaser, lenticular and sand-clay alternating bedding. *Sedimentology,* 19, 85-98.

Thanh, T. D., 1995. Coastal morphological changes concerning the management of coastal zone in Vietnam. Coastal Change 95 Conf. Bordeaux. Feb. 1995, 1-12.

Thanh. T. D., D. V. Huy and Ng. H. Cu, 1991. Tidal creek-channel system in the western coastal zone of Tonkin Gulf. Haiphong Mar. Res. Center: Marine Environment and resources. Hanoi. Sc. Techn. Publ. House, 26-32. (in Vietnamese).

Thannheiser, D., 1975. Beobachtungen zur Küstenvegetation auf dem westlichen kanadischen Arctis-Archipel. *Polarforschung* 45, 1-16.

Thom, B. G., 1967. Mangrove ecology and deltaic geomorphology: Tabasco, Mexico. *J. Ecol.* 55: 301-343.

Thom, B. G., 1975. Mangrove ecology from a geomorphic viewpoint, in G. Walsh, S. Snedaker and H. Teas, eds.: Proc. Int. Symp. Biol. Management of Mangroves, Honolulu Oct. 1974: 469-481.

Thom, B. G., 1982. Mangrove ecology: a geomorphological perspective, in B. F. Clough ed.: *Mangrove Ecosystems in Australia, Structure, Function and Management.* Austr. Natl. University Press, Canberra, 3-17.

Thom, B. G. 1984. Coastal landforms and geomorphic processes, in S. C. and J. G. Snedaker eds.: *The Mangrove Ecosystem: Research Methods,* ch. 1: 3-17.

Thom, B. G., ed., 1984. *Coastal Geomorphology in Australia.* Academic Press, Sidney, 290 p.

Thom, B. G., 1985. New South Wales, in E. C. F. Bird and M. L. Schwartz eds.: *The World's Coastline.* Van Nostrand Reinhold, New York, 969-974.

Thom, B. G., and L. D. Wright, 1983. Geomorphology of the Purari delta, in T. Petr ed.: *The Purari — Tropical Environment of a High Rainfall River Basin, Junk,* Den Haag, 47-65.

Thom, B. G., L. D. Wright, and J. M. Coleman, 1975. Mangrove ecology and deltaic-estuarine geomorphology: Cambridge Gulf — Ord River, western Australia. *J. Ecol.* 63, 203-232.

Thom, B. G. and L. D. Wright, 1983. Geomorphology of the Purari delta, in T. Petr ed.: *The Purari — Tropical Environment of a High Rainfall River Basin.* Junk Publ., Den Haag, 47-65.

Thompson, R. W., 1968. Tidal flat sedimentation on the Colorado river delta, northwestern Gulf of California. *Geol. Soc. America Memoir* 107, 133 p.

Thompson, R. W., 1975. Tidal flat sediments of the Colorado river delta, northwestern Gulf of California, in R. N. Ginsburg ed.: *Tidal Deposits,* ch. 7, 57-65.

Thorez, J., E. Goemaere and R. Dreesen, 1988. Tide- and wave-influenced depositional environments in the Psammites du Condroz (Upper Famennian) in Belgium. In.: P. L. de Boer, A. van Gelder and S. D. Nio, eds.: *Tide-Influenced Sedimentary Environments and Facies.* Reidel Dordrecht. 389-415.

Tinley, K. L., 1985. Mocambique, in E. C. F. Bird and M. L. Schwartz eds.: *The World's Coastline.* Van Nostrand Reinhold. New York, 660-677.

Tomlinson, P. B., 1986. *The Botany of Mangroves.* Cambridge University Press, Cambridge, 413 p.

Trewin, N. H. and W. Welsh, 1976. Formation and composition of a graded estuarine shell bed. *Palaeogeogr. Palaeoclimatol. Palaeocecol.* 19, 219-230.

Trueman, E. R. 1968. The mechanism of burrowing of some naticid gastropods in comparison with that of other molluscs. *J. Exp. Biol.* 48, 663-678.

Tucker, M. E., 1973. The sedimentary environments of tropical african estuaries: Freetown peninsula, Sierra Leone. *Geol. Mijnb.* 52, 203-215.

Tufail, A., P. S. Meadows and P. McLaughlin, 1989. The influence of waves and seagrass communities on suspended particles in an estuarine embayment. *Mar. Geol.* 59, 85-103.

Turmell, R. J., and R. G. Swanson, 1976. The development of Rodriguez Bank, a Holocene mudbank in the Florida Reef Tract. *J. Sed. Petrol.* 46, 497-518.

Twidale, C. R., 1985. South Australia, in E. C. F. Bird and M. L. Schwartz eds.: *The World's Coastline.* Van Nostrand Reinhold. New York, 913-927.

UNESCO, 1981. Bibliography on mangrove research, 1600-1975. Compiled by B. Rollet: 479 pp.

Usoro, E. J., 1985. Nigeria, in E. C. F. Bird and M. L. Schwartz eds.: *The World's Coastline.* Van Nostrand Reinhold, New York, 607-613.

Vale, C., and B. Sundby, 1987. Suspended sediment fluctuations in the Tagus estuary on semi-diurnal and fortnightly time scales. *Est. Coast. Shelf Sci.* 33, 495-508.

van Bennekom, A. J., E. Krijgsman-van Hartingsveld, G. C. M. van der Veer and H. F. J. van Voorst. The seasonal cycle of reactive silicate and suspended diatoms in the Dutch Wadden Sea. *Neth. J. Sea Res.* 8, 174-207.

van den Berg, J. H., 1981. Rhythmic seasonal layering in a mesotidal channel fill sequence, Oosterschelde Mouth, the Netherlands. In.: S. D. Nio, R. T. E. Schüttenhelm, and T. C. E. van Weering, eds.: *Holocene Marine Sedimentation in the North Sea Basin.* Int. Assoc. Sedim. Spec. Publ. 5, 147-159.

van der Spek, A. J. F., 1994. Large-scale evolution of Holocene tidal basins in the Netherlands. Thesis Utrecht University 191 p.

van der Werff, A. 1960. Die Diatomeen des Dollart-Emsgebiet. Verh. Kon. Nedf. Geol. Mijnb. k. Gen. *Geol. Serie* 19, 163- 201

van Dongeren, A. R., and H. J. de Vriend, 1994. A model of morphological behaviour of tidal basins. *Coastal Engineering* 22, 287-310.

van Heerden, I. Ll., J. T. Wells, and H. H. Roberts, 1981. Evolution and morphology of sedimentary environments, Atchafalaya delta, Louisiana. *Trans. Gulf Coast Assoc. Geol. Soc.* 31, 399-408.

van Heerden, I. Li., J. T. Wells, and H. H. Roberts, 1983. River-dominated suspended-sediment deposition in a new Mississippi delta. *Can. J. Fish. Aquat. Sci.* 40 (Suppl. 1), 60-71.

van Pelt, J., M. J. Woldenberg, and R. W. H. Verwer, 1989. Two generalized models of stream network growth. *J. Geol.* 97, 281-299.

van Straaten, L. M. J. U., 1951. Quelques particularités du relief sousmarin de la Mer de Wadden (Hollande). *C. R. Congres Sedim. Quat. France, Bordeaux,* 139-145.

van Straaten, L. M. J. U. 1952. Biogene textures and the formation of shell beds in the Dutch Wadden Sea.- *Proc. K. Ned. Akad. Wet.* (B) 55(5): 500-516.

van Straaten, L. M. J. U. 1954. Composition and structure of Recent marine sediments in the Netherlands.- *Leidse Geol. Meded.* 19: 1-108.

van Straaten, L. M. J. U., 1964. De bodem der Waddenzee, in W. F. Anderson, J. Abrahamse, J. D. Buwalda, and L. M. J. U. van Straaten eds.: *Het Waddenboek,* Thieme, Zutphen, 75-151.

van Straaten, L. M. J. U., and Ph. H. Kuenen, 1957. Accumulation of fine-grained sediments in the Dutch Wadden Sea. *Geol. Mijnb.* 19, 329-354.

van Straaten, L. M. J. U., and Ph. H. Kuenen, 1958. Tidal action as a cause of clay accumu-lation. *J. Sed. Petrol.* 28, 406-413.

Vance, D. J., M. D. E. Haywood, and D. J. Staples, 1990. Use of a mangrove estuary as a nursery area by postlarval and juvenile Banana Prawns, Penaeus merguiensis de Man, in northern Australia. *Est. Coast. Shelf Sci.* 31, 689-701.

Vann, J., 1959. Landform-vegetation relationships in the Atrato delta. Ann. Ass. Am. Geogr. 49: 345-360.

Vann, J. H., 1959. The geomorphology of the Guyana coast, in *Second Coastal Geography Conference,* April 5-9, 1959 Louisiana. Washington D. C., 153-187.

Verger, F., 1988. *Marais et Wadden du littoral francais.* 3e ed. Paradigme, Caen. 549 p.

Verstappen, H. Th., 1964. Geomorphology in delta studies. Publ. Int. Inst. Aereal Survey and Earst Sciences (ITC), Delft, ser. B, nr. 24.

Verstappen, H. Th., 1973. *A Geomorphological Reconnaissance of Sumatra and Adjacent Islands* (Indonesia). Wolters-Noordhoff Publ. Groningen, 182 p.

Verweij, J. 1930. Einiges über die Biologie Ost-Indischer Mangrovekrabben. *Treubia* 12, 167-257.

Verweij, J. 1952. On the ecology of distribution of cockle and mussel in the Dutch Wadden Sea, their role in sedimentation and the source of their food supply. *Arch. Néerl. Zool.* 10, 171-239.

Vilas Martin, F., 1983. Medios sedimentarios de transicion en la Ria de Vigo: sequencias progradantes. *Thalassas* 1983, 49-55.

Vilas, F., and M. A. Nombela, 1985. Las zonas estuarinas de las costas de Galicia y sus medios asociados, N. W. de la peninsula iberica. *Thalassas* 3, 7-15.

Viles, H., and T. Spencer, 1995. Coastal Problems. Edward Arnold, London, 350 p.

Visser, M. J., 1980. Neap-spring cycles reflected in Holocene subtidal large-scale bedform deposits.: a preliminary note. *Geology* 8, 543-546.

Visser. W. A., ed. 1980. Geological Nomenclature. Scheltema en Holkema, Utrecht 540 pp.

Volker. A., 1965. The deltaic area of the Irrawaddy river in Birma, in *Proc. Dacca Symposium* Febr.-March 1964. UNESCO, 373-379.

von der Borch, C. C., 1976. Stratigraphy of stromatolite occurrences in carbonate lakes of the Coorong Lagoon area, south Australia. In M. R. Walter, ed.: *Stromatolites Developments in Sedimentology* 20, Elsevier, ch. 8. 3, 413-420.

Vos, P. C., P. L. de Boer and R. Misdorp, 1988. Sediment stabilization by benthic diatoms in intertidal sandy shoals; qualitative and quantitative observations, in P. L. De Boer ed. *Tide-Influenced Sedimentary Environments and Facies.* Reidel Publ. Dordrecht, 511-526.

Walker, H. J., 1985. Alaska, in E. C. F. Bird and M. L. Schwartz eds.: *The World's Coastline.* Van Nostrand Reinhold, New York, 1-10.

Walker, H. J., J. M. Coleman, H. H. Roberts and R. S. Tye, 1987. Wetland loss in Louisiana. *Geografiska Annaler* 69A, 189-200.

Walter, M. R. ed., 1976. Stromatolites. *Developments in Sedimentology* 20, Elsevier, 790 p.

Wang, B. C., and D. Eisma, 1988. Mudflat deposition along the Wenzhou coastal plain in southern Zhejiang, China, in P. L. de Boer, A. van Gelder, and S. D. Nio eds.: *Tide-Influenced Sedimentary Environments and Facies.* Reidel Publ., 265-274.

Wang, B. C., and D. Eisma, 1990. Supply and deposition of sediment along the north bank of Hangzhou Bay, China. *Neth. J. Sea Res.* 25, 337-390.

Wang, Y., 1983. The mudflat system of China. *Can. J. Fish Aquat Sci.* 40 (suppl 1), 160-171.

Wanless, H. R., and J. J. Dravis, 1984. Comparison of two Holocene tidal flats — Andros Island, Bahamas, and Caicos, British West Indies. *Bull. Am. Assoc. Petrol. Geol.* 68, 577.

Wanless, H., K. M. Tyrrell, L. P. Tedesco and J. J. Dravis, 1988. Tidal-flat sedimentation from hurricane Kate, Caicos platform, British West Indies. *J. Sed. Petrol.* 58, 724-738.

Ward, L. G., 1981. Suspended-material transport in marsh tidal channels, Kiawah Island, South Carolina. *Mar. Geol.* 40, 139-154.

Warme, J. E., 1967. Graded bedding in the Recent sediments of Mugu Lagoon, California. *J. Sed. Petrol.* 37, 540, 547.

Warren, R. S., and W. A. Niering, 1993. Vegetation change on a northeast tidal marsh: interaction of sea-level rise and marsh accretion. *Ecology.* 74, 96-103.

Watson, J. D., 1928. Mangrove forests of the Malay Peninsula. *Malay Forest Res.* 6: 1-275.

Weart, S. R., 1997. The discovery of the risk of global warming. *Physics Today,* 50, 34-40.

Wells, J. T., 1983. Dynamics of coastal fluid muds in low-, moderate and high- tide-range environments. *Can. J. Fish. Aquat. Sci.* 40 (Suppl. 1), 130-142.

Wells, J. T., and J. M. Coleman, 1977. Nearshore suspended sediment variations, central Surinam coast. *Mar. Geol.* 24, M47-M54.

Wells, J. T., and J. M. Coleman, 1978. Inshore transport of mud by waves: northeastern coast of South America. *Geol. Mijnb.* 57, 353-359.

Wells, J. T. and S.-Y. Kim, 1989. Sedimentation in the Albemarle-Pamlico lagoonal system: synthesis and hypotheses. *Mar. Geol.* 88, 263-284.

Wells, J. T., and G. P. Kemp, 1981. Atchafalaya mud stream and recent mudflat progradation: Louisiana chenier plain. *Trans. Gulf Coast Assoc. Geol. Soc.* 31, 409-416.

Wells, J. T., and G. P. Kemp, 1981. Physical processes and fine-grained sediment dynamics, coast of Surinam, South America. *J. Sed. Petrol.* 51, 1053-1068.

Wells, J. T., and H. H. Roberts, 1981. Fluid mud dynamics and shoreline stabilization: Louisiana chenier plain. Proc. 17th Conf. Coastal Eng. Sydney, 1382-1401.

Wells, J. T., and J. M. Coleman, 1984. Deltaic morphology and sedimentology, with special reference to the Indus River delta, in B. U. Haq and J. D. Milliman eds.: *Marine Geology and Oceanography of Arabian Sea and Coastal Pakistan.* Van Nostrand Reinhold, New York, 85-100.

Wells, J. T., and G. P. Kemp, 1986. Interaction of surface waves and cohesive sediments: field observations and geologic significance, in A. J. Mehta ed.: *Estuarine Cohesive Sediment Dynamics.* Springer Verlag, 43-65.

Wells, J. T., J. M. Coleman, and Wm. J. Wiseman, Jr. 1979. Suspension and transportation of fluid mud by solitary-like waves. Proc. 16th Conf. Coastal Eng. Hamburg, 1932-1952.

Wells, J. T., Y. A. Park and J. H. Choi, 1984-1985. Storm-induced fine-sediment transport, west coast of South Korea. *Geo-Marine Letters.* 4, 177-180.

Wells, J. T., C. E. Adams Jr., Y-A. Park, and E. W. Frankenberg, 1990. Morphology, sedimentology and tidal channel processes on a high-tide-range mudflat, west coast of South Korea. *Mar. Geol.* 95, 111-130.

West Coast of India Pilot, 1975. 11th ed. Hydrographer of the Navy, 203 p.

West, R. C., 1956. Mangrove swamps of the Pacific coast of Columbia. *Ann. Ass. Am. Geogr.* 46: 98-121.

Wilhelmy, H., 1968. Indus delta und Rann of Kutch. *Erdkunde*, 22 (3), 177-191.

Williams, H. F. L., and M. C. Roberts, 1989. Holocene sea-level change and delta growth: Fraser River delta, British Columbia. *Can. J. Earth Sci.* 26, 1657-1666.

Williams, H. F. L., and T. S. Hamilton, 1995. Sedimentary dynamics of an eroding tidal marsh derived from stratigraphic record of 137-Cs fallout, Fraser Delta, British Columbia, Canada. *J. Coast. Res.* 11, 1145-1156.

Williams, S. L., 1990. Experimental studies of Caribbean seagrass bed development. *Ecol. Monogr.* 60, 449-469.

Wingfield, R. T. R., C. D. R. Evans, S. E. Deegan and R. Floyd, 1978. Geological and geophysical survey of The Wash. Inst. Geol. Sc. NERC Report 78/18, 23 p.

Winkelmolen, A. M., 1971. Rollability, a functional shape property of sand grains. *J. Sed. Petrol.* 41, 703-714.

Winkelmolen, A. M. and H. J. Veenstra, 1974. Size and shape sorting in a Dutch tidal inlet. *Sedimentology* 21, 107-126.

Wohlenberg, E. 1937. Die Wattenmeer-Lebensgemeinschaften im Königshafen von Sylt. Helgoländer wiss. Meeresunters. 1, 1- 92.

Wolanski, E., 1988. Circulation anomalies in tropical australian estuaries, in B. Kjerve, ed.: *Hydrodynamics of Estuaries.* CRC Press, Boca Raton, Fl., 53-60.

Wolanski, E., 1992. Hydrodynamics of mangrove swamps and their coastal waters, in V. Jaccarini and E. Martens, eds.: *The Ecology of Mangrove and their Related Ecosystems.* Hydrobiologia. 247, 141-161.

Wolanski, E. and P. Ridd, 1986. Tidal mixing and trapping in mangrove swamps. *Est. Coast. Shelf. Sci.* 23: 759-771.

Wolanski, E., Y. Mazda, B. King, and S. Gay, 1990. Dynamics, flushing and trapping in Hinchinbrook Channel, a giant mangrove swamp, *Australia. Est. Coast. Shelf Sci.* 31, 555-580.

Wolanski, E., Y. Mazda and P. Ridd, 1992. Mangrove hydrodynamics, in A. I. Robertson and D. M. Alongi eds.: *Tropical Mangrove Ecosystems.* Coast. East. Studies 41, Amer. Geophys. Union, Washington D. C., 43-62.

Wolanski, E., and R. J. Gibbs, 1995. Flocculation of suspended sediment in the Fly river estuary, Papua New Guinea. *J. Coast. Res.* 11, 754-762.

Wolaver, Th. G., R. F. Dame, J. D. Spurrier, and A. B. Miller, 1988a. Sediment exchange between a euhaline salt marsh in South Carolina and the adjacent tidal creek. *J. Coastal Res.* 4, 17-26.

Wolaver, Th. G., R. F. Dame, J. D. Spurrier, and A. B. Miller, 1988b. Bly creek ecosystem study — inorganic sediment transport within an euhaline salt marsh basin, North Inlet, South Carolina. *J. Coastal Res.* 4, 607-615.

Wolff, W. J., ed. 1988. De internationale betekenis van de nederlandse natuur. RIN-rapport 88/32, 55 p. (in Dutch)

Wolff, W. J., and C. J. Smit, 1990. The Banc d'Arguin, Mauretania, as an environment for coastal birds. Ardea 78, 17-38.

Wood, M. E., J. T. Kelley, and D. F. Belknap, 1989. Patterns of sediment accumulation in the tidal marshes of Maine. *Estuaries* 12, 237-246.

Woodroffe, C. D., 1981. Mangrove swamp stratigraphy and Holocene transgression, Grand Cayman Island, West Indies. *Mar. Geol.* 41, 271-294.

Woodroffe, C. D., 1985a. Studies of a mangrove basin, Tuff Crater, New Zealand. II. Comparison of volumetric and velocity-area methods of estimating tidal flux. *Est. Coast. Shelf Sci.* 20, 431-445.

Woodroffe, C. D., 1985b. Studies of a mangrove basin, Tuff Crater, New Zealand. III. The flux of organic and inorganic particulate matter. *Est. Coast. Shelf Sci.* 20, 447-461.

Woodroffe, C. D., 1987. Pacific island mangroves: distribution and environmental settings. *Pacific Sc.* 41, 166-185.

Woodroffe, C., 1992. Mangrove sediments and geomorphology, in A. I. Robertson and D. M. Alongi eds.: Tropical Mangrove Ecosystems. Coast. Est. Studies 41, Amer. Geoph. Union, Washington D. C., 7-41.

Woodroffe, C. D., 1993. Late Quaternary evolution of coastal and lowland riverine plains of Southeast Asia and northern Australia: an overview. *Sedim. Geol.* 83, 163-175.

Woodroffe, C. D., J. Chappell, B. G. Thom, and E. Wallensky, 1989. Depositional model of a macrotidal estuary and floodplain, South Alligator River, northern Australia. *Sedimentol.* 36, 737-756.

Woodroffe, C. D., R. J. Curtis and R. F. McLean, 1983. Development of a chenier plain, Firth of Thames, New Zealand. *Mar. Geol.* 53, 1-22.

Woods, P. J., M. J. Webb and I. G. Eliot, 1985. Western Australia, in E. C. F. Bird and M. L. Schwartz eds.: *The World's Coastline.* Van Nostrand Reinhold. New York, 929-947.

Wright, L. D., 1975. Sediment transport and deposition in a macrotidal river channel: Ord river, western Australia, in L. E. Cronin ed.: *Estuarine Research,* II, Academic Press, 309-321.

Wright, L. D., 1976. Nearshore wave-power dissipation and the coastal energy regime of the Sydney-Jervis Bay region, New South Wales: a comparison. *Aust. J. Mar. Freshwater Res.* 27, 633-640.

Wright, L. D., J. M. Coleman, and B. G. Thom, 1973. Processes of channel development in a high-tide-range environment: Cambridge Gulf-Ord River delta, Western Australia. *J. Geol.* 81, 15-41.

Wright, L. D., J. M. Coleman, and M. W. Erickson, 1974. Analysis of major river systems and their deltas: morphologic and process comparisons. Coastal Studies Institute, Louisiana Stata University Techn. Rep. 156, 114 p.

Xu Shi-yuan, Shao Xu-sheng, Chen Zhong-yuan, and Yan Qin-shang, 1990. Storm deposits in the Changjiang delta. *Science in China (Series B)* 33, 1242-1250.

Xu, Zh. M., 1985. The eastern flat sedimentation of the Chongming Island. *Oceanol. Limnol. Sinica* 16, 231-239.

Yan, L., E. Wolanski, and Q. Xie, 1993. Coagulation and settling of suspended sediment in the Jiaojiang river estuary, *China. J. Coastal Res.* 9, 390-402.

Yang Shilun, 1991. Impact of wind driven wave on short-period erosion and deposition of tidal flat. *Marine Sciences* 2, March 1991, 58-64. (in Chinese with English abstract)

Yang Shilun, and Chen Jiyu, 1994. The role of vegetation in mud coast processes. *Oceanol. Limnol. Sinica* 25, 631-635. (in Chinese with English abstract)

Yang Shilun, and Xu Haigen, 1994. Tidal flat sediments and sedimentation on the Changxin and Hengsa islands at the mouth of Changjiang river. *Acta Geogr. Sinica* 49, 449-456.

Yin, F., 1984. Tides around Taiwan. In T. Ichiye ed.: Ocean hydrodynamics of the Japan and East China Seas. *Elsevier Oceanogr. Series* 39, 301-316.

Young, B. M., and L. E. Harvey, 1996. A spatial analysis of the relationship between mangrove (Avicennia marina var. australasica) physiognomy and sediment accretion in the Hauraki Plains, New Zealand. *Est. Coast. Shelf Sci.* 42, 231-246.

Yun Caixing, 1983. Scouring and siltation processes of tidal flats of the Yangtze River estuary and sediment exchange between flats and channels. *J. Sed. Res.* 4, 43-52.

Zarillo, G. A., 1985. Tidal dynamics and substrate response in a saltmarsh estuary. *Mar. Geol.* 67, 13-35.

Zeff, M. L., 1988. Sedimentation in a salt marsh-tidal channel system, southern New Jersey. *Mar. Geol.* 82, 33-48.

Zenkevitch, L., 1963. *Biology of the Seas of the U.S.S.R.* Allen and Unwin, 955 p.

Zenkovich, V. P., 1962. *Coastal Zone Development.* Moscow, Publ. House Academic Sci. USSR, 710 p.

Zenkovich, V. P., 1985. Arctic USSR, in E. C. F. Bird and M. L. Schwartz eds.: *The World's Coastline.* Van Nostrand Reinhold. New York, 863-871.

Zernitz, E. R., 1932. Drainage patterns and their significance. *J. Geol.* 40, 498-521.

Zhang Guodong, Wang Yiyou, Zhu Jingchang, Dong Rongxin, and Wu Ping, 1984. Modern tidal flat sedimentation in Jianggang, north Jiangsu. *Acta Sedimentol. Sinica* 2, 39-51.

Zhang, G. D., Wang, Y. Y., Zhu J. Ch., and Dong R. X., 1987. Modern tidal flat sedimentation in Jianggang, northern Jiangsu province. *Acta Oceanol. Sin.* 6, suppl. 11, 216-224.

Zhang Keqi, Jin Qingxiang, and Wang Baocan, 1993. Seasonal changes of the tidal flat from Jinhuigang to Caojing along the north bank of Hangzhou Bay. *Chin. J. Oceanol. Limnol.* 11, 321-332.

Zhang Renshun, and Wang Xueyu, 1991. Tidal creek system on tidal mudflat of Jiangsu province. *Acta Geogr. Sinica* 46, 195-206.

Zhang Yong, 1988. The sinuous channel system of the tidal flats in the Dollard, The Netherlands. Report NIOZ, 31 p.

Zhu Yongkang, 1986. Some characteristics of the Jiao jiang mountain river estuary under strong tides in Zhejiang province. *Geogr. Res.* 5, 21-31.

Zieman, J. C., 1972. Origin of circular beds of Thalassia (Spermatophyta: Hydrocharitaceae) in South Biscayne Bay, Florida, and their relationship to mangrove hammocks. *Bull. Mar. Sci.* 22 (3), 559-574.

Zimmerman, J. T. F., 1974. Circulation and water exchange near a tidal watershed in the Dutch Wadden Sea. *Neth. J. Sea Res.* 8, 126-138.

Zimmerman, J. T. F., 1986. The tidal Whirlpool: a review of horizontal dispersion by tidal and residual currents. *Neth. J. Sea Res.* 20, 133-154.

Zunica, M., 1985. Italy, in E. C. F. Bird and M. L. Schwartz eds.: *The World's Coastline*. Van Nostrand Reinhold. New York, 419-429.

Author Index

Subject Index

Index of Geographical Names

Latin Fauna Names

Latin Flora Names